Biodiversity Dynamics

Biodiversity Dynamics
Turnover of Populations, Taxa, and Communities

Michael L. McKinney and James A. Drake

Editors

COLUMBIA UNIVERSITY PRESS ▲ NEW YORK

Columbia University Press
Publishers Since 1893
New York   Chichester, West Sussex
Copyright © 1998 Columbia University Press
All rights reserved

Library of Congress Cataloging-in-Publication Data

Biodiversity dynamics : turnover of populations, taxa, and communities / Michael L.
　McKinney and James A. Drake, editors.
　　　p.　cm.
　Includes bibliographical references and index.
　ISBN 978-0-231-10414-2 (cloth)
　ISBN 978-0-231-10415-9 (paper)

　1. biological diversity.　2. Evolution (Biology)　3. Population biology.
　I. McKinney, Michael L.　II. James A. Drake
　QH541.15.B56　1998
　577.8′8—dc21　　　　　　　　　　　　　　　　　　　　　　　　98–17973

Casebound editions of Columbia University Press books are printed on permanent and
durable acid-free paper.

Printed in the United States of America

This book is dedicated to Michael L. Rosenzweig and James H. Brown for their decades of work to promote interdisciplinary discoveries between ecology and paleontology.

Contents

Introduction

Michael L. McKinney

*Ecologists must take a more active role in investigating the
processes of species production and extinction.*
                                        —Ricklefs and Schluter 1993b

The biodiversity crisis has had at least one positive outcome: It has forced
biologists from many disciplines to interact and exchange data, which generally improves our overall understanding of ecology and evolution. *Biodiversity dynamics* refers to the turnover of biological units across all temporal and
spatial scales (chapter 1). Like most of the recent literature on biodiversity,
this book represents a synthesis and distillation of data derived from a variety
of disparate fields that have traditionally had little interaction. In this case,
data from population biology are presented with data from community ecology, comparative biology, and paleontology. Major theoretical and practical
gains can be made from such a synthetic view.

 This book has its roots in a symposium jointly sponsored by the Ecological Society of America and the American Institute of Biological Sciences, at
their 1994 national meeting. Many of the book contributors gave papers at
that symposium. However, as the scope and goal of the book became clearer,
other contributors were invited to submit papers to help satisfy gaps or deficiencies. In many ways, this book can be seen as an extension of the recent
books by James Brown (1995) and Michael Rosenzweig (1995), which seek
to extrapolate ecological dynamics to large scales of time and space.

Theoretical Importance of Turnover Across Many Scales

The need for a broad view of biodiversity dynamics has often been expressed by both ecologists (e.g., Pimm 1991) and paleontologists (e.g., Morris 1995a). Ecologists have tended toward an ahistorical focus on general principles of current biotic interactions at relatively small scales of time and space. Paleontologists (and evolutionary biologists in general) have tended toward the other extreme, taking a historical view of biotic interactions at very coarse scales of time and space. Also, ecologists have often emphasized generality dynamics, whereas evolutionists have tended to emphasize the lack of generality found in contingent "random" events (e.g., Gould 1989a).

This book seeks a middle ground, to relate processes occurring at fine and coarse scales while acknowledging that both general dynamics and contingent events are important. To understand the processes controlling diversity at any scale, one must examine both origination and extinction (see also McGhee 1996). Furthermore, we should ideally try to interrelate origination and extinction across many scales. Unlike the theories of some (e.g., Gould 1985), the theme of this book does not emphasize the independence of turnover across various scales of space and time, which would hold, for example, that mass extinctions select victims in ways unrelated to "normal" background extinctions.

To the contrary, much of this book provides evidence that there is often a direct correlation of turnover dynamics across many scales. It is important to realize that much of this evidence is empirical because, in theory, there are many ways that independence among scales could occur. Metapopulation theory, for instance, predicts that a species composed of many small local populations that undergo frequent extinction and recolonization could persist longer than a species composed of just one or a few long-lasting populations (Hanski and Gilpin 1997). In this case, high population extinction rate does not correlate with high species extinction rate. Yet Susan Harrison's chapter, discussed later, reviews empirical data that suggest this is not so. Real taxa do not persist as classic metapopulations but show dynamics that often link increasing population persistence to increasing species persistence. Similar evidence for extrapolated turnover patterns is also found when examining communities. Chapters by Russell and Aronson and Plotnick show how many patterns of turnover are extrapolated from local communities to the biosphere.

Practical Application of Turnover to Biodiversity Issues

The study of turnover has taken on a new urgency with the rise of conservation biology. We need to examine how human impacts at small scales translate into biodiversity loss at coarser temporal and spatial scales (Meffe and Carroll 1994). Many papers in this volume provide practical information that can be of immediate use to management of biodiversity at many scales.

At the fine scale of population dynamics, Maurer and Nott show evidence from birds detailing why rare species are especially prone to extinction: They have not only smaller ranges but also more fragmented ranges. Conservation management must therefore specifically focus on counteracting the effects of enhanced range fragmentation in rare species.

Data discussed by Cutler, at the coarser scale of community turnover over thousands of years, show a surprising determinism in extinction selectivity among component species. Populations of the same species seem most prone to extinction at many spatial and temporal scales.

At still coarser scales of species turnover in evolutionary time, Gittleman and others, for example, use phylogenetic reconstruction to show that carnivores have tended to experience relatively high extinction rates in the recent past, indicating that they are currently exceptionally vulnerable to extinction. This can help justify placing very high priority on immediate efforts toward their preservation.

At even coarser scales, such as North American mammalian biodiversity in evolutionary time, Alroy reviews considerable evidence that evolutionary diversification will slow down as niche saturation sets limits on the number of species that can be sustained in a region. This has very important implications for making large-scale predictions of biodiversity conservation. For example, long-term regional impacts of introduced species and habitat loss can be based on species–area considerations to predict the final equilibrial diversity that a region can sustain.

Overview of the Book

In the first chapter of this book, I present a broad overview of what is meant by *biodiversity dynamics*. My basic view is that it represents turnover across all scales of time and space. How, for example, is turnover (colonization/extinction) of populations related to turnover (speciation/extinction) of species and higher taxa? The focus is on turnover because it is the only way to gain a full understanding of ecological and evolutionary processes. Information that is limited to either extinction or speciation alone can be very misleading. For one thing, extinction and speciation (and colonization) are often strongly interdependent. Mass extinction may precede major speciation events by removing incumbent species, for example. Or the same physical environmental change may stimulate both speciation and extinction. Even aside from extrinsic causation, there is an intrinsic (biotic) correlation, in that groups with high speciation rates also tend to have high extinction rates (e.g., Stanley 1990c).

Part One: Phylogenetic Turnover: From Populations Through Higher Taxa

The relation of population turnover to species turnover is most directly addressed by Susan Harrison. Her work has often reviewed empirical data showing that some of the basic assumptions of metapopulation theory do not apply to real species (e.g., Harrison 1994). In her chapter, she thus argues that taxa are not likely to persist in evolutionary time as metapopulations in the classic sense. She discusses how species, instead of persisting as populations in an extinction–colonization equilibrium, tend to show mainland–island, nonequilibrium, or very patchy distributions that show much less or much more colonization among patches than in the restrictive assumptions of metapopulation theory (see also Harrison and Hastings 1996). In the case of mainland–island and nonequilibrium distributions, the persistence of the species is often strongly correlated with the persistence of single (large) populations.

The relationship between geographic range and evolution is examined by Brian Maurer and Philip Nott. They show, using data from North American insectivorous birds, that rare species, with small geographic ranges, also tend to have more fragmented ranges. This synergism between localization and fragmentation means that rare species have relatively higher extinction rates and lower net diversification rates than abundant species. This is supported by fossil data demonstrating that globigerinids had consistently higher net species diversity with lower extinction and speciation than globorotaliids. They discuss how the origin of this pattern may lie in Darwin's idea that more ecologically generalized species, such as the globigerinids, are more locally abundant and widespread (also see Brown 1995). Such species have lower speciation and extinction rates than rare species, with diversification rate (which equals speciation rate minus extinction rate) being greater in more generalized (more common) species.

The next two chapters address turnover at the species level, using the fast-growing methods of phylogeny reconstruction based on living species. John Gittleman and coauthors present a very stimulating overview that covers a wide range of key issues. They show that molecular phylogenies have major practical applications too. A basic theme of their paper is that ecological, life history, and morphological traits are not free to evolve as needed but are phylogenetically constrained to varying degrees (also see Harvey 1996 for review). Using a large database of mammalian molecular phylogenies, they show that morphological and life history traits are generally less evolutionarily labile than behavioral and ecological traits such as population density and day range. This implies that the latter traits are less constrained by developmental, genetic, or other correlated limitations.

Jody Hey and coauthors test specific models of speciation and extinction with phylogenetic trees of living taxa. This approach expands on the basic random cladogenetic models produced by David Raup beginning in the early 1970s. A key finding is that 10 of the 11 data sets analyzed are best fit by a growth model with an extinction rate of zero. This constitutes tentative evidence that these small young clades are in a growth phase of cladogenesis, although this is not conclusive because of the difficulty in distinguishing extinction from sampling bias and other artifacts.

In the past few years, Daniel McShea has written some very influential articles on evolutionary trends, especially the evolution of complexity (McShea 1996). In his chapter, he extends earlier work to develop a conceptual scheme on clade diversification in state space (where *state space* is any feature of a species such as size, geographic range, or complexity). He shows that large-scale behavior of a clade, such as a trend in the mean or maximum of a trait, is the product of two factors: (1) the clade's small-scale behavior, i.e., rules governing the individual lineages of the clade, and (2) the structuring of those rules in state space. If structuring is minimal, then large-scale clade patterns will be the direct result of small-scale dynamics. An example is a "driven" trend where all component lineages are biased toward size increase. But as structuring increases, large-scale behavior will be increasingly independent of small-scale dynamics. An example is a barrier on attainable small size that produces an asymmetrical "passive" trend wherein the maximum size of the clade increases while minimum size is unchanged. In extreme cases, structuring could control many details of clade behavior, almost irrespective of small-scale dynamics.

A classic example of clade diversification is body size evolution. This is explored by Douglas Kelt and James Brown, who expand a model proposed by Brown and others (1993) based on PEF (potential energetic fitness). The PEF model predicts that the peak body size of a clade reflects the optimal size for most readily converting available energy resources into offspring. For mammals, this is about 100 g. Kelt and Brown show evidence that PEF is a major factor underlying the distribution of body sizes at local and regional scales across ecological and evolutionary time scales (see also Brown 1995).

Small-scale and large-scale dynamics of clade diversification can result from either intrinsic properties, such as developmental constraints, or extrinsic forces of environmental selection. Or, most often, such dynamics result from the interaction of both intrinsic and extrinsic forces. This interaction is explored by Gunther Eble, who uses fossil data to test Kauffman's (1993) rugged fitness landscape model. Although Kauffman's application of self-organization to biodiversity evolution has received much theoretical attention, Eble shows that Kauffman's model does not provide a good fit to the patterns of origination of higher taxa in the fossil record. While certain regularities and similarities to the rugged fitness model do occur, Eble shows that the model may need to be refined.

The final chapter in Part One addresses the very coarse scales of turnover of higher taxa in geological time. Norman Gilinsky presents evidence of long-term decline of background family origination and extinction. This represents a decline in turnover "volatility." What causes this pattern? Part of the explanation, suggested by John Sepkoski, who has pioneered such work, is that extinction-prone clades such as trilobites have become extinct. Thus, if family turnover translates directly into species turnover, then Gilinsky infers that niche breadth may be a factor. Citing a number of previous suggestions, he notes that more specialized species tend to have higher turnover rates.

Part Two: Community Turnover: From Populations Through Global Diversity

Ecosystems and communities experience turnover at many scales of time and space (Brown 1995; Rosenzweig 1995). Kenneth Schopf and Linda Ivany present a novel view of ecosystem stasis and change that is explicitly hierarchical. Reviewing extensive data from the fossil record, they discuss how finer scales of observation tend to show evidence for fluctuating species composition of ecosystems. Such finer scales include temporal scales ranging from a few years though a few hundred thousand years. In contrast, coarser scales of study have tended to yield patterns interpreted as showing stasis. Such coarser scales include time spans on the order of a few to many millions of years. Examples of such coarse-scale stasis are Boucot's (1990a) ecological-evolutionary units, and subunits of it identified by Brett and Baird (1995). The main question is whether this stasis is real, being caused by such processes as "ecological locking," or whether it is simply an artifact from coarser scales of resolution filtering out much of the small-scale turnover (McKinney et al. 1996; Alroy, chapter 12).

Species are not randomly distributed across the earth. Alan Cutler discusses an important nonrandom pattern that may eventually reveal much about the underlying processes of origination and extinction. This is the very common pattern of "nested subsets": less species-rich biotas are composed of subsets of species of more species-rich biotas. A species absent from one biota will tend to be absent from all smaller (less species-rich) biotas. If it is present in a biota, it will tend to be present in all of the more species-rich biotas. Cutler shows how three nonexclusive processes can produce such nestedness: (1) passive sampling, (2) nested habitat distributions, and (3) colonization/extinction. Of special interest to biodiversity dynamics is the third process. Selective extinction can produce nestedness as can differences in colonization ability. Broadly adapted species that can both resist extinction and colonize a range of habitats will be found in many subsets, of all sizes. In contrast, species specialized to a narrow range of habitats will be found only in a few subsets. Can differences in speciation and extinction, on an evolutionary time scale, explain nestedness on continental spatial scales?

The crucial issue of diversity equilibrium is supported by John Alroy's thorough study. Using a huge database, he finds cogent evidence that North American mammal diversity has been relatively constant throughout much of the Cenozoic, indicating the existence of a static equilibrial point. Origination rate is inversely related to diversity, whereas extinction rate is not, so Alroy infers that evolutionary niche saturation is the key process underlying the logistic diversity pattern observed. This conforms to the suggestion of Rosenzweig and McCord (1991) of evolutionary incumbency and niche-preemption as a main control producing equilibrial diversity patterns: As new species evolve to fill niches after a mass extinction, the number of available niches diminishes proportionately with time. The pattern seen in Alroy's data also fits data from other studies showing that origination, and not extinction, is the main control on regional and global diversity. Extinction rate plays little role, as, at geological scales, it is often relatively constant through time. Importantly, Alroy's data seem to contradict a number of other currently popular theories about diversity dynamics, such as the Red Queen, Raup's "kill curve," Vrba's "turnover pulse" hypothesis, and coordinated community stasis.

Another example of a logistic diversity pattern is also seen in Ordovician marine genera as shown by Arnold Miller and Shuguang Mao. The Ordovician experienced major evolutionary radiations of both Paleozoic and Modern marine faunas, with both genus and family diversity increasing by threefold or more. This logistic growth apparently occurred quite rapidly, even faster than previously thought (Miller and Foote 1996). Miller and Mao's work is especially illuminating because they attempt to relate this global increase in diversity to processes that were occurring at smaller spatial scales. They conclude that some scales of observation show unique patterns not visible at other scales. Thus, the global diversity signal was not simply the summed result of community-level diversification. Their interesting suggestion is that biotic factors may have been more important at local, community scales, whereas abiotic diversity controls may have dominated at larger, paleocontinental scales.

Equilibrial community diversity patterns are also a theme in Michael Rosenzweig's stimulating and provocative discussion of species accumulation in space and time. This is an expansion of Preston's concept that a "species–time" curve is a theoretical analog to the much better known species–area curve. If true, it would imply a dynamic where horizontal (spatial) division of habitat is similar to vertical (temporal) division of habitat. Rosenzweig discusses evidence that species do in fact temporally accumulate in an apparently regular way, at both regional and global spatial scales. Based on the rate of turnover in the accumulation curves, Rosenzweig infers that species turnover in evolutionary time is similar to the rate of species turnover seen spatially among major geographic provinces. In contrast, space and time do not seem to be as interchangeable at smaller, ecological scales. However, there is evidence for a regular temporal accumulation in paleocommunities in local stratigraphic sections (see McKinney 1996a; McKinney et al. 1996). One of the most contentious of Rosenzweig's findings is that long-term speciation rates over the last 500 million years may have been relatively constant [see e.g., Gilinsky (chapter 9) and Alroy (chapter 12)].

Evolutionary equilibrium indicates that there is a balance between speciation and extinction. A main benefit of studying evolutionary turnover is that we are in a much better position to understand the origins of biodiversity rather than just its maintenance. Compared to studies of biodiversity maintenance in a community, the origins of that biodiversity have been greatly neglected. This is the central point made by Warren Allmon, Paul Morris, and Michael McKinney in their intermediate disturbance hypothesis of maximal speciation. They discuss both fossil and modeling evidence for this hypothesis, which proposes that maximum rates of speciation will be produced at intermediate levels of disturbance. Very high levels of disturbance will result in extinction, stress, and depauperate faunas, whereas very low levels will not provide the environmental stimuli that drive natural selection. This hypothesis has an obvious connection to the older notion that intermediate disturbance maintains higher levels of diversity (e.g., Petraitis et al. 1989), but it explicitly addresses the very different process of diversity origination. The turnover dynamic of extinction is thus balanced by speciation and not immigration.

Turnover at many scales has generally been underappreciated by both ecologists and evolutionary biologists. Gareth Russell suggests that similar approaches can be used to study turnover at any level in the hierarchy of life, including turnover in time and space. The general principles that emerge are an important link between ecology and evolutionary biology. While ecologists have focused on turnover at fine spatiotemporal scales and evolutionary biologists have focused on coarser spatiotemporal scales, Russell shows that similar patterns and processes occur at both ecological and evolutionary scales. While it has long been recognized that species origination and extinction are analogous to population immigration and local extinction (e.g., MacArthur and Wilson 1967), Russell rigorously quantifies interpretation and modeling of this analogy. The result is that spatiotemporal turnover in "time-averaged" paleocommunities can be studied to describe how cumulative ecological turnover is translated into long-term ecological and evolutionary turnover (also see McKinney and Allmon 1995).

We might expect that turnover at many scales can be related to fluctuations in nutrient levels, biomass, productivity, and other ecosystemic properties. Evidence from fossil and living marine biota, as Ronald Martin discusses, indicates that changes in nutrient levels may have played a major role in many mass (and minor) extinctions. A counterintuitive insight of his view is that increased productivity can accompany extinction at many scales. This period of destabilization is followed by reequilibration to new conditions, which is often characterized by increased biomass and biodiversity. The result, as Martin discusses, has been a long-term global increase in biodiversity, the complexity of ecosystems, and such ecosystemic properties as biomass and productivity. Increased nutrient input, at many scales, may thus cause a temporary surplus of extinction over origination, but leading to a greater accumulation of species in the long term.

Patterns of change are often related to scale of observation, as discussed in Schopf and Ivany's paper above. A basic message of Richard Aronson and Roy Plotnick is that both scale-dependent and scale-independent patterns and processes occur in biological dynamics. However, in seeking emergent properties unique to just one level, scale-independent processes have often been overlooked. Physical disturbances, for instance, have a strong influence on community, and taxic turnover may operate at a similar fashion at many or all scales. Some biotic traits such as extinction resistance may operate similarly at all scales from that of individual death, through population, and up through species and even higher taxa (see also Harrison's chapter and the preceding discussion of it). Aronson and Plotnick note that, in such cases, long-term community and taxic patterns may simply result from the summed, additive effects of small-scale processes rather than any synergistic interactions such as "ecological locking" within communities.

Contributors

Warren D. Allmon, Paleontological Research Institution, 1259 Trumansburg Road, Ithaca, NY 14850

John Alroy, National Center for Ecological Analysis and Synthesis, 735 State Street, Santa Barbara, CA 93101

Richard B. Aronson, Dauphin Island Sea Lab, P.O. Box 369, Dauphin Island, AL 36528

James H. Brown, Department of Biology, University of New Mexico, Albuquerque, NM 87131

Alan H. Cutler, Department of Paleobiology, National Museum of Natural History, Smithsonian Institution, Washington, D.C. 20560

Gunther J. Eble, Department of Paleobiology, National Museum of Natural History, Smithsonian Institution, Washington, D.C. 20560

Norman L. Gilinsky, Department of Geological Sciences, Virginia Polytechnic and State University, Blacksburg, VA 24061

John L. Gittleman, Department of Ecology and Evolutionary Biology, University of Tennessee, Knoxville, TN 37996

Susan Harrison, Division of Environmental Studies, University of California, Davis, CA 95616

Jody Hey, Department of Microbiology and Genetics, Rutgers University, Nelson Laboratories, 604 Allison Road, Piscataway, NJ 08854

Linda C. Ivany, Museum of Paleontology, University of Michigan, Ann Arbor, MI 48109

Douglas A. Kelt, Department of Wildlife, Fish, and Conservation Biology, University of California, Davis, CA 95616

Shuguang Mao, Department of Geology, University of Cincinnati, Cincinnati, OH 45221

Ronald E. Martin, Department of Geology, University of Delaware, Newark, DE 19716

Brian A. Maurer, Department of Zoology, Brigham Young University, Provo, UT 84602

Michael L. McKinney, Department of Geology, University of Tennessee, Knoxville, TN 37996

Daniel W. McShea, Department of Zoology, Duke University, Durham, NC 27708

Arnold I. Miller, Department of Geology, University of Cincinnati, Cincinnati, OH 45221

Paul Morris, Department of Organismic and Evolutionary Biology, University of Massachusetts, Amherst, MA 01003

M. Philip Nott, Department of Ecology and Evolutionary Biology, University of Tennessee, Knoxville, TN 37996

Roy E. Plotnick, Department of Earth and Environmental Sciences, University of Illinois, Chicago, IL 60607

Michael L. Rosenzweig, Department of Ecology and Evolutionary Biology, University of Arizona, Tucson, AZ 85721

Gareth J. Russell, Department of Ecology and Evolutionary Biology, University of Tennessee, Knoxville, TN 37996.

Kenneth M. Schopf, Museum of Comparative Zoology, Harvard University, Cambridge, MA 02138

Biodiversity Dynamics: Niche Preemption and Saturation in Diversity Equilibria

Michael L. McKinney

*The rapid build-up of diversity to near original levels . . .
suggests that many systems spend most of their time near the
equilibrium species richness.*   —Brown, 1988

*We might expect abundant, widely distributed species to
have large litters, younger ages of reproduction, and so on.*
                    —Brown, 1995

Biodiversity dynamics, as used here, refers to the turnover of populations, taxa, and communities, where turnover is origination and extinction. Analyzing only origination or only extinction is inadequate for most theoretical and practical goals. Theoretically, origination and extinction, at many scales, influence one another and are often influenced by the same external forces. In practice, long-term loss of biodiversity can occur by decreasing origination as well as by increasing extinction (McGhee 1996). Furthermore, conservation requires more than knowing extinction proneness of taxa. The differing abilities of taxa to recover from extinction by origination of new populations and species is also important and can play a huge role in the composition of the future biosphere (Kerr 1994).

It is often argued that populations are the most basic unit of biodiversity dynamics. Eldredge (1992), for example, notes that populations form the basis of both genealogical units (e.g., species and higher taxa) and ecological units (communities). Given this, the question of understanding biodiversity dynamics becomes one of understanding how population turnover is translated into taxic and community turnover.

My chapter draws much of its theoretical impetus from two ecologists, Michael Rosenzweig and James Brown, who have each spent decades attempting to extrapolate ecological dynamics to evolutionary timescales. Many of their theories are in fact supported by emerging evidence, including many chapters of this book. Rosenzweig has long argued (1995; also chapter 14 of this volume) that species diversity at regional and global spatial scales and through evolutionary time show equilibrial patterns. The first part of my chapter is devoted to the emerging evidence for species diversity equilibria, and the role of niche saturation in causing it.

Brown's recent book *Macroecology* (1995) summarizes much of his work attempting to extrapolate abundance dynamics to large spatial and temporal scales. Of special relevance here is Brown's reliance on the niche concept as the basis for understanding diversity patterns. The second part of my chapter explores this further, especially the view that broad-niched, generalized species are both ecologically and evolutionarily dominant. Such species are more ecologically abundant and trophically dominant. I suggest that they also tend to evolutionarily expand into more diverse clades because their adaptations allow them to preemptively evolve into niche space in a finite world.

Both parts of my chapter conclude with a brief application to practical issues of conserving biodiversity. Equilibrial diversity dynamics, for instance, have many applications for predicting future biodiversity loss at large spatial scales. Similarly, and less discussed in the conservation literature, the ecological and evolutionary dominance of some taxa provides valuable predictive data on both short-term extinction proneness and long-term recovery. A few dominant clades may, for example, preferentially suffer fewer extinctions and also have better prospects for evolutionary recovery (McKinney 1997).

Diversity Equilibria

MacArthur and Wilson (1967) initiated an enormous investigation into equilibrial models of diversity regulation at ecological spatial and temporal scales. These models have been extended to larger scales, of continents and evolutionary time, by a number of workers, most notably Rosenzweig (1975, 1995), Brown (1988, 1995), and Sepkoski (many articles; e.g., 1991a, 1992c, 1996). Such large-scale models have dynamics similar to those of island biogeography, with the following general traits that are usually applied to continental and global scales of community (and often clade) diversity:

1. Diversity, usually defined as species richness, is assumed to stochastically fluctuate around an equilibrium value.

2. The equilibrium is a dynamic state. Species richness is roughly constant, but there is changing species composition driven by similar rates of speciation and extinction.

3. The equilibrium is temporary when viewed over geological time.

Trends in plate tectonics and other geological processes alter available habitat. Continental fragmentation, for instance, has probably helped to increase the equilibrium diversity of marine invertebrates since the early Mesozoic Era (e.g., Sepkoski 1992c).

4. The equilibrium value is largely determined by competition for the available finite resources. Speciation is thus density dependent, with decreasing species origination as diversity increases. In contrast, considerable data support the view that extinction is often relatively constant and density independent (Alroy, chapter 12).

In recent years, there has been increasing discussion that equilibrial models are inadequate descriptions of species diversity at ecological spatial and temporal scales (e.g., DeAngelis and Waterhouse 1987; Botkin 1990). Yet the opposite seems to be true of species diversity at continental spatial and evolutionary temporal scales. While there have been a few dissenters (e.g., Hoffman 1989; Benton 1995, 1996), there is a growing body of evidence reviewed by Sepkoski (1992c, 1996), Brown (1995), Rosenzweig (1995), Alroy (chapter 12), and Courtillot and Gaudemer (1996) that species diversity at continental and evolutionary scales is characterized by equilibrial dynamics.

That diversity equilibria are more evident at continental spatial and evolutionary temporal scales (than at smaller scales) may validate theoretical findings that "the stable equilibrium state should not be viewed as a fundamental property of ecological systems, but as a property that can emerge asymptotically from extrapolation to sufficiently large spatial scales" (DeAngelis and Waterhouse 1987). The underlying mathematical relationships for this are explored by DeAngelis and Waterhouse (1987) and by Levin (1992), who show how variability tends to decay as scale of observation increases. The fluctuations of diversity (via immigration and extinction) seen at local scales should thus average out to produce a more constant pattern of diversity at larger scales of time and space.

This does not mean, of course, that rapid, major deviations from diversity equilibrium never occur at continental or global scales. Mass extinctions clearly do occur (Sepkoski 1992c). But such major changes in "equilibrium" (standing) diversity should (1) be progressively less common, and (2) occur more slowly, with increasing spatial and temporal scale. Indeed, these theoretical predictions of scale effects are apparently borne out. Major (large) fluctuations in global diversity, including mass extinctions, are relatively rare and have been relatively gradual on ecological time scales (e.g., Erwin 1993a; Prothero 1994).

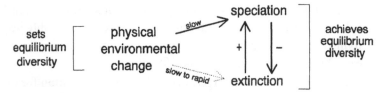

Figure 1.1. Physical environmental changes drive origination and extinction of species and higher taxa. Origination and extinction mutually influence each other. Only extinction by physical change can be rapid on ecological time scales.

A Graphical Model of Diversity Equilibrium

Figure 1.1 is a graphical summary of many basic ideas traditionally proposed about diversity equilibrium at many scales. While much is self-explanatory (and not new), the following key points are noted because they may clarify some contentious issues:

1. Ultimate Cause of Diversity Change

It has been debated, for over 100 years, whether abiotic or biotic factors are more important in producing and maintaining biodiversity (Gould 1977; Miller and Mao, chapter 13). As with equilibrium itself, a main source of misunderstanding seems to be a failure to specify the scale that is being analyzed. Ecologists, with their focus on local spatial and very short temporal scales, have tended to emphasize biotic diversity controls and an ahistorical view (Ricklefs and Shluter 1993a). The fossil record, on the other hand, documents a strong influence of abiotic diversity controls when continental and global spatial scales are viewed through geological time. A major review by Valentine (1990) concludes that "for massive changes in diversity . . . and the waxing and waning of marine clades, the chief driving factors are abiotic." Cracraft (1992) discusses a general model of diversity controls ultimately driven by such abiotic factors. Miller and Mao (chapter 13) discuss a specific example of the role of geological activity.

Figure 1.1 thus depicts abiotic environmental change as the ultimate driving force influencing both origination and extinction species. Global cooling, for example, has caused the extinction of some tropical groups but has directly caused the evolution of new ones (e.g., Prothero 1994; McGhee 1996). Allmon (chapter 15) and others extend this idea to argue that mild abiotic changes produce the highest ratio of origination rate to extinction rate.

2. Mechanisms of Proximate Diversity Regulation

Figure 1.1 also shows that origination and extinction are influenced not only by abiotic changes but also by each other. Rapid extinction of many species creates ecological vacuums that are filled by relatively rapid origination of new species (e.g., Patzkowsky 1995). Conversely, origination of new species may actually drive extinction of other species where competitive replacement occurs (e.g., Roy 1996). The mutual influence of origination and extinction are seen as *proximate* regulators that keep diversity fluctuating around the stable diversity equilibrium level that is ultimately determined by long-term abiotic changes, as noted above, such as global and geological processes.

3. Rate of Diversity Change

A final key point of figure 1.1 is the different rates of change among the diversity controls. Most are slow in terms of ecological time: speciation, whether promoted by abiotic processes or extinction, and extinction promoted by speciation of other species, are relatively slow. In contrast, extinction driven by abiotic processes can sometimes occur very rapidly, as with bolide impacts (Archibald 1996). This may explain why there is greater variation in fossil extinction rates than in origination rates (Foote 1994). This asymmetry means that rapid major deviations from equilibria can only reduce (and not increase) diversity. Despite other similarities to population curves, there is thus no diversity "overshoot" where diversity temporarily exceeds carrying capacity, as can happen with populations. Regional and global diversity are therefore either near a slowly changing equilibrium (see Brown's quote at the chapter beginning), or (much more rarely) they are recovering from rapid abiotic perturbations. Since abiotic disturbance is increasingly more common at progressively smaller scales (McKinney and Allmon 1995), we may infer that regional and global equilibrium diversity represents something less than the actual maximum diversity that could be sustained in the absence of ongoing local disturbances.

Diversity Equilibrium via Niche Saturation

Despite decades of criticism, especially about its measurability, the niche concept is still widely used in ecology (e.g., Ricklefs 1991). A main reason is that niche dimensions provide an excellent way of uniting traits of the individual, such as diet and physiology, with population traits such as mean abundance and intrinsic rate of growth (Brown 1995). This is critical for efforts (such as this book) that seek to explain cladogenetic processes in terms of population dynamics.

The idea that niche saturation is a main cause of equilibrial diversity patterns is an old one. Alroy (chapter 12), Brown (1988; 1995), Rosenzweig (1995), and Sepkoski (1992c, 1996) review the considerable literature on this and evidence for it. Among the most cogent evidence is Alroy's finding that origination is density dependent, whereas extinction is not, indicating that niche space is being "filled." New species evolve until most available niches are filled, at which point speciation rate begins to approximate extinction rate. Other lines of evidence are (1) logistic patterns of diversification in many clades (Wagner 1995a; Patzkowsky 1995), (2) patterns of clade interactions including competitive replacement (Patzkowsky 1995), (3) larger morphological "jumps" during radiations just after mass extinctions indicating the occupation of more open adaptive zones (Patzkowsky 1995; Eble, chapter 8), and (4) niche-filling in ecological time as indicated by community assembly models (Drake 1990) and increased exotic invasion resistance with greater native biodiversity (Brown 1995).

Application: Global Biodiversity Predictions

The equilibrial theories of island biogeography have already been widely applied in efforts to design regional preserves, although with considerable debate (Beeby 1993). A main reason for the debate is that island biogeography, and its derivative species-area curves, treats all species alike. I return to this problem later.

A less explored application is the extension of island biogeography to make predictions about continental and global biodiversity changes from human impacts. A notable exception is the use of species-area curves to predict the ultimate impact of massive human introduction of exotic species onto all the continents. Figure 1.2 shows a prediction by Vitousek and others (1996) that the current global diversity of 4200 mammal species would decrease to about 2000 species if all the continents were reunited into a single land mass (see also Brown 1995; McKinney 1997, 1998b). This may approximate the ultimate impact of extensive human transportation of mammal species to the point where geographic isolation no longer exists.

A similar approach has been used by paleontologists to test for the effects of continental assembly on past mass extinctions. An important lesson has been that global species–area relationships will not explain past mass extinctions unless the details of areal changes are considered. Loss of continental shelf marine habitat by lower sea level, for example, is partly offset by a gain in shallow water habitat around oceanic islands (discussion in McGhee 1996). Also, reduction in biogeographic provincialization (heterogeneity) can cause extinction even if no loss of area occurs (Schopf 1979). When such

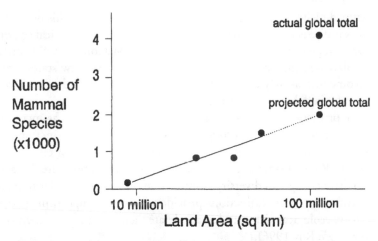

Figure 1.2. Projected ultimate effect of human transport of exotic mammal species and homogenization of the global biota. In effect, this is the same as uniting all land into a giant supercontinent, as occurred in the late Paleozoic Era (Erwin 1993a). Projected number based on extrapolation of species–area curve, where solid line regression uses known data from areal surveys. Actual global total is currently much higher because of high endemism, which could disappear if increasing global transportation overwhelms the effects of geographic isolation. *–Redrawn from Vitousek et al. 1996.*

details are included, data from modern marine species show that species–area relationships seem to be good descriptors of past and modern marine biodiversity patterns (Briggs 1995; McKinney 1998b).

Yet another application from paleontology is that biodiversity can actually increase even though equilibrial global biodiversity, as measured by number of taxa, is constant. In a number of papers, Foote (review in Foote 1996a) has documented how morphological diversity, or "disparity," among taxa in a group can increase by evolutionary divergence although the number of taxa is constant or even decreasing. A clade that is near an equilibrial state in one type of diversity metric (number of species) can be undergoing expansion (or contraction) in another diversity metric. In theory, of course, a clade can (and sometimes does) simultaneously maintain an approximate equilibrium in both metrics for a long time (Foote 1996a).

Diversity Dominance via Niche Preemption

A major implication of niche saturation as a determinant of diversity levels is the pivotal role of niche preemption in determining the composition of biotas: Niches are occupied on a "first come, first served" basis, so that groups that immigrate (on ecological scales) or evolve (on geological scales) into unoccupied niche space become "entrenched." Evidence for this on ecological scales is discussed by Vermeij's (1991) "hypothesis of ecological opportunity" and on geological scales by Rosenzweig and McCord's (1991) "incumbent replacement model." Indeed, Hallam's (1994) overview states that "it has become increasingly clear that in general the fossil record of both vertebrates and invertebrates supports the preemptive model."

A key prediction of the niche preemption model is that, as incumbent occupants of niches are not dislodged by competition, then extinction of the incumbents by disturbances provides the main opportunity for replacement. As Roy (1996) discusses (and shows evidence for), speciation-rate disparities tend to drive changes in diversity composition during both background and mass extinctions. Mass extinctions provide widespread opportunities to occupy many ecological niches and so accelerate incumbent replacement (Patzkowsky 1995; Roy 1996).

Are there any traits that enhance a group's capacity for niche preemption, to occupy a disproportionate number of niches before another group? Or, is it just a matter of chance, as Gould, for example, has argued for so many years (e.g., Gould 1996)? I review evidence for the former, that some groups are *supertaxa*, to use Stanley's (1990c) term. I will argue that evolutionary supertaxa, defined as clades with many species, tend to be ecological supertaxa, defined as clades containing many abundant and broadly adapted species. The same traits that promote population dominance may also tend to promote high long-term diversification potential.

Clade-Level Traits

The logistic pattern of clade diversification (e.g., Wagner 1995a) has led to the use of population models as a tool for analyzing clade evolution (see especially Sepkoski 1991c; Patzkowsky 1995). Clades can thus be described as having a characteristic intrinsic rate of growth, r, and carrying capacity, K. [I use *clade* in a very general sense, to refer to any phylogenetic group of significant size, as represented by such taxonomic proxies as classes, orders, or families (see Foote 1996b)]. In this case, r represents the per species rate of species (or genus or family) diversification and K is the equilibrial species (or genus or family) level (e.g., Alroy, chapter 12; Maurer and Nott, chapter 3). Clades with high r would have high net speciation (speciation minus extinction); those with high K would be able to speciate into a large number of niches.

Regardless of its causes, the empirical fact of logistical patterns of clade evolution in the fossil record opens up a wide range of potential applications of population models to clade models that have yet to be explored. Patzkowsky (1995) shows how the basic logistic pattern of each clade can be modified by interactions with other clades. Thus, it appears that, just as interactions between populations of different species alter the various population curves of those species, so clade interactions occur, with similarly predictable results.

Extension of this model to clade dynamics permits theoretical predictions about clades. Using population models, Lande (1993), for example, rigorously shows that average time to extinction, under stochastically fluctuating conditions, increases almost exponentially with K where population size is low, and it increases as a power of K where population size is large. By analogy, the same increases in time to extinction with K are predicted for clades when species richness is substituted for population size in Lande's (1993) estimates (see chapter 9). Evidence for this includes the longer persistence of fossil higher taxa, such as genera that have higher species richness, during background times (Jablonski 1991) and sometimes even during mass extinctions (Erwin 1993a). High K clades should have certain traits, discernible at the population level, that can be identified and that provide an explanation for their clade-level dynamics. We should, for example, be able to specify why some clades have a high K and other clades do not. I now address this key issue.

Ecological Supertaxa: Are They Evolutionary Supertaxa?

The idea of ecologically dominant species is at least as old as Darwin. He called them dominants because they are both locally abundant and geographically widespread. There is now extensive evidence at many spatial scales in support of this, as discussed by Brown (1995), who also reviews evidence for why such dominants exist. The most common explanation is that dominant species are generalists, occupying broad niches, that are able to utilize a wide range of abundant resources (Lawton et al. 1994; Brown 1995). This also follows from the observation that species with wide tolerances in one niche dimension also tend to have wide tolerances in other dimensions (Futuyama and Moreno 1988). This apparently produces a synergistic effect because dominant species are not only more widespread but they tend to be more locally abundant where they coexist with other species. As Brown (1995) put it, "the jack-of-all-trades appears to be master of all."

There is some evidence that such ecological dominance, as measured by abundance and range, is also related to evolutionary dominance, as measured by species diversity.

1. *Clade diversification patterns:* Generalist species have more net species diversification (r) [e.g., planktic foraminifera (Maurer and Nott, chapter 3), flowering plants (Ricklefs and Renner 1993)]. Fossil gastropods show that

sessile filter feeders have higher turnover rates and a lower equilibrial diversity than deposit feeders, which are mobile and have a wider variety of diet and environmental tolerances (Wagner 1995a and references within).

2. *Life history in diversification:* The most species-diverse subtaxa (e.g., genera) within a taxon (e.g., family) nonrandomly tend to have species with opportunistic life history traits such as short generation times and high abundance (Marzluff and Dial 1991).

3. *Phylogenetic heritability of abundance and niche-breadth:* Abundant species are nonrandomly clustered in certain genera and other higher taxa (Gaston 1994); generalist species give rise to specialist species in the absence of continual selection pressures (Brown 1995). This heritability is a prerequisite to permit any differences in clade-level diversification.

4. *Body size in diversification:* The most species-diverse subtaxa (e.g., genera) within a taxon (e.g., family) nonrandomly tend to have small body sizes (Dial and Marzluff 1989; McKinney 1990b; Nee et al. 1992; Blackburn and Gaston 1995). Given that increasing abundance correlates with small body size (Currie 1993), this follows the expected pattern. It also follows the suggestion that smaller species in a clade tend to be more generalized (Stanley 1973b).

Accumulating Species in Clades of Many Scales

To summarize, available evidence tentatively conforms to the idea that ecological supertaxa exist. These broadly adapted abundant species are widespread and have high rates of population growth. I have also suggested, based on the four lines of evidence described, that such ecological supertaxa are often evolutionary supertaxa because they, on average, have a greater positive long-term disparity between speciation and extinction rates. This produces the interesting analogy that species with high r at the population level will tend to belong to clades that have high K at the clade level of species diversity.

There is evidence of considerable consistency of selectivity between mass and background extinctions (Raup and Boyajian 1988; Sepkoski 1990). Furthermore, there is evidence that being widespread and abundant may be a factor in this selectivity, promoting survival during both background and mass extinctions (Erwin 1993a; Van Valkenburgh and Janis 1993; McKinney 1996b). If true, then mass extinctions often just accelerate the process of replacement of high turnover groups by low turnover groups. Gilinsky (chapter 9) has referred to this as the replacement of "high-volatility" groups being replaced by "low-volatility" groups. This has a long history in population biology, where populations with high volatility tend to go extinct sooner because their abundance fluctuations take them near zero abundance more often (Pimm 1991).

Clade turnover has been explored for many years by Sepkoski (e.g., 1992c, 1996) in efforts to explain why the low turnover "modern" marine faunas have replaced earlier, higher turnover marine faunas. It is notable that plants show the opposite pattern, of replacement of low-turnover by high-turnover groups (Valentine et al. 1991b). This may be expected given the suggestion that plant evolution is driven more by competition than animal evolution (Knoll 1984). If true, animal species occupy niches that may be difficult to dislodge by competition, so that incumbency plays a major role (discussion and evidence in Roy 1996; Sepkoski 1996), even at the population level.

A few years ago, it would have been acceptable to formulate the supertaxa concept in terms of r-K theory. However, as reviewed by Stearns (1992), classic r-K theory is a gross oversimplification that does not accurately describe population dynamics among closely related species. In fact, the very existence of ecological supertaxa helps explain why: Widely abundant generalist species are not limited to inhabiting disturbed environments. As Brown (1995) says, such species are usually more abundant than the specialist (K-selected) species with which they coexist.

This does not, however, by itself invalidate the incumbency ideas discussed here. The problem with r-K theory is that it tried to explain too much (Stearns 1992), leaving open the possibility that parts of it are correct. In this case, the notion that species with higher rates of population growth can preempt other species after a disturbance (of any scale) seems valid in both theory and observation (Ricklefs 1991). One place where r-K theory seems to have failed is in its suggestion that these incumbent abundant generalists would be replaced by more competitive specialists.

Brown (1995) has pointed out that the evidence for ecological dominants is most clear in comparisons between closely related taxa. For example, a positive correlation between high local abundance and wide geographic abundance is most clearly seen when comparing many species in the same genus as opposed to species in different families (Gaston 1994). This is critical to the ecological supertaxa concept because we might expect species accumulation at small scales to scale upward to accumulate in larger phylogenetic groups so that the same pattern of incumbency is visible at many scales of clade-level comparisons. The compilation by Marzluff and Dial (1991) showing opportunistic traits in the more diverse subtaxa within many levels of taxa supports this. Similarly, the smaller, more abundant passerines are the most diverse clade of birds (Nee et al. 1992).

At a larger scale, we might ask if clades characterized by relatively high rates of population growth and abundance tend to be more diverse. Indeed, Nee and others (1996) calculate that arthropods, considered as a clade, are enormously more diverse than would be expected from random expectations.

Labandeira and Sepkoski (1993) attribute the high diversification of insects to low extinction rates, as predicted here, by the importance of incumbency.

Taxonomic Allometry: Describing Subtaxic Superdiversity

A more rigorous view of these patterns is found in the study by Ricklefs and Renner (1994) of taxonomic richness in living flowering plants. They use a method that one might call taxonomic allometry because they regress the number of genera and species per family using the same power function used in traditional allometric studies. Ricklefs and Renner (1994) regress log(genera) = k log(species) + log b, so that k represents the allometric rate of relative change that species accumulate in genera of flowering plants. They find that, for 365 families of flowering plants, this equation is a highly significant fit ($r^2 = 0.78$) where $k = 0.60$ (standard error = 0.02), indicating "negative allometry": as families increase in species richness, species accumulate much faster than genera in a highly regular way. Examination of their data shows that this occurs because of supertaxic patterns: some genera are exceptionally rich in species.

To extend (and further test) the analysis of Ricklefs and Renner (1994), I analyzed another data set of well-studied living organisms, the mammals. Family, genus, and species diversity data were obtained for all known living mammals from *Walker's Mammals of the World* (Nowak 1991) for a total of 127 families and over 4000 species. Regression of these data using the negative taxonomic equation in the preceding paragraph yielded surprisingly similar results: $k = 0.63$ (standard error = 0.03; $r^2 = 0.75$).

I also used the taxonomic allometry method on living irregular echinoids. Using Ghiold's (1988) global compilation of 414 known species (in 28 families), I calculate a negatively allometric slope (k) of 0.44 (standard error = 0.08; $r^2 = 0.56$). About 60 percent of all known living species are spatangoids (Ghiold 1988), and the two most species-rich families of all living irregular echinoids (Brissidae, Schizasteridae) are also spatangoids.

Ricklefs (1995) notes that the increasing species-to-genus ratios of negative taxonomic allometries reflect "relatively high rates of speciation and more recent derivation" of superdiverse genera (see also chapter 5). Negative taxonomic allometry conforms to the previously stated predictions that the biosphere tends to accumulate extinction-resistant species through time, and these species differently accumulate in superdiverse, low-turnover taxa and subtaxa.

A recent, very important substantiation of this has been the findings of Gaston and Blackburn (1997), using phylogenetic analysis of birds, that younger (more recently derived) taxa tend to contain a lower proportion of threatened species. They suggest that the most probable mechanism for this is that these older taxa are in a phase of long-term decline. Again, this conforms to the fossil patterns of replacement by lower-turnover groups, and declining origination rates. In ecological terms, this also conforms to the body-size and life-history findings (e.g., Marzluff and Dial 1991) that more opportunistic traits characterize superdiverse taxa. Indeed, Bennett and Owens (1997) have shown, also using phylogenetic analysis, that extinction-proneness in the selective extinction of bird species is correlated with large body size and low reproductive rates.

Supertaxa as Distortions in the Fossil Record

The fossil record is highly biased toward preservation of abundant species (Sepkoski 1994). Given the proposal here, that abundant subtaxa tend to have qualitatively different ecological and evolutionary dynamics, our perception of historical events in the fossil record could be seriously distorted. The majority of rare species, with their higher turnover rates and lower rates of diversification, may be completely or largely invisible and thus unaccounted for in the major studies of fossil biodiversity (e.g., Sepkoski 1992c).

It is difficult to prove this because we cannot quantify what we do not see (McKinney 1996b). However, initial efforts to examine the effects of the bias on paleocommunity patterns, such as "stasis," indicate that there may be a strong artifactual (and thus misleading) contribution (McKinney et al. 1996). At coarser scales, such as extinction rates of higher taxa, such biases may lead to substantial underestimation of true extinction (and origination) rates (Jablonski 1995; McKinney et al. 1996). Fortunately, statistical methods are emerging that allow paleontologists to adjust for rarity biases and make probability estimates of true turnover rates (Foote and Raup 1996).

Mass Extinction Recovery and Generalists

While seemingly in conflict, the evidence discussed here may actually complement the evidence in Harries and others (1996) that mass extinction survivors consist of far more than generalist or opportunistic species. I readily agree with Harries and others (1996) that refugia, chance, preadaptations, and other factors also play a role in species survivorship of catastrophic events. However, my main concern here has been with the very long term patterns of evolutionary radiation, and the ecological (life-history) correlates of those patterns. As a result, my view has been more explicitly phylogenetic than that of Harries and others (1996), whose survivor categories cut across a wide range of ecological and taxonomic groupings. In contrast, this chapter has focused on more specific issues: identifying the life-history (and other) traits that promote long-term diversification within single groups (clades, taxa, and subtaxa).

Chapter Overview and Conservation Implications

While the concept of dynamic equilibria has been oversimplified in ecology (Botkin 1990), there is a large body of evidence that global biodiversity (usually measured by taxon richness) shows equilibrial patterns through geological time. The physical environment seems to generally determine biodiversity levels that are achieved by speciation and extinction. The finite availability of resources translates into eventual near-saturation of available niches (Ricklefs and Schluter 1993). Key evidence favoring this is the role of incumbency in influencing the composition of biota at many scales, ranging from local communities to global (Jablonski and Sepkoski 1996). Other evidence is that global diversity equilibria seem to be attained by decreasing origination rates (accompanied by relatively constant extinction rates) during evolutionary time (Alroy, chapter 12) and that high rates of diversification occur after a mass extinction (Patzkowsky 1995).

When synthesized, these (and other) lines of fossil evidence produce a tentative view of a biosphere that gradually accumulates more extinction-resistant species during geological time (Gilinsky, chapter 9; Sepkoski 1992b; Gaston and Blackburn 1997), except for brief catastrophic intervals when at least some of the "usual" rules of survival are broken (Jablonski 1995). This restarts the process of accumulating more extinction-resistant species.

Brown (1995), Rosenzweig (1995), and Lawton (1995) are among those who have argued that large-scale biodiversity patterns can be more usefully understood if explained in terms of population dynamics (see also chapter 9). This chapter has attempted to explain taxic incumbency at the population level. The extinction-resistant species that accumulate during most of geological time may thus also tend to be extinction resistant at the population level. Such extinction resistance in a species may be promoted by high local abundances, wide geographic ranges, and high intrinsic rates of population growth (Holt et al. 1997). These traits would promote incumbency at many geographic scales because disturbances could lead to replacement of extinction-prone populations at many scales. It has been suggested, for instance, that species with higher local carrying capacities have lower extinction rates and/or higher colonization rates (Holt et al. 1997 and references therein). A specific example is that living mammal taxa with relatively high extinction-prone populations also tend to have disproportionately more species that are threatened. Thus primates and perissodactyls have higher rates of local extinction than artiodactyls and rodents (Bodmer et al. 1997) and also tend to have relatively more threatened species globally (McKinney 1998a).

The biological causes of such extinction resistance are often explained in terms of niche breadth. Brown (1995) defines a broad-niched (or generalist) species as one that "can tolerate a wide range of physical conditions, use many kinds of resources, and survive in the presence of many potential enemies." Lawton (1995) notes that such tolerances are genetically influenced (physiology, diet, and so on), so that they are likely to be phylogenetically heritable. Specific examples are the role of large body size and low reproductive rates in promoting selective extinction among bird (Bennett and Owens 1997) and mammal (Bodmer et al. 1997) taxa. In other words, extinction resistance can be passed on to descendent species and is nonrandomly clustered among taxa (McKinney 1997). Furthermore, younger clades seem to have more extinction-resistant species, for reasons as yet unclear (Gaston and Blackburn 1997).

In sum, the available evidence conforms to (but does not prove) the hypothesis that extinction-resistant species accumulate through most of geological time, including the last 60 million years. Furthermore, they accumulate nonrandomly among taxa, to produce the taxonomic allometric patterns that we see today, with steadily diversifying subtaxa such as rodents (among mammals), passerines (among birds), and spatangoids (among echinoids). The last example suggests the important role of key innovations in producing extinction resistance. Spatangoid success is often attributed to their key adaptations for deep burrowing in a variety of sediments (Smith 1984), allowing them to exploit food resources not available to other irregular echinoids, and to avoid predation more effectively. Similarly, Ricklefs and Renner (1994) explain exceptional species richness in plant subtaxa as occurring from "the ability to enter new habitats," which allows the persistence of nascent species and other advantages that accrue when preempting available niches.

Conservation Implications

Any evolutionary trend toward increasing extinction resistance should not be misinterpreted to mean that the current biosphere is buffered from the looming specter of massive human disturbance. A more sophisticated interpretation of the fossil patterns for conservation biology would be as follows. One, while species may have become, on average, generally less prone to extinction, there are clearly many species that remain very prone to extinction. Two, differences in extinction proneness among taxa should be a focus of conservation efforts, a consideration that is often omitted (see Gaston and Blackburn 1997 for discussion). For example, the evidence here suggests that species in older, more unique taxa may (unfortunately) be more prone to extinction than species in younger taxa that are less phylogenetically unusual and thus contribute less to net biodiversity.

Three, extinction-prone taxa tend to be more prone at many scales. Taxa that are more threatened at global scales (with relatively more threatened species) also tend to be more threatened at local scales (with more easily extinguished populations). Four, the projected result of current highly selective extinction patterns is to accelerate biodiversity loss faster than would occur if population and species extinction were taxonomically random.

Part One

Phylogenetic Turnover:
From Populations Through Higher Taxa

2

Do Taxa Persist as Metapopulations in Evolutionary Time?

Susan Harrison

Studying the history of diversity involves trying to explain variation among taxa in their rates of extinction and speciation (e.g., Simpson 1953; Stanley 1975; Sepkoski 1992b; Cracraft 1992; Gilinsky and Good 1991). For species or genera within particular groups, correlates of either or both of these rates include physical traits such as body size (Stanley 1987; Jablonski 1991), temperature tolerance (Jablonski et al. 1985), or armor (Vermeij 1987b); ecological traits such as the degree of trophic specialization (Vrba 1980); and environmental variables such as productivity (Vermeij 1993), latitude (Jablonski 1991), or ocean depth (Jackson 1974; Jablonski and Lutz 1983; Signor 1990). However, two interrelated factors are considered to be virtually universal influences on both extinction and speciation, namely geographic range structure and dispersal ability (Jablonski and Lutz 1983; Valentine and Jablonski 1983; Vermeij 1987b; Cracraft 1992). Patchy distributions and poor dispersal are expected to increase the rates of both extinction and speciation, whereas continuous distributions and high dispersal should generally have the opposite effects.

In an interesting parallel, ecologists have for the past several decades been developing a theory that relates the persistence of species in contemporary time to their spatial distributions and dispersal abilities. This theory, known as metapopulation dynamics, has been the subject of considerable mathematical modeling, a steadily growing number of empirical studies, and a widening range of applications in ecology, evolution, and conservation (reviewed by Hastings and Harrison 1994; by Hanski and Gilpin 1997; and later in this chapter).

The purpose of this essay is to evaluate the usefulness of metapopulation theory in understanding patterns in macroevolution (also see McKinney and Allmon 1995). I will first define metapopulation dynamics; then, leaving aside speciation for the moment, I will ask whether the persistence of taxa in geologic time can be understood in terms of extinction–colonization balance, and what the implications of this might be. Then, attempting to answer this question, I will draw on indirect evidence from ecological studies and a modest amount of direct evidence from paleobiology. Finally, I will conclude by speculating about the net effects of metapopulation structure on diversification, as a product of its potentially opposing effects on persistence and speciation.

What Are Metapopulations?

In the broadest sense, metapopulations are simply sets of subdivided populations, created by the fact that patches of a species' habitat are farther apart than the species typically disperses. In this sense, it is possible to say that virtually every species is a metapopulation at some spatial scale. Most metapopulation models and theories, however, use the term in a narrower sense that carries many more potential implications. The typical framework for metapopulation models is that of a "population of populations," undergoing a process analogous to birth and death: each population is subject to local (population) extinction, new populations are founded when dispersing individuals reach empty patches of habitat, and the metapopulation persists as long as the rate of recolonization balances that of local extinction.

In such a "classic" metapopulation (figure 2.1a), for any given rate of local extinction, the persistence of the metapopulation depends critically on the rate of dispersal among patches. Also, although the simplest models feature an infinite number of equidistant patches, more realistic models show that regional persistence depends on two other factors as well: metapopulation size, i.e. the number of local populations and suitable habitat patches, and the distances among habitat patches, which influence rates of colonization.

Empirically, this spatially and temporally dynamic view of populations arose in large part from observations of species in naturally patchy and transient environments, e.g. plants and insects in successional habitats (Harrison and Taylor 1996). However, metapopulation models have since been applied to a far broader range of phenomena. Much theory has been devoted to showing that pairs of species incapable of coexisting locally, such as strong

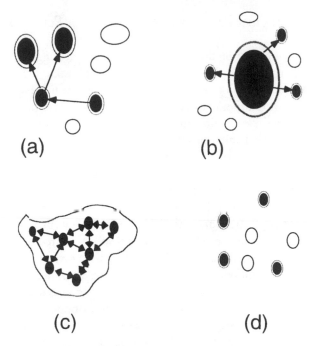

Figure 2.1.   Four types of metapopulation structure: classic (a), mainland–island (b), patchy populations (c), and nonequilibrium (d). *Filled circles* represent habitat patches that currently support populations; *empty circles* are temporarily vacant habitat patches; *dotted lines* represent the lifetime movement radii of individual organisms; *arrows* represent dispersal between patches. See text for further explanation. *–From Harrison 1991.*

competitors, predators and prey, or pathogens and hosts, can coexist in sub-divided environments with limited dispersal (see reviews in Hanski and Gil-pin 1996). Recently, metapopulation models have been developed to predict the fates of species in habitats becoming fragmented by humans (reviewed in Harrison 1994).

Metapopulation ideas have also been important in evolution, ever since Wright (1940, 1978) proposed the existence of a "shifting balance" between processes acting within and among populations. In fact, the term *metapopulation,* and formal metapopulation models, originated Levins' (1970) theo-retical study of interdemic selection. More recent theoretical and empirical work has examined the effects of extinction and colonization on the parti-tioning of genetic variation among populations (see reviews in McCauley 1991; Harrison and Hastings 1996), and the implications of metapopulation processes for coevolution have been explored by Thompson (1994).

From Ecological Persistence to Geological Persistence

The elegant simplicity of the metapopulation model suggests an equally simple way to view the evolutionary persistence of taxa. Do higher taxa persist as metapopulations in geologic time? That is, can the life span of a family or genus (or any multispecies clade) be usefully viewed in terms of a dynamic balance between the extinction and colonization of populations, played out on a field of suitable habitat patches? Here, I am defining *local extinction* as the extirpation of any spatially continuous segment of a species, up to and including the entire species. Speciation, unlike extinction, has no analogy in ecological time, and so I will ignore it for the present, while focusing on extinction and persistence. Therefore, it does not matter here whether the newly founded populations are of the same species that went extinct, another (existing) species of the same clade, or a novel species in that clade.

The conceptual model just described would have several interesting implications. First, evolutionary biology has seen decades of debate on whether natural selection acts at levels higher than those of individuals and kin (Stanley 1975). If local extinction and recolonization of populations have been omnipresent in the history of taxa, it adds support to the idea that selection can act on population-level traits. Second, recent metacommunity models (e.g., Holt 1994) have proposed that entire biotas are structured geographically into local subsets of their component species, through the interplay of metapopulation processes (extinction and colonization) with local ones (competition and predation). For this to be true, the component species must have behaved as metapopulations throughout the history of the community. Third, if extinction and recolonization are ubiquitous in the histories of many taxa, it suggests that the search for patterns in the history of diversity should focus on traits conducive to success in metapopulation terms, in particular, dispersal.

Before evaluating the evidence for this metapopulation view of evolutionary persistence, it may be useful to describe the ecological methods for inferring metapopulation dynamics—excluding experiments and direct measurements of extinction and colonization, since these have little relevance to paleobiology. Ecologists who have used species distributions to evaluate metapopulation dynamics have done so by mapping the locations of all local populations and habitat patches within a geographic area; by showing that suitable but unoccupied habitat patches exist; and by showing that habitat patches are more likely to be occupied the nearer they are to other occupied patches (e.g., Harrison et al. 1988; Opdam 1990; Laan and Verboom 1990; Lawton and Woodroffe 1991; Thomas et al. 1992).

What patterns can we seek in the fossil record, which generally consists of vertical records of species' distributions at single points in space, to determine whether taxa fit the metapopulation model described above? Some necessary, although perhaps not sufficient, conditions are as follows. First, a taxon should show a highly dynamic distribution through time, such that there is no single geographic location at which it is present throughout its known history. Second, for taxa whose longevity in the fossil record can be estimated, some of the variation in longevity should be explainable by traits that influence susceptibility to local extinction. Third, some of the variation among taxa in their longevity should be explained by their relative dispersal abilities.

Unfortunately, the first prediction is unlikely ever to be directly testable, given the biases and incompleteness of the fossil record (Signor 1990). Obtaining an adequate profile of the spatial distribution through time of any taxon, let alone many taxa as would be required, is an almost hopeless task. It is clearly impossible to obtain evidence on the existence of suitable and unoccupied habitat, or on the spatial relationships among empty and occupied patches, for fossil taxa. As an alternative approach, I will draw upon ecological studies of metapopulations, and ask whether available evidence supports the idea that species persist in an extinction–colonization balance in contemporary time. I will then argue that the answer has parallels in evolutionary time.

The second prediction is difficult to approach for purely logical reasons. It is hard to conceive of traits of species or taxa that would influence their susceptibility to local extinction, without also affecting their likelihood of global extinction. Any predictions would risk circularity ("taxa persist better because they are less likely to go extinct"). The third prediction yields more hope of direct evaluation, however. For certain groups, the link between dispersal abilities and taxon longevity has been examined, although the evidence is far from complete. I will review this evidence and discuss its possible meaning and limitations.

Do Species Persist as Metapopulations in Ecological Time?

Two defining features of classic metapopulations, as discussed above, are that all local populations are subject to extinction, yet the metapopulation persists through recolonization. These features are the basis for most of the unique predictions made by metapopulation models, such as that the species will become locally and regionally extinct if dispersal rates decline or if patches become too few or too isolated. However, a review of the empirical literature casts some doubt on how frequently these basic assumptions of metapopulation models are met, and it suggests several alternative metapopulation structures that may be more common than the classic kind (also see Hastings and Harrison 1994; Harrison and Taylor 1996).

Natural systems in which local populations are patchily distributed, and in which extinction and colonization are frequent, often include one or several large or otherwise fairly stable populations (e.g., Schoener and Spiller 1987; Harrison et al. 1988; Pimm et al. 1988; Schoener 1991; Harrison and Taylor 1996). These extinction-resistant populations may act as the major sources of the migrants that recolonize the smaller, shorter-lived populations. In such mainland–island metapopulations (figure 2.1b), local extinction and recolonization occur but need not be in balance, since the metapopulation will persist as long as the mainland populations survive.

Species that naturally occupy short-lived patches of habitat, such as early successional plant species or the insects that feed on them, are almost by definition excellent dispersers. Individuals of such species inhabit and leave progeny on several to many patches in their lifetimes. Such species tend to form patchy populations (figure 2.1c) in which there is little demographic subdivision among patches, and therefore no true local extinction or recolonization on patches (Harrison et al. 1995; Harrison and Taylor 1996). This implies that persistence is not sensitive to small differences in dispersal rate or interpatch distances, and that suitable habitats are rarely unoccupied. More generally, these conclusions apply to any species that disperses well relative to the distances among units of its habitat.

Conversely, when habitats are separated by very large distances relative to a species' dispersal abilities, the result is a set of conspecific populations that are completely isolated from one another. In such nonequilibrium metapopulations (Brown 1971; Harrison and Taylor 1996; figure 2.1d), local extinctions are seldom or never offset by recolonization, and they are therefore merely steps on the way to regional extinction.

Finally, there are some systems that appear to exhibit classic metapopulation dynamics (see figure 2.1a). Notable examples include recent studies of pool frogs (Sjogren 1994) and butterflies (Hanski et al. 1994). However, existence in a balance between extinction and recolonization may be a less common condition than it would at first appear; habitats may often be subdivided too unequally, too finely or too coarsely (relative to dispersal) to meet the basic assumptions of metapopulation models (figure 2.2). This is not just a semantic distinction, because only in classic metapopulations is regional persistence sensitively dependent on the among-population processes of dispersal and recolonization. In the other cases, persistence is largely a function of processes acting within individual populations.

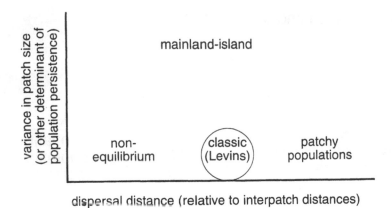

Figure 2.2.   Relationships between different types of metapopulation.

A qualitatively similar conclusion has often been drawn about evolution through interdemic selection: Although it is theoretically possible, it can occur only under a narrow range of parameter values that permit a high level of genetic differentiation among populations to coexist with a high rate of population turnover (see, e.g., Slatkin 1985). In defense of interdemic selection, it has been shown that turnover may enhance differentiation among populations, if populations are founded by small and homogeneous groups (Wade and McCauley 1988; McCauley 1991; Whitlock 1992). Counteracting this effect, however, are two processes that erode differentiation among populations within metapopulations: the high rates of dispersal (therefore gene flow) exhibited by species that undergo high rates of local extinction, and the long-term loss of total genetic diversity from the metapopulation because of the accelerated genetic drift caused by turnover (see Harrison and Hastings 1996; and figure 2.3).

Returning to the persistence of higher taxa in geologic time, some parallels may now be drawn. There are obvious similarities between mainland–island metapopulation dynamics and the contraction of taxa to refugia during periods of glaciation or other adverse environmental events (e.g., Jablonski 1991; Vermeij 1993). Likewise, by analogy with patchy populations, there may be some taxa that disperse so well that geographic isolation, local extinction, and suitable but vacant habitats have not featured importantly in their histories (see next section). Finally, taxa may form nonequilibrium metapopulations; one possibility is that once founded, populations remain completely isolated from one another and never experience gene flow or recolonization.

Another possible nonequilibrium mode of persistence in evolutionary time is through metapopulation dynamics that are driven by habitat change.

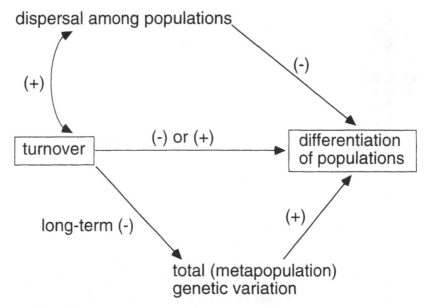

Figure 2.3. Relationships between population turnover (extinction and colonization) and genetic differentiation among populations. *Straight lines* represent cause and effect; *curved arrow* represents correlation. Directly, turnover can either increase or decrease differentiation, depending on propagule size and composition (*central arrow*). However, taxa with high population turnover tend to be good dispersers, leading to low differentiation (*curved arrow*). Also, turnover causes the long-term loss of variation from the entire metapopulation, eventually leading to homogeneity among populations as well (*bottom two arrows*). See Harrison and Hastings (1996).

Thomas (1994) argues that in real metapopulations, turnover in ecological time is seldom stochastic as most theory implicitly assumes. Rather, extinctions are usually a result of habitat loss or degradation, and colonization occurs when new habitats are created near existing populations. The important implication is that equilibrium at the level of the metapopulation, which arises (in models) from the assumption of a fixed universe of habitat patches, does not exist. The analogy on an evolutionary time scale would be a taxon that spreads and colonizes, then declines and becomes extinct, as its habitat spreads and then declines (McKinney and Allmon 1995).

Of course, in evolutionary time, suitable habitat for a taxon must be defined relative to its evolving adaptations and its evolving biotic environment. The universe of patches may therefore be just as volatile as species distributions themselves. This represents a general limitation on applying metapopulation theory to macroevolutionary processes.

It is interesting to note the strong similarities between the three alternative types of metapopulation structure described above, and three major modes of speciation that have been proposed (see Valentine and Jablonski 1983). Mainland–island population structures are considered optimal for promoting founder speciation (Mayr 1942); sets of local populations that are not panmictic, but are linked by consistent gene flow, create the conditions for clinal evolution (Endler 1977); and sets of isolated (but not small) populations are the precursor to vicariant speciation (Hennig 1966).

To recapitulate the above arguments, I suggest that simply having a patchy habitat, or exhibiting some local extinctions and recolonizations, is not enough to make species fit the classic metapopulation model. Ecological evidence suggests that many species have these traits, yet few can be described as existing in a dynamic balance between extinction and colonization. More often, they exist either by virtue of a persistent mainland population, or by a degree of dispersal that prevents population subdivision, or in a process that resembles a random walk more than a dynamic equilibrium. By analogy, the same may be true for taxa in evolutionary time.

Do Better-Dispersing Taxa Persist Longer in Evolutionary Time?

The paleobiological evidence most relevant to the questions posed here comes from studies of fossil marine invertebrates. Benthic marine invertebrates may have planktotrophic (PT) larvae that feed in the water column for an extended period, or nonplanktotrophic (non-PT) larvae that either never enter the plankton or enter it for a short time and do not feed until they settle (see Jackson 1974; Jablonski and Lutz 1983 for detailed descriptions of these dispersal modes and their taxonomic and ecological correlates). Species with PT larvae presumably disperse their propagules over vastly greater distances than non-PT species, as is generally confirmed by electrophoretic studies on extant taxa (Jablonski and Lutz 1983; Ward 1990; Hunt 1993; Hellberg 1994).

For numerous taxa in the fossil record, PT species show greater longevity in the fossil record than do non-PT species (Scheltema 1971; Shuto 1974; Hansen 1978; Jablonski and Lutz 1983; Valentine and Jablonski 1983; Vermeij 1987b; Jablonski 1987, 1991). This is the clearest direct evidence linking a key metapopulation characteristic, namely dispersal ability, to persistence in evolutionary time.

There are several limitations on the interpretation of this result, however. The most important, for our purposes, is that the link between dispersal and longevity appears to be mediated by geographic range size: better dispersers persist longer because they are more widely distributed (Jackson 1974; Jablonski and Lutz 1983; Valentine and Jablonski 1983; Vermeij 1987b, 1989; Jablonski 1987, 1991). Under a classic metapopulation scenario, in which taxa persist through a balance between local extinction and recolonization, it should also be true that dispersal promotes longevity even when the total size of the geographic range is held constant. This issue has never, to my knowledge, been examined.

Moreover, the generality of the dispersal–longevity connection can be questioned on several grounds. First, other traits besides dispersal ability may lead to large geographic ranges, for example broad temperature tolerance (Jablonski et al. 1985). Second, in some groups of marine invertebrates, range size is not a good predictor of taxon longevity (Stanley 1987). Third, it is unclear whether this pattern can be generalized to other groups of organisms, few of which manifest the huge range in dispersal abilities represented by PT and non-PT marine invertebrates.

Few other groups are as well represented in the fossil record as marine invertebrates, but once again, some indirect ecological evidence may be useful. In a study of spatial patterns of diversity among terrestrial species in the United Kingdom, Harrison et al. (1992) found that, contrary to conventional wisdom, there were no differences in beta diversity between small- and large-seeded plants, flying and nonflying animals, or plants and animals. Beta diversity is the spatial turnover component of regional diversity, and it is inversely related to the sizes of species' geographic ranges. This suggests that at least at that spatial scale, organisms had dispersed relatively well during their histories of postglacial colonization, so that their dispersal abilities did not represent a constraint on their geographic distributions. In related work, Harrison (1992) found that dispersal constrains the distributions of Pacific bird species only at scales of greater than 1000 km across ocean barriers (see Downes 1980 for related evidence on insects). If limited dispersal constrains ranges only at these very large scales, it casts some doubt on the expectation that dispersal ability will be a strong, general correlate of taxon longevity. However, more work on this is needed.

Conclusions: Alternative Types of Metapopulation Structure, and Their Implications for Persistence and Diversification

Because neither species nor higher taxa can persist for long if extinctions of populations outpace the founding of new ones, it is sometimes taken as self-evident that species exist in the balance between extinction and colonization portrayed in metapopulation models. However, I have argued that this is untrue; in both ecological and evolutionary time, metapopulations have a variety of structures, not all of which require or imply an extinction–colonization balance. Important alternative ways to persist include having refugia, in which certain populations never go extinct (Jablonski 1991; Vermeij 1993); and nonequilibrium dynamics, in which the species or taxon shifts its geographic distribution in response to changing availability of habitat, and there is no tendency toward a steady-state distribution.

Returning to some of the original reasons for interest in the metapopulation model, we can now examine the implications of having concluded that local extinction, colonization, and narrow-sense metapopulation dynamics do not appear ubiquitous. The conclusions drawn here give little support to the plausibility of selection at the level of populations; the traits of individuals inhabiting highly persistent (mainland) populations may be more important than traits of extinction-prone (island) populations. Nor is there any support here for metacommunity models of community assembly, which propose that geographical patterns of diversity are shaped importantly by the chance appearances and disappearances of species on habitat patches. Instead, it appears likelier that species are found in most of their habitats most of the time, at least within the bounds set by absolute (e.g., oceanic) barriers to dispersal.

Most importantly for our present purposes, rejection of the classic metapopulation model implies less reason to believe that dispersal ability will have a universally large effect on the persistence of taxa. Good dispersal may lead to greater persistence for some groups of organisms, but it does so mainly through increasing the sizes of their ranges, which is not a metapopulation effect. For many taxa, persistence will have nothing to do with dispersal ability, but instead it will be related to other traits influencing geographic range size, or by extrinsic factors such as the proneness of different types of habitat (e.g., shallow versus deep marine, tropical versus temperate) to major environmental changes.

These speculative conclusions set the stage for returning briefly to the subjects of speciation and net diversification. Because geographic distribution and dispersal abilities affect the rates of extinction and speciation in opposite directions, it remains unclear which type of taxon should diversify faster. But from the arguments just presented, the influence of these factors on the rates of extinction of taxa should be weak and inconsistent. If, in contrast, their influences on speciation are generally quite strong, as argued by Valentine and Jablonski (1983), then we can predict faster diversification in more patchily distributed and poorly dispersing groups. Direct evidence is once again limited, but at least for some groups this prediction appears to be met (Futuyma 1986).

Acknowledgments

I thank Phil Ward, Geerat Vermeij, Phil Signor, James Drake, Chad Hewitt, and Michael McKinney for insightful comments, and Ted Grosholz for introducing me to the world of fossil marine invertebrates.

3

Geographic Range Fragmentation and the Evolution of Biological Diversity

Brian A. Maurer and M. Philip Nott

That which biologists measure as biological diversity is the consequence of a great number of complicated processes operating over numerous spatial and temporal scales. One important component of biological diversity is species richness, or the number of species found in a given place during a specific period of time. Species richness is measured at a variety of scales, from that of the local community in a given season to an entire continent over evolutionary time. In the past, it has been somewhat difficult to relate processes occurring in local communities to those operating on much larger scales, but recently a number of promising approaches provide new ways to integrate ecological and evolutionary perspectives of species diversity (Brown and Maurer 1989; Ricklefs and Schluter 1993b).

In this chapter, we explore how ecological processes affecting diversity might be related to evolutionary ones. The basic idea is relatively simple: Every species has a spatially discontinuous geographic range and the more discontinuous the range, the more likely a species is to go extinct. The degree of discontinuity in a species' range is thought to ultimately be related to its ecological needs and their spatial distribution in the environment (Brown 1984, 1995; Lawton et al. 1994; Maurer and Villard 1994). A consequence of this view is that the likelihood of extinction of species is related to its ecology through the degree of fragmentation of its geographic range. Species that have rare or widely dispersed resources, or are limited more severely by climate or food, will have more fragmented ranges. Fragmentation of a range increases the number of isolated populations, and hence it decreases the likelihood of "rescue effects" preventing local extinctions. This should lead to increased rates of extinction of local populations that should in turn increase the risk of global extinction. This proposed mechanism has implications for the evolution of species diversity: For two sister clades (groups of related species) sharing similar ecologies, the rate of species extinction should be higher in the clade whose species have more fragmented geographic ranges.

Although these ideas are relatively straightforward, actually measuring the hypothesized processes presents an enormous empirical challenge. Our intent here is to show how current distributional information might be used to indirectly examine aspects of the diversification process. Our argument begins with a theoretical discussion of how species diversity might be regulated over evolutionary time. This, of course, assumes that regulation of diversity occurs. Alroy (chapter 12) examines evidence of such regulation by examining the dynamics of species diversity in Cenozoic mammals in North America; we examine a data set on Neogene foraminifera that provides further evidence for diversity regulation. The theory of diversity regulation does not specify explicit mechanisms, however. We use data on the current geographic distributions of North American passerine birds to examine how diversity might be regulated in part by nonrandom extinction.

The Regulation of Diversity

There are a number of ways to model the regulation of diversity of a clade over evolutionary time. Generally, rates of speciation and extinction can be regulated by factors that do not depend directly on the ecological characteristics of species in a clade (external factors) and by factors that are a direct consequence of those characteristics (internal factors). External factors include the geological activity of a continent, changing patterns of climate, the diversity and ecological "efficiency" of natural enemies such as predators, competitors, and diseases, and the availability of appropriate food and habitat. Clearly, these factors are linked with one another, so it is generally difficult to separate out their effects. Furthermore, they operate over a number of spatial and temporal scales. Competition, for example, may play different roles at different scales (Maurer 1985). At local scales, it may be complexly linked to habitat quality and food abundance, while at continental scales it may determine the ability of a newly formed species to successfully establish itself. Internal factors that regulate diversity include the genetically determined morphological, physiological, and behavioral attributes of species that determine their abilities to use resources, interact with other species, and maintain a characteristic spatial structure of populations distributed throughout their geographic ranges. Additionally, there may be certain aspects of a species' genetic system that may predispose it to speciate more readily.

These internal and external factors interact with each other in complex ways to determine rates of speciation and extinction in clades. The rate at which a clade diversifies has been expressed as a function of species number by several authors (MacArthur 1969; Rosenzweig 1975; Sepkoski 1978, 1979, 1984; Carr and Kitchell 1980; Maurer 1989), but this rate is also a function of the many external factors that define the ecological successes of species in evolutionary time. In what follows, we assume that the mathematical functions used to describe speciation and extinction rates are implicitly functions of the geological and ecological context in which a clade evolves. Clades evolving in different geological circumstances, for example, may have functions that differ both quantitatively (i.e., in magnitude) and qualitatively (i.e., in shape).

Walker (1985) called mathematical functions used to describe the evolution of biological diversity, diversification functions. A diversification function gives the instantaneous per species rate of change as a function of the number of species in a clade. If dS/dt is the rate of change in species number, then the diversification function governing that rate of change, $\rho(S)$, is given by

$$\rho(S) = \frac{1}{S}\frac{dS}{dt} \qquad [1]$$

so that the rate of change in species number is

$$\frac{dS}{dt} = S\ \rho(S) \qquad [2]$$

Commonly, $\rho(S)$ is decomposed into per species speciation and extinction rates, such that $\rho(S) = \zeta(S) - \xi(S)$, where $\zeta(S)$ is the per species rate of speciation and $\xi(S)$ is the per species rate of extinction. Thus, the rate of species change is

$$\frac{dS}{dt} = S\ [\zeta(S) - \xi(S)] \qquad [3]$$

The biological interpretation of the function $\rho(S)$ is that it expresses how the internal and external factors that regulate diversification depend on the number of species in the clade (S). Clades with exceptionally high speciation rates are not necessarily more diverse than those with low speciation rates, because they may also experience high extinction rates. Similarly, clades with exceptionally low extinction rates may not be more diverse than those with high extinction rates. Rates of speciation and extinction are often correlated (Stanley 1990c; Gilinsky, chapter 9), so species with large speciation rates often have large extinction rates as well. Equation 3 implies that it is the difference between speciation and extinction rates that is most important in regulating diversity.

This general theoretical framework for modeling diversification suggests that examining the diversification functions for sister clades in the fossil record will provide insights into how the external and internal factors regulating diversity operate. Sister clades, by virtue of their shared plesiomorphies that have ecological consequences, should have similar diversification functions, differing in only a few key attributes (i.e., internal factors that affect diversity). These key attributes should predispose members of one clade to be more likely to speciate, less likely to go extinct, or both. Hence, over a range of diversities (S), one clade will have a greater net rate of diversification than the other (figure 3.1). Since by definition sister clades originated at roughly the same time, the clade with the greatest net rate of diversification will diversify faster and maintain a higher number of extant species at any one point in time.

An example illustrates the complexities of inferring mechanisms of diversity regulation from speciation and extinction rates alone. Stanley et al. (1988) analyzed the diversity of two clades of foraminifera during the Neogene. One clade, the globigerinids, was made up of an ecologically diverse group of species, while the second, the globorotaliids, consisted of more specialized species. With the exception of the last million years or so of the Neogene, globigerinids were more diverse than globorotaliids (figure 3.2). Globorotaliids had a higher average number of speciation events per 1-million-year (my) interval than the globigerinids, but they also had a higher average number of extinction events (table 3.1). Average percent speciation and extinctions were also higher for the globorotaliids. The difference between average percent speciations and extinctions was higher for the globorotaliids as well (see table 3.1). Therefore, examining speciation and extinction rates as they are commonly calculated by paleontologists does not explain the differences in diversity between to the two clades. Based on the data in table 3.1, the expectation is that the globorotaliids should have maintained the highest diversity during the Neogene, but they did not.

Examination of $\rho(S)$, on the other hand, makes it quite clear why the globigerinids were the more diverse clade. Average per species rates of change

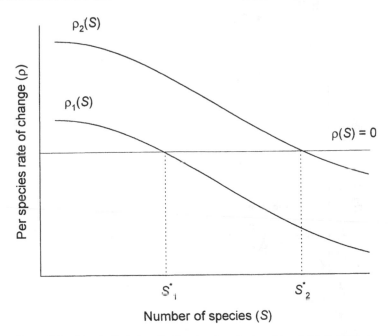

Figure 3.1. Hypothetical diversification functions for two sister clades that differ in ecological tolerances. Because the clades are sister clades, they started as the consequence of a cladogenetic event that produced two different species from a single common ancestor. Clade 2 is derived from the more ecologically generalized species of the pair. Its per species speciation rate exceeds its per species extinction rate by a larger margin than that of the more specialized species from which clade 1 is derived. Thus, clade 2 diversifies at a faster rate than clade 1 when there are few species, and it has a higher equilibrium species number. The shapes of the diversification functions are arbitrary (they would be parabolic if the diversification functions were logistic), but they imply that diversity is regulated (i.e., diversification is not exponential).

during a 1-my interval were calculated from the number of speciations (n_s) and extinctions (n_e) per interval as

$$r = \ln\left(1 + \frac{n_s - n_e}{S}\right) \qquad [4]$$

where r is an estimate of the average value of $\rho(S)$ over any given 1-my interval, and S is the number of species present at the start of that interval (Stanley 1979; Maurer 1989). Note that Equation 4 does not necessarily assume that diversification during an interval was exponential. It represents the average rate of change over the interval regardless of the form of $\rho(S)$.

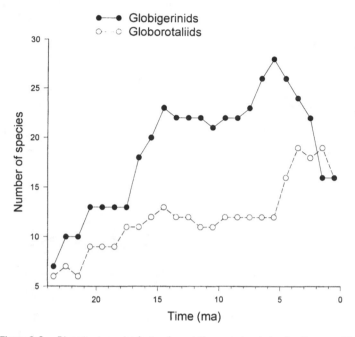

Figure 3.2. Diversity dynamics for two foraminiferan clades during the Neogene. The globigerinids (*filled circles*) were ecological generalists and the globorotaliids (*open circles*) were ecological specialists.

TABLE 3.1 Averages for variables describing changes in species number and the component processes of speciation and extinction for two foraminiferan clades across twenty-four 1-my intervals during the Neogene

Variable	Clade	
	Globigerinids	*Globorotaliids*
Number of species	19	12
Number of speciations	1.54	1.88
Number of extinctions	1.21	1.41
Percent speciation	10.23	16.45
Percent extinction	5.98	11.12
Percent speciation − percent extinction	4.25	5.33
Rate of diversification (my^{-1})	0.03	0.04

Note: Although the average difference between percent speciation and extinction is larger for globorotaliids, they maintained a lower average number of species than globigerinids.

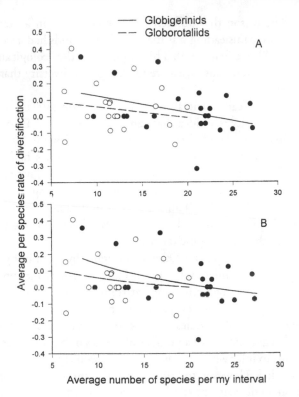

Figure 3.3. Estimated diversification functions for two foraminiferan clades during the Neogene using linear **(A)** and nonlinear **(B)** diversification functions. Note that the ecologically generalized globigerinids (*solid lines*) had a higher diversification function for a given number of species than the more specialized globorotaliids (*broken lines*). Curves for the linear model were estimated by least squares regression; those for the nonlinear model of Maurer (1989) were estimated from linear regression of diversification rates on log species number.

To examine the nature of $\rho(S)$ for the two foraminiferan clades, we plotted values of r against S. Although the average value for r was higher for the globorotaliids (see table 3.1), r was consistently higher for the globigerinids when plotted against S (figure 3.3). There was considerable scatter in r among the twenty-four 1-my periods for both clades, but the trend estimated from linear regression for different kinds of diversification functions (see Maurer 1989) indicated that globigerinid rs exceed globorotaliid rs for any given value of S (see figure 3.3). Hence, the standing diversity of the generalized globigerinids always exceeded that of the globorotaliids, even though both clades were roughly the same age. This means that at any given point in time, the globigerinids diversified at a faster per species rate than the globorotaliids. The reason that differences between clades in average rates of diversification were misleading was that the globigerinids had a steeper negative slope for $\rho(S)$ (see figure 3.3). This implies that the ecologically generalized globigerinids were more sensitive to changes in diversity than their less generalized relatives.

That the ecologically generalized clade of foraminifera maintained a higher species diversity is an idea that can be traced back to Darwin's model of evolution in chapter IV of *On the Origin of Species* (Maurer 1998). Darwin felt that species with a wider range of ecological abilities had larger geographic ranges, and that this generally enabled them to give rise to more descendant species than those with smaller geographic ranges and narrower ranges of ecological abilities. Thus, a key to understanding differences in diversity among groups of related species, in Darwin's mind, was to be found in the nature of their geographic ranges.

In what follows, we pursue Darwin's insights by examining how geographic range structure might relate to the likelihood of extinction. We argue that certain kinds of geographic range attributes increase the likelihood of extinction for any species possessing them, and that these characteristics are related to the ecology of a species. We base our argument on descriptions of the degree of fragmentation of a number of species of North American birds, and we use this to infer which kinds of species are likely to experience higher rates of extinction. However, to fully explain the effects of geographic range fragmentation on diversity, its relation to speciation must be understood. We consider briefly, possible relationships between geographic range fragmentation and speciation rate.

Geographic Range Fragmentation and Extinction

Brown and Maurer (1987) showed that species of North American terrestrial birds with small geographic ranges almost always had low densities. One striking exception to this pattern was the tricolored blackbird (*Agelaius tricolor*). This species has a geographic range that is restricted to western California, yet it is very abundant where it is found. Brown and Maurer's (1987) data on geographic range size were based on rather crude estimates taken from planimetry of geographic range maps in Robbins et al. (1983). We recalculated geographic range sizes for 242 species of insectivorous birds (Maurer 1994) using quantitative data on patterns of abundance taken from the North American Breeding Bird Survey (BBS), and we found virtually the same pattern that Brown and Maurer (1987) obtained for the original 380 species (figure 3.4).

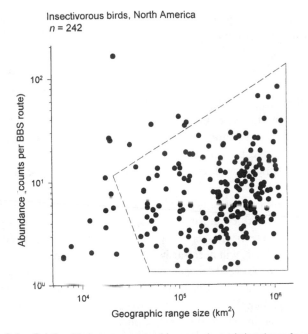

Insectivorous birds, North America
n = 242

Figure 3.4. Relationship between geographic range size and abundance for species
(N = 242) of insectivorous birds in North America. Note the paucity of species with
high average abundance and small geographic range size. Each point represents a
single species. The *polygon* represents boundaries on the relationship between these
variables, hypothesized by Brown and Maurer (1987) to be the result of species dynamics. The *broken lines* indicate boundaries that represent combinations of variables
more likely to result in the extinction of species, whereas the *solid lines* are more absolute constraints (e.g., the size of the North American continent). BBS, North American
Breeding Bird Survey.

Of interest to the present discussion is Brown and Maurer's (1987) suggestion that the pattern in Figure 3.4 implies that species with small geographic ranges rarely build up high average densities because they are more
likely to go extinct than species with larger geographic ranges. Occasionally,
species such as the tricolored blackbird are able to exploit spatially restricted
but highly productive habitats (the tricolored blackbird inhabits the highly
productive marshes and agricultural areas in the central valleys of California),
but on the average, a species with a small geographic range will be limited by
a sufficient number of environmental factors that its average density will be
low (Maurer 1989; Maurer 1990; Brown 1995). Because its average density
is low, the total number of individuals in a species with a small geographic
range will be low, and the species as a whole will have a relatively high probability of going extinct (i.e., a relatively short expected time to extinction).

Can the general mechanism hypothesized by Brown and Maurer (1987) be related to geographic range fragmentation? As it stands, the hypothesis implies that a species goes extinct as a single, spatially continuous population. This certainly seems to underlie much of the thinking behind early attempts to relate population size to extinction likelihood (MacArthur and Wilson 1967; Leigh 1981). If geographic populations are actually spatially fragmented collections of subpopulations, then ignoring the effects of the spatial distribution of subpopulations, and their degree of connectedness, may limit the application of such models to global extinction of species. This is where geographic range fragmentation may provide additional insight into the extinction process.

To see the importance of geographic range fragmentation in determining the likelihood of extinction for a species, consider the relationship between the area and perimeter of a geographic range. A small geographic range, by virtue of its small size, will necessarily have more perimeter per unit area than a large geographic range (Maurer 1994). For example, suppose we have two geographic ranges that are perfect circles, and one geographic range has twice the area of the other. The larger of the two geographic ranges will have only 1.41 times the perimeter of the smaller one. This means that if subpopulations are uniformly distributed within the boundaries of the geographic ranges of the two species, the species with the smaller geographic range would have a larger proportion of subpopulations in contact with its range boundary than the species with the larger range. The boundary of a geographic range exists because ecological conditions cannot support viable populations of the species beyond it (Brown and Maurer 1989). Consequently, a species with a small geographic range would have a relatively high proportion of its subpopulations facing likely extinction. The consequence of this would be a higher rate of extinction of subpopulations for species with small geographic ranges. All else being equal, a higher rate of extinction of subpopulations should translate into a higher probability of global extinction. Therefore, species with small geographic ranges should face a higher likelihood of global extinction than species with large ones. The effect on extinction of having a higher proportion of local populations experiencing environmental extremes in a small geographic range must also be added to the fact that such a species will have a smaller total number of local populations than a species with a large geographic range. Thus, such a species will experience a double jeopardy: (1) fewer local populations to go extinct, and (2) a higher rate of local extinctions. We will come back to this point in a moment. First, we consider the possibility that geographic range shape may be different in species with small and large geographic ranges.

The above argument holds only for geographic ranges with smooth boundaries. On the one hand, if small geographic ranges have smoother boundaries than large ones, then small geographic ranges may possibly have a lower proportion of subpopulations in contact with conditions that are extreme for the species than large ones. This depends, of course, on the degree of "jaggedness" of large geographic ranges relative to small ones. On the other hand, small geographic ranges may have more jagged or fragmented boundaries than large ones. If so, then the effect discussed in the previous paragraph would be exaggerated.

Geographic range fragmentation, then, can influence the likelihood of extinction by affecting the rate of extinction of local populations. The higher the degree of fragmentation, the more likely a species is to go extinct. The important question implied by the previous paragraph is how fragmentation is related to geographic range size. Are small geographic ranges more or less fragmented than large ones? Maurer (1994) suggested that geographic range fragmentation could be measured using concepts from fractal geometry. Basically, a fractal dimension of a shape is related to the degree of smoothness of the shape. The box dimension is a measurement of a fractal shape taken by counting the numbers of boxes of various sizes that cover the shape (in this case, a geographic range). The slope of the logarithm of the number of boxes that cover a shape regressed against the logarithm of the inverse of box size is called the box dimension of the shape (Maurer 1994). The more jagged the shape, the lower the box dimension. Thus, geographic ranges with high box dimensions are relatively smooth. Importantly, the box dimension is a measure of *relative* geographic range shape, so that there is no size bias built into its method of calculation. A semilogarithmic plot of box dimension against geographic range size for 242 species of insectivorous birds in North America indicated that there was a positive relationship between range size and box dimension (figure 3.5). Therefore, at least for North American insectivorous birds, species with large geographic ranges have relatively smooth range boundaries. Hence, the effect on extinction of having a small geographic range is exaggerated yet once again by the tendency of species with small geographic ranges to also have more fragmented ones. This suggests that a species with a small geographic range will have a much shorter expected time to extinction than a species with a large one for at least three reasons: (1) it has fewer populations; (2) it has a higher proportion of those populations in contact with environmental extremes; and (3) there is a greater degree of spatial discontinuity among those populations exposed to environmental extremes.

It is important to note that there are some potential benefits of a frag-

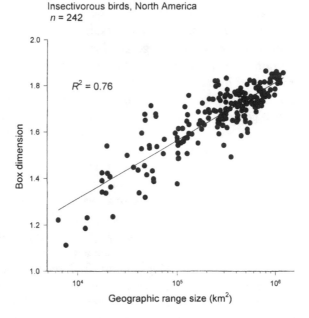

Figure 3.5. Relationship between geographic range size and degree of geographic range fragmentation for North American insectivorous birds. Species that have low box dimensions have high degrees of fragmentation. Note that species with small geographic ranges have high degrees of range fragmentation, whereas species with large ranges have relatively smooth range boundaries. Each point represents a single species ($N = 242$).

mented geographic range. Most notably, fragmentation may allow isolation of negative effects such as disease or catastrophe to a small number of local populations. However, there are additional complications that argue against such effects in geographic ranges. Density is not uniformly distributed across a geographic range; rather, it is highest in central populations (Brown 1984, 1995; Hengeveld 1990; Maurer 1994; Maurer and Villard 1994; Brown et al. 1995). Furthermore, peripheral populations are relatively more variable than central populations (Curnutt et al. 1996). Data such as these suggest that in most species, peripheral populations represent demographic sinks maintained primarily by immigration. The important point to note here is that in a species with a small geographic range, there is a higher proportion of these peripheral, sink populations than in a species with a large geographic range. Hence, risk spreading among sink populations may have a relatively small effect on the global persistence of a species.

The Mechanism of Extinction of Geographic Populations

No species has a constant geographic range size. Over evolutionary time, it is likely that a species' geographic range fluctuates considerably. Ultimately, every species will sustain a sufficiently long period of time during which its geographic range is shrinking and its total population size is declining so that it will eventually suffer extinction. This was clearly part of Darwin's view of evolution (Darwin 1859). As we argued in the preceding section, it is the species with the large geographic ranges that persist for the longest period of time, and that have the highest probability of leaving descendant species. Is there any further insight into the extinction process that we can gain from considering geographic range structure? We believe there is.

If species with large geographic ranges have low degrees of fragmentation, we should expect that there is a direct relationship between fragmentation and the rate of extinction of local subpopulations. This hypothesis can be tested indirectly with the BBS data set. The BBS consists of about 4000 individual survey routes. In any given year, about 2500 of those routes are surveyed. Routes surveyed in consecutive years can be used to estimate the average rate of detection of a given species across all routes.

Consider for a moment the dynamics of occupancy of a species on the BBS routes on which it is observed. Let p be the proportion of routes that the species is observed on in any given year, then the rate function governing the dynamics of p, say $\pi(p)$, can be modeled as

$$\pi(p) = \frac{1}{p}\frac{dp}{dt} \qquad [5]$$

so that the dynamics of route occupancy is

$$\frac{dp}{dt} = p\ \pi(p) \qquad [6]$$

The rate function $\pi(p)$ has two components. First, when a census route is not occupied in a given year and is in the following year, a local "colonization" has occurred. This colonization is a result of the species being either entirely absent from the route in the first year or simply rare enough on the route to escape detection, and building up a sufficient population size in the subsequent year to be detected. Either way, it is not unreasonable to assume that the detection in the second year is the result of an increase in abundance between years, and thus should reflect population increases on the route. Of course, the colonization may be a consequence of random sampling. This will affect the nature of $\pi(p)$ as considered below. Second, a census route on which the species is detected in one year may be unoccupied in the following year. If so, a local "extinction" has occurred. As with the colonization, this extinction may be a result of decreases in local abundance or simply of random detections. Below, we consider differentiating between changes in p that are the consequence of random detections, and those caused by changing local abundances.

The rate function $\pi(p)$, then, is the difference between the rate of local colonizations and local extinctions. If colonizations and extinctions are simply a result of stochastic variation in detections that do not correspond to changes in abundance at local sites, $\pi(p)$ should be independent of p. If there is a statistically significant relationship between $\pi(p)$ and p, then it is unlikely that changes in the proportion of routes on which a species is detected are a result of chance. In fact, we can specify the form that $\pi(p)$ should take on. When p is relatively small, then we expect that the species will be common only on those routes that contain the best habitat conditions for that species. These are most likely to be routes located in central portions of the species' geographic range (Brown 1984), where abundance is more stable (Maurer and Villard 1994). Consequently, changes in p will most likely occur through increases in abundance on routes located in peripheral portions of the species' geographic range, leading to increased colonizations. Therefore, local colonizations will exceed local extinctions. When p is relatively large, then many peripheral routes will already be occupied. Populations on these routes are less stable than those on central ones (Curnutt et al. 1996), so there should be more local extinctions than colonizations. These arguments suggest that $\pi(p)$ should be positive when p is small and negative when p is large. Again, we emphasize that there should be no negative relationship between $\pi(p)$ and p if changes in p are a result of random detections.

We obtained estimates of $\pi(p)$ for several species from the BBS data by calculating the proportion of routes that are colonized and suffer extinctions, and plotting these against the proportion of routes occupied. These calculations are illustrated for two species of wood warblers (*Dendroica*) in figure 3.6. For both species, the difference between the proportion of colonizations and extinctions is negatively related to the proportion of routes on which each species is detected (see figure 3.6A). For both species, when p exceeded a certain value, $\pi(p)$ tended to be less than zero, and when p was smaller than that value, $\pi(p)$ was positive. Linear regression was used to approximate this "equilibrium" value for p for each species. This was done by fitting a straight line to the relationship between p and the estimated value of $\pi(p)$,

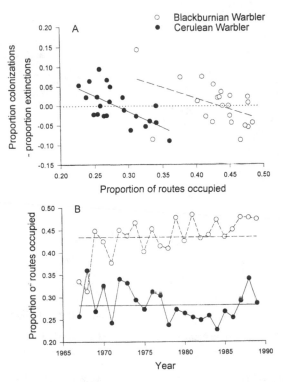

Figure 3.6. **(A)** The difference between the proportion of BBS routes colonized and the proportion that suffered extinction, plotted against the proportion of routes occupied in a given year for two species of wood warblers [Blackburnian (*open circles*) and Cerulean (*filled circles*)]. Each point represents a single year. **(B)** Dynamics of range occupancy for the two species of wood warblers in A. Note that the proportion of routes occupied fluctuates about the proportion where local colonizations equal local extinctions.

and dividing the intercept of the resulting regression line by the absolute value of its slope. The dynamics of p for both species oscillated about this value (see figure 3.6B). Let this value be p^*, then $\pi(p^*) = 0$. The quantity p^* can be considered a dynamical average of the proportion of censuses a species could occupy in its geographic range that it actually does. Thus, a species with a small value for p^* will be found on relatively few of the censuses each year than it has historically been recorded on. There should be an indirect relationship between p^* and the likelihood of global extinction for a species. The fewer censuses a species is detected on in any given year, the more likely it maintains relatively small populations on those census sites, and therefore, the more likely major perturbations will affect these small

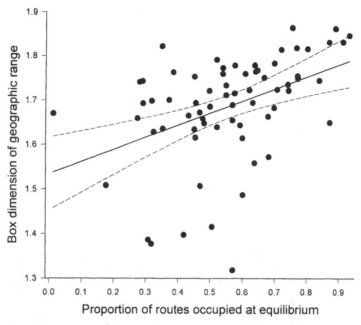

Figure 3.7. Relationship between the proportion of the geographic range occupied by a species for which local colonizations equal local extinction, and the geographic range fragmentation for 68 species of small insectivorous birds in North America. Note that species with high degrees of geographic range fragmentation (low box dimension) tend to have a smaller proportion of their geographic range occupied than species with low degrees of range fragmentation. The *solid line* is the least squares regression line, and *broken lines* are 95 percent confidence intervals about the line. The regression was significant (slope = 0.27, $t = 3.87$, $df = 66$, $P = .0003$, $R^2 = 0.18$).

populations negatively. A species with a small value for p^* has fewer populations that can serve as sources should a major catastrophe occur.

How is p^* related to geographic range fragmentation? Recall that species with small geographic ranges have relatively fragmented ones. If this has dynamical consequences, then we would expect that the box dimension would be positively correlated with estimates of p^*; that is, species with high degrees of fragmentation (low box dimensions) would also be found on a relatively small proportion of censuses. This is indeed the case for the North America birds (all are passerines) for which these calculations have been done (figure 3.7). We conclude, then, that the reason species with small geographic ranges have highly fragmented ones is that dynamically, they maintain relatively fewer stable populations than species with large geographic ranges. We hypothesize that this makes species with small geographic ranges much more likely to go extinct over geological time than those with large geographic ranges.

Geographic Range Fragmentation, Speciation, and Diversity

Up to this point, we have argued that geographic range fragmentation has definite consequences for global extinction of a species. Species with highly fragmented ranges will tend to have small ranges, and they will occupy fewer local sites throughout their range than species with large geographic ranges. Recall that fossil evidence suggests that there is a correlation between speciation rate and extinction rate (Stanley 1990c; Gilinsky, chapter 9). If species with fragmented ranges are more likely to go extinct, are they more likely to speciate as well? This is an exceedingly complicated problem to which we cannot give its deserved attention here. However, we hope to at least outline some of the issues that must be addressed to answer the question.

The basic problem in examining the relationship between speciation and geographic range fragmentation is twofold: Any hypothesized relationship depends on (1) what one chooses to call a species, and (2) how one believes species form. Unfortunately, there is no consensus on either of these points (see, e.g., papers in Otte and Endler 1989). A variety of criteria may be used to recognize species (Cracraft 1989; Templeton 1989), and species may form by a number of possible mechanisms (White 1978; Lynch 1989; Brooks and McLennan 1991). A detailed discussion of how these mechanisms and species definitions affect the relationship between formation of new species and geographic range fragmentation is beyond what we can accomplish here. Suffice it to say that some speciation mechanisms may be enhanced in species with small, highly fragmented geographic ranges (e.g., peripheral isolates speciation) and others enhanced in species with large geographic ranges (e.g., vicariant speciation).

If we take the empirical observation from the fossil record of a correlation between per species rates of speciation and extinction as our starting point, then we could conclude that species with small, fragmented geographic ranges have higher per species rates of speciation (as a consequence of the argument in the previous section). But from the foraminifera example cited above (see table 3.1, figures 3.2, 3.3), it does not follow that clades consisting of species with small, fragmented geographic ranges will necessarily be more (or less) diverse than clades composed of species with large geographic ranges. Certainly, we would expect turnover in species to be greater for the clade with small geographic ranges, but this does not help us to predict diversity. There is an additional assumption needed to argue which of the two kinds of clades should be more diverse. We believe that assumption was provided first by Darwin (1859). Darwin was greatly impressed with geographic range size, and he argued that it followed from the success of individual organisms. Species with large geographic ranges, according to Darwin, had been shaped by natural selection to use a wide variety of "resources," or in his words, "many and widely diversified places in the polity of nature" (Darwin 1859, p. 112), whereas species with small geographic ranges were necessarily more restricted in the places where they could find appropriate ecological conditions. The ecological advantages of species with large geographic ranges allow them to persist at the expenses of rarer species, in Darwin's view. In the present context, we suggest that the additional assumption needed to link diversity with geographic range fragmentation is this: Species with less fragmented geographic ranges are capable of maintaining viable populations over a wider variety of ecological conditions than related species with more fragmented ranges, and hence they suffer less extinction relative to speciation. Thus, clades composed of widespread species will maintain higher diversity because the difference between per species speciation and extinction rates is greater for such a clade than for one composed of narrowly distributed species. This is borne out by the foraminiferan clades: the more diverse globigerinids occurred across more biogeographic provinces than the more narrowly distributed globorotaliids (Stanley et al. 1988).

Are Geographic Ranges Metapopulations?

We have argued that geographic range structure is related to extinction, so that species with small, fragmented geographic ranges are more likely to suffer extinction. Can what we know about geographic range structure be explained by existing metapopulation theory? There are many kinds of metapopulation structures (Harrison, chapter 2), but metapopulation theory is most clearly developed for—and carries the most significance in—structures that assume that patches of appropriate habitat are more or less the same and that the population persists in a given landscape by continual colonization of patches not occupied (Hanski 1991). If colonizations keep pace with local extinctions on patches, then the metapopulation is able to persist in the landscape; otherwise it goes globally extinct. All nonoccupied patches are assumed to have equal likelihood of receiving colonists; all occupied patches are assumed to have equal likelihood of local extinction. These assumptions can be relaxed somewhat by assuming variation among patches in the density they can support (Hastings 1991), but the assumption of homogeneity within patch types is still necessary.

There are undoubtedly many population systems that approximate the structure required to apply metapopulation theory. But can it be argued that a geographic range is a metapopulation? We suggest there are several factors that must be considered before accepting such a proposition. First, geographic ranges, except for perhaps those of highly restricted species or island endemics, have a very well defined spatial structure. Generally, places where a species is common are clustered together in one or a few centers of abundance (Hengeveld and Haeck 1981, 1982; Brown 1984, 1995; Hengeveld 1990; Maurer and Villard 1994). Areas peripheral to these centers of abundance sustain lower populations. Therefore, sites at the fringes of a species' geographic range are less likely to receive colonists than sites near the center, and this violates one of the basic assumptions of many metapopulation models. Second, geographic ranges are often much larger than the average dispersal distance of individuals. Even in species that migrate long distances between breeding and nonbreeding seasons, individuals often are philopatric to natal areas or nest sites where they were previously successful as breeders. Philopatry acts to effectively limit the breeding dispersal of individuals to distances smaller than the span of the breeding range. Metapopulation models that assume constant rates of colonization assume that the rate of movement between patches is relatively low, and that they are spatially unstructured; that is, movement from one patch to a neighboring patch is the same for all patches. If suitable habitats are more dispersed, for example, at the range periphery, there may be less movement among peripheral patches than among central ones. Third, there may be mass movements of individuals within the boundaries of the geographic range that occur over many generations (Shmida and Wilson 1985). The breeding ranges of many North American birds surveyed over the 30 years of the BBS seem to be relatively stable (see, e.g., Maurer and Villard 1994), but major range expansions of several introduced and native species (Wing 1943) occurred over the first five decades of the twentieth century. Geographic range dynamics may occur on time scales appropriately measured in decades or even centuries. Hence, assumptions regarding constant rates of colonization and extinction will be violated.

From the preceding considerations, it seems that metapopulation theory will require further modifications and elaborations to be useful in describing geographic range dynamics. For example, peripheral populations in a geographic range may approximate traditional metapopulation structure, and equations defining colonization and extinction dynamics could incorporate additional terms to account for the unidirectional flow of individuals into these populations from more central source populations. We suggest that making such modifications in metapopulation theory should provide a better understanding of the complexities of geographic range dynamics that might ultimately prove useful in understanding the role of extinction, and to some degree speciation, in determining the dynamics of biodiversity.

Summary

Diversity changes over time within lineages as a function of the difference between per species rates of speciation and extinction. Although these rates are often correlated, diversification depends on the extent to which speciation exceeds extinction for a given number of species. Clades composed of species with wide ecological tolerances or that can use a wide variety of resources should be more diverse at any given point in time than those that are more specialized or restricted. The extinction component of diversification is related to geographic range structure. Species with small geographic ranges also have more fragmented ranges. These two factors act in concert to increase the rate of local population extinction in restricted species. Measurements of geographic range fragmentation and of the dynamics of local colonization and extinction for species of North American insectivorous birds support this interpretation. Because geographic ranges are more complex spatially than current metapopulation theories assume, we suggest that current metapopulation theory requires modification before further theoretical insights into the evolutionary dynamics of diversity can be obtained. Such modifications should include expansion or generalization of current metapopulation theory to the geographic scale.

Acknowledgments

We thank Dan McShea, Susan Harrison, and Kevin Gaston for many excellent and helpful comments on this chapter. Any errors that remain after their excellent comments are our responsibility. Funding for this research was provided by the United States Environmental Protection Agency and the National Science Foundation.

4

Detecting Ecological Pattern in Phylogenies

**J. L. Gittleman, C. G. Anderson, S. E. Cates,
H.-K. Luh, and J. D. Smith**

The Oxford English Dictionary defines *diversity* as "the condition or quality of being different in character or quality." Biodiversity is, in simple terms, a catchall for describing levels of being biologically different or variable. Biodiversity is a reflection of certain patterns of turnover, as measured by differential speciation and extinction. Ecological aspects of turnover are related to specific ecological characteristics such as geographic range, life histories, or habitat structure. Evolutionarily, the questions are, how do these characteristics vary at different taxonomic levels such as populations, species, and so forth, and do they seemingly have an impact on biodiversity and turnover? That is, as species become extinct and are phylogenetically replaced by their descendants, do these descendants retain the same ecological characters or do they evolve new characters that lead to origination? Certainly, given heritability and coefficients of selection, it is expected that across monophyletic groups some traits (e.g., body size, metabolic rate) should have a closer relationship with phylogenetic change than other traits (e.g., home range movements, population density). In this paper, we ask the following questions: (1) Are such expected phylogenetic differences indeed observed with various kinds of traits? (2) How do we look for these differences? and (3) Do they help us in learning about origination and extinction of biodiversity? We emphasize that answers to these problems are preliminary, mainly because the opportunities for understanding differential phylogenetic patterns have only recently become available with modern analytical tools and molecular phylogenies. Further, a fundamental assumption behind our analyses is that phylogenies reflect actual evolutionary patterns, as also assumed in parallel work that use phylogenies to test models of evolution (Hey et al., chapter 5) and turnover rates of higher taxa (Gilinsky, chapter 9). Support for this assumption is receiving ever-increasing support (Harvey et al. 1996) and, in the long term, our analytical abilities to use phylogenies will progress more rapidly by using them now.

Why Should Phylogeny Be Important Ecologically?

Until recently, paleontologists and ecologists have seemingly been quite comfortable ignoring one another. One reason for this is that history, especially phylogenetic history, has been important in paleontology but relatively insignificant in ecology. In the last few years, the awareness of the need to consider phylogenetic pattern in ecology has dramatically increased (e.g., Gorman 1992; Cadle and Greene 1993; Brooks and McLennan 1991, 1992, 1993; Farrell and Mitter 1993; Arnold 1994, 1995). Two recent books include phylogeny as an integral part of a larger agenda to better understand ecological and evolutionary processes. In one, Brooks and McLennan (1991, p. ix) state, "Throughout this book, we will attempt to convince you that the integration of phylogeny with ecology and behavior produces exciting new research." In the other, Harvey and Pagel (1991, p. 204) claim, "A proper understanding of community structure will not be achieved independently of phylogenetic structure." But why should phylogeny be important in ecology, especially in studying traits such as population density or geographic range, which are thought to be flexible, or labile, in response to environmental selection? Certainly, there exists in the ecological literature a "tension" (Ricklefs and Schluter 1993b, p. 242) between those advocating that organisms conform to ecological conditions and those who feel that organisms largely reflect speciation and unique historical events.

In the last decade, phylogeny has appeared in a variety of contexts under the headings of *constraints, niche conservatism,* or *alternatives to adaptation;* indeed, almost any phylogenetic pattern in phenotypic variation has received some label such as *phylogenetic inertia* (Wilson 1975), *local constraint* (Maynard Smith et al. 1985), *history* (Brooks and McLennan 1993), *phylogenetic constraint* (McKitrick 1993), or *lineage effects* (Arnold 1994, 1995).

Generally, phylogenetic pattern in a trait is thought to serve as a sort of governor, limiting the extent and direction an organism can change evolutionarily. But explaining trait variation in terms of phylogenetic effects can often lapse into simply a restatement of the facts: phylogeny constrains traits and traits show phylogenetic constraints! The most difficult aspect of analyzing phylogenetic pattern is, indeed, pinpointing causal relations (Antonovics and van Tienderen 1991; Lord et al. 1995). Basically, we know very little about what factors cause phylogenetic effects. In a sense, this is unsurprising because we are dealing with historical patterns that occurred long ago (or at least before some of us realized that phylogeny might be important) and are rarely open to experimental tests. Harvey and Pagel (1991) and Futuyma (1992) have identified six salient (non–mutually exclusive) factors that will produce phylogenetic correlation: (1) stabilizing selection imposed by ecological factors or habitat fidelity; (2) biomechanical constraints among characters, allowing some structures to evolve more slowly or more quickly than others; (3) nonlinear relations between genotype and phenotype, such as in "threshold characters" where a continuous character is expressed discontinuously (e.g., number of digits in mammals); (4) pleiotropic effects, frequently related to developmental constraints; (5) lack of genetic variance to act upon, thus revealing no character evolution; and (6) adaptation, whereby species move into new niches or respond similarly to the same selection pressures.

At this stage, we do not know whether any or all of these factors actually produce phylogenetic effects. Undoubtedly, experimental systems with relatively fast reproductive turnovers and ecological changes will usher in the next phase of testing causal relations. Two examples illustrate the kind of data collection that will be helpful. Futuyma et al. (1993) assessed whether genetic variation in an insect (*Ophraella communa*), which has coevolved with a particular host (*Ambrosia*), affects the insect's ability to feed on new hosts. In experiments that offered an array of hosts as food, *Ophraella* did not show sufficient genetic variation to feed and survive on different hosts. Thus, referring to the preceding list, factor 5 and possibly factor 3 (as a result of a threshold feeding response) are consistent with the phylogenetic patterns in this insect–host interaction. In the context of behavioral evolution, McLennan (1991) performed some elegant experiments based on hypotheses generated from phylogenies of gasterosteid (stickleback) fishes. The phylogenies suggested that the origin and elaboration of nuptial coloration is related to (1) male–male interactions, (2) parental care, and (3) male–female interactions. McLennan tested these macroevolutionary hypotheses with choice experiments in which *Gasterosteus aculeatus* females were allowed to select males based on color, breeding conditions, or behavioral affinities. Results indicated that females mainly cue on male body color rather than patterns of parental care or male–male interactions, suggesting that observed phylogenetic patterns relate to adaptive responses (factor 6) or some form of stabilizing selection on reproductive characters (factor 1); genetic variation was not examined. Both of these studies clearly indicate that causal factors related to phylogenetic patterns can be tested experimentally.

Even so, there are many taxa and problems at the macroecological and evolutionary levels not open to such experimental treatment. With these, it is helpful to show precisely when and how phylogenetic patterns emerge in different kinds of traits and across quite disparate taxa. For example, do species at higher trophic levels, those having to find unpredictable and widely distributed foods, reveal less phylogenetic pattern because they must adapt to rapidly changing conditions? Or do relatively young taxa, because they are still experimenting with new characters ("adaptive mutations"), fail to show phylogenetic signal compared to more "mature" taxa (see also Gilinsky, chapter 9)? We will show that combining quantitative trait data with statistical comparative methods and molecular phylogenies allows us to see more clearly these types of emergent patterns.

The Promise of Molecular Phylogenies

Phylogenetics is "the study of the evolutionary routes followed by particular organisms" (Desmond 1982, p. 148). Information conveyed through phylogenetic study is typically presented in some graphical form such as a phylogenetic tree, cladogram, or branching diagram, all of which have analytical variants (e.g., rooted or unrooted) and reflect different types of hierarchical structure (Eggleton and Vane-Wright 1994). Such illustrations represent transformation, but in a tricky sense: observations are rarely known for the actual transformations. Usually, transformations are inferred in various ways, including the use of different types of characters (morphology, ontogeny, DNA, RNA, etc.) and evolutionary assumptions (polarity of characters, parsimony, etc.), all of which will have a significant bearing on a phylogeny. A phylogeny thus constructed is admittedly only an estimate (or hypothesis) of the actual historical events (or transformations) but nevertheless a hypothesis that allows us to consider a number of important questions.

Apart from the mathematics of generating and verifying a phylogenetic tree (see Penny et al. 1992), a phylogenetic tree usually contains four elements that are useful for examining character evolution (figure 4.1): a root, nodes, branches, and tips. Roots are branching points at the base of a tree. Nodes represent putative ancestors at branching points. Branches connect internal nodes or, at the terminal branches of tree, a node to a tip. And, tips are the taxa of study, most often species, occurring at the terminal branches. With this information in hand, we can ask questions about (1) the overall relationship between a trait and a given phylogenetic tree [i.e., "phylogenetic correlation" (Gittleman and Luh 1992)], (2) changes in trait variation at

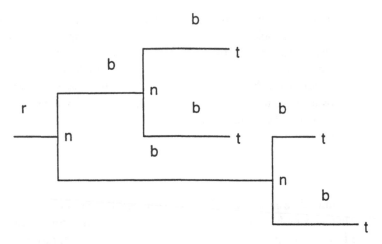

Figure 4.1. Information contained in phylogenetic trees. r, root; n, node; b, branch; t, tip.

precise phylogenetic time intervals (i.e., are traits evolving in certain ways at nodal splits?), and (3) the relationship between the extent of trait evolution and branch lengths (i.e., rates of trait evolution). For our purposes here, we want to use phylogenetic trees to examine whether different kinds of traits (morphologic, ecologic, behavioral), presumably those that receive more emphasis in studies of biodiversity, reveal different evolutionary patterns.

Phenotypic traits, including measures of size, growth, or morphological specialization, are typically used to reconstruct phylogenies. Recently, phylogeny reconstruction has entered a new era. In terms of studying interrelationships of ecology and morphology, and indeed the evolution of any traits, it is critical to ask whether the characters under study are also ensconced in the phylogeny reconstruction. Such circularity is obviously problematic in searching for relative differences among traits. For example, if a phylogeny were based on some morphological trait and we then used this phylogeny to evaluate whether body size or geographic distribution were phylogenetically correlated, it would be expected that the analysis would be biased toward the morphological trait. Now, with molecular phylogenies, this circularity can be broken.

Mammalian Phylogeny and Ecology: An Empirical Example

In this section, we demonstrate how molecular phylogenies, specifically those that contain information about evolutionary time scale and topology, can be used to estimate the relative lability of phenotypic trait evolution. We show that various ecological traits (home range size, day range movements, population density) are more labile than morphological or life history (brain size, body size, gestation length) traits across six mammalian taxa. We end with a discussion about how such phylogenetic analyses suggest that trait lability across taxa should be included as an important index for measuring biodiversity, ecological changes in taxonomic turnover, and ecological factors associated with extinction.

The Data

Out of 97 published molecular phylogenies of mammalian taxa, we identified six that were usable for our purposes here, because they contained time frame information, species-level phylogenetic analysis, and appropriate genetic assays for the examined taxa (see Gittleman et al. 1996 for further discussion of the restrictions and applications of molecular trees). The taxa (and species sample sizes) included Primates (26), Carnivora (39), Ceboidea (19), Canidae (28), Bovidae (27), and Cervidae (9). For species represented in the molecular phylogenies of these taxa, trait values were derived from published literature for two morphological (brain weight, body weight), two life history (gestation length, birth weight), and four behavioral/ecological (home-range size, population group size, population density, day range) traits. [Complete definitions and quantitative (phylogenetic) analyses of these traits are given in Gittleman et al. (1996); the present analysis is a partial summary of this work, with the addition of population density and day range length.] Population density is defined, and measured accordingly, as the number of individuals per square kilometer; day range is the daily path length of a population group (note, however, that day range was not analyzed in Bovidae and Cervidae because there were few available data). Prior to analysis, all of the trait data were logarithmically transformed because cross-species values approximate a lognormal distribution and logarithmic transformations are necessary to reduce skew. The original data files, including both the molecular distance matrices and the species trait values, are available from the authors.

A Statistical Tool

To measure how different types of phenotypic traits change with phylogenetic distance, we borrow concepts and methods from the extensive literature on spatial autocorrelation (Cliff and Ord 1973, 1981; Upton and Fingleton 1985; see also Legendre 1993 for review of biological applications). Moran's (1950) I statistic is used to examine the relationship between phylogenetic distance and phenotypic variation. We denote the observed phenotypic trait for species i by y_i and write $\bar{y}$ for the average of y_i over the n species of the data set. Using this notation, Moran's coefficient is given as

$$I = \frac{n}{S_0} \frac{\sum\limits_{i=1}^{n}\sum\limits_{j=1}^{n} w_{ij}(y_i - \bar{y})(y_j - \bar{y})}{\sum\limits_{i=1}^{n} (y_i - \bar{y})^2} \qquad [1]$$

where

$$S_0 = \sum_{i=1}^{n}\sum_{j=1}^{n} w_{ij} \qquad [2]$$

Moran's I is, in essence, an (estimated) autocorrelation coefficient: The numerator is a measure of covariance among the (y_i) and the denominator is a measure of variance. At the heart of this statistic lies the weighting matrix **W**, $\mathbf{W} = [w_{ij}]$. In searching out phylogenetic autocorrelation at some level, I will compare the phenotypic trait of a species with a weighted average of the trait over a set of neighbors. The ijth element of the **W** matrix, w_{ij}, is the weight assigned to the jth species in computing the weighted average for species i. The value for w_{ii} is always set to 0 (as a species is clearly always its own best predictor), and because we average over neighbors, we will, for non-zero rows, transform the matrix such that

$$\sum_{j=1}^{n} w_{ij} = 1, \qquad i = 1, 2, \ldots, n \qquad [3,4]$$

The w_{ij} are determined by the correlation we are trying to ascertain. With specified (molecular) branch lengths, as in the present study, we average over all species within some finite interval of distance, typically set in accord with the frequency distribution of branch length intervals (see below). We use **W**, in effect, to flag species that are to be averaged.

Because we refer to I as an autocorrelation, it is natural to assume that $-1 \leq I \leq 1$, with values ± 1 indicating a strong phylogenetic correlation. To standardize values of I to compare among traits, we scale by the maximum observed value. To interpret the values of I, we tabulate associated standard deviates, z, in which z-scores greater than 1.96 may be used to reject the null hypothesis of no phylogenetic correlation at the 0.05 level.

To assess relative phylogenetic change among traits, we employ a "phylogenetic correlogram" (see Gittleman and Kot 1990). This is simply a graph showing how autocorrelation (observed I's or associated z's) varies with phylogenetic distance, as hypothesized in figure 4.1. With detailed, complete phylogenetic information (i.e., branching topology), we first calculate a distance set for each phylogeny (see Purvis et al. 1994). The distance set is formed by arranging frequency distributions of the distances observed for each phylogeny and then selecting those distances with the highest frequency ("D"). This *maximizes* the chance of finding trait change at a given distance in a phylogeny (i.e., phylogenetic correlation); distances with the highest frequency also correspond to those portions of the tree that reflect high speciation rates. Weighting matrices are then defined by applying Moran's I as follows:

w_{ij} = 1 if the distance of species i and j belongs to distance set D,
    = 0 otherwise

We then use the correlogram to show where phylogenetic correlation occurs with divergence and branching patterns in the phylogeny.

From these correlograms, we can detect four hypothetical patterns (figure 4.2; see also Gittleman et al. 1996):

1. No relationship with phylogeny, as possibly observed under strong stabilizing selection.

2. A specific autoregressive ("directional") form with correlation falling-off with phyletic distance, a pattern possibly observed from a simple Brownian motion random walk (i.e., pure random genetic drift).

3. A "stepwise" function with phylogenetic distance, representing stasis followed by change between nodes (note that this is not, in actuality, an alternative hypothesis to the second hypothesis, because the overall pattern of change is in proportion to phyletic distance).

4. A "piecewise" function with phylogenetic distance, representing linear change of significant positive and negative (phylogenetic) correlation and, importantly, greater phenotypic plasticity than in the "directional" or "stepwise" patterns; this pattern is particularly instructive because it suggests that characteristics have been replaced by new or different characteristics between speciation events. Our aim is to empirically observe these "phylogenetic snapshots" to see whether behav-

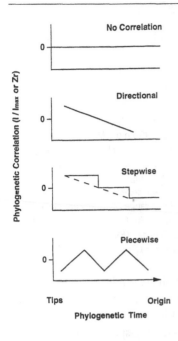

Figure 4.2. Hypothetical patterns (shown in four correlograms) of phylogenetic change in quantitative traits. Patterns are measured using a spatial autocorrelation statistic (Moran's *I*) and are indicative of various forms of evolutionary lability among traits (from Gittleman et al. 1995; see text for further details). The range of phylogenetic pattern is from no change, expressed by zero, to strong phylogenetic change, represented by ±1.

ioral/ecological traits (e.g., home range size, population density) indeed are changing more phylogenetically than morphological and possibly life history traits.

Analytically, it is important to recognize that the observed differences in (phylogenetic) autocorrelation among traits are not caused by differences in trait variance as this is factored out in Moran's *I* equation.

Empirical Results

To view phylogenetic differences among traits, we present a phylogenetic correlogram for each taxonomic group (figure 4.3). Five general patterns emerge that apply consistently to all examined taxa.

1. The morphological traits (body weight, brain weight) and the life history traits (birth weight, gestation length) are significantly correlated with phylogeny and in an expected "autoregressive" manner; that is, taxa more closely related (species tips) tend to be more similar phenotypically, with similarity falling off with phylogenetic distance. This is expected for more conservative, heritable traits that are under directional selection (see figure 4.2). Interestingly, however, the morphological traits show slightly higher correlations than the life history traits.

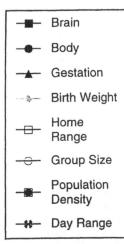

— Brain

— Body

— Gestation

— Birth Weight

— Home Range

— Group Size

— Population Density

— Day Range

Figure 4.3. Phylogenetic correlograms showing observed patterns of z-scores (from the Moran's *I*) for the quantitative traits of brain weight (*filled squares*), body weight (*filled circles*), gestation length (*triangles*), birth weight (*diamonds*), home range size (*open squares*), population group size (*open circles*), population density (*circle in square*), and day range (*X*). Traits are plotted for Primates **(A)**, Carnivora **(B)**, Ceboidea **(C)**, Canidae **(D)**, Bovidae **(E)**, and Cervidae **(F)**. Observed z-scores are significant at values of ±1.96.

2. Most of the behavioral and ecological traits reveal little if any phylogenetic pattern, thus reflecting greater evolutionary lability (i.e., no phylogenetic correlation) than the morphological or life history traits. This suggests that such traits are replaced by different traits during speciation.

3. Home range size is an exception to the overall trend because as an ecological trait it shows phylogenetic pattern in Primates, Ceboidea, and possibly Canidae and Bovidae. This might be the result of allometric effects of phylogenetic changes in size.

4. Relative differences of lability among traits do not relate to the age of the taxon, as comparatively old (e.g., Carnivora) and new (e.g., Canidae) taxa contain similar patterns or variation in speciation patterns.

5. While most of the observed phylogenetic correlation among all traits converges around zero at the root of the tree, in some cases it does not. As examples, home range size in the primates and the bovids, four traits in the ceboids, and three traits in the cervids show significant negative correlation.

Discussion

General Trends Across Mammals

The above example of applying a phylogenetic comparative method to different kinds of quantitative traits across mammals highlights a number of points related to ecology. Generally, although they remain essentially hypotheses, molecular phylogenies allow us to precisely reconstruct phyloge-

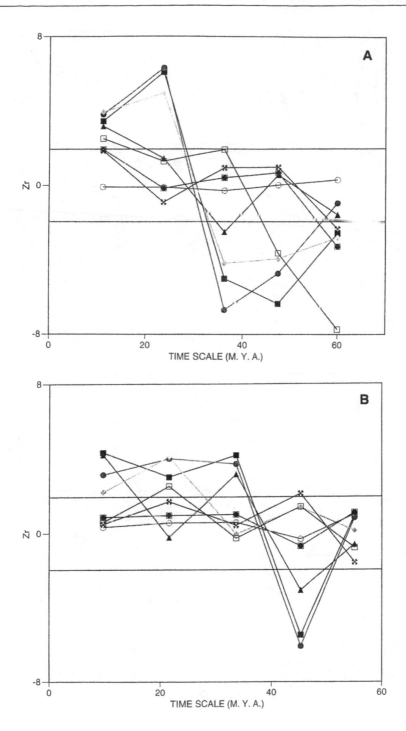

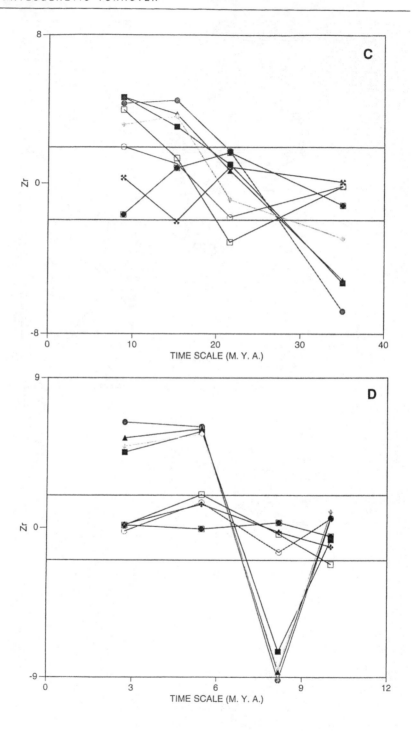

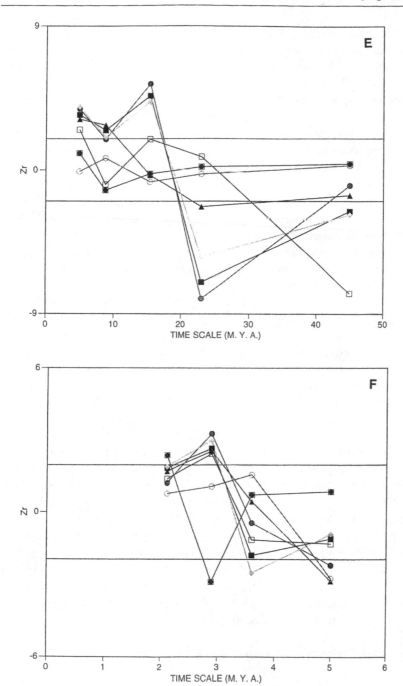

netic history. Hanging different kinds of ecological and other traits onto phylogenies permits not only a general assessment of phylogenetic pattern among traits, but also a precise view of when traits change with respect to speciation. This is a clear method for analyzing both origination and diversification of traits across divergent taxa, the only prerequisite for such an analysis being some information on phylogenetic topology and time scale.

More specifically, the mammal example reveals a number of interesting patterns. First, in terms of maintaining biodiversity, some traits may be more manageable than others. Clearly, traits such as population density and day range movements are more phylogenetically labile, thus potentially responding to immediate environmental selection, which should be helpful for conservation management programs. Also, in terms of simply describing biodiversity, statistical comparative methods can be used to compare relative phylogenetic correlation of single traits (e.g., population density) among diverse taxa to evaluate whether traits in some taxa are more variable than in others; a reduction in variability following a speciation event may indeed contribute to extinction. Second, there appear to be certain moments in the history of a lineage when dramatic phenotypic change occurs. Across mammals, traits such as body size and life histories are very similar in closely related taxa, but as one moves back in phylogenetic time a point is reached at which traits are very dissimilar. We hypothesize that such phenotypic changes reflect a major speciation event. However, an equally plausible explanation is that this pattern is a direct extension of a molecular phylogeny, perhaps reflecting idiosyncrasies of phylogenetic reconstruction rather than anything about ecological or evolutionary pattern (Gittleman et al. 1996). We think that, overall, evolutionary history does reflect phenotypic variability, showing quite divergent patterns via mechanisms of adaptive change (convergence, parallelism, etc.) whereby sister taxa are becoming more and more similar in a remarkably consistent manner. On the other hand, if a given phylogeny is unbalanced (e.g., Ceboidea) or one species at the root of the tree has very different trait values than the remaining species in a group (e.g., white-tailed deer compared to other cervids), then the phylogenetic correlation will be significantly negative to show this divergence at the base of the phylogeny. Only repeated analyses across many taxa with many traits and independent molecular phylogenies will tease apart how consistent the phylogenetic trends are as described here.

Using Phylogenies for Problems of Biodiversity

The luxury of now having more basic natural history data across taxa, at the same time that our reconstructions of phylogenetic history are precise and reliable, almost gives optimism for countering daily extinction rates (see Pimm et al. 1995). We end with two suggestions for future directions in using phylogenies to study biodiversity.

1. Developmental canalization and conservation biology. Phylogenetic analyses underscore the features of constraint, often represented by a reduction in phenotypic variability through evolutionary history. One important mechanism of this is canalization or developmental stability (McKinney and Gittleman 1995). In relation to biodiversity and conservation biology, analyzing development can identify patterns of population stress, developmental effects on fitness, and developmental effects on captive breeding populations (Clarke 1995). For example, in a phylogenetic analysis of life history patterns across carnivores (Gittleman 1994), the giant panda (*Ailuropoda melanoleuca*) compared to other carnivores showed clear developmental protraction in growth rate from birth to weaning. It is likely that such growth is directly linked to unusually high juvenile mortality, both of which may be a consequence of the dietary inefficiency of being a bamboo-eating carnivore. An important question now is what (causal) developmental factors bring about this pattern. Obviously, the giant panda cannot be studied experimentally; however, the red panda (*Ailurus fulgens*), a smaller bamboo-eating relative that shows similar life history abnormalities, can be used as an experimental model. Future work should tease apart when and why in early development growth rate trajectory begins to depart from expected carnivore patterns.

2. Traits, cladogenesis, and biodiversity. Another phylogenetic issue, one that is tied into the empirical patterns described above, is cladogenesis. Do some taxa produce more clades than others? Or, conversely, do some taxa go extinct at higher rates than others? An important diagnostic step in answering these questions is seen in the kind of analyses described by Gilinsky (1994; chapter 9), in which it is shown that the correlation between origination and extinction among higher taxonomic levels is remarkably high. Another approach is to examine evolutionary models against phylogenetic branching patterns: by posing evolutionary null models against molecular phylogenies, it is possible to reveal patterns of cladogenesis, thus showing increases (speciation) and decreases (extinction) in species numbers, i.e., biodiversity (Harvey et al. 1994; Nee et al. 1994a; Kubo and Iwasa 1995). For example, we can compare the difference between actual (from fossil evidence) and reconstructed phylogenies (from molecular phylogenies) to reveal important information about biodiversity: the relative number of taxa at any given time intervals, whether the taxa increase or decrease in number through history, and the precise moments of time in which lineage splits occur. In addition to describing the extent to which diversity changes among

taxa, it is possible to search for causal factors, which is where the above analyses come in. For example, if certain lineages show high cladogenesis, one can ask whether this relates to various ecological (population density, geographic range) or morphological (small body size, big brain size) traits (see Gittleman and Purvis 1997). The phylogenetic comparative analyses shown here will help to pinpoint when and to what degree trait differences emerge phylogenetically.

Patterns of biodiversity and extinction can also be studied when molecular phylogenies are placed against the backdrop of explicit theoretical models of evolution. A simple (null) model may be constructed from a constant-rate birth–death process (Nee et al. 1994a) in which each lineage, at each point in time, has the same probability of giving rise to a new lineage (birth) or going extinct (death) as any other lineage. Using this model, we can compare reconstructed (molecular) phylogenies against real phylogenies to see whether lineages have no extinction (identical to the birth–death curve) or, under more realistic conditions, extinction pulls apart the lines representing the number of lineages through time (figure 4.4); each line also has a characteristic curvature, which Nee et al. (1994a) call the "push of the past" (initial rate of cladogenesis) and the "pull of the present" (more recent taxa have had less time to go extinct).

So, what do reconstructed phylogenies look like against this null model? Of all the available molecular phylogenies, by far the most complete representation of any taxon at the ordinal level and above is Sibley and Ahlquist's (1990) phylogeny of 1700 living bird species. Assuming that intervals of time and branch lengths are correct—always a questionable assumption—it appears that either rates of branching patterns have been decreasing or extinction rates have been increasing (figure 4.5a). Recent analyses of mitochondrial DNA fortunately accord with the former explanation that cladogenesis has been slowing down across avian taxa (Nee et al. 1992). But what would a high extinction rate look like? The order Carnivora, a group comprising many top predators that seem quite vulnerable to human disturbance, might be expected to reveal variable patterns of cladogenesis and more recent extinction rates (Gittleman and Purvis 1997). Two patterns emerge when using a partial molecular phylogeny (Wayne et al. 1989) representing 30 carnivore species (figure 4.5b). First, during the heyday of carnivore radiation around 40 to 50 million years ago, there was a massive upswing in biodiversity that produced several distinct lineages in the canoid (canids, procyonids, mustelids, ursids) and feloid (viverrids, hyeanids, felids) carnivores. Second, there is a steep upturn in the recent past. The difference between the slope in the recent past and that in the more distant past is a rough

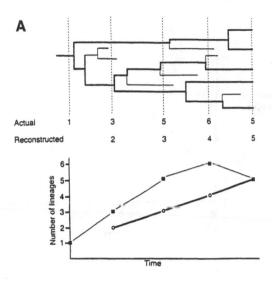

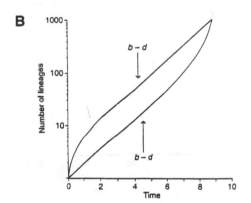

Figure 4.4. **(A)** Detecting patterns of cladogenesis through comparing actual (fossil) and reconstructed (molecular) phylogenies using lineage-through-time plots. If there has been any extinction, the reconstructed phylogeny (*lower line*) will tend to underrepresent the number of lineages in the past. **(B)** Expected growth of numbers of lineages of actual (*upper line*) and reconstructed (*lower line*) phylogenies under a constant-rate birth–death process. The slopes of the lines are the speciation rate minus the extinction rate at each point in time, according to each phylogeny. (Both from Nee et al. 1995.)

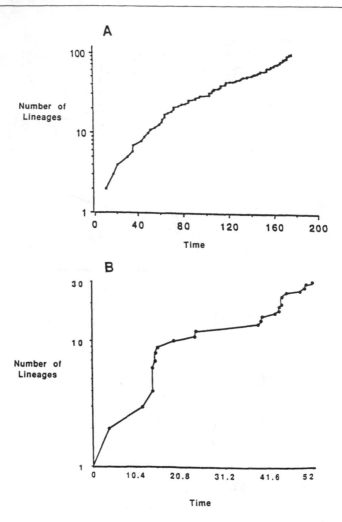

Figure 4.5.   Lineage-through-time plots for **(A)** Sibley and Ahlquist's (1990) phylogeny of the birds (after Harvey et al. 1994) and **(B)** Wayne et al.'s (1989) phylogeny of the carnivores. Assuming the phylogenies are correct, cladogenesis appears to be slowing down in birds, and a high extinction rate may be occurring in carnivores (see text for further explanation). Both plots are constructed using analytical programs of Rambaut, Harvey, and Nee (in press).

estimate of the extinction rate that has historically characterized the group (Nee et al. 1994a). Thus, the rather steep upturn in the carnivore plot, probably resulting from a historically high extinction rate, could indicate that carnivores are currently vulnerable to extinction.

Molecular phylogenies will clearly be important for estimating patterns of biodiversity and extinction, especially when ecological and evolutionary processes are lost to history and molecular data provide the only means for detecting these processes. Future applications will undoubtedly involve improving the quality of phylogenetic trees, the statistical procedures for analyzing tree structure, and the analysis of how particular ecological factors correlate with the structure of phylogenies across taxa.

Acknowledgments

We are grateful to two anonymous referees and Mike McKinney for comments that improved the content and presentation of this chapter. This work was partially funded by a Howard Hughes Medical Institute Undergraduate Biological Sciences Education Program Grant #71195-539601 to the Division of Biology at The University of Tennessee.

5

Testing Models of Speciation and Extinction with Phylogenetic Trees of Extant Taxa

**Jody Hey, Holly Hilton, Nicholas Leahy,
and Rong-Lin Wang**

Could the diversity of species have been caused by simple random processes of speciation and extinction? This question, asked repeatedly in recent decades (Raup et al. 1973; Gould et al. 1977; Stanley et al. 1981; Raup 1985; Hey 1992; Nee et al. 1992) and in this chapter, might seem irrelevant to investigators of specific cases of speciation or extinction. Clearly, every case of speciation or extinction has ecological, biogeographic, and genetic causes, and the idea of "random causes" may have little meaning to students of the mechanisms of speciation and extinction. However, scientists have few tools to study the causes of diversity for the vast majority of speciations and extinctions that happened long ago. The kinds of data that can be brought to bear on the causes of ancient events are mostly limited to just one type: assessments of the similarities of and differences between organisms. If the comparisons are among fossils of organisms, then some information on the age of fossils is also available. In general, these data offer no information on individual ancient speciation and extinction events, but they do present limited scope for inference on general causes of diversity. For instance, comparative phenotypic data (or genotypic data if species are extant) can be used to estimate a phylogenetic tree. We may then ask what kinds of processes could have given rise to a particular phylogenetic tree. Again, there is no room to inquire of the details of specific events, so we must find a way to think about the collection of all processes that have determined the shape of the tree. As in other historical sciences (such as archaeology and cosmology), we begin with a simple model that includes a rudimentary representation of the processes that create history, and we assess how well the model fits our picture of history. If the model fits poorly, then it is probably wrong and should be modified or replaced.

The general approach of comparing phylogeny estimates with the predictions of simple models of speciation and extinction was popularized by Raup et al. (1973) and Gould et al. (1977). These authors used models of random branching with damped equilibrium to generate phylogenetic trees and compared these with trees built from actual fossil data from a variety of taxa. Comparisons of diversity and clade shape revealed that many patterns in the fossil record were similar to those expected from random processes.

Recently, models have been developed that permit the study of speciation and extinction in the history of extant species (Hey 1992; Nee et al. 1994a,b). This approach offers two clear advantages over the use of fossils, at least in terms of data collection. First, the sampling of extant species can be completed to a high degree, without the presence of arbitrary gaps such as those caused by the vagaries of fossilization. This does not mean that sampling is either easy or thorough, and our knowledge of the diversity of extant taxa of large organisms is much better than for small organisms. Second, extant organisms have genotypic characters and a wider array of phenotypic characters than fossils, permitting better assessments of phylogenetic trees. However, data on extant taxa come from essentially a single point in time, and they offer no information akin to the procession of changes and the record of appearance and disappearance of forms that can be seen with a good fossil record.

We present a review of available methods for extant taxon studies, together with an evaluation of their utility. Two major points emerge from this analysis:

1. Phylogenetic tree models, and appropriate data, can reveal historical processes of speciation *and* extinction, even though the data come only from extant species.
2. The statistical power of this approach is limited, so that insights on extinction can be expected only when extinction has played a large role in the history of a group of extant species or when the species group is large.

We also apply several phylogenetic models to 11 data sets taken from the primary literature. These analyses conform, in general, to points 1 and 2. Most taxa reveal little evidence of extinction, with some informative exceptions.

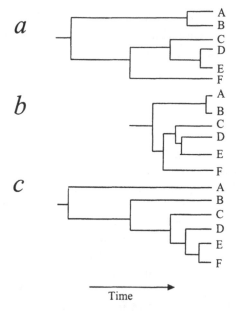

Figure 5.1. Hypothetical phylogenetic trees for groups of six species. Each species is indicated by an uppercase letter. Each tree consists of branches representing the persistence of species through time, and nodes (junctions of branches) that represent speciation, when one species splits into two. Three trees are shown. Both *a* and *b* have the same topology, but *b* has shorter distances between nodes. The topology of *c* is different, but here there are similar distances between nodes.

The Models

Phylogenies, hereafter often called trees, can be described as having two general attributes: length and shape (figure 5.1). The majority of this chapter focuses on questions about tree length, specifically within the context of rooted bifurcating trees, such as those in figure 5.1. These trees are called rooted because time has a direction such that one of the nodes is older than all of the others. Unrooted trees are not considered because they are not explicitly historical (i.e., time lacks direction). This review is also not concerned with multifurcating trees (i.e., more than two descendant species from a single node), primarily for simplicity and because the models that have been developed described bifurcating histories.

Almost without exception, the models of speciation and extinction that have been used to generate phylogeny predictions are Markov chains. By way of example, consider that at some point in time there is some number of species, N, and consider a sequence of short time intervals, within each of which there is some chance that speciation or extinction may occur. After some random number of time intervals, t, a speciation event may occur or an extinction may occur. If speciation has occurred, then the number of species becomes $N + 1$; if extinction, then the number of species is, of course, $N - 1$. We may specify that this model has the Markov property, which means that the value of t does not depend in any way on the time since the previous speciation or extinction event. Models of this type are also usually homogeneous, meaning that the values of the extinction and speciation parameters are the same for each species and constant over time.

Discrete Markov chains have been widely used for the study of species turnover and taxon sizes (Raup et al. 1973; Gould et al. 1977; Stanley et al. 1981; Gilinsky and Good 1991). These models usually begin with a single species and then consider what can happen at successive discrete time intervals. In the very general model of Gilinsky and Good (1991, p. 150), at each time interval it is possible for each species to undergo speciation, or for a species to go extinct, or for nothing to happen. The important parameters under this model are the probability of speciation and the probability of extinction (the probability that neither occurs being 1 less the sum of these values). The principal shortcoming of these discrete models in a phylogenetic context is that time, per se, is not included. The random component of the model is the occurrence (or absence) of a speciation or extinction event within an interval. If the probabilities of speciation and extinction are low, then more intervals will pass without a change in N, on average. Thus, the number of intervals can be considered an indication of the passing of time, but the meaning of an interval remains unclear. Discrete Markov chains have a more appropriate application, for instance, in the study of the population dynamics of organisms that have discrete generations.

Continuous Markov chains, also widely used, differ from discrete models in that the time between events is an explicit random variable. Under a discrete model, the central question is whether or not a speciation or extinction event has occurred within an interval; under a continuous model, the central question is how much time has passed before the next speciation or extinction event occurs. In the parlance of continuous-time, homogenous Markov chains, speciation occurs at a specified rate (usually the Greek letter λ is used) and extinction occurs at a specified rate (usually μ). A discrete model with very low values for the speciation and extinction probabilities will generate results that are effectively indistinguishable from a continuous model. Continuous models are often more analytically tractable, relying on differential rather than difference equations. They are also usually easier to simulate, especially under low speciation and extinction rates, because they require fewer random numbers. Only continuous Markov chain models have been used to address questions on the structure of trees for extant species.

Tree Length

Perhaps the simplest model is one in which there is no extinction, but only speciation. This model, called a pure birth process or a Yule process, was first described in a phylogenetic context by Yule (1924) and was further developed for statistical tests on phylogenies by Hey (1992). Consider a single species that undergoes speciation at rate λ. Then the probability that the time until speciation takes place, time t, is given by

$$P(t) = \lambda\, e^{-\lambda t}. \tag{1}$$

This is an exponential distribution, and it can be shown that only an exponential probability distribution fits a continuous homogeneous process under the Markov property. The expected value of t is $1/\lambda$, meaning that the average time until speciation is the inverse of the speciation rate. After speciation, there are two species. Because we are assuming that the speciation rate per species is constant, the time until the next speciation event is also an exponential distribution:

$$P(t_2) = 2\lambda e^{-2\lambda t_2}. \tag{2}$$

On average, the time until the next speciation event is $1/(2\lambda)$, exactly half of the average time for the case of one species. At a time when there are N species, the time until the next speciation event is

$$P(t_N) = N\lambda e^{-N\lambda t_N}. \tag{3}$$

In general, we see that the waiting time until the next event is less if there are more species that can undergo speciation.

Figure 5.2 provides an example of a phylogenetic tree under this model. In this example, the times between speciation events have been set to the expected values. This example reveals a phenomenon common to Markov chain phylogenetic models, as well as to phylogenetic trees estimated from actual data: the time between the deepest nodes of the tree (those farthest in the past) is greater than the times between the recent nodes of the tree.

A more complex model includes an extinction rate, μ. If there are N species, the probability that the time until either speciation or extinction occurs is given by

$$P(t_N) = N(\lambda + \mu)e^{-N(\lambda+\mu)t_N}. \tag{4}$$

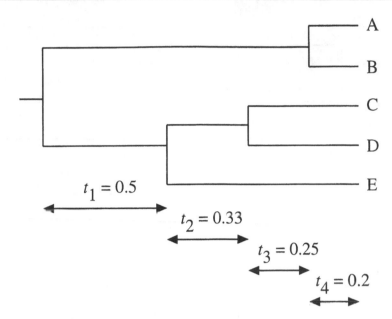

Figure 5.2. A hypothetical phylogenetic tree under a model of no extinction. The values of *t* represent the times between successive nodes and have been set to the expected values for a speciation rate of 1.0 per species. In general, there are $I + 1$ species persisting between node I and node $I + 1$, so the expected time between node I and node $I + 1$ is $1/(I + 1)$ (see text for further explanation).

The probability that the event is speciation is $\lambda/(\mu + \lambda)$ and the probability that it is extinction is one minus the probability of speciation, that is $\mu/(\mu + \lambda)$. This model, also called a pure birth and death process, has been employed in a phylogenetic context (Raup 1985; Nee et al. 1994b). Unlike the pure birth process, in which the number of species is equal to the number of events plus one, the number of species after some time of a birth and death process may have a wide distribution, and it will include the case of all species having gone extinct.

When extinction is included, an important distinction emerges between the true history of a set of species and the history that is perceived with a phylogenetic tree. A bifurcating phylogenetic tree of N species will contain exactly $N - 1$ nodes, each representing a case of speciation. However, if the true history includes speciation *and* extinction, then some of the speciation events will not be represented in the phylogenetic tree. Only those speciation events for which both daughter species are ancestors of the extant species at the tips of the tree will be represented by the phylogenetic tree (figure 5.3) (Hey 1992; Nee et al. 1992). In a recent series of papers, Nee and colleagues (1992, 1994a,b) have described the relationship between the true history of

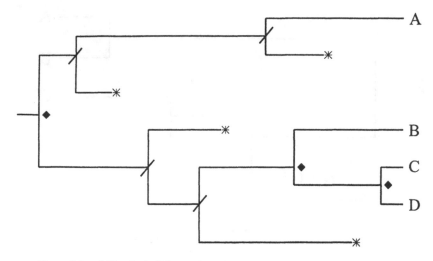

Figure 5.3. A hypothetical history of speciation and extinction. Four species (A, B, C, and D) are extant. Branches indicate all species that existed for some period of time. Only the four labeled branches (A through D) extend to the far right. Cases of extinction are indicated by a branch terminating in an asterisk (*). *Black diamonds* indicate those three nodes for which both descendant species were ancestors of extant species. These are the only speciation events that show up in a phylogenetic tree. Other speciation events are indicated with a *slash*.

speciation and extinction (which includes branches that end in extinction and do not reach the present) and the history perceived with a phylogeny estimated from extant species. If one considers the number of species present at times extending into the past, then this value will usually be lower (it cannot be greater) for the perceived history than for the actual history, because of extinction. Furthermore, this discrepancy is expected to be small for recent times and greater for times further in the past.

A special case of the birth and death process that is of special interest is when the speciation rate equals the extinction rate ($\lambda = \mu$). Like the pure birth model, this model is attractive for having just a single parameter. Also, most groups of organisms that have existed on earth have gone extinct (Raup 1994). This means that most groups of organisms have experienced identical levels of speciation and extinction (although the process may not have been homogeneous; see, e.g., Pearson 1992). A model in which speciation and extinction occur with equal rates is one tool to address the question of whether extant groups of organisms have similar processes of diversification and turnover as did extinct groups.

Yet another Markov chain model that has been employed within a phylogenetic context is one in which speciation and extinction always occur together (Hey 1992). Consider a group of N species, and suppose that whenever one undergoes speciation, another species becomes extinct, so that the number of species remains constant. This model was developed by Moran (1958) and is often used in a population genetic context for the case when population size is constant. It turns out that the waiting time between nodes in the perceived phylogeny of a group of N species under this model has a form that is very similar to those expected of a pure birth process. The probability distribution of the time between node I and node $I + 1$ is given by

$$P(t_i) = NB\gamma_i e^{-NB\gamma_i t_i}, \qquad [5]$$

where B is the speciation/extinction rate per species and γ_i is equal to $I(I - 1)/[N(N - 1)]$ (Kingman 1982; Tavare 1984; Hey 1992).

Hey (1992) developed the Yule model and the Moran model in parallel so that both could be compared with phylogenetic trees estimated from molecular data. In that paper, the Yule model was called model G, for growth, and the Moran model was called model C for constant. These various models can be arrayed as a sort of continuum with respect to the amount of extinction they include. Model G has no extinction; the general birth and death model has arbitrary rates for each process, but the case of equal rates is of special interest; and model C has speciation and extinction at equal rates, but these two processes always occur together.

To better compare the pure birth and death model with equal rates to models G and C, a computer simulation was developed. The simulation proceeded exactly as the birth and death process was described earlier. Furthermore, the number of species was recorded at the time of each event, as was the pattern of ancestor–descendant relationships that spanned events. Figure 5.4 shows the average time between successive nodes of phylogenetic trees under all three models. All three models predict increasingly greater time intervals further in the past, as one approaches the base of a phylogenetic tree.

Tree Shape

Two trees may have the same number of species, and identical time intervals between nodes, yet have very different topologies. In general, the time intervals between nodes can vary independently from the actual branching pattern. All of the models that have been described so far employ the assumption that the species that undergo speciation and extinction at any point in

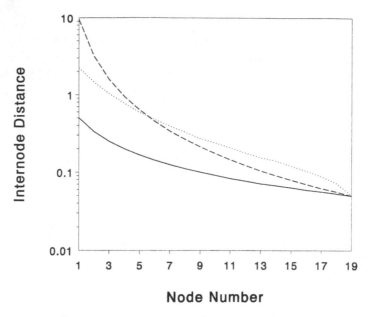

Figure 5.4. The expected distances between successive nodes for the phylogenetic tree of a monophyletic group of 20 species. Node 19 is the most recent; node 1 is the oldest. The speciation rate per species was set to 1.0 for models G (*solid line*) and C (*broken line*). The expected distance between node l and $l + 1$ for model G is equal to $1/(l + 1)$ and equal to $1/[l(l + 1)/19]$ for model C. For the model of equal but independent birth and death [$\lambda = \mu$ (*dotted line*)], both rates were set to 0.5 for each species (so that the time for node 19 would be equal to that for models G and C) and the internode distances were found via simulation. The simulations began with a single species and then proceeded as described in the text. Under this model, the number of species fluctuates randomly. To ensure that the results were not biased by recording the first trajectory that reached 20 species, the following protocol was followed. Each run included a large number of events (up to 500, although often all species became extinct prior to this time). If during a simulation, the number of species reached 20, then the simulation was used; if not, the simulation was discarded. During each run, all time points when the number of species was exactly 20 were recorded. After the conclusion of the run, one of these points was randomly selected and the history (which had been saved) was determined backwards from that point. The values indicated by the *dotted line* are the average internode distances found from 4000 successful runs.

time are randomly selected from those that exist at that point in time. This means that at a time when one of N species is undergoing speciation, there are N differently shaped histories that can result from that speciation event. If extinction is a separate possibility, then there are an additional distinct N number of trees as well. In general, the number of possible branching patterns of a phylogenetic tree is very large (Felsenstein 1978), and the value is even greater if nodes are ordered in time (Hey 1991), as they are in the models discussed here. For N equal to 5, there are 105 distinct rooted trees and 180 distinct rooted trees with ordered nodes. For N equal to 10, there are 3.45×10^7 distinct rooted trees and 2.57×10^9 distinct rooted trees with ordered nodes.

What kinds of tree shapes are predicted by the Markov chain models? To answer this question, it is first necessary to define *tree shape*. In the trivial case where tree shape means a topology with ordered nodes, then all possible outcomes are equally likely under the Markov chain models. However, tree shape may be defined in such a way that some shapes are more likely than others. The most common meaning of shape is balance or symmetry. Trees that are more balanced have roughly equal numbers of descendants extending from each node (e.g., compare figure 5.1a and figure 5.1c). Balance is most easily assessed for a single node, in which case the numbers of species descendant from the left and right sides of the node (L and R, respectively) are expected to be consistent with a random branching model (Slowinski and Guyer 1989; Sanderson and Donoghue 1994). However, for a node with M descendant species, such that $L + R = M$, all possible values of L ($L = 1$ to $M - 1$) are equally likely (Harding 1971; Slowinski 1990). This means that the models do not predict balance in the common vernacular of "symmetry," any more than they predict a lack of balance. It is still possible to ask whether the difference between L and R is bigger than expected by chance, although significant results are not possible without large values of N (Slowinski and Guyer 1989; Sanderson and Donoghue 1994). It is also possible to develop metrics for overall tree balance or symmetry, and several measures have been described (see Kirkpatrick and Slatkin 1993).

The specific assumption of the Markov chain models that is examined by tests of balance is the idea that all species are equally likely to undergo speciation or extinction. If there exist nodes such that the taxa on one side have different rates of speciation and/or extinction from those species on the other side, then the tree is expected to lack balance. A related but distinct method focuses specifically on whether the species that descend from each node are more or less likely than other species to be connected to the next node (Hey 1992). In effect, this test examines whether new species are more or less likely to undergo speciation than old species.

The remainder of this chapter is limited to tree length issues. The various Markov chain models generate predictions about lengths, yet they differ in the degree to which extinction is included and thus in the predicted pattern of variation. None of the tree shape analyses that have been developed are capable of distinguishing among the different Markov chain models.

The Data

In practice, fitting the Markov chain models to data on extant groups imposes significant practical constraints that limit the appropriateness of many data sets. One of the most problematic discrepancies between the models and comparative data is that the models generate predictions about time, whereas the data come in units of character change or divergence. Thus, to apply the models it must be roughly true that the amount of divergence that has occurred between species be proportional to the amount of time that has passed since their most recent common ancestral species. It is this requirement, that divergence be clocklike, that has the effect of limiting comparisons between models and data to those studies using molecular genetic data. For closely related species, the divergence of DNA is often roughly clocklike. Furthermore, the mutational process for DNA is fairly simplistic, relative to that for other traits, such that adjustments can be made to account for multiple mutations that can occur under high levels of divergence (see, e.g., Nei 1987). If divergence is clocklike, then the data are also very suitable for tests of tree shape. Any of a variety of methods can be used to generate accurate estimates of tree shape under these circumstances (Rohlf and Wooten 1988; Rohlf et al. 1990).

A second constraint arises from the fact that all of the models generate predictions about monophyletic groups: i.e., groups for which all of the extant descendants of the most ancestral species (represented by the basal node) are included. It is necessary, then, that data sets used for testing models also completely represent monophyletic groups. Model C, in which speciation and extinction are coupled, is a partial exception. It turns out that equation 5 applies for a random sample of species from a monophyletic group. At any rate, the requirement of monophyly presents three difficulties. First, monophyly is a statement about phylogenetic history, so that the monophyletic status of a group of species selected for phylogenetic study is usually not well known prior to the collection of data. Second, the requirement that all species of a monophyletic group be included has the effect of ruling out groups of organisms for which thorough global diversity surveys have not been made. Last, many phylogenetic studies that employ molecular genetic methods, even those focused on small taxonomic groups, have not included all of the known species for the taxon under investigation.

Testing Models

In practice, the conditions of monophyly and clocklike evolution are likely to be approximated only for relatively recent taxa containing small numbers of species. Typically, the monophyletic status of small groups is more likely to be known, because they are more often studied as part of a larger group. Also, recently diverged species are more likely to accumulate mutations at similar rates, so that the assumption of clocklike divergence is more likely to fit for recent groups than for older species groups.

It is also important to emphasize that the models that have been described do not incorporate the uncertainty that comes with actual data sets. In particular, phylogenetic tree estimates typically have considerable uncertainty associated with branch lengths and with topology. Similarly, the models assume monophyly, but in practice this is difficult to know with certainty. In short, the models ignore sources of variation that are inherent to phylogenetic estimates.

A thorough literature search has returned 11 data sets for which DNA divergence data have been estimated from all of the species of a taxon. Six of these data sets were included in Hey (1992), with the remainder having appeared or been found since then. Three types of DNA studies have been included: studies on restriction fragment length polymorphism (RFLP), DNA-DNA hybridization studies, and DNA sequencing studies.

Table 5.1 presents the results of several analyses on these data sets. For each data set, models G and C were applied, and for each of these models, the likelihood ratio statistic, Λ, and the probability of the likelihood ratio statistic, p, are presented. Also shown are the estimates of the speciation rate for model G, $\hat{a}$ (this notation is in keeping with Hey 1992, although this parameter is a speciation rate similar to λ), the speciation/extinction parameter for model C, $\hat{b}$, and the 95% confidence intervals of these estimates. The model of Nee et al. (1994b) was also applied to these data. Expression 21 of Nee et al. (1994b) was maximized, under the data, to generate estimates of λ and μ, and these are also given in table 5.1.

An error was found in Hey (1992). In that paper, it was reported that the most recent node should not be included in the analyses for models G and C. In fact, this node is suitable, and the arguments against using it are a fine example of the Bus Stop fallacy (Feller 1971, p.12). Correcting this error causes some small changes in Table 5.1 relative to the analyses on the same data sets reported in Hey (1992).

In Hey (1992), the principal finding from an analysis similar to that presented in Table 5.1 was that model G (no speciation) fit the data better than model C. This pattern persists with the addition of more species groups; 10 of the 11 data sets generate a lower likelihood ratio statistic under model G than under model C (and concomitantly, a higher value of p). A striking parallel to this pattern is that in all ten of these cases, the estimate of μ, under the Nee et al. model, is zero and the estimate of λ is very similar to $\hat{a}$. It

TABLE 5.1 Fitting models to data

Data set	Data type[a]	N[b]	Model G					Model C				Independent Birth and Death[g]	
			â[c]	95% CI limit[d]	λ[c]	p[f]	b[c]	b[c]	95% CI limit	λ	p	λ	μ
Subgroup *melanogaster*	DNA-DNA	8	0.59	0.31–1.4	3.25	0.91	0.13	0.13	0.07–0.33	7.04	0.54	0.63	0.0
Tribe Hadenoecini	DNA-DNA	9	0.26	0.15–0.63	4.94	0.75	0.04	0.04	0.02–0.10	12.6	0.15	0.26	0.0
Family Gruidae	DNA-DNA	14	1.52	0.95–2.8	13.5	0.49	0.23	0.23	0.14–0.43	17.6	0.22	1.79	0.0
Genus *Equus*	RFLP	6	0.29	0.14–0.85	4.76	0.41	0.07	0.07	0.03–0.21	9.59	0.09	0.25	0.0
Sect. *Peripetasma*	RFLP	9	2.27	1.27–5.4	5.4	0.71	0.42	0.42	0.23–0.96	11.2	0.21	2.38	0.0
Subgroup *nasuta*	RFLP	10	0.33	0.19–0.73	5.08	0.83	0.07	0.07	0.04–0.16	7.82	0.58	0.39	0.0
Complex Narbonensis	RFLP	6	0.37	0.18–1.17	9.75	0.08	0.10	0.10	0.05–0.30	13.4	0.02*	0.38	0.0
Genus *Krigia*	RFLP	7	2.0	1.0–5.3	2.36	0.85	0.66	0.66	0.34–1.8	2.55	0.83	2.3	0.0
Genus *Graptemys*	DNAseq	13	0.87	0.53–1.66	13.0	0.43	0.13	0.13	0.08–0.25	18.6	0.13	0.97	0.0
Genus *Microseris*	RFLP	8	4.14	2.23–10.3	13.8	0.06	1.91	1.91	1.0–4.8	7.55	0.37	13.9	11.9
Genus *Calycadenia*	DNAseq	12	0.55	0.33–1.1	6.4	0.86	0.08	0.08	0.05–0.16	14.0	0.29	0.61	0.0

[a]DNA-DNA, DNA hybridization; RFLP (restriction fragment length polymorphism), restriction enzyme assays of variation; DNAseq, DNA sequencing studies.

[b]N, number of species in the data set. The number of internode distances used for models G and C is equal to $N-1$.

[c]Values of $\hat{a}$ and $\hat{b}$ were calculated using expressions 3 and 10 of Hey (1992), respectively, with the exception that all $N-1$ internode distances were used.

[a] The 95 percent confidence interval (CI) limits of $\hat{a}$ were calculated as follows: Using the observed value of $\hat{a}$, 5000 simulated data sets were created; for each simulated data set, $\hat{a}$ was calculated; the simulated values were ranked; and the upper and lower limits were taken from the 97.5 percent and 2.5 percent positions in the ranking, respectively. The same procedure was done for the 95 percent CI limits of $\hat{b}$.

[b] Λ is the likelihood ratio statistic, generated using expression (13) of Hey (1992), again with the exception that all $N-1$ internodes were used.

[c] f_p is the probability of getting a likelihood ratio statistic equal to or greater than that observed under the model; it was generated using the same simulations generated for the CI intervals.

[d] λ and μ are the birth and death rate parameter estimates under the independent birth and death model and were calculated using expression 21 of Nee et al. (1994). The internode distances required for the calculations of all rate parameters were generated by applying the UPGMA protocol in the SAHN program of NTSYS (Rohlf 1985) to the distance data reported in the references.

[e] Subgroup *melanogaster* (Caccone et al. 1988): ΔT_m data for all species of these members of the genus *Drosophila*.

Hadenoecini (Caccone and Powell 1987): ΔT_m data from all species of this tribe of the family Raphidophoridae.

Gruidae (Krajewski 1990): ΔT_m data for all 14 species of this family of class Aves. Data for the four subspecies of *D. sulfurigaster* were pooled because of zero divergence among them.

Equus (George and Ryder 1986): sequence divergence based on restriction site analysis of mitochondria DNA of family Equidae.

Section *Peripetasma* (Systma and Gottlieb 1986): sequence divergence based on restriction site analysis of chloroplast DNA for all nine species of this section of the genus *Clarkia* (Onagraceae).

Subgroup *nasuta* (Chang et al. 1989): sequence divergence based on restriction site analysis of mitochondria DNA for these members of genus *Drosophila*.

Complex Narbonensis of genus *Vicia* (Raina and Ogihara 1994): chloroplast sequence diversity based on restriction site analysis on a six-member group of legumes.

Genus *Krigia* (Kim et al. 1992): chloroplast sequence divergence based on restriction site analysis of seven species of dwarf dandelions.

Genus *Graptemys* (Lamb et al. 1994): combined cytochrome *b* and control region sequences from the mitochondria of 13 members of this genus of turtles.

Genus *Microseris* (Asteraceae) (Wallace and Jansen 1990): chloroplast sequence divergence based on restriction site analysis. The full data set included all 16 species of the genus, but several species exhibited no net interspecific variation. These species were collapsed on one another, with the effect that recent internode distances of length 0 were ignored.

Genus *Calycadenia* (Compositae) (Baldwin 1993): DNA sequences of nuclear ribosomal gene spacers. The full data set included all 11 species, but some exhibited no net divergence. These species were collapsed on one another, with the effect that recent internode distances of length 0 were ignored.

seems that the predominant history among small monophyletic groups of extant species includes speciation but not extinction. The one exception to this pattern, the data for *Microseris*, is better fit by model C than model G, and it generates high and similar estimates of λ and μ.

However, before accepting the conclusion of zero or low extinction, it is important to inquire of the sensitivity of these tests to the presence of extinction. There are three distinct causes for the concern that extinction has not been well measured: insensitive tests, sampling bias, and estimator bias. These topics are discussed in turn.

Insensitive tests: It has been shown that a simple model of history without extinction is expected to have increasingly greater node intervals as one looks deeper in a phylogenetic tree. This pattern is also expected of histories with extinction, only more so. The difficulty is that a history of equal and independent processes of speciation *and* extinction does not have a pattern of node intervals that differs sharply from the case of no extinction. In figure 5.4, the shapes of the curves under model G and the independent birth and death model are very similar. Figure 5.4 suggests that a history of extinction and speciation might generate phylogenetic trees similar to those with only speciation, albeit with lower apparent rates of speciation.

Sample bias: All of the Markov chain models that have been described are intended to approximate the history of extant species groups, and these groups necessarily have experienced more cases of speciation than extinction—a group of N species has had exactly N more cases of speciation than extinction. However, it is possible that the underlying speciation and extinction rates are no different for extant groups than for extinct groups (which have equal amounts of both). Thus, when we study a group at a time prior to the extinction of all species within the group, we have selected a history that may underrepresent any underlying propensity to extinction. This condition is realized in the method of Nee et al. (1994b), which requires that the estimate of the speciation rate be greater than that for the extinction rate ($\lambda > \mu$).

Estimator bias: It is possible that there is an actual statistical bias in the estimators. In other words, the expected value of a parameter estimate may differ from the true value of the parameter, even when samples are not biased and when evolution has occurred as assumed and the model is completely correct. The estimates $\hat{a}$ and $\hat{b}$ in table 5.1 have a slight bias of this type (Hey 1992).

To examine the difficulty of distinguishing extinction, we have employed a simulation in which birth and death rates are equal. In some respects, this is the most attractive null model for tests of the structure of phylogenetic trees: the model has only a single parameter, and it enjoys the justification

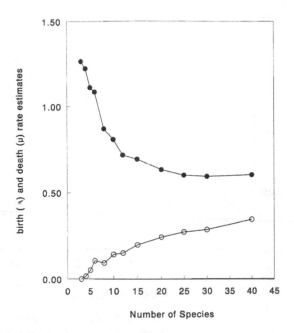

Figure 5.5. Assessing the bias of estimates of speciation and extinction. Average values, from 4000 simulations, are shown for λ (*filled circles*) and μ (*open circles*), estimated by the method of Nee et al. 1994b. For both λ and μ, the true values were set to 0.5. Simulations were conducted under the model of independent speciation and extinction as described in the text and in the figure 5.4 legend.

that birth and death rates have been equal for most groups of organism (those that have gone extinct). We used the model of Nee et al. (1994b) to inquire of the bias in estimates of the parameters λ and μ. Figure 5.5 shows the average estimates of λ and μ for a wide range of differently sized monophyletic groups. The simulations were run with λ = μ = 0.5, with 1000 replicates per sample size. It is clear that both parameters are highly biased, even for species groups as large as 40. It is expected that most of this bias is the result of sampling bias, and is not estimator bias. Although λ and μ are estimated by maximum likelihood, a method that often generates biased estimates, the similar values found for λ and $\hat{a}$ in table 5.1 argue against a high estimator bias for λ. It is known that the estimator bias for $\hat{a}$ is slight (Hey 1992).

We next asked about the statistical power of these methods. Statistical power is the probability of making the correct decision to reject the null model when the null model is false and the alternative model is true. To assess power, it is also necessary to specify an alternative model. Because our concern is that these methods may have little power to detect extinction, we

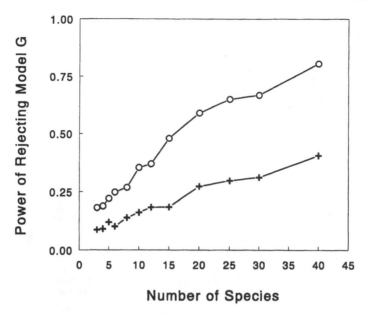

Figure 5.6.   Statistical power of tests of model G. Data sets were simulated under model C (*open circles*) and the model of equal but independent birth and death rates [λ = μ (*pluses*)]. For each simulated data set, model G was applied and the likelihood ratio test (Hey 1992) used to assess whether model G could be rejected. Each point represents the proportion of 1000 independent simulated data sets that were rejected, under model G, at the 95 percent level.

will use the growth model (no extinction) as the null hypothesis, and models that include extinction as alternative hypotheses. In other words, we simulated phylogenetic trees under a model that includes extinction (i.e., the alternative hypothesis is true) and then fit the growth model to the simulated data and asked if the growth model was rejected. Two alternative hypotheses were considered: the independent birth and death model with λ = μ, and the constant model with complete linkage of speciation and extinction. The results over a range of sample sizes are given in figure 5.6.

As expected, power increases with sample size. However, power is quite low under both alternative models for sample sizes in the range of those reported in table 5. 1. Perhaps of greatest concern is that the model of independent but equal speciation and extinction processes has very low power across the entire range of sample sizes considered. In short, if speciation and extinction have been going on independently under a homogeneous Markov process, then most of the phylogenies are going to be found consistent with a model that lacks extinction. The phylogenies will certainly be deeper with the extinction than without, but the spacing of the nodes will not usually depart much from the pattern expected without extinction. If this is the case, then the estimated speciation rate, assuming no extinction, will be lower than the true rate.

Although most species groups in table 5.1 exhibit patterns that are apparently well fit by model G, consistent with a lack of extinction, there are some noteworthy departures. One taxon [genus *Microseris* (family Asteraceae, tribe Lactuceae)] is fit poorly by model G and better by model C, and it has high and similar estimates for λ and μ. The estimate of λ (13.9) is over 3 times that for $\hat{a}$ (4.14). The poor fit of model G, and the discrepancy between the parameters $\hat{a}$ and λ, suggest a history with frequent speciation and nearly as much extinction. The final phylogenetic tree suggests a much lower rate of speciation when fit to model G, but the fit is poor. The UPGMA tree for this species group is shown in figure 5.7a, and it can be seen that the intervals between the deepest nodes are much greater than those for the shallowest nodes. A second taxon that does not fit the overall pattern is the Narbonensis complex in the legume genus *Vicia*. Both model G and C fit poorly, and model C is rejected (table 5.1). Figure 5.7b shows a UPGMA tree for this group, in which it can be seen that there is a pattern of short internodes near the base of the tree and near the tips of the tree, with the longest internodes in the middle. This type of pattern is not predicted by any of the models presented here.

Conclusions

Extinction and speciation are the yin and yang of life's diversification. The current state of the web of life that occurs on our planet can be viewed as a function of those organisms that have persisted or in terms of those organisms that have perished. However, biologists and paleontologists tend to overlook extinction. Despite some initial emphasis by Darwin, "extinction largely dropped out of the consciousness of evolutionary biologists and paleobiologists" (Raup 1994). Without dwelling on the psyche of scientists, some of this "extinction gap" can be explained by the differences between extinct and extant organisms. For biologists who study diversity at the present moment, or for paleobiologists who study diversity at time points in the past, information on those organisms that exist at a specific point in time is more accessible than for those that are absent. The sampling bias described in this chapter for studies on extant species is essentially a manifestation of the difference in accessibility between extinct and extant species. In fact, extinction is expected to leave a mark on the phylogenetic history of those organisms that have not gone extinct, and this chapter has reviewed several models that can be used to assess the role of extinction. Unfortunately, the

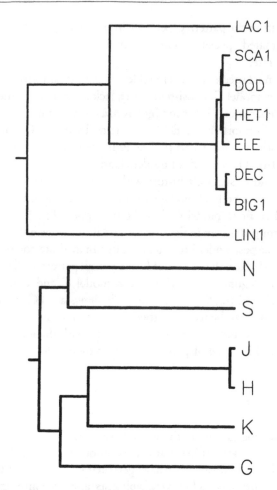

Figure 5.7. UPGMA trees (Sneath and Sokal 1973). **(A)** Genus *Microseris* (Astera-ceae) (Wallace and Jensen 1990). Data presented in table 2 of Wallace and Jensen (1990) were converted to estimates of nucleotide substitutions and adjusted for multi-ple hits (Nei and Li 1979), and they were used to generate the distance matrix from which the UPGMA tree was made. The full data set included 16 species, but several recently diverged species exhibited no net interspecific variation (see table 5.1 legend). **(B)** The Narbonensis complex of genus *Vicia* (Raina and Ogihara 1994). The data presented in table 3 of Raina and Ogihara (1994) were first converted to estimates of the numbers of nucleotide substitutions and adjusted for multiple hits (Nei and Li 1979).

mark that extinction leaves is subtle, and usually there is little statistical power to distinguish between models that include extinction and those that do not. Extinction in a phylogenetic history causes longer internode distances deep in a tree, but it can be difficult to distinguish between a model with extinction, and a model without extinction at a lower rate of speciation.

Our analysis of 11 data sets has revealed some interesting patterns. First, 10 out of the 11 data sets fit the growth model best and return an estimated extinction rate of zero. Is this surprising? The species groups listed in table 5.1 share the traits of being small and young (typically less than 10 million years since the last common ancestor of the group). It is possible that extinction has not been common in these groups. This conclusion necessarily entails a corollary that extinction is more common among older groups of species. At present, we cannot say if the apparent absence of extinction is because of a true lack of extinction or whether it is caused by sampling bias, and the lack of resolution with small species groups, with respect to the tested models.

The data from *Microseris,* a genus of annual and perennial herbs (Asteraceae: Lactuceae) (Wallace and Jansen 1990) is consistent with a history that includes extinction. This exception to the general pattern suggests that we are able to detect extinction at least in some cases and that not all small extant groups are in a phase of growth without extinction. Another exception is the Narbonensis complex, a group of species that are thought to be the closest wild relatives of the broad bean (*Vicia faba*) (Raina and Oghihara 1994). These species have a history that did not closely match any of the models tested. In short, the longest internodes are not deep in the tree, but rather at an intermediate position, as if caused by some non-homogeneous process of speciation and/or extinction.

The ability to detect speciation and extinction with phylogenetic trees expands as the number of species within a monophyletic taxa increases. The current bloom in systematic research using DNA tools should increase the number of appropriate data sets and their size. Given some, albeit limited, resolution of speciation rates and extinction, we can look forward to larger data sets and the research questions that may become accessible.

Some concepts discussed elsewhere in this volume suggest a few a priori distinctions that could become interesting contrasts as more data become available:

1. Good versus poor dispersers. As suggested by Harrison (chapter 2), if taxa exist as metapopulations, then well-dispersing taxa should persist longer than taxa that disperse poorly.

2. Abundant versus rare species. Abundant species with large ranges should persist longer and have more chances to speciate (Maurer and Nott, chapter 3).

One question of special interest from our findings would be a contrast between older versus younger clades (focusing on the time in the past of the oldest node). As suggested by our analyses, very recent groups may be in growth-only mode, while older species may be holdouts in shrinking groups.

6

Dynamics of Diversification in State Space

Daniel W. McShea

America spread west in the mid to late nineteenth century. In our own century, the story of the mass migration has been told and retold in countless ballads and novels. And in now-classic movie westerns, it has been epitomized in scenes of long wagon trains winding across the prairie toward the setting sun.

Census data from that period confirm the demographic aspect of the story: the mean location of Americans—the country's center of gravity, so to speak—shifted westward. However, the data also show that from two broad belts occupying the middle third of the country, the west-central states and the mountain states, about the same number of people moved *east* as moved west, at least during the first part of the migration from 1850 to 1880 (data from Mitchell 1983). Actually, slightly *more* moved east during this period.

The pattern is puzzling at first. The country as a whole spread west, while throughout the West, slightly more people migrated east. But it is consistent with a simple explanation. The westward shift of the mean could be explained by short, local westward movements by people in the East; unlike other regions of the country, migration there is necessarily biased westward, because migration eastward is blocked by the Atlantic Ocean. The principle is illustrated abstractly in figure 6.1A. The horizontal axis is location in space, reduced to a single dimension, east and west, and the vertical axis is time. Suppose for simplicity that we are dealing with an asexually reproducing spe-

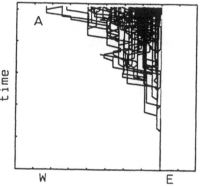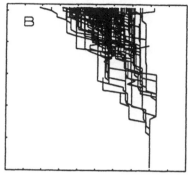

Figure 6.1. **(A)** Westward diffusion of a population. Individuals migrate east and west equally often, except at a boundary in the east—an ocean—where migration further east is blocked. **(B)** Migration is biased so that individuals migrate west more often than east. The ocean boundary is still present, but it is no longer necessary to account for the westward trend.

cies instead of people. The population begins as a single individual. As time passes, the individual reproduces or splits, and its progeny do likewise, causing the population to grow. Across the country, individuals migrate east slightly more often than they migrate west, except at one point in space where migration to the east is blocked by a boundary, the Atlantic Ocean. The result is that, despite the slight eastward bias in the movement of individuals from certain regions, the population as a whole diffuses slowly to the west. More generally, the behavior of the system at the large scale seems to be independent of at least one major feature of its dynamics at the small scale.

The pattern after 1880 was different. The mean continued to shift west, but more people moved west than moved east from all regions [except the Pacific states (Mitchell 1983)], encouraged no doubt by the west's great natural resources and by the development of a transcontinental railroad. Figure 6.1B shows that the new dynamic produces a pattern broadly similar to figure 6.1A. However, in this new regime, large-scale behavior—the shift of the mean westward—and small-scale dynamics—the westward bias—are more closely connected. Indeed, the large-scale behavior seems to be a direct result of the small-scale dynamics.

The same disparity (figure 6.1A), and congruence (figure 6.1B), between large-scale behavior and small-scale dynamics can arise in a wide variety of contexts. In the above example, the diversifying group was a population, and the lineages consisted of individuals. But the group could just as well be a monophyletic group of species, a clade, and the lineages could be individual species. Also, the horizontal axis was physical space. But it could equally well have been almost any "state variable" (McKinney 1990a), such as size, fitness, complexity, metabolic rate, geographic range, temperature tolerance, or speciation or extinction rate—in principle, any feature of a species. (To emphasize the generality of the principles developed here, the one-dimensional space within which diversification occurs will be called simply a state space, and in figures, the horizontal state-space axis will not be labeled.)

Suppose, for example, that the state space in figures 6.1A and 6.1B were a "size space." (Size increases to the left, contrary to the convention.) The two figures could then be understood to show two possible explanations for Cope's Rule, the well-known tendency for mean size to increase in many clades. In figure 6.1B, selection favors large size in all or most lineages (Newell 1949), creating a bias toward increase and, correspondingly, an increase in the mean. In figure 6.1A, a group originates near a lower bound, a lower limit on size, perhaps the result of a design constraint (Stanley 1973b). No bias is present (increases and decreases occur equally often), but mean size increases anyway as the group diversifies.

This chapter addresses two questions. First, how shall we understand the causes of large-scale behavior of clades diversifying in state space? The causes must have something to do with the small-scale dynamics of lineages, but such dynamics seem to come in many forms, producing not just biases and boundaries, but local optima and other attractors, and variations in speciation and extinction probability. To clutter the conceptual picture further, large-scale behavior for a clade includes not just trends in the mean, but stasis in the mean, and also trends and stasis in the minimum and maximum (that is, in the locations of a clade's extremes in state space). It also includes changes and stasis in the distributions of species in state space, that is, how many species there are in each range of size, temperature tolerance, extinction rate, and so on. At present, we have no systematic way to think about all of these factors or how they interact.

Second, how shall we understand the causes of independence between the small and the large scale when it occurs? What determines whether a change in dynamics produces a corresponding change in large-scale behavior; for example, does an increase in speciation rate in some region of state space produce an increase in diversity there? (It may not, if diversity is limited in that region.) What factors need to be taken into account to justify extrapolation across hierarchical levels, that is, to justify inferring the behavior of a clade from the behavior of its component lineages, or vice versa? If mean temperature tolerance in a clade as a whole has been rising, does that mean that any tendency to increase exists among most individual lineages? (Not necessarily: see figure 6.1A.)

The overall problem can be posed in other terms. What factors control whether the distribution of species in some state space is historical or equilibrial? In other words, what controls whether a clade's distribution will be the historically contingent result of its small-scale dynamics (its intrinsic probabilities of speciation, extinction, and morphological change) or the result of its dynamics conforming to some structuring imposed by ahistorical constraints?

For example, the present distribution of people in America could be nearly equilibrial. In other words, it might reflect the prior structuring of state space (the country), which is some function of the distribution of natural and commercial resources. (Both change somewhat as technologies and industries change, of course, so no true equilibrium is possible.) In contrast, it is clear that no equilibrium had been reached by the late nineteenth century, when the distribution of people at any given time reflected more the small-scale dynamics, the frequency of birth, death, and migration, than the actual distribution of usable resources. In retrospect, it seems clear that the early dominance of this system by its small-scale dynamics was (or is still) temporary. It could continue only until the limitations imposed by the structuring were engaged, in other words, until the country and the use of its resources reached saturation.

In the diversification of clades in state space on evolutionary time scales, the possible sources, or causes, of both the dynamics and the structuring are less easy to grasp, and the distinctions and relationships among them are less clear. This chapter is an attempt to understand and make explicit those sources, distinctions, and relationships.

I first present a conceptual scheme for organizing and thinking about the ways in which structuring can affect clade behavior and the degree of independence between small-scale dynamics and clade behavior. The focus is on the behavior of the mean, and there is some ancillary discussion of the behavior of maxima and of distribution shape. (The behavior of minima and of clade skewness were discussed in McShea 1994b.) The scheme will show that the behavior of a clade and the degree of independence between a clade's small-scale dynamics and its large-scale behavior depend partly on the nature or type of small-scale dynamics at work, but also on differences in the degree of "structuring" of those dynamics. [The word is pirated from Fisher's (1986) apt description of systems such as figure 6.1A as "diffusion within a structured design space."]

I then describe three major grades of structuring, classic cases that are intended as useful reference points for discussion of structuring and independence. Finally, I try to clarify the conceptual scheme further with a discussion of two cryptic cases, two systems in which the role of structuring and its relation to the small-scale dynamics is apparently (and in one case, actually) problematic.

This analysis is not concerned with why systems are structured the way they are, but only with the effect of structuring of various kinds on large-scale behavior and on the degree of independence between the small and the large scale. Further, only some of the many possible small-scale dynamics are explored, and only some of the many possible ways of structuring, enough to enable readers who find the approach agreeable to extend the reasoning to other cases on their own. Finally, the treatment is conceptual: the point is to study the relationship between structuring and hierarchical independence in principle, to help train our intuitions and condition our expectations. Empirical studies are an essential next step to verify, to refine, or perhaps to reveal as false those intuitions.

This treatment attempts to build on the foundational work on hierarchical evolutionary dynamics as done by Stanley (1973b, 1979), Raup and Gould (1974), Raup (1977a), Salthe (1985, 1993), Fisher (1986), Gould (1988, 1990), and McKinney (1990a), and it extends the train of thought begun in McShea (1994b).

The Structuring of Dynamical Rules

A Computer Model

Figures 6.1A and 6.1B are outputs of a computer model of clade diversification. In the model, a clade begins as a single lineage, a single species, at a value of zero in an unspecified state space. In each time step, each iteration of the model, the lineage moves left in state space with probability pi (probability of increase) and right with probability pd (probability of decrease). For both left and right movement, the magnitude of the step taken in state space is the absolute value of a normally distributed random variable with mean zero. (The standard deviation is varied to scale the output appropriately for the figures.)

Also in each time step, the lineage has the opportunity to generate a new lineage, to speciate or branch, which it does with probability pb (probability of branching), and to terminate, which it does with probability pe (probability of extinction). When branching occurs, new lineages follow the same movement, branching, and extinction protocols as the original. Finally, the vertical line at zero in figure 6.1A is a "cushioning" boundary. Lineages that try to cross it are instead assigned their original value before the attempt occurred. These protocols correspond to an "anagenetic" model of evolution, in that movement (or anagenetic change) occurs between as well as during branching events. In a cladogenetic or punctuated equilibrial version (see McShea 1994b), lineage movement would be confined to branching events.

In figure 6.1B, there is a strong bias ($pi = 0.15$; $pd = 0.01$) and a trend in the mean results; such trends might be called driven (McShea 1994b) to emphasize the fact that they occur as a result of a driving force, or bias, among lineages. In figure 6.1A, there is no bias ($pi = pd = 0.08$); such trends are called passive (McShea 1994b) to emphasize that they occur in the absence of any driving forces. In both, the branching probability was set higher than the extinction probability ($pb = 0.15$; $pe = 0.05$) to make diversification likely. Also in both, the model was run for 50 time steps.

The model is additive in that movement of lineages in the state space is accomplished by adding and subtracting numbers in each time step. A multiplicative model might seem more appropriate for some spaces, especially for size or other aspects of morphology, but the additive model is more versatile. It appropriately models change that is truly additive and also can be transformed to a multiplicative model simply by reconceiving the state space as a log state space (McShea 1994b).

The parameters of a system are assumed to be clade specific. That is, pi, pd, pb, and pe, as well as any structuring of these, will differ from one clade to the next, so that, for example, a boundary at some value in size space for one clade may not impede the movement of lineages in another clade at all. Finally, the model is reasonable and appropriate only in cases in which we can assume that parameters of a system remain stochastically constant. Notice that for some morphological features, at some time scales, this assumption may be decidedly unrealistic. For example, the suggestion has been made that certain parts of developmental programs become increasingly resistant to change as they become increasingly elaborated, or burdened (e.g., Riedl 1978). To accommodate such cases in which parameters change directionally, the model and the reasoning developed here would have to be modified. Further caveats are discussed in McShea (1994b).

The use of probabilistic biases and diversification parameters in the model does not deny that change or branching in each real lineage would have its own unique, fully deterministic explanation. In a series of coin flips, the trajectory and outcome of each flip has a unique deterministic explanation, but the series as a whole can be treated statistically. Similarly, change or branching in each lineage is the deterministic result of a unique concatenation of structural and ecological factors, but an ensemble of lineages, a clade, may nevertheless have robust, statistical properties. It is those statistical properties that are captured by the model parameters.

The model gives us a wide range of parameter values and value combinations to consider, and some of these are explored in McShea (1994b). Here it is not a tool for exploration, however, but a heuristic, a device for visualizing clade dynamics and structuring. The model is allied with a class of Monte Carlo models developed by Raup and Gould (1974) and Raup (1977a) for simulating morphological change in growing systems. Some analytical studies of such systems have also been done (e.g., Skellam 1951; Slatkin 1981; Toft and Mangel 1991).

Definitions

The small-scale dynamics of a diversifying clade are the rules that govern the behavior of a lineage in the state space. A lineage is an ancestor–descendant sequence of species, and its behavior includes movement, branching, and extinction. Figure 6.2A shows the two rules that underlie the passive trend in figure 6.1A. One rule—shown as a vertical line with a reflected arrow— is a blocking rule, which dictates that lineages attempting to traverse a place in state space where the rule applies are blocked and returned to their starting positions. The other—shown as a pair of equal and opposing arrows—dictates that movements left and right are equally probable in any place in state space where the rule applies. Figure 6.2B shows the single rule that accounts for the driven trend in figure 6.1B: movement to the left is more probable than to the right.

The structuring of the small-scale dynamics is the arrangement of the regions of application of the dynamical rules in state space. So, in figure 6.2A, the blocking rule is structured so that it applies to all lineages, arriving at any time, at a certain fixed value in state space, creating a boundary. The equal-probability rule is not structured at all, and it applies uniformly everywhere to the left of the boundary. In figure 6.2B, none of the rules is struc-

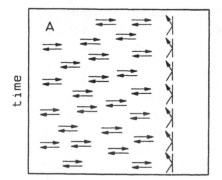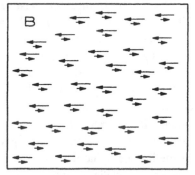

Figure 6.2. The small-scale dynamics, and their structuring (A only) for the systems in figure 6.1. See text.

tured. There is only one rule—greater probability of moving left—and it applies equally to all lineages, at all times, throughout the space.

Large-scale behavior refers to the behavior of certain properties of what I have been calling the clade, but it would more precisely be described as a temporal cross section of a clade. The clade properties of interest—mean, maximum, etc.—are the summary statistics of, or distributions for, the members of a clade that exist at a given time. And the behavior of the clade refers to changes in those properties over time, such as trends in the mean.

Hierarchy

The hierarchical relationship here can be described using Salthe's (1985) basic triad of levels: focal, lower, and upper. The focal level is the clade, the lower level is the lineage (along with its dynamics), and the upper level consists of the (relatively) invariant features of the "context" (see later) that the clade occupies and that produces structures. A clade is composed of lineages, and thus together the two constitute what has been called a scalar (Salthe 1993), structural (O'Neil et al. 1986), or aggregative (Valentine and May 1996) hierarchy.

The role of the "context" is less clear, because it may include features internal to the organisms themselves. For example, in the explanations offered earlier for Cope's Rule, the boundary could have been the result of external selection that acts only against organisms of some critical small size, but it could also have been the result of an internal developmental constraint. If external selection, then the context is some combination of factors in the external abiotic and biotic environment, perhaps competition from small species in another clade occupying the same habitat. The selective environment qualifies as context not because it is external to the lineages, but because it is stable or invariant, relative to the lineage dynamics (Allen and Starr 1982; Valentine and May 1996), and because it is largely unaffected by the dynamics—at least at the focal time scale. In this case, because the context *is* external, it could be argued that the clade is physically contained within its context, and that the context is also part of an aggregative hierarchy.

If the boundary is produced by an internal developmental constraint, the constraint must apply to all or most lineages in order to constitute a boundary, and it is therefore a relatively invariant feature of the clade body plan. It qualifies as a context for the lineages in that it changes slowly, at least relative to the lineage dynamics, and the lineage dynamics do not influence it. In this case, the clade is contained within its context only metaphorically.

Notice that the causes of the small-scale dynamics can also be either internal or external. For example, if a bias accounts for Cope's Rule, it could be the result of external selection favoring large size in all or most lineages, but it could just as well be produced by internal factors, such as directed speciation (Stanley 1979) or developmental channeling (Alberch 1980). [Internal drives (Lamarck 1809)—another possibility—are not now fashionable in evolutionary discourse, although there is no reason in principle why biases toward size increase, for example, could not be a general property of developmental systems.] In sum, both the internal-versus-external and the dynamics-versus-structuring distinctions have to do with the causes of clade behavior, but they carve the space of possible causes in different ways.

Independence

Notice that in figure 6.1B, the small-scale dynamics (see figure 6.2B) completely account for the behavior of the clade as a whole, in particular, for the trend in the mean to the left. In other words, the behavior of the clade as a whole is a direct and obvious expression of the dynamics. In figure 6.1A, however, the small-scale dynamics alone are insufficient to account for the trend. It is not enough to know what the dynamics *are;* we must also specify their structuring, for example the fact that one of them is organized in space and time so as to produce a boundary. Compare the system in figure 6.3A to

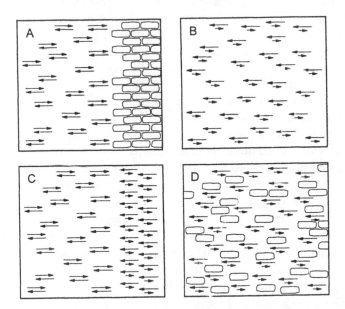

Figure 6.3. Structurally equivalent variations on the systems in figure 6.1. (Here, the blocking rule is represented as a brick.) C is equivalent to A; in C, the boundary is produced by a strong local bias. D is equivalent to B; in D, the state space contains many scattered local blocking rules. Because the rules are not structured, however, there is no net effect on clade behavior.

that in 6.3D; both contain many blocking rules, but only in 6.3A are they structured in any way. This structuring supersedes the dynamics, in a sense, producing a trend that is not inherent in the dynamics alone. In other words, the trend is to some extent *independent* of the dynamics.

The large-scale behavior in figure 6.1A is also independent of the dynamics in another sense. Any of a number of different dynamics could have been structured to produce a boundary. The blocking rule might have dictated lineage repulsion or adsorption, rather than cushioning. Or it might have been an absorbing boundary; for example, if the state space were geographic range, then lineages with small ranges might have a higher extinction probability (Maurer and Nott, chapter 3). A trend to the left would occur with all of these dynamics, provided they were structured properly. Thus, to some extent, the structuring renders the small-scale dynamics invisible, or in other words, it "screens off" clade behavior from the dynamics (Brandon 1985).

Of course, a single boundary does not insulate the behavior of the clade as a whole from the small-scale dynamics under all circumstances. If the small-scale dynamics consisted of a *very* strong bias to the right, no trend would be produced. Initial position can also make a difference. If the clade started much further to the left in figure 6.1A, far from the boundary, no trend would be produced, at least initially. Thus, a boundary at a single value in state space is a fairly weak form of structuring that produces independence only under a restricted set of conditions. Greater degrees of structuring produce independence more reliably, under a wider range of possible small-scale dynamics (and of initial positions, as well). Some of the more structured cases are considered later.

The suggestion that systems can be independent, to some extent, of their small-scale dynamics does not imply that they are somehow able to transcend, override, or contradict those dynamics. On the contrary, at the small scale, all behavior is a direct result of the small-scale dynamics. The expansion of a gas inside a closed box is governed by the rules of interaction of gas molecules with each other and the rules of interaction with the box. At the scale of the molecules, nothing more is going on.

Viewed at a larger scale, however, different behaviors emerge. The final configuration of the gas as a whole is different in boxes of different shapes, and the behavior of the gas is different in a box with no lid. For all of these systems, the list of applicable dynamical rules is the same, but the systems differ in the structuring of their rules, in the placement of the rules in space.

Unlike gases in boxes, the applicable dynamical rules for lineages vary a great deal among diversifying clades. And the variation among rules accounts for much of the variation in large-scale behavior among clades. But as in the case of the gas, the structuring of the dynamical rules is also significant,

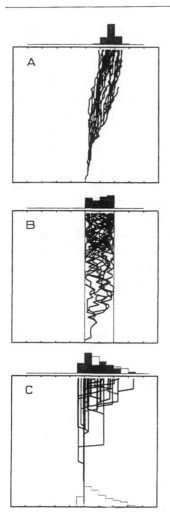

Figure 6.4. Three classic cases (see text). **A:** An unstructured system. **B:** A system that is somewhat structured by the presence of two boundaries that limit the movement of lineages in state space. **C:** A highly structured system, in which a preexisting distribution limits both movement and diversity of lineages in each region of state space.

strongly influencing the behavior at the larger scale, that at the scale of the clade as a whole, and therefore also the degree of independence between the rules and clade behavior.

Three Grades of Structuring

Three grades, or levels, of intensity of structuring can be identified. The grades are exemplified in the three runs of the computer model shown in figure 6.4. The grades are not equally spaced points on a linear scale of degree of structuring. (The first grade is an extreme, a completely unstructured system, but the third is not, because more highly structured systems can be imagined.) They should instead be viewed as classic cases, a series of instances showing successively greater degrees of structuring, in which the behavior of the system as a whole is increasingly constrained.

1. The system in figure 6.4A is *unstructured*. A bias was present throughout the space ($pi = 0.54$; $pd = 0.06$), no boundary was present, and both anagenetic change and branching ($pb = 0.10$; $pe = 0.05$) were allowed. Examples of other unstructured systems would be those in which the direction of lineage movement is unbiased but movements or steps in one direction are larger than those in the other (which would also produce a trend), as well as those in which both direction and step size are unbiased (so that no trend occurs at all).

One possible example of such a system is the diversification of the horses, the family Equidae, in size space during the Cenozoic. The mean size for the clade as a whole increased, and a strong bias toward increase is evident among lineages. In a random sample of ancestor–descendant comparisons, increases outnumbered decreases nine to zero (McShea 1994b; data from MacFadden 1986). In addition, minimum size, or the size of the smallest horse in existence at a given time, increased as well, a pattern consistent with a bias-driven system (McShea 1994b). In present terms, no structuring seems to be present, and the trend could well be a direct result of the small-scale dynamics, consisting entirely of the bias rule, distributed uniformly throughout size space. A likely cause of the dynamic is selection for greater size acting on all or most horse lineages.

2. The system in figure 6.4B is *somewhat structured*, on account of the presence of two (blocking-rule) boundaries at fixed values in state space. No bias was present ($pi = pd = 0.50$), and again lineages move and branch ($pb = 0.10$; $pe = 0.05$).

Dynamics other than the blocking rule might have been used to construct the boundaries. For example, biases might have been used, perhaps biases that engage abruptly at some fixed value in state space (see figure 6.3C) or perhaps graded biases that increase continuously as some fixed value is approached. Alternatively, boundaries might have been produced by abrupt increases in branching or extinction probability. Some have suggested that the right skew of most size distributions might be the result of higher than average branching (speciation) rates for smaller organisms (e.g., Stanley 1973b; Dial and Marzluff 1988; Brown and Nicoletto 1991). Such structuring is effectively a soft boundary. For any clade diversifying in the high-branching region of state space, the low-branching region adjacent would constitute a significant limit on clade movement.

A similar kind of structuring might be one that takes the form of one or more attractors, perhaps adaptive optima, rather than blockades. An attractor might be absolute, so that any lineage arriving at it is prevented from leaving, a kind of black hole in state space. Or it might be graded, so the probability of movement in the direction of the optimum increases as the optimum is approached.

Notice that all of these forms of structuring amount to controls on lineages that are dependent only on their position in state space. Absent at this intermediate level of structuring is any sensitivity to standing diversity, that is, to the number of other species already present at each position in state space. Such diversity-dependent controls are introduced in the next section.

Here, the added structure—whether consisting of boundaries, biases, or attractors—means greater independence between the small-scale dynamics and clade behavior. In figure 6.4B, a brief (passive) trend in the mean occurs until movement and diversification have brought the clade into full contact with both boundaries and the space between them has been filled. The rate at which the space is filled depends mainly on the frequency of change ($pi + pd$), or on the phylogenetic lability of the trait in question (Gittleman et al., chapter 4). In any case, once the space is filled, opportunities for change in clade mean, maximum, and minimum become fairly limited. Sandwiched between the two boundaries, there is little room for increase or decrease, so long as diversity increases. The point is that clade behavior is largely, although not entirely, a product of the structuring, rather than of the small-scale dynamics.

A possible example of such a doubly bounded system is the diversification of aquatic, free-living arthropods in "tagmosis" space (Cisne 1974). Tagmosis is the degree of differentiation among leg pairs within an individual, and it is computed as a function of number of leg-pair types and the number of each type, using a formula from information theory (Cisne 1974). Figure 6.5 shows the trajectory for minimum, mean, and maximum tagmosis in the diversification of the world fauna of aquatic arthropods over the Phanerozoic (data from Cisne 1974).

The existence of a lower bound can be inferred. The lowest tagmosis value attained, 0.5 for trilobites, represents an extremely undifferentiated condition, close to the analytical minimum of zero, which would correspond to an arthropod in which all leg pairs were identical. The increase in the minimum at about 230 million years is probably not significant, as it corresponds to a decrease in diversity, the extinction of the trilobites in the end-Permian mass extinction. Arthropod diversity increased after that, but the mean stopped increasing and, significantly, so did the maximum. The leveling of the maximum suggests an upper bound, a limit to the limb- pair complexity of aquatic arthropods. Any number of causes for such a limit can be imagined. For example, if each leg pair corresponds to an organismal function,

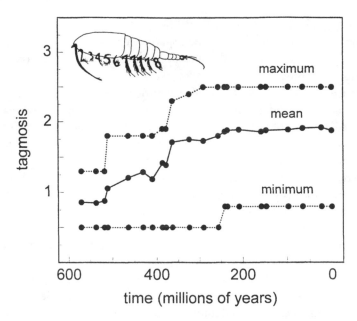

Figure 6.5. Behavior of mean, maximum, and minimum tagmosis for the world fauna of aquatic, free-living arthropods. Tagmosis = $(1/N) \log_2(N!/N_1!N_2! \ldots N_i!)$, where N = total number of leg pairs and N_i = number of the ith type. Data are from Cisne (1974); Burgess Shale problematic taxa are omitted. Inset: Figure reproduced with permission.

then the number might be limited by the number of functions to be performed. Wilson (1975) makes an analogous argument for the existence of an upper bound in number of castes in certain social insects.

The small-scale dynamics are unknown in this case, but the system could be substantially insulated from them anyway. If both upper and lower boundaries were indeed present, then together they made up a highly structured system in which movement of the mean, maximum, and minimum were severely limited—a system in which large-scale behavior attained a substantial degree of independence from the underlying small-scale dynamics.

3. Figure 6.4C shows a yet *more structured* system in which large-scale clade behavior is limited by a preexisting distribution, outlined with a thin line at time zero. No bias was present and a purely cladogenetic model was used ($pi = pd = 0$, except at branching events, where $pi = pd = 0.50$). For clarity, only surviving lineages are shown ($pb = 0.20$, $pe = 0$; when diversity limits are exceeded, $pe = 1.0$). At time zero, the distribution "contains" no species, but it represents the number of species that could exist—in principle, given existing ecological controls, perhaps—in each of several ranges, or bins, in state space. In this system, a dramatic increase in extinction proba-

bility ($pe = 1.0$) was the rule introduced to limit diversity. For example, the first bin can hold only five species, so if a sixth species was introduced by a branching event or if a lineage moved into that bin from an adjacent one, it became extinct in the next time step.

Other small-scale dynamics could have been used to produce the same sort of structuring, such as a diversity-dependent blocking rule, so that lineages attempting to enter a saturated region of state space would be turned back. A diversity-dependent bias might have been used, so that lineages entering a saturated region would tend to move back out. Whatever rules are used, the structuring here is more severe than in figure 6.4B, in that here it includes controls on diversity, as well as on the locations of lineages in state space.

The structure of the distribution exerts considerable control over the clade mean, as well as the ultimate shape of the clade distribution in state space, and to this extent releases these large-scale properties from the underlying dynamics. This is not to say that a small-scale dynamics cannot be invented that would thwart to some extent the control exerted by the structuring. For example, a very strong bias to the right would fill the right bin of the distribution, pinning those lineages in that bin, and the rest of the distribution would not fill up. But under a wide range of possible small-scale dynamics, essentially any that allow substantial movement in both directions, the clade "diffuses" to fill the distribution, as a gas expands to fill a box.

Brown et al. (1993) developed an equation for computing the "reproductive power" of a species, which they construed as a measure of fitness. Reproductive power is a function of a species' average body mass and of the rate at which it converts resources to offspring. Plugging in reasonable rate values for various mammal species allows them to predict fitness as a function of body mass alone; the resulting fitness curve is right-skewed, similar to the distribution in figure 6.4C. They note that the shape of the distribution of mammalian body masses is very similar to the shape of the fitness curve, with the body mass mode falling very nearly at the calculated optimum. And they suggest that the distribution may be the result of—in present terms—the structuring of two small-scale dynamics: first, an attractor at the optimum body mass, and second, diversity-dependent competition (Brown et al. 1993; Kelt and Brown, chapter 7), which sets limits on diversity at each body mass that are proportional to the fitness of that body mass. The mode of the distribution is determined by the modal mammalian rate constant, and the shape is determined by the dynamics of competition. If these two factors can be treated as fixed, perhaps embedded somehow in the constraints of the mammalian body plan, then the body mass distribution was preexisting in the sense that it was determined before the diversification of mammals occurred, and therefore it constrained or structured that diversification considerably as it occurred.

Cryptic and Complex Cases

The following cases appear to raise difficulties for interpretation in the preceding scheme. I address these difficulties as a device for explaining further how the scheme is to be understood.

Case 1

Colloquially, in discussions of trends, a distinction is often made between occasional and necessary increase. Specifically, the suggestion is that trends in size or complexity or some other state variable occur occasionally, accounting for the apparent rise in the maximum over the history of life, but that increase is not necessary, accounting for a (roughly) stable minimum. One such system would be that in figure 6.6, in which almost all diversification and movement occur within groups in locally structured spaces (in figure 6.6, the local structuring consists only of single boundaries), but on rare occasions when new groups arise, they do so at higher values in state space. For example, Stebbins (1969) imagined that increases in complexity in life as a whole occurred essentially in this fashion, with little significant change in complexity occurring within levels of complexity, but each new level of

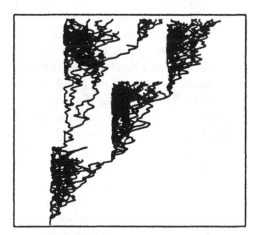

Figure 6.6. A system in which structures (boundaries) are present locally, causing passive trends within groups, but no structure is present globally. A bias operates, so that increase occurs in the origin of new groups, but it operates uniformly—in the origins of all groups, throughout the space—and therefore the system is unstructured at the scale of the clade as a whole (i.e., the clade that includes all five groups).

complexity providing a foundation—or perhaps a springboard—for further advances.

The system in figure 6.6 contains a number of structures (boundaries), but at the scale of the clade as a whole, it is not structured. The source of the trend is a kind of small-scale dynamic not yet discussed, a bias to the right that engages only in the origin of new groups. The bias is a strong one, because all new groups arise to the right of their ancestral groups, even though it engages rarely. The situation is similar to that illustrated in figure 6.3D; both boundaries and biases are present, but they are distributed uniformly, so the state space is homogeneous or unstructured. (Actually, the bias is reversed in figure 6.3D.) In effect, this is a weakly driven trend (McShea 1994b).

Case 2

The probability of a new lineage remaining in a given part of state space might depend *only* on the number of lineages already present in the range and not on its particular location in state space. In other words, the filling of state space might be purely density dependent. This differs from the filling of a preexisting distribution described earlier in that no structuring is present initially. Instead, the structuring is produced directly and entirely by the small-scale dynamics and is therefore historically contingent. Any densely filled region of state space acts like a boundary, excluding other lineages, but the boundary is a contingent result of the diversification and movement that produced the region of high density in the first place. The small-scale dynamics are structured, but the structuring is itself dynamic, changing as dense regions appear, dissolve, and move.

Of course, if the structuring changes slowly relative to diversification and movement, perhaps if the bulk of the lineages occupying the space belong to other, more slowly changing clades, then it can be treated as static. But otherwise, the behavior of the clade as a whole would be extremely complex and difficult to predict, placing it beyond the scope of this analysis.

Summary

This chapter is an attempt to develop a conceptual scheme for thinking about hierarchical aspects of the diversification of clades in state space. The central insight offered is that the large-scale behavior of a clade, such as a trend in the mean or maximum, is a product of two factors: (1) the clade's small-scale dynamics, or the set of rules governing the behavior of the lineages that make up the clade; and (2) the structuring of those rules in state space. Importantly, the scheme is agnostic on the subject of internal versus external causation. For understanding the relationship between scalar levels, it makes no difference whether the small-scale dynamics and the large-scale structuring are caused by selection, developmental constraints, or both.

If structuring is minimal, large-scale behavior will be a direct result of the small-scale dynamics, but as structuring increases, the degree of independence between the two increases. Three grades of structuring can be identified, namely, systems in which (1) structuring is absent and the same small-scale rules apply everywhere uniformly, throughout the space; (2) structuring is present in the form of one or more boundaries that limit the movement of lineages arriving at certain values in state space; and (3) structuring takes the form of controls on diversity as well as movement. In the extreme case, structuring is sufficient to determine not just the behavior of clade parameters, such as mean and maximum, but the shape and location of the distribution of species in state space, almost irrespective of the small-scale dynamics.

Acknowledgments

I thank M. McKinney, D. Ritchie, and S. Salthe for careful readings of the manuscript.

7

Diversification of Body Sizes: Patterns and Processes in the Assembly of Terrestrial Mammal Faunas

Douglas A. Kelt and James H. Brown

It is only a slight overstatement to say that the most important attribute of an animal, both physiologically and ecologically, is its size. Size constrains virtually every aspect of structure and function and strongly influences the nature of most inter- and intraspecific interactions.

—G. A. BARTHOLOMEW 1981:46

The diversity of life is composed of two parts. In part it is made up of species diversity, which is the number of discrete evolutionary lineages. Over 4700 species of mammals are known to science, for example, and new forms, even new genera, are being described routinely (Wilson and Reeder 1993). Diversity is also reflected in the tremendous variety of characteristics that these organisms exhibit. This is particularly true with respect to features of structure and function. One particularly important characteristic, and the most visible feature of living organisms, is body size. The variation in body sizes is impressive: it ranges over 20 orders of magnitude, from the smallest microplasms weighing only 10^{-13} g to the largest known organisms, whales, some of which weigh more than 10^8 g. Even within most large clades, size diversification spans many orders of magnitude. As an example, the class Mammalia spans 10^8 g, from the tiniest shrews (e.g., *Microsorex*, 2.5 g) to the blue whale (*Balaenoptera musculus*, about 1.6×10^8 g). Within this class, sizes within the order Primates range from about 100 g (e.g., mouse lemur, *Microcebus murinus*, and pygmy marmoset, *Cebuella pygmaea*) to over 200 kg (*Gorilla gorilla*), the rodents from about 6 g (e.g., pygmy mouse, *Baiomys taylori*) to over 60 kg (capybara, *Hydrochaeris hydrochaeris*), and the Carnivora from 35 g (least weasel, *Mustela nivalis*) to nearly 800 kg (brown bear, *Ursus arctos*). Furthermore, within the Carnivora, the family Mustelidae exhibit body sizes from about 35 g (*M. nivalis*) up to 45 kg (otters, *Pteronura* and *Enhydra*), and the Felidae range from about 1.5 kg (e.g., flat-headed cat, *Felis planiceps;* margay, *Felis wiedii*) to over 300 kg (Siberian tigers, *Panthera tigris*). It is clear that this diversification in sizes is not simply a result of the domination of particular regions of the body size range by specific clades. Rather, we see that successively smaller subclades span a large portion of the total range in body sizes of the larger clade, and the overall pattern is gradually built up by the accumulation of nested groups.

In part because of the constraints of allometry, this diversity in body sizes is reflected in a similarly high diversity of other characteristics, ranging in spatial scale from characteristics of individual anatomy, to characteristics of home ranges, territories, and geographic distributions, and in temporal scale from physiological rates and patterns of daily activity to evolutionary probabilities of speciation and extinction.

If we are to understand the diversity of body sizes, we need to understand the structural and functional constraints that operate at the level of the individual. Additionally, we need to understand how these constraints influence other features that operate at greater spatial and temporal scales. To this end, we will borrow a metaphor from Hutchinson (1965), who referred to species as being actors in an evolutionary play being acted out in an ecological theater (figure 7.1). In our adaptation of Hutchinson's metaphor, organisms of

Figure 7.1. (Opposite)   A conceptualization of "the ecological theater and the evolutionary play" (Hutchinson 1965) as they interact to influence the evolution of body size. Local communities are represented by *boxes with permeable borders;* the *solid arrows* represent immigration (*i*) and local extinction (*e*) of species. The local ecological theater includes both the abiotic conditions (the stage scenery, e.g., soils, physiognomy, weather, etc.) and the biotic interactions present (other actors in the play). The local evolutionary play involves microevolutionary shifts in body size to adjust to the structure and dynamics of the local theater. Local communities are nested within regional assemblages, which also exchange individuals and species to some degree, but much less so than communities (indicated by *less permeable boundaries* and *dashed immigration arrows*). The ecological theater varies from place to place (i.e., riparian woodlands versus desert shrub land versus sand dunes within the Great Basin biome), but it still includes the biotic players and the abiotic stage scenery. As a result, the evolutionary play at the regional scale produces geographic variations in body size, which we perceive as races, varieties, or subspecies. Regional assemblages are nested within continental faunas, whose physical boundaries are even less permeable (although some intercontinental movements do occur). Inter-regional variations in the ecological theater (i.e., Boreal forest versus Chihuahuan desert) are similar in kind, but they are functionally coarser than the intraregional variations mentioned. Continental dynamics (e.g., area, shape, spatial heterogeneity) influence the macroevolutionary processes of speciation (both cladogenetic and anagenetic) and extinction of species possessing particular traits such as large versus small body sizes. Islands support small-scale communities that are relatively isolated and therefore receive limited immigration (hence the dashed immigration arrows).

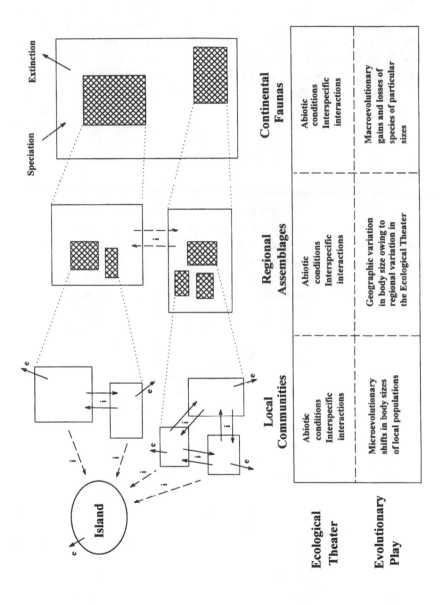

	Local Communities	Regional Assemblages	Continental Faunas
Ecological Theater	Abiotic conditions Interspecific interactions	Abiotic conditions Interspecific interactions	Abiotic conditions Interspecific interactions
Evolutionary Play	Microevolutionary shifts in body sizes of local populations	Geographic variation in body size owing to regional variation in the Ecological Theater	Macroevolutionary gains and losses of particular species of particular sizes

different sizes are subject to the rules of allometry and the constraints of phylogeny. Different actors bring different sizes to the stage, and this in turn influences how they interact with other actors and with the abiotic environment. The stage scenery (the abiotic environment), as well as the number and kinds of other actors, varies over both space and time, and this in turn causes changes in other actors. These changes occur through the processes of microevolution and macroevolution: the modification of species by gradual adjustments to their environment, and the gain and loss of species by speciation or by extinction. In the present chapter, we consider how actors respond to changes in scenery and to changes in the other actors that are on stage, by modifying their body sizes and consequently their roles in the evolutionary play.

There are three approaches to the study of how actors in our theater change roles. One approach is to watch the play through time. However, we have only one good snapshot of the play (the present), and we must rely on paleontological records to deduce the condition of the actors and the stage in the past. Such records are very useful, but to varying degrees they are inevitably fraught with problems of incomplete preservation, and consequent difficulties in interpretation. A second approach is to reconstruct the play by means of phylogenetic reconstruction (e.g., Felsenstein 1985; Brooks and McLennan 1991). This approach has been useful in evaluating the evolution of community structure and body size in Caribbean *Anolis* (Losos 1995), tropical diversity of cockroaches (Grandcolas 1993), and parental care in fishes (Gittleman 1982). A third alternative is to observe the actors and their performances and how these vary over different parts of the stage. By understanding how players' roles vary in ecological time and in different ecological settings, we can learn much about the evolutionary processes that have molded them.

We will focus our attention on terrestrial mammals. This diverse, conspicuous, and charismatic group of animals has received much attention by neontologists. As a result they are well understood ecologically. Additionally, their hard endoskeleton has resulted in a good fossil record with which evolutionary (and ecological) hypotheses may be tested. Hence they represent a good group for relating inferences from the paleontological record to insights from neontological studies.

Our approach will involve making comparisons of the roles of contemporary players over different spatial scales and in different environmental settings. We will employ a combination of inductive inference and mechanistic models to begin to evaluate hypotheses regarding the ecological and evolutionary processes that have produced the observed patterns of body size in terrestrial mammals. We will begin by outlining what we believe are the major patterns in body size. Then we will review three classes of processes that may produce these patterns. Third, we will compare several attempts to explain the evolution of body sizes. Finally, we will discuss the use of models in evaluating processes operating in body size evolution, and we will suggest several avenues of future work.

The Patterns

There are four principal patterns in body sizes that we attempt to explain. First, frequency distributions of body sizes at continental scales are highly modal and strongly right-skewed when plotted on a logarithmic scale (figure 7.2A; Maurer et al. 1992; Brown et al. 1993). This also is true at the generic level for all of the major continents (Rusler 1987; Maurer et al. 1992). Body size distributions on smaller continents (Australia, New Guinea, Madagascar) and islands are strongly modal but are relatively symmetrical (Rusler 1987; Maurer et al. 1992). This pattern has been known for many years (e.g., Hutchinson and MacArthur 1959; May 1978; Bonner 1988) and has been documented for a variety of taxa in addition to mammals. As such, this may be a very general characteristic of most or all large clades, in which case the arguments that we will present may be generally applicable to the cladogenetic evolution of body sizes.

Second, local ecological communities exhibit nearly uniform distributions of body sizes on a logarithmic scale (see figure 7.2C). Thus, species that coexist in local habitats are significantly nonrandom subsets of the continental faunas (Brown and Nicoletto 1991). The spacing between species, however, is not constant, and as a result there are clumps and gaps in local body size distributions (Holling 1993). Intermediate spatial scales, such as biomes or geographic regions within a continent, demonstrate correspondingly intermediate patterns. Hence, body size distributions for mammals in all of the principal biomes in North America are modal and right-skewed, but these distributions are much flatter than for the continent as a whole (see figure 7.2B; Brown and Nicoletto 1991).

Third, there is a macroevolutionary reduction in extreme body sizes as area of an isolated land mass decreases (figure 7.3A; Brown et al. 1993; Marquet and Taper ms). The largest species are generally smaller, and the smallest species are generally larger, as the size of the land mass decreases from the largest continents to the smallest islands. Note that the frequency distributions of body sizes among species is completely different for isolated land masses and for nonisolated sample areas within larger land masses. In the former case, the modal shape is retained but the range of body sizes becomes

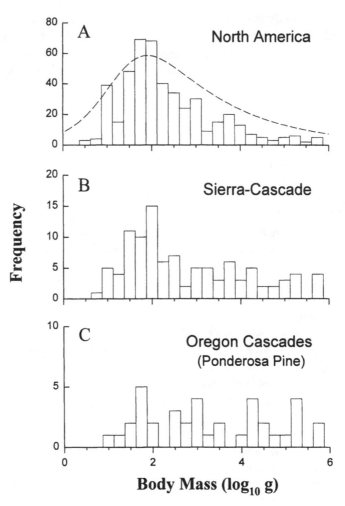

Figure 7.2. Principal patterns in the frequency distributions of body size. **(A)** At the scale of continents or the entire globe, the distribution of body sizes of mammals is highly modal and strongly right-skewed on a logarithmic axis. The *dashed line* gives potential energetic fitness for mammals (after Brown et al. 1993). **(B)** Body size distributions within biomes also are skewed, but they are much flatter than those at continental scales. **(C)** At the scale of local communities, body size distributions are generally log-uniform and are significantly nonrandom subsets of larger spatial scales (all data from Brown and Nicoletto 1991).

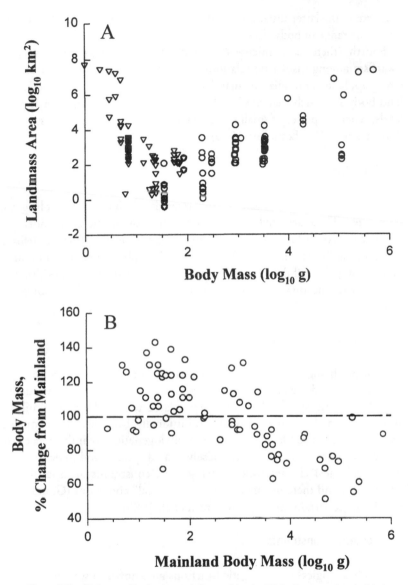

Figure 7.3.   Patterns of body size evolution on islands. **(A)** As the area of an isolated landmass decreases, there is a corresponding macroevolutionary loss of the smallest and the largest species. As a result, the smallest landmasses possess species of intermediate body size (about 100 g). **(B)** Island populations of small species exhibit microevolutionary trends toward gigantism, whereas those of large species exhibit dwarfism, with respect to the most closely related continental and ancestral populations (from Brown et al. 1993).

reduced; in the latter situation, the distribution becomes less modal but preserves the range in body sizes.

Fourth, there is a microevolutionary trend towards gigantism and dwarfism among insular populations of species when compared to conspecific populations on adjacent mainlands. This trend is a function of mainland body size, such that species of large size tend to develop dwarfs on islands, whereas species of small size develop insular races of giants (see figure 7.3B; Foster 1964; Lomolino 1985).

The Processes

There are three primary classes of processes that influence the evolution of body sizes. These are phylogenetic constraints, allometric constraints, and ecological relationships. We follow Maynard Smith et al. (1985) in defining constraint as "a bias on the production of variant phenotypes or a limitation on phenotypic variability caused by the structure, character, composition, or dynamics of the developmental system," and we discuss both categories below.

Phylogenetic Constraints

Because of phylogenetic descent and inheritance of characters from common ancestors, related species tend to preserve a legacy of their phylogenetic history. As a result, species that are closely related tend to be more similar in size than more distantly related taxa. Additionally, because of the allometric constraints discussed later, and because phylogenetic proximity reflects recency of evolutionary divergence, closely related species tend to be similar in many ecological characteristics. Phylogenetic constraints are specific to a given taxon and therefore are a type of "historical" constraint (Gould 1994; the "local" constraint of Maynard Smith et al. 1985).

Allometric Constraints

Many physiological and energetic functions are known to scale to body size (Peters 1983; Schmidt-Nielson 1984; Calder 1984). Such allometric relationships reflect structural and functional constraints on the size and shape of the individual organism. As such, these are "universal" constraints (Maynard Smith et al. 1985; the "formal" constraint of Gould 1994).

If different organisms were merely isomorphic versions of a single body plan, then the physical laws of diffusion, and the known relations of surface area to volume, would result in three kinds of relatively straightforward allometries. In such an idealized set of organisms, functions of volume would scale linearly with body mass, M, as M^1, while functions of surface area would scale as $M^{0.66}$, and linear relationships as $M^{0.33}$ (see, e.g., Schmidt-Nielson 1984).

Real organisms are not isomorphic, however, and generally do not scale in this simple fashion. Whole-organism rate processes, including such processes as metabolic rate, growth rate, milk production, requirements for energy and limiting nutrients such as nitrogen, phosphorus, etc., generally scale as $M^{0.75}$. Conversely, mass-specific rate processes, including rates of cellular metabolism, clearance of toxins from the body, etc., scale as $M^{-0.25}$. Finally, biological times, including such features as gestation period, life span, the time required for blood to circulate through the body, etc., scale as $M^{0.25}$.

Because of these allometric constraints on structure and function, including energetic and nutritional requirements and timing and allocation in life histories, body size ultimately influences many features of ecology, biogeography, and evolution (Peters 1983; Calder 1984). While not all of these allometric relationships are well understood, size appears to place important constraints on such variables as dietary specialization, niche breadth, home range and territory size, population density, clutch size, area of geographic range, and the dynamics of extinction and speciation (Brown 1995 and included references).

The result of these allometric relations is that small organisms [but probably not the smallest (see later)] require fewer total resources but have large mass-specific requirements. They live fast and have large reproductive allocations, as reflected in short life spans, large and frequent litters, and high intrinsic rate of increase, r_{max}. They have high densities and small territories and home ranges, which they exploit over short time scales, and as a result they are able to avoid extinction and persist in small geographic ranges. Finally, small organisms likely have relatively high speciation rates. Larger organisms tend to have opposite patterns; importantly, they are ecologically constrained to low population densities and large geographic ranges, and they often appear to have high extinction rates during periods of catastrophic environmental change.

Ecological Interactions

The local environment in which an organism lives may act as an agent of selection, resulting in adaptive evolution of body size. Two processes appear to be involved. Organisms may be selected to alter their ecological functions in ways that can be accomplished by changing size, thereby reaping the benefits that are consequences of allometry. For example, because age at first reproduction scales as $M^{0.25}$, selection for reduction in the age of first reproduction, perhaps because of altered predation, may be achieved by evolving towards smaller size. In addition, however, interactions with the environment may cause organisms to be selected to break the constraints of allometry. Thus, selection on weasels to follow prey into burrows and refuges has apparently resulted in the evolution of a long and thin body shape. The result of this is that weasels differ from other mammals of similar mass by having greater body surface area, rates of heat loss, and food requirements at low ambient temperatures (Brown and Lasiewski 1972).

Because of the influence of phylogeny, related organisms will tend to be similar in size and other characteristics (Harvey and Pagel 1991). This tendency towards similarity will tend to be reinforced by the relative difficulties in breaking the constraints of allometry. Ecological pressures, however, will often tend to promote the divergence of closely related species. To the extent that speciation is allopatric, related organisms will tend to occur in somewhat different environments, to be subject to different selective pressures, and therefore to diverge in body size and other characteristics. Alternatively, to the extent that closely related allopatric species occur in similar environments, they may be selected to remain similar in body size and other ecological characteristics. Allopatric species that occur in similar environments but are not closely related may be selected to converge. To the extent that speciation may be sympatric, or that closely related species have secondarily come to coexist, they may be subject to selection to diverge in body size and/or other characteristics. The most obvious selective pressure promoting such divergence is interspecific competition, and its evolutionary consequences have been referred to as character displacement (Brown and Wilson 1956; Taper and Case 1992). It is also possible that other ecological interactions, such as predation or even mutualism, could produce selection for divergence in body size or related characteristics.

Explanations of Observed Patterns

Modal and Right-Skewed Distribution of Clades and of Faunas at Continental Scales

Hypotheses

Three main explanations have been proposed to explain the continental pattern of body sizes.

Gould (1988) presented the null hypothesis, that the observed distribution reflects random divergence of species of a clade from a common ancestor. Because of Gould's advocacy of punctuated equilibrium (e.g., Eldredge and Gould 1972; Gould and Eldredge 1977), he considered natural selection operating differentially on individuals to be a less significant factor in the evolutionary diversification of body sizes than the large changes that occur during cladogenetic and extinction events.

Hutchinson and MacArthur (1959) argued that the high diversity of small body sizes can be attributed to the fact that these smaller species are more specialized than larger species, and that they operate at spatial and temporal scales at which the environment is inherently more heterogeneous. May (1978, 1986; see also Morse et al. 1988; Lawton 1990) recast Hutchinson and MacArthur's (1959) hypothesis in terms of fractal geometry. If smaller species occupy narrower niches, and the environment exhibits a fractal structure so that niches are more abundant and more spatially fragmented, then it follows that smaller species should be more abundant.

Finally, Brown et al. (1993) hypothesized that the modal and right-skewed distribution of body sizes reflected an underlying energetic dynamic, and that the peak in diversity of body sizes occurred at an optimal mass, at which species could most readily obtain energetic resources from the environment and convert them into offspring. Brown et al. suggested that allometric constraints on energy and resource acquisition and allocation limit the reproductive power or potential energetic fitness (PEF) that can be realized by individuals of different size. They developed an allometric model of PEF that predicts the optimal size of a species within a clade, and that should correspond closely to the mode of the observed frequency distribution.

Evidence

Gould's hypothesis has been subjected to a computer simulation (Maurer et al. 1992) in which faunas of 350 species evolved in response to either cladogenetic or anagenetic processes, or both. Additionally, speciation and/or extinction were incorporated and were modeled as unbiased or as biased towards either large or small species. Maurer et al. (1992) demonstrated that random cladogenetic events alone did not produce skewed frequency distributions, but when extinction probabilities were biased towards small or (especially) large body sizes, the resulting distributions were significantly skewed. These results were accentuated when anagenetic evolution was incorporated. Hence, the null hypothesis, that the evolution of body sizes is a product of random cladogenesis, was rejected, and some kind of deterministic process was implied to have produced the distributions of sizes observed empirically.

McShea (1994a) also used computer simulation to explore possible mechanisms affecting the distributions of quantitative traits in proliferating lineages. He compared different combinations of three kinds of models: (1) equiprobable random divergence, equivalent to the null model of Gould

(1988) and Maurer et al. (1992); (2) passive divergence with a boundary on one side that could be approached but not crossed, as might occur, for example, if some functional constraint placed an absolute limit on the smallest possible body size; and (3) divergence biased in one direction (i.e., driven) by differential speciation and/or extinction rates. McShea found that the null model of equiprobable divergence did not generally produce highly modal and skewed distributions, but that the passive and especially the driven models readily did so. He also compared the distributions generated by his simulation models with empirical distributions of body sizes and skeletal traits.

Hutchinson and MacArthur's (1959) argument provides a good fit to observed distributions, but it is not very testable. Fractal versions (May 1986, 1988a; Morse et al. 1988; Lawton 1990) predict that the largest number of niches should be available to the smallest species. Thus, they fail to account for the decline in frequency of species of extremely small size. All of these hypotheses based on size-related specialization are difficult to evaluate empirically, because they require an operational way of measuring specialization and the number of "unfilled niches" available to species of various sizes.

Finally, the distribution of PEF as a function of mass is unimodal and right skewed, and it is very similar in form to the empirical distribution of body sizes for the 464 species of North American mammals (see figure 7.2A). PEF predicts an optimal body size for mammals at about 100 g, which corresponds well with the peak of the observed distribution. The model of Brown et al. also makes predictions about the evolution of insular populations, as well as of life history and demographic traits. This model does not explicitly explain the diversity of body sizes. Given that an optimum exists, why are not all species of this size? How do we account for the diversity of sizes found in ecological communities and continental faunas? To address this, we have been working with models of local community assembly that incorporate characteristics of PEF (see later). We hope that these may be extended to operate at greater spatial scales, and that they may provide testable hypotheses for the evolution of regional and continental faunas.

The Nearly Uniform Size Distribution Within Local Communities

Hypotheses

As described, we can envision three possible explanations for this local pattern. The most obvious null hypothesis is that local communities are a random selection from the pool of species that are present on the continent (Brown and Nicoletto 1991).

Alternatively, the distribution may reflect some form of local, ecological displacement, most likely to avoid competition between species of similar size. Ecological displacement may arise from two mechanisms: ecological assortment or evolutionary adjustment, the dynamics of which are manifested at different spatial scales. Ecological assortment reflects the local compatibility of species and is a result of the daily interactions among species living in close proximity. It is mediated by local colonization/extinction dynamics, the capacities of species to immigrate into and persist in local communities depending on the extent to which they are similar to and compete with the other species that are present. In contrast, evolutionary adjustment influences the regional evolution of body sizes, producing races and subspecies, and reflects microevolutionary adjustments by species to different environmental conditions and to different faunal composition. The most common form of such adjustment is character displacement, the tendency for evolutionary divergence in body size and other characteristics, in response to selection to reduce competition with coexisting species (see Taper and Case 1992).

Finally, Holling (1993) has suggested that the environment presents a given array of niches, the dynamics and distribution of which are dictated by both biotic and abiotic environmental characteristics. As a result, species have sorted themselves into discrete body size categories that reflect the parameters of the available niches. Presumably, both ecological assortment and evolutionary adjustment can act as mechanisms to produce a match between the available niches and the body sizes of the species that fill them. Holling's explanation places more emphasis on the presence of gaps and clumps in local body size distributions than on the overall, relatively uniform distribution.

An alternative to Holling (1993) invokes the overriding dominance of certain key taxa (Brown 1995). Certain widespread, abundant species occur in many local communities and habitats over large spatial scales, and they form the clumped nodes of body sizes over these scales. Such "continental keystone species" will also often be members of diverse and widespread lineages (e.g., *Sorex cinereus, Peromyscus maniculatus, Canis latrans*), and, therefore, where they are absent they will often be replaced by closely related taxa of similar sizes, reinforcing the observed clumping of body sizes at certain nodes. While Holling's explanation suggests that niches are set in an a priori manner and hence should be relatively constant across biome types both within and among continents, our alternative suggests that they are dependent on the particular organisms present. Therefore, we would predict that the modal sizes depend on phylogenetic constraints and would often differ among communities in similar habitats, or biomes on different continents.

Evidence

The null hypothesis, that local communities are random subsets of the continental species pool, was tested and disproven by Brown and Nicoletto (1991), who found that random selection from the North American species pool produced simulated assemblages that had much more skewed body size distributions than real communities. It follows from this that some local process operates to inhibit local coexistence of similar-sized species. Competition is the most logical candidate, although we can imagine scenarios in which predation or mutualism might also produce such patterns (i.e., apparent competition; Holt 1977). Even if Holling's (1993) cross-scale explanation is right, some process operates to limit the number of species that occur within the nodal niches. We do not suggest that all similar-sized species interact competitively [e.g., least weasels, *Mustela nivalis* (45 g), do not compete with voles, *Microtus* (40 to 80 g)], but if competition operates within guilds, and the number of guilds is relatively small, then competition may spread species out within guilds.

While the observed local body-size distributions are not highly modal, and they are more dispersed than a random subsample of the North American pool, they do exhibit clumps and gaps at local and biome scale. Some mechanism appears to aggregate species into relatively discrete clumps. It will be very difficult to separate Holling's explanation from the alternative that we have presented above. Intercontinental comparisons will be difficult in many cases (e.g., are woodlands really comparable habitats in North America and Australia?) but may be feasible in others (*Larrea* shrublands are very similar in North America and South America).

Macroevolutionary Reduction in Extreme Sizes as Area of Isolated Land Mass Decreases

Hypotheses

The null hypothesis for this pattern is that the distribution of body sizes on smaller land masses is a random sample of the body sizes that are present in the global pool.

Alternatively, Damuth (1971) demonstrated that the probability of extinction scales positively with body size. Because local population density varies inversely, and area of geographic range varies directly, with body size, and because the probability of extinction is strongly influenced by total population size (e.g., MacArthur and Wilson 1967), it follows that extinction probability should be correlated directly with body size.

Finally, PEF (Brown et al. 1993) may explain the spatially dependent distribution of body sizes. According to this argument, as species richness decreases, species that are divergent from the optimum should be least likely to persist, so both maximal and minimal body sizes should converge towards the optimal body size for a taxon. For mammals, this body size is approximately 100 g (Brown et al. 1993). Hence, on the smallest islands, where only a single species is present, it should approximate this size.

Evidence

Marquet and Taper (ms) developed a simple randomization protocol to draw a given number of species (without replacement) from the global pool. They found that the smallest and largest body sizes on land masses with low richness were significantly larger and smaller, respectively, than those predicted by a random draw of the same number of species. We conclude that the reduction in extreme body sizes on smaller land masses is not a product of random selection.

Damuth's argument accounts for the decrease in large species as land area decreases, but because his allometric relationships are linear, they do not account for the loss of the smallest species. In fact, a strict application of Damuth's argument would suggest that as the number of species on a land mass is reduced to one, the final species present would be in the smallest size class (about 2 to 3 g for mammals), which is not observed (Brown et al. 1993).

The observed shift in the distribution of body sizes with decreasing richness agrees with the predictions of the PEF model of Brown et al. (1993). We note that their argument predicts the convergence on optimal size when only a single species is present, but it must invoke mechanisms of faunal buildup and species packing in order to develop larger faunas. Extensions of the PEF model (Kelt 1997; see later), which incorporate competitive interactions and character displacement, demonstrated remarkably good fit to empirical data.

Microevolutionary Trends Towards Gigantism and Dwarfism in Insular Populations

Hypotheses

The null hypothesis is that changes in body size in insular faunas reflect random changes (e.g., founder effects, genetic drift) with respect to the ancestral populations. In contrast, Brown et al. (1993) argued that insular species will converge towards the optimal size for a taxon. Because islands have fewer species, the number of competitors and predators likely is reduced, relative to the mainland. With fewer competitors, species should experience micro-evolutionary shifts in body size towards the optimum. Species of large body size that encounter reduced predation pressure on islands also may become dwarfed on islands. We suggest that in both cases, selection for insular dwarfism involves allometric changes in life history traits, such as the production of offspring at earlier ages and a reduced interval between litters, which increase PEF. Selection for insular gigantism in populations of small original body size also involves allometric changes to increase PEF, but these would be attained by increasing the capacity for energy acquisition, rather than conversion.

Evidence

In contrast to the null model, Lomolino (1985) demonstrated that the direction of body size shifts among insular taxa are allometrically related, such that smaller species exhibit gigantism and larger species become dwarfed. A regression through these data is significant and crosses the zero-change isocline at approximately 200 g. Given the wide confidence intervals around the regression, this intersection is in reasonably close agreement with the 100 g predicted from PEF, and it further supports the underlying importance of an energetic definition of fitness in the evolution of body size. We emphasize that there is a substantial stochastic element in how insular colonization and extinction affect the species composition of insular faunas. The predicted direction and magnitude of body-size adjustment will depend on the time that the species has been on the island and on the number and sizes of other species that are present. As a result, we would predict a fair amount of variation around the predicted shift, and this is observed.

Implications for Spatial and Temporal Patterns of Species Abundance

Potential energetic fitness implicitly predicts that many ecological and physiological allometric relationships should change slope and perhaps even sign at approximately the optimal body size (Brown et al. 1993). In support of this, Brown et al. (1993; see also Brown 1995) noted that while both home range size and maximal population density varied with body size in mammals, these relationships are not well fit by standard allometric power functions. Instead, both parameters appear to have their maximal values at some intermediate size, which is very close to the modal size and the optimal size predicted by the model of PEF. These nonlinear relationships appear to have important consequences at geographic spatial scales and evolutionary time scales. Geographic range size and ecological specialization as reflected in habitat specificity are correlated with home range size and population density, so they also vary nonlinearly with body size (Brown 1995). A consequence is that species near the modal body size tend to exhibit the greatest habitat specificity, have the smallest geographic ranges, and show the highest spatial turnover or beta diversity (Brown and Nicoletto 1991). At the same time, however, other species of near-modal size have very large geographic ranges and are habitat generalists. These species, together with ones of more extreme size, are widely distributed and form the core species of mammal communities across the North American continent.

We can speculate on the pattern of temporal turnover we might expect to see in the fossil record. On the one hand, since we expect extinction rates to be negatively correlated with geographic range size and habitat nonspecificity—and perhaps for speciation rates to show the same pattern—specialized species of modal size should turn over rapidly. On the other hand, the generalized species of near-modal size with dense populations and wide geographic ranges should exhibit the lowest rates of turnover. Species of extreme size, especially very large, should have high rates of extinction because their very low population densities do not compensate for their large geographic ranges.

Models of Size Diversification During Faunal Assembly

The simplest approach to modeling faunal diversification involves random cladogenesis. This was explored for the general case by computer simulations (Raup et al. 1973; Raup and Gould 1974; Gould et al. 1977). Gould (1988) suggested that such random diversification from a common ancestor could also account for the distribution of body sizes within clades, but he did not develop explicit models. Maurer et al. (1992) modeled the evolution of body sizes in continental faunas with a simulation approach and found that random cladogenetic processes alone were unlikely to account for the right-skewed distributions that are almost universally observed. They showed that incorporation of size-biased speciation or extinction into the model could generate right-skewed distributions, but their models did not incorporate any specific mechanisms through which body size could cause these biases. McShea (1994a) simulated lineage diversification in passive and directed and in bounded and unbounded conditions, with the goal of determining which gave resulting frequency distributions of quantitative morphological traits most similar to those observed empirically. He found that highly modal and skewed distributions rarely resulted from unbounded passive divergence, but that they could be produced either by passive divergence in the presence of a boundary or by directional forces. He found that the distributions of body sizes in rodents and vertebral complexity in mammals resembled the simulations of random bounded divergence, while the distribution of body size in horses resembled his directional simulations. McShea's results supported Stanley's (1973b) earlier suggestion that the evolutionary trend towards larger body sizes seen in many lineages and taken to exemplify Cope's rule is more likely to reflect the greater diversity and higher speciation rates of lineages of small body size than directional evolution towards larger sizes.

None of these approaches have incorporated allometric relationships of body size. Recently, Kelt (1997) combined the model of PEF (Brown et al. 1993) with a general model of character displacement (Rummel and Roughgarden 1985; Taper and Case 1992) to simulate assembly of ecological communities. Kelt's model simulates the development of local communities as a consequence of immigration from a larger pool of species, followed by anagenetic shifts in body size to maximize fitness. Fitness is a function of both PEF and competitive displacement. Selection operates to shift body sizes towards the energetically optimal size. Opposing this, competitive interactions operate to shift species away from each other, thereby spreading them out along the mass axis. As species richness increases, the fitness landscape is gradually transformed from a unimodal humped distribution to one with multiple peaks and troughs. This model produces size-structured communities as a result of both evolutionary character adjustment and ecological sorting. Simulated local communities of up to 50 species were generated. These resembled real communities in having similar extreme sizes, and log uniform distributions of species between these limits. Simulated communities differed from real ones by lacking the clumps and gaps along the size axis noted by Holling (1993).

Kelt (unpublished) modified his original model to simulate cladogenetic faunal evolution, by incorporating both speciation and extinction. Assembly is initiated with a single species of random size, which shifts body mass to maximize fitness, as described. This species is allowed to speciate, and the resulting daughter species evolve anagenetically to maximize PEF and minimize competition (figure 7.4A). Cladogenetic diversification continues until the extinction rate, resulting from ecological carrying capacity, equals the speciation rate. As species richness increases, the fitness landscape is transformed from a unimodal to a polymodal distribution, and ultimately to a unimodal trough in which fitness is greatest at extreme sizes. Like the ecological model, the cladogenetic model produces faunas with species uniformly distributed along a logarithmic axis of body sizes.

An interesting outcome of the cladogenetic model is that it generates Cope's rule (Stanley 1973b). This occurs because, at any given point in time, species tend to be clustered under the peak in PEF, and competition for ener-

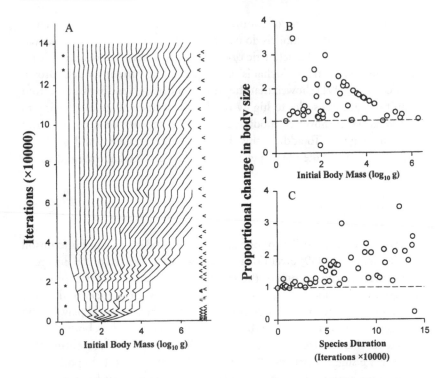

Figure 7.4. Results of a model of interactive faunal assembly involving both cladogenetic speciation and anagenetic body size shifts to maximize local fitness (see text). In the model presented in this figure, the probability of stochastic extinction upon reaching stability was set at 20 percent; in other words, there was a 20 percent probability that a randomly chosen species would go extinct once all species masses had stabilized. **(A)** Trajectories of body size change for 50 species; speciation events are marked with *arrowheads* (<), while stochastic extinctions (not including extinctions resulting from evolution to extreme body sizes) are noted with *asterisks* (*). **(B)** Proportional change in body size for 50 species, as a function of their initial body size. Even very small species may experience substantial increases in their mass. **(C)** Proportional change in body size for 50 species as a function of their duration. The majority of species clearly evolved towards greater body size, although one species in this simulation underwent dwarfism. Regression statistics: body size change = (7.415×10^{-6}) (species duration) + 1.071, $P < .0001$, $r^2 = 0.303$.

getic resources is therefore greatest among species at modal body sizes. Because the slope of the fitness landscape is less steep at larger than at small body sizes, when speciation occurs there is a greater opportunity for anagenetic shifts to larger body sizes (see figure 7.4B,C).

Kelt's models further support the potential roles of energetics and interspecific interactions in the evolutionary diversification of faunas, as well as the ecological organization of communities. At their present stage of development, however, these models do not generate the modal and right-skewed distributions of sizes characteristic of continental faunas. We believe this reflects the limited spatial scale that is considered in the current models. Brown and Nicoletto (1991) showed that the shift in the body size distribution from log-uniform at local scales to highly modal at continental scales is owing to the differential turnover of modal-sized species between habitats and over geographic regions. To address this, Kelt's assembly models could be developed further to include the dispersion of species among habitats and across geographic space.

Prospectus

We have explored the utility of one approach to understanding Hutchinson and MacArthur's (1959) evolutionary play. We have focused on an energetic fitness currency and on one of the most obvious features of its expression, body size, to describe how species respond to the stage scenery and to the other actors present, at any given moment during the ecological play. We do not suggest that life is so simple that a single parameter such as body size can be used to characterize the dialogue and choreography of the entire play. However, we suggest that the allometry of PEF offers testable hypotheses to explain some aspects of this dialogue and choreography that have remained elusive until now.

It is always reassuring when a single model or mechanism appears to be able to explain empirical observations made over broad spatial and temporal scales. Such was the general appeal of MacArthur and Wilson's (1967) theory of island biogeography. PEF appears to hold much promise in explaining the distribution and evolution of body sizes, and energetics may provide a currency to synthesize and unify heretofore poorly integrated fields such as physiology, ecology, ethology, evolution, and paleontology. The PEF model predicts an optimal body size for a given taxon, and it provides a unified explanation for body size distributions at various spatial and temporal scales. Additionally, most features of local community assembly are predicted by simple models that incorporate PEF and ecological character displacement.

However, we stress that such models are only the first step of a research program, certainly not the final word on the evolution of body size. We do, however, believe that the generality of this modeling approach makes it a good starting point. We encourage the development and testing of alternative explanations for these patterns and of further extensions of our own models' analyses. Some potentially fruitful avenues to pursue include:

- Alternative niche axes. Our analyses have emphasized body mass as a single axis of primary importance in ecological assembly and evolutionary diversification. Certainly, other factors, including trophic specialization, habitat structure and regional heterogeneity, interspecific interactions, microhabitat utilization, and climatic regimes, influence the assembly and diversification of ecological communities and continental faunas. Further work incorporating these factors, and considering their relationships to body size, should prove insightful.

- Incorporation of the fossil record. Our analyses provide predictions that may be tested by detailed investigation of the fossil record. There may not be a large number of fossil communities preserved with sufficient fidelity to test our ecological predictions, but the evolutionary patterns, such as Cope's rule, should be widely testable.

- Generalization to other taxa and clades. Our analyses have focused almost exclusively on the evolution and diversification of extant terrestrial mammals. Brown et al. (1993) presented data suggesting that their PEF model applies to other extant vertebrate groups. We urge the application and evaluation of these ideas with other taxa, both extinct and extant. Cases in which the model fails to provide a good fit to empirical data will provide useful insights into the limitations of the present model, and we can hope they will lead to greater understanding.

- Interactions between taxa or clades. Recognition that players share the ecological stage and interact with other kinds of organisms may have important influences on the patterns of body sizes at both ecological and evolutionary scales. For example, large insects may interact with the smallest mammals, influencing the lower size limit for the latter group. Similarly, the diminutive size and inconspicuous nature of mammals throughout the Mesozoic may have reflected competitive or other interactions with the several groups of giant reptiles (sensu lato) that were dominant members of ecosystems. And the diversification and dramatic increase in the body sizes of mammals may reflect altered interactions as a result of the extinction of the dinosaurs at the Cretaceous/Tertiary boundary about 65 million years ago.

Conclusion

The evolutionary diversification of body sizes within and among clades has produced a remarkable array of species, but understanding the dynamics of this diversification has been an elusive task for evolutionary biologists. We suggest that this is partly because fundamentally different currencies are used by the various subdisciplines of biology. Physiological and ecosystem ecologists study the dynamics of energetic changes (e.g., dE/dt), whereas population and community ecologists investigate the dynamics of numerical changes (e.g., dN/dt, Brown 1994). We have attempted to bridge this gap with the use of PEF, which uses allometric functions to characterize the ability of an organism to acquire energetic resources from the environment, and to convert these resources to reproductive functions (Brown et al. 1993).

This approach has allowed for the prediction of an optimal body size for a clade; this was calculated to be approximately 100 g for mammals (Brown et al. 1993). Subsequent investigations have incorporated character displacement to model ecological assembly (Kelt 1997), and speciation and extinction dynamics to model evolutionary diversification of ecological communities (Kelt, unpublished). These studies support the suggestion that PEF is an important factor underlying the distribution of body sizes at both local and regional spatial scales, and over both ecological and evolutionary time scales.

PEF appears to provide testable hypotheses for important influences of ecology and allometry on the evolutionary diversification of species. We urge further explorations of this as well as alternative approaches. We tentatively suggest that at one level of abstraction, Hutchinson's play may be easier to understand than we may have thought.

Summary

Using terrestrial mammals as an example, we review four patterns in the frequency distributions of body sizes among species: (1) the modal and right-skewed distribution (on a logarithmic scale) characteristic of the faunas of large continents and the entire globe; (2) the log-uniform distribution at the scale of local ecological communities within continents; (3) the macroevolutionary tendency to lose species of extreme body sizes as the area of an isolated land mass decreases; (4) the microevolutionary trends towards gigantism in small species, and dwarfism in large species, when isolated on islands. We evaluate alternative hypotheses to explain these patterns.

A model of potential energetic fitness (PEF) appears to provide a single, unifying explanation for all of these patterns. As initially modeled by Brown et al. (1993), PEF predicts an optimal body size based on the allometry of energy acquisition and allocation to reproduction. Kelt (1997; unpublished) has extended this model, incorporating both ecological character displacement and cladogenetic evolution, to simulate the assembly of local communities. Kelt's models not only produce simulated communities similar in many respects to real ones, but the cladogenetic models also generate the evolutionary trend toward increasing body size, known as Cope's rule.

Acknowledgments

We thank Mike McKinney for the invitation to contribute to this volume, Mark Taper and Pablo Marquet for valuable discussions and insights during the development of this approach, and Brian Maurer and George Stevens for reading the manuscript. This research was partially funded by NSF grants DEB 9221238 and DEB 9318096.

8

The Role of Development in Evolutionary Radiations

Gunther J. Eble

How are biodiversity dynamics developmentally constrained at various hierarchical levels? Patterns of diversity and associated process theories have conventionally been treated in extrinsic, particularly ecological, terms, and development has not been sufficiently integrated into discussions of diversity change (e.g., Rosenzweig 1995). This may be due, in part, to a reluctance of many students of development or ecology to delve into the interplay between allegedly ahistorical principles of form generation and their historical realization in the process of evolution. It was not until recently, when the reality of the Cambrian explosion of metazoan body plans became clear, that developmental explanations have been called on to account for the apparent temporal asymmetry in the turnover of higher taxa (and associated morphological innovations).

Changes in developmental integration and regulation are now an interesting alternative hypothesis to the usual ecological explanations (e.g., Erwin and Valentine 1984; Jablonski and Bottjer 1990a; Hall 1992; Erwin 1993b, 1994; Valentine 1995; Raff 1996). An array of macroevolutionary problems associated with changes in diversity through time has not yet been fully explored in explicitly developmental terms. Is diversification at lower levels ever developmentally constrained? What is the developmental context of key innovations? Can developmental constraints underlie failure to radiate and macroevolutionary lags [low-diversity delays to diversification (Jablonski and

Bottjer 1990b)]? Are diversity-dependent phenomena always played out in ecospace? Null hypotheses for diversification have usually been based on the assumption of randomness, but given nonrandom pattern we must logically pose development-based explanations as alternative hypotheses to ecology-based ones. Testing such explanations is a basic challenge faced by neontologists and paleobiologists alike.

I address the role of development in the context of evolutionary radiations, not only because they have traditionally attracted the attention of both paleobiologists and neontologists, but also because problems of testability become more evident. However, much of the discussion is potentially applicable to background diversification times as well. Additionally, I present a case study on a recent hypothesis for the early introduction of evolutionary innovations (as in the Cambrian explosion) that purportedly includes contributions from both ecology and development: Kauffman's model of rugged fitness landscapes (Kauffman 1989, 1993, 1995).

Evolutionary Radiations, Development, and Paleontology

Evolutionary radiations constitute a very important aspect of evolution because they generate much of the biosphere's taxonomic and morphological diversity. We identify evolutionary radiations as interesting subjects for study by comparison with previous and later quiescence (background diversification). Thus, temporal asymmetries, i.e., unusually early or late radiation in the history of a clade or clades, will attract the most attention. However, exponential branching processes (Stanley 1975, 1979; Sepkoski 1978, 1991c; Hey 1992; Patzkowsky 1995) will produce apparent radiations even under constant diversification rates, and ultimate decline in standing diversity is the expectation over very long periods of time (Raup 1985). These null hypotheses should be rejected before looking for legitimate processes underlying asymmetry.

Evolutionary radiations can either be adaptive or exaptive [after terminology of Gould and Vrba (1982)]. Following Stanley (1990a), adaptive radiations should bear a direct relationship with beneficial ecological traits (true key innovations) that enhance occupation of empty ecospace. But radiations can also be a result of an inherent propensity to generate novelties and taxa that is not in itself related to the radiation (see Vrba 1983; Stanley 1990a; Skelton 1993), via incidental cooptation of previous or novel traits. Here we can talk about *exaptive radiations,* and this is an important realm in which to evaluate the contributions of development. An intrinsic bias to generate new taxa could be based either on a conserved trait [such as nonplanktotrophic development or small body size (Jablonski 1986c; Brown and Maurer 1989;

Brown 1995)] that happens to incidentally enhance the chances of speciation, or on developmental flexibility that, for structural reasons, allows for greater production of variant phenotypes (Wake and Larson 1987; Buss 1988; Lauder and Liem 1989; Liem 1990). Whether a conserved trait or a developmental bias appeared by previous selection is irrelevant in this context. Once they are fixed (with heritable variation absent), their maintenance is not the product of stabilizing selection but of developmental dynamics (Maynard Smith et al. 1985). Stasis at deeper levels in the hierarchy is almost synonymous with fixation: development must then operate via mechanisms of self-organization and self-regulation, generation after generation, to produce stable morphologies. Hence, even adaptive radiations are prone to be influenced by the limitations of development, to the extent that key innovations become entrenched in the developmental system.

Thus, whenever stasis holds and can be understood as the partial result of developmental constraints, development is liable to have a direct impact on processes of clade sorting. Whether development can be a clade-level property (McKinney 1988a; see Rieppel 1986, 1991; Løvtrup 1987; Nelson 1989), justifying its involvement in clade selection sensu stricto, or must rather be viewed as an organismal property conducive to incidental effects via emergent fitnesses (see Vrba and Gould 1986), the macroevolutionary implications of clade sorting, such as the generation of trends, can be tied in directly with development. Therefore, the study of the role of development in evolutionary radiations would benefit from the identification and description of radiations that are exaptive and can thus better be understood by reference to morphogenetic, rather than adaptive, opportunities; it would also benefit from the placement of key innovation hypotheses in an explicit developmental context that helps understand not only origin (the materials available) but also maintenance (fixation and developmental entrenchment). This means that in order to dissect radiations empirically, as much attention should be paid to the dynamics of diversification in developmental spaces as in ecospaces. Developmental spaces [the among-individual analogue of epigenetic landscapes (see Ellers 1993)] imply stability and preferred pathways of form generation and transformation; adaptedness arises from the interplay of structural possibilities and the set of realizable functions (ecospace). There is no necessary one-to-one mapping of an adaptedness function onto a stability or complexity function, although some attempts in this direction have been made (Kauffman 1993, 1995).

The challenge is to investigate developmental spaces in a paleobiological context, where most radiations are ultimately described, while acknowledging that paleontological data are often consistent with both ecological and developmental arguments. As all organisms develop and have an ecology at the same time, answers are not likely to come in simple form. The framing of appropriate questions with diverse kinds of data should be the target of continual evaluation. In what follows, I try to provide an outline of research options relevant to the evolution of biodiversity.

Data

Development has many aspects, but it is through its morphological results that most paleobiological and neontological interpretations at various hierarchical levels and time scales are made. Morphological data are usually represented as characters or character combinations, and from them three partially overlapping kinds of data are assembled: cladistic or phylogenetic, taxonomic, and phenetic. Character combinations may involve synapomorphies only, as in cladistics, revealing primarily a branching hierarchy of nested sets to be explained by history. Or they may involve a mixture of synapomorphies, symplesiomorphies, and autapomorphies, as in conventional taxonomy and phenetics, revealing relatively discrete groupings in morphospace, although not necessarily hierarchically nested. On genealogical grounds, a cladistic hierarchy clearly assumes precedence. Ultimately, though, cladistics, taxonomy, and phenetics would reflect a (nonbranching) hierarchy of levels of inclusiveness or distinctness (disparity) that is partially independent from the cladistic hierarchy of relationships. The role of development, if any, depends on the level of morphological distinctness considered. Therefore, all three kinds of data can provide meaningful information about the way a hierarchy of disparity is structured by ontogeny.

The deeper the hierarchical level, the more likely that similarities among taxa involve constraints imposed by development, because of invariance in the face of different environments (Levinton 1988; McKinney and McNamara 1991). Morphological distinctness is mainly a phenetic issue, but we are faced with the paradox that our ability to use phenetic data is inversely related to the level of analysis in the hierarchy: as dissimilarity increases, homologous features disappear. Although attempts have been made to circumvent the problem by sidestepping the issue of homology (e.g., the skeleton space of Thomas and Reif 1993), such approaches may diminish the potential developmental meaning of observed patterns of morphological distribution.

In the absence of phenetic measures, taxonomic and cladistic ranks have been accepted as proxies for morphological distinctness (e.g., Bambach 1985; Bambach and Sepkoski 1992; Erwin 1994; Erwin et al. 1987; Jablonski and Bottjer 1990a,b,c, 1991; Jacobs 1990; Løvtrup 1988; Patzkowsky 1995; Raup 1983, 1985; Raup and Boyajian 1988; Valentine 1990b; Valentine and Erwin 1987; Valentine et al. 1991a). This approach avoids the problem of dwindling homologies. However, taxonomic and cladistic information should not be viewed as surrogates for phenetic data, but as complementary data that can stand as a signal on their own (Foote 1996b). Taxonomic and cladistic data can be useful in describing evolutionary radiations and in suggesting (1) whether coherent patterns are present, (2) whether phenetic data might be fruitfully added to the analysis, and (3) whether development or any other explanatory process should be considered. A null hypothesis of exponential diversification, for example, provides a baseline against which deviations can be measured and interpreted. On its own, an exponential pattern may imply no absolute constraint to diversification, either ecological or developmental (but see Miller and Sepkoski 1988). However, different exponential patterns suggest differences in rates of speciation and extinction as well as taxonomic structure [the ratio of lineages to paraclades (see Patzkowsky 1995)] and, accordingly, possible underlying ecological and morphological correlates. A logistic pattern, in turn, almost automatically suggests the existence of constraints. Although the expected increase in extinction rate and decrease in origination rate with time presupposes ecological mediation of diversity dependence, the relative constancy of extinction rates (Walker and Valentine 1984; Van Valen 1985b) makes "density dependence" of origination rates as reasonably compatible with dwindling of possibilities in developmental space as with crowding in ecospace (Gilinsky and Bambach 1987). This implies the existence of developmental "carrying capacities" (finite sets of feasible morphologies), as opposed to ecological ones. This is a possibility rarely explored but maybe appropriate over macroevolutionary time scales.

Nevertheless, "taxon-free" studies of overall morphological variation (see Foote 1991, 1993a,b, 1995, 1997) are important, by potentially permitting analyses that directly link developmental constraints or allowances to morphological consequences without the confounding effects of the genealogical hierarchy. It is usually a very tricky exercise to use characters at the organismal level to explain proliferation of taxa (Lauder and Liem 1989; Cracraft 1990; but see Carlson 1992; Baumiller 1993). All in all, in studying the role of development in evolution it seems best to experiment with as many different kinds of data as possible, while keeping in mind the methodological limitations inherent in the very production of empirical knowledge.

The Nature of the Questions

Consistency arguments are an important part of historical explanations, and consilience of inductions has been an acceptable strategy of investigation at least since Darwin (Ruse 1986). Different kinds of data sometimes inductively suggest or prohibit certain questions, but we may also be interested in the inverse, deductive problem: posing a priori hypotheses that stand on their own and await testing in a variety of ways. There are general and specific approaches, depending on whether developmental information or its proxies are used. General (indirect) approaches use adult phenotypic data, and they implicitly assume that constraints on origination and extinction of adult form that are not consistent with ecological arguments must be due to development. Specific approaches attempt to test for the role of development explicitly, by direct inclusion of developmentally relevant information (growth series, larval characters, etc.) or its proxies in evolutionary studies. Current controversies can serve as a guide for causal investigation of radiation and diversification in general, and I will draw on them as appropriate. In particular, the Cambrian explosion, however unique, has become a focus for discussions pertaining to ecological and developmental causality behind radiations (Valentine and Erwin 1987; Gould 1991, 1993; McShea 1993; Ridley 1993; Erwin 1994; Valentine 1995; Valentine et al. 1996). Below, and without claiming exhaustiveness, I review some previous approaches (see also McShea 1993; Erwin 1994) while suggesting a few additional possibilities that have been generally neglected.

General Approaches

Study of Origination and Extinction

Data on origination are naturally tied in with developmental arguments. By looking at origination in different ways, insight can be gained into the nature of innovation across levels, taxa, and time (Eble 1995a). To a first approximation, the role of development can be studied by interpreting variation in rates vis-à-vis ecological hypotheses. Arguments for occupation of empty ecospace invariably rely on an extrinsic change in the rate of success of novelties in time or space (see table 8.1), and they imply that the rate of production should vary with environmental changes. Rebounds from mass extinctions provide an ideal situation for testing such covariation; bursts after mass extinctions have been shown for genera (Sepkoski 1995a), families (Sepkoski 1984; Erwin et al. 1987), and, to a lesser extent, orders (Sepkoski 1978; Erwin et al. 1987). Although this general association between disturbance and production of novelties is amenable to ecological interpretation, it is worth emphasizing that many rebounds would be an expected consequence of diversification even if it was stochastic or if it was primarily driven by occupation of developmental space. Analysis of origination through cumulative origination functions (Eble 1995a) might be useful here, for its ability to test for long-term regularities in change of rates that might indicate particular constraints on novelty introduction. Rosenzweig (1995, pp. 63 to 69) discussed cumulative origination functions in time at the species level, looking primarily for interpretability in terms of species–area relationships. Later in this chapter, I address cumulative origination functions at higher taxonomic levels.

More studies of rate of origination in the context of functional/ecological data are needed [e.g., trophic categories (Erwin et al. 1987); degrees of environmental disturbance (Jablonski and Bottjer 1990a,b,c, 1991); body size (Brown and Maurer 1989; Brown 1995)], as well as in conjunction with developmental information [e.g., frequency of heterochrony (McNamara 1986, 1988); timing of germ-line determination (Buss 1988); developmental architecture (Jacobs 1990); developmental control mechanisms (Valentine et al. 1996)], while refining the level of analysis by detailed study of individual clades. Jacobs (1990), for example, analyzed the pattern of ordinal origination in the Phanerozoic for two different developmental Baupläne (serial and nonserial), and he argued that the more dramatic decline in origination of serial forms is not predicted by ecological considerations, being instead associated with the underlying regulatory hierarchy in development.

Intrinsic factors are also important in extinction, however, as in the following examples:

1. Variation in degree of developmental constraint (and thus degree of plasticity) across groups might play a role analogous to that of genetic variation in extinction. In the face of environmental change, highly constrained species may be unable to adapt, and therefore go extinct. Studies of morphological variability in preextinction intervals should be promising in testing for extinction selectivity, and in assessing to what extent such selectivity may be blind to fitness in normal environments and a contingent result of fixation of developmental pathways—Raup's (1991b) "wanton extinction."

2. Also, immediately after an extinction event, higher developmental or growth rate may favor rebounding capabilities within a group (see Lidgard 1986 on bryozoans).

3. Further, a certain pattern of development leading to a particular adult morphology might become frozen simply by reason of extinction: Paleozoic echinoids showed much variation in numbers of plate columns in ambulacra and interambulacra, but the survival of only two lineages at the end-Permian (with two plate columns per ambulacrum/interambulacrum) historically constrained the whole radiation of post-Paleozoic echinoids (all with the same arrangement). Extinction here led to the fixation of a previously variable morphogenetic sequence.

4. Finally, in coevolutionary avalanches of extinction, small perturbations in internally cohesive ecosystems can cascade into varying degrees of breakdown (Kauffman 1993, 1995; Plotnick and McKinney 1993). To the extent that perturbations can be purely endogenous, arising for example from speciation events affecting the connectedness of ecosystems, the nature and timing of avalanches has to depend on the ultimate raw material for origination provided by development. Support for such coevolutionary, biotically driven avalanches is at present based on modeling and on claims for ecosystem cohesiveness (Plotnick and McKinney 1993; Schopf and Ivany, chapter 10; but see Alroy, chapter 12; Jablonski and Sepkoski 1996). It must be balanced against the extensively documented extrinsic forcing factors that underlie extinction in the history of life. Nevertheless, *susceptibility* to disturbance, biotic or abiotic, must partially depend on taxon-specific traits that translate into differential metapopulation dynamics, speciation, and extinction (McKinney and Allmon 1995; Brown 1995; Allmon et al., chapter 15). Taxon-specific traits are a function of the developmental context of their generation and maintenance. Origination and extinction propensities can be concurrently interpreted in such context, and they might be even predicted when specific developmental information is available.

How might development be involved in producing the observed temporal asymmetry in turnover of taxa at higher levels of the taxonomic hierarchy (phyla, classes, orders)? The Vendian-Cambrian history of diversity at such taxonomic levels is marked by great increases, with origination outweighing extinction (Valentine et al. 1991a). Turnover at lower levels (families, genera) follows the same pattern (Sepkoski 1992c), but the ratio of higher taxa to families and genera is at a Phanerozoic peak during the Cambrian (Valentine et al. 1991a). Regardless of the relative importance of ecological opportunity in triggering the Cambrian explosion, the uniqueness of taxonomic structure in this period is in retrospect an expression of the way constructional themes (Baupläne) became available at least at the same rate as variations within themes. Such constructional themes, being very conservative, retrospectively imply a high degree of developmental or generative entrenchment (Wimsatt 1986). But when such themes were being established, characters likely to become more entrenched (with more "entrenchability," or capacity for integration) were probably as readily produced as any other character (provided they were ecologically relevant or at least neutral). The contrast between relatively high diversity of higher taxa and relatively low diversity of lower taxa in the early Paleozoic may then be a simple result of the inherently smaller number of all possible more entrenchable characters (corresponding to higher taxa) in comparison with that of less entrenchable ones (corresponding to lower taxa). Thus, the burst in originations at the highest taxonomic levels in the early Paleozoic might reflect the exploration of a given range of possibilities for entrenchment. Concomitantly, other characters would build up through time and move integrated phenotypes further and further away from the opportunity of incorporating highly entrenchable characters (see also Erwin et al. 1987; Levinton 1988; Stearns 1992). This would reduce the chances of higher-level origination, but at the same time it would reduce the chances of extinction (which was high in the Cambrian) by accumulation of subtaxa within clades. Such decline in turnover ultimately equilibrates into a relative constancy in number of phyla, classes, and orders observed after the Cambrian and Ordovician (Sepkoski 1978; Erwin et al. 1987; Valentine 1990a), although a small burst in origination of orders after the end-Permian mass extinction may demand special explanation (e.g., availability of developmental or ecological space). In any case, studies of rate of origination and extinction of characters with different degrees of entrenchment should help test such a developmental hypothesis. Note that this treatment of turnover decline in terms of entrenchment is distinct from debates concerning the decline in origination and extinction at the family and lower levels (see Gilinsky 1994), although it does rely on a temporal change in taxonomic structure (see Flessa and Jablonski 1985) and it does attempt to address origination and extinction in coordinated fashion (Gilinsky 1994, chapter 9). The potential role of developmental factors in the decline in turnover at lower levels deserves further examination, however.

Analysis of Morphological Disparity

The quantitative study of morphological variation through time is becoming a major research agenda in paleobiology (Foote 1991, 1993a,b, 1995, 1997; Gould 1991; Jablonski 1995; Wagner 1995b). Consistent regularities in phenotypic covariance through time and across taxa, despite variation in environment, are strongly suggestive of ontogenetic constraints. Crinoids expanded their occupation of morphospace only slightly after their initial radiation, despite 250 million years (my) of evolution in the Paleozoic (Foote 1995). Burgess Shale arthropods have at least the same (Briggs et al. 1992; Wills et al. 1994), and arguably more (Foote and Gould 1992), morphological disparity than Recent ones, despite more than 500 my separating the two samples. Natural selection is opportunistic, with diffusion in morphospace being a null expectation (Fisher 1986; Foote, 1993b; McShea, chapter 6). However, quiescence over such long spans of time strongly suggests a decrease in the generation of opportunities themselves. *Reduction* in variance through time, despite mass extinctions clearing out ecospace, also points to morphogenetic entrenchment (Gould et al. 1987; Valentine et al. 1996). We are not dealing with definitive tests here, because "environment" is a complex parameter (Lewontin 1983; Brandon 1992) and variability in the intensity of natural selection is almost an irrefutable alternative (see McShea 1993).

Although a particular pattern of character covariance may arise by selection or chance, the continual *persistence* of a pattern of character covariance across taxa can be viewed as a potential result of developmental constraint. This approach is implicit in the standard definition of developmental constraint (Maynard Smith et al. 1985), and in genetic terms it implies pleiotropy and gene regulatory interactions. A role for selection is by no means discarded, but, operationally, intrinsic correlations of growth are bound to be expressed as regularities in the covariance matrix (see Lande 1986; Arnold 1992). An unconstrained phenotype will show equal variance for all characters, and no significant covariance (Wagner 1988). More entrenched characters should have higher covariances with a wider range of other characters (McShea 1993). Developmentally linked or serially homologous characters, with no obvious functional coupling, should display significantly higher correlation than nonlinked or nonhomologous characters (Bader and Hall 1960; Lande 1986).

Thus, Gould (1984) argues that covariance sets in *Cerion* reflect many automatic correlations in growth of a constrained structure, and that they stand as evidence for nonadaptation. In the same vein, Hughes (1990) suggests that isometric growth is evidence of a more canalized development, and conversely that allometry implies less canalization and more mosaicism; degree of isometry and allometry could thus be compared with phylogeny and stratigraphy for tests of hypotheses of developmental flexibility.

Test of Model Predictions

Explicit, a priori model predictions can facilitate testing by avoiding the retrospective fallacy, the danger of imposing a particular explanation on a historical pattern. Lauder and Liem (1989) cogently discussed a phylogenetic procedure to allow testing of historical factors such as developmental constraints and allowances, which I summarize here. Given a certain morphological novelty (e.g., metamery), one might explicitly predict certain consequences (e.g., a bias toward greater morphological variation as a result of semi-independent units, the segments), and test such consequences via multivariate analysis of taxa in a phylogenetically defined ingroup (metameric clade) and outgroup (nonmetameric clade). The prediction can be further tested in other ingroup/outgroup comparisons, assuming that the feature in question is homoplastic. By analogy with the subclade test (McShea 1994b), one might also wish to analyze morphological variation in component *subclades* (e.g., annelids, arthropods) to check whether an increase in morpho-

TABLE 8.I   Major hypotheses to account for evolutionary radiation and accompanying temporal asymmetry, expressed as later quiescence

Hypothesis	Radiation	Quiescence	Process
Empty ecospace	Ecology, development	Ecology	Extrinsic change in rate of success
Development	Development (higher flexibility)	Development (lower flexibility)	Intrinsic change in rate of production
Rugged fitness landscapes	Development + ecology	Development + ecology	Intrinsic change in rate of success in rugged fitness landscapes

Processes involved in the generation of asymmetry are briefly described. Mutation-driven and genome hypotheses (Erwin 1994) are subsumed under the development hypothesis.

logical variation over time is indeed a necessary correlate of the structural novelty.

Table 8.1 presents more general models of evolutionary radiation. These have been very refined from a theoretical point of view, generating a plethora of predictions in different empirical circumstances. They provide a platform for testing of specific hypothesis. Wagner (1995b), for example, tested the empty ecospace and the development models against data on the diversification of early gastropods. He partitioned his data into "internal" characters (more prone to developmental constraints) and "external" characters (reflecting general trophic strategies), and he asked whether the associated composite variables changed in magnitude through time in agreement with the observed decrease in disparity. A marked decrease was observed in the magnitude of composite variables associated with internal characters, which could be interpreted as support for the development model and the associated prediction of increasing developmental constraint.

Kauffman's model of rugged fitness landscapes predicts early radiation and later quiescence of higher taxa (Kauffman 1989, 1993, 1995). The model relies on long-jump adaptation achieved by early ontogenetic mutants under a tight interplay of developmental constraints and selection. Although some workers have commented on the (in principle) untestability of the model (e.g., Charlesworth 1995; Levinton 1995), certain predictions *are* testable (Eble 1995a,b). Further below I present a case study in the context of the Cambro-Ordovician radiation of phyla, classes, and orders.

Specific Approaches

Description and Study of Heterochrony

In many ways, discussions about the importance of heterochrony are reminiscent of the debate about the importance of mutation in evolution. Mutation is ubiquitous, it varies in effect, and it interacts with selection; so does heterochrony (McKinney and McNamara 1991). But heterochrony is clearly nonrandom (in a statistical sense), and this is what makes it so interesting and relevant from the perspective of development and evolution. Not all evolutionary changes are reducible to heterochrony (David 1989; Müller 1990; McKinney and McNamara 1991), but many are (see contributions in McKinney 1988b). Much attention has been paid to the inference of heterochrony from phylogeny and the characterization of heterochronoclines. From the point of view of evolutionary radiations, data on heterochrony could be further explored to allow testing of hypotheses of epigenetic entrenchment through time (Valentine and Campbell 1975; Levinton 1988; Wimsatt and Schank 1988): the frequency of heterochrony, regardless of mode, can be an index of developmental lability (e.g., McNamara 1988). Such an index could also be used in testing hypotheses about the dynamics of novelty in space; for example, the post-Paleozoic onshore ordinal origination bias (Jablonski and Bottjer 1990a,b,c, 1991) may or may not be a *direct* result of involvement of heterochrony in the origin of orders (for contrasting views, see Jablonski and Bottjer 1990a,b,c; McKinney and McNamara 1991).

Study of Fluctuating Asymmetry

As suggested by Jablonski and Bottjer (1990a), and more recently by Ridley (1993) and McShea (1993), decrease in levels of fluctuating asymmetry through time could indicate more labile development early on, to the extent that fluctuating asymmetry is a proxy for developmental buffering (see Palmer 1986; Palmer and Strobeck 1986). Studies in a paleontological context include those by L. Smith (1994b), who suggests that high fluctuating asymmetry is not typical in a sample of Cambrian trilobite species.

Inferences from Mode of Larval Development

Mode of larval development in marine invertebrates can affect both origination and extinction: planktotrophic development is conducive to wider geographic ranges and higher resistance to extinction; nonplanktotrophic development leads to the reverse, with associated higher probabilities of origination (Jablonski 1986a,c). Although such correlations may break down during mass extinctions (Jablonski 1986a), they provide important opportunities to investigate the direct impact of a developmental feature on rates of diversification, as shown for mollusks (e.g., Jablonski 1986a), cheilostome bryozoans (Taylor 1988), and echinoids (Emlet 1989, 1995).

Mapping of Developmental Information onto Cladograms

There is no necessary correspondence between biological homologies [shared developmental constraints (Wagner 1989, 1994); see also Roth 1984, 1994] and historical homologies [synapomorphies (see Rieppel 1992)], but the finding of *matches* between these different kinds of homology points to development as a control on innovation or extinction over evolutionary time. One can map developmental characters on an independently derived cladogram and test whether phylogenetically defined clades can also be defined by developmental features (Wray 1992; Erwin 1993b, 1994; Meyer et al. 1995), whether particular adult features are consistently associated with developmental ones (Wray and Bely 1994), whether rampant homoplasy has a developmental basis (Wake and Larson 1987; Wake 1991), or whether extinct clades or living fossils (clades experiencing little speciation) might owe a turnover bias to go extinct or originate to developmental apomorphies.

Wray (1992) found that most orders and families of echinoids (but not less inclusive taxa) have synapomorphies in larval morphology, suggesting a possible control of development on origination at least at those levels. Valentine et al. (1996) assembled molecular, developmental, and fossil evidence to show that each distinct metazoan phylum or class for which the *Hox* homeobox cluster has been described is associated with a distinct pattern of gene duplication and loss. This is an interesting result that must be interpreted in terms of different models for the timing of homeobox gene diversification relative to the timing of splitting of lineages in the context of the Cambrian explosion (see Valentine et al. 1996; Wray et al. 1996). It does indicate that developmental flexibility was a major determinant in the radiation of higher taxa, at least as far as the homeobox evidence is concerned.

Rather than mapping the distribution of developmental characters or character-states on independently constructed trees, one may choose to optimize such data on trees constructed using a total-evidence approach. Smith et al. (1995) argued that there is no consistent congruence among molecular, adult morphological, and developmental (larval) data in echinoids. This does not rule out the potential importance of development in echinoid evolution, but it *does* suggest that changes in larval morphology are partly decoupled from changes in adult morphology, and that partly decoupled radiations within the same genealogical nexus must have taken place.

Alternatively, one can hypothesize that the evolution of development *must* be isomorphic with the evolution of, say, body plans (Davidson et al. 1995) and then concentrate on detailed interpretation of developmental characters in total-evidence trees and their implications for the timing and structure of radiations (for an example involving the metazoan radiation, see Peterson and Marshall 1995). At the limit, one can concentrate exclusively on developmental characters and the reconstruction of developmental trees either through parsimony or through direct study of bifurcations in the developmental process (Ho 1990, 1992). In this case, the goal is to understand the diversity of forms purely in terms of the diversity of developmental transitions (see Goodwin 1990, 1994).

Morphospace Comparisons

While the significance of development to systematics is still hotly debated, among-group developmental transformations and hierarchies are not always recoverable through parsimony (Alberch 1985; Ho 1990, 1992; Marshall et al. 1994). One may expect the orderliness of ontogeny to underlie the order recovered by phylogeny, but the developmental and the genealogical hierarchies are logically separate—developmental constraints can affect both homology and homoplasy. The developmental hierarchy is one of levels of distinctness and of overall morphological relatedness. This should naturally invite the use of phenetic methods in the investigation of the role of development in evolution (Gould 1991, 1993).

Comparison of morphospaces constructed with and without developmental information provides a way of consistently studying the impact of development in constraining or facilitating changes in diversity [in a manner analogous to Hickman's (1993) design spaces, where the focus is on functional causation]. Congruence in range and/or location in "developmental" and "nondevelopmental" morphospaces, when properly interpreted to account for possible stochastic effects, is powerful evidence for a controlling influence of development. To allow comparison, homologous features must be used. This restricts our choices (with qualifications) to three kinds of contrasts: theoretical/empirical, abnormal/normal, and juvenile/adult morphospaces.

Theoretical morphospaces (Raup 1966, 1968; see McGhee 1991) rely on models that generate form on the basis of a few fundamental parameters. Models vary in how explicitly they incorporate actual processes of morphogenesis (e.g., McGhee 1991; Savazzi 1995). Naturally, the more elaborate the model, the more likely it is to accurately reflect the morphogenetic processes involved. But simplicity has its appeal. If a certain range of variation can be successfully accounted for by a simple model, this means that the actual biological system and pathways of control need not be more complicated than the model itself (Raup 1968). From a morphogenetic point of view, we should be interested in constraints of both physical and biological nature:

they are equally expressed in development (Gould 1989b) and as such have equivalent causal status relative to functionally based extrinsic controls. Seilacher's (1994) candle experiments, where a simple process of wax accretion at the interface of two media effectively simulates certain shell forms, is an example of such reasoning; the exercise is purely physical, but it underscores the potential simplicity of morphogenetic processes.

The classical example of comparison of theoretical and empirical distributions is the mapping of ammonoid cephalopods in simple coiling space (Raup 1967). That the actual range and locations are mostly consistent with theoretical possibilities of the model used is telling in itself: whatever the causes of particular phylogenetic transitions among ammonoids, their radiation was fundamentally limited by the developmental constraint of coiling (Maynard Smith et al. 1985). The *density* of occupation of particular regions is a separate problem (Ward 1980; Hickman 1993). More attention should thus be paid to realistic modeling of domains in morphospace, where different parameter values result in the same phenotype, the stability of phenotypes being proportional to the area of a domain (Alberch 1989).

Discontinuities in morphospace are usually reflected in the existence of teratologies, which brings us to the potential usefulness of comparisons of abnormalities with normal forms. Teratologies, however defined, are highly unfit. Their appearance cannot be the result of natural selection and must reflect internal constraints and opportunities (Alberch 1989). Under proper sampling, the set of teratologies and the directions of transformations should represent a space of possible forms that is purely developmental and empirically derived. Trends in the generation of teratologies can then be scrutinized for similarities with trends in actual origination. If abnormal variation mimics normal variation, a case is made for a direct impact of the logic of development (and its preferred channels of transformation) in the dynamics of origination. Examples of such approaches include the match of patterns of limb variation between intraspecific teratologies and evolutionarily established genera (Alberch 1989), in conformity with Shubin and Alberch's (1986) morphogenetic model for tetrapod limb diversity; the extensive similarity between *Drosophila melanogaster* mutants and variation in the Drosophilidae as a whole (DeSalle and Carew 1992); and the remarkable resemblance between pollutional teratologies of regular sea urchins and a variety of irregular forms typical of other orders (Dafni 1986, 1988). These studies are all neontological and amenable to experimental manipulation. Teratologies in the fossil record are very rare, but abundant taxa should be tractable. In addition, teratologies could be more operationally defined in terms of morphological outliers; if properly characterized, outliers become data to allow comparison of normal and abnormal forms much in the same way as described previously, and with the advantage of allowing paleobiological insights.

Finally, juvenile/adult morphospace comparisons can be especially relevant in assessing the role of development in radiations. Similarities in morphological distributions of pooled juveniles and pooled adults can indicate if changes in ontogenetic trajectories are likely to be involved in the shaping of general trends. Alternatively, differences can indicate whether more or less intergroup variation is present earlier in development, thus testing von Baerian recapitulation (which states that more specific characters are developed from more general ones). David (1989) does exactly that in an analysis of diversity in deep-sea echinoids of the family Pourtalesiidae, showing morphoclines from generalized juveniles to divergent adults. Eble (1996, 1997), in a multivariate study of echinoids of the order Spatangoida, finds conflicting patterns in different dimensions.

A Case Study: Testing Higher-Taxon Innovation in Rugged Fitness Landscapes

The following empirical study relies on two previously outlined general approaches to the investigation of the role of development in evolutionary radiations: study of origination and test of model predictions.

Explanations for Temporal Asymmetry

Two main explanations for evolutionary radiations pervade the debate of the past 10 years, at least as far as the problem of the Cambrian explosion and later quiescence is concerned. According to the empty ecospace hypothesis, there is an extrinsic, ecologically driven change in the rate of success of evolutionary innovations. According to the development/genomic hypothesis, there is an intrinsic change in the rate of production of novelties (see table 8.1). This dichotomy has dominated discussion, but recently a third alternative was proposed: Kauffman's rugged fitness landscapes model (Kauffman 1989, 1993, 1995), where a change in the rate of success in an intrinsically constrained fitness landscape occurs during diversification (see table 8.1).

Data

The work here relies on compilations from compendia of global stratigraphic ranges, including Sepkoski's familial (1992a) and generic (unpublished) database for marine invertebrates and Benton's familial compilation (1993).

The data were culled to minimize taxonomic and sampling biases by removing monotypic, single-stage, incertae sedis, problematic, or poorly preserved taxa from the analysis. The observed patterns thus represent a conservative case, with stronger biological signal.

Evaluating the Pattern of Temporal Asymmetry

The temporally asymmetric pattern of origination for phyla, classes, and orders, if not for lower levels, is almost universally agreed upon (Valentine 1969; Erwin et al. 1987). Taken at face value, plots of total origination through time indicate that a major burst indeed happened by the Cambrian and Ordovician, with later reduction in the intensity of originations (figure 8.1). I performed bootstrap analyses to test the statistical significance of the Cambro-Ordovician burst, against the null hypothesis that the burst could have happened even if different magnitudes of origination were equally likely through time. Origination increments were sampled at random with replacement from the whole of the Phanerozoic plus upper Vendian, and a bootstrap distribution (based on a thousand replications) for number of originations up to the end of the Ordovician constructed and compared with the actual number (figure 8.2). For phyla ($P = .006$), classes ($P < .001$), and orders ($P < .001$), there is a highly significant burst, which can then serve as a reliable baseline for additional analyses. Heuristically, wherever possible one should test for the statistical significance of supposed patterns of radiation (see also Guyer and Slowinski 1993; Sanderson and Donoghue 1996).

Kauffman's Model

The rugged fitness landscapes model is at the core of Kauffman's attempt to achieve a marriage of self-organization and selection in the context of the evolution of complexity. Over many years, Kauffman (see 1985) has drawn attention to the power of developmental constraints imposed by the structure and dynamical behavior of genetic regulatory systems. The self-organizing complexity of such systems would lead to ensemble properties (universals) that represent limits that selection cannot overcome. His work has attracted significant attention, and quite a bit of excitement (Burian 1988; Grene 1990; Wimsatt and Schank 1988). Now, he explicitly calls for a synthetic approach, where the contributions of selection and ecology are incorporated by reference to rugged fitness landscapes (Kauffman 1993, 1995; Kauffman and Levin 1987).

Landscapes can vary from highly smooth to highly rugged, with a whole series of intermediate possibilities. Microevolution should occur in

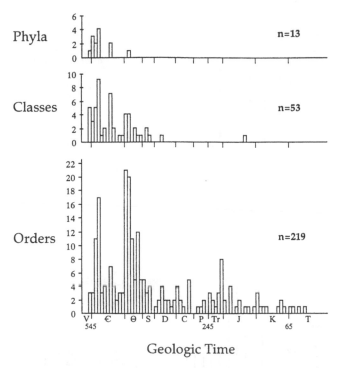

Figure 8.1.   Total number of originations per stratigraphic interval for phyla, classes, and orders of marine invertebrates through geologic time, from Vendian (base at 610 my ago) to the Pleistocene. Each histogram bar corresponds to a geologic stage. Letters on x-axis represent major divisions of the stratigraphic record (from left to right: Vendian, Cambrian, Ordovician, Silurian, Devonian, Carboniferous, Permian, Triassic, Jurassic, Cretaceous, and Tertiary). Numbers on x-axis correspond to the base of the Paleozoic, Mesozoic, and Cenozoic eras, respectively, in millions of years before present. Data (see text for sources) were culled to account for taxonomic and sampling biases. No monotypic, single-stage, incertae sedis, problematic, or poor preservation taxa were included.

smoother, correlated landscapes, where small jumps in the vicinity of one's position accrue slight changes in fitness. Macroevolution (e.g., the origin of Baupläne) should occur in more rugged landscapes. Kauffman argues that long-jump adaptation via mutants in early ontogeny, leading to major innovations conducive to phyla, classes, and orders, should occur in highly mountainous landscapes, where fitness values are uncorrelated. By a variety of simulation experiments and theoretical considerations that partly rely on Feller's theory of records (Feller 1971), Kauffman defines a "universal law" of long-jump adaptation, in which the number of improvement opportunities decreases exponentially, and the waiting time for fitter variants doubles after

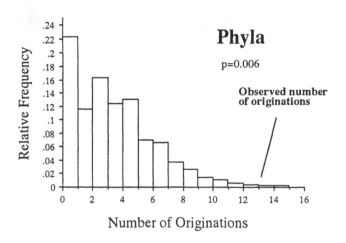

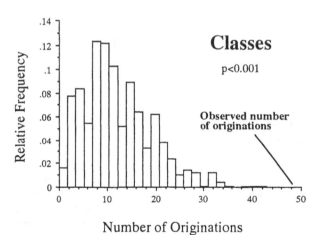

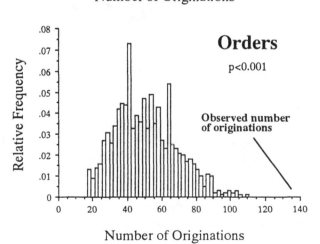

each successful step. This leads to the prediction of a logarithmic relationship between the cumulative number of improvements (S) and the cumulative number of tries (G). Thus, $S = \ln G$ (see Kauffman 1993).

Under the rugged fitness landscapes model, early radiation and later quiescence are robust inevitabilities of long-jump adaptation. The burst of originations of phyla, classes, and orders in Cambrian and Ordovician times (about 540 to 440 my ago; see figure 8.1) contrasts starkly with later quiescence, as discussed earlier. Thus, Kauffman (1989, 1993, 1995) argues that the cumulative pattern of originations (assumed to be improvements) conducive to such higher taxa should be compatible with the logarithmic prediction of the rugged fitness landscapes model. At the higher taxonomic levels to which this relationship is supposed to apply (phyla, classes, orders), deformation of the landscape through time by reason of environmental changes is expected to be only minor.

A Test of the Model of Rugged Fitness Landscapes

I test Kauffman's prediction by looking at the cumulative pattern of origination of phyla, classes, and orders in the marine invertebrate fossil record, as documented by their times of first appearance. A more realistic prediction (incorporating taxonomic and sampling biases), which still preserves Kauffman's prediction while allowing for "measurement" error, is $S = a \ln G + b$ (figure 8.3). The structure of the relationship is more important here than the magnitude of the regression coefficients (see Blalock 1964, pp. 50–52). In a conventional plot, one expects a monotonically increasing (but decelerating) asymptotic curve. A convenient way to express (and test for) such a relationship, which mimics Weber-Fechner's law and dose-response relationship in bioassay—where the independent variable must increase by a constant proportion to produce successive increments in the dependent variable—(Batschelet 1979; Sokal and Rohlf 1995), is by a logarithmic transformation of the x-axis (see figure 8.3). The test of Kauffman's prediction then becomes equivalent to an analysis of the degree of linearity in such transformed plots, or of particular assumptions of a linear model.

In testing the rugged fitness landscapes model, I treat the logarithmic expectation as the null hypothesis, and I make the assumption (implicit in Kauffman's model, and thus biasing the test against its rejection) that origi-

Figure 8.2. (Opposite) Bootstrap distribution of cumulative originations for phyla, classes, and orders of marine invertebrates through Cambrian and Ordovician, under random sampling of origination increments through time. Observed number of originations and statistical significance (based on 1000 replications) are indicated.

THEORETICAL PREDICTION: S=a lnG +b

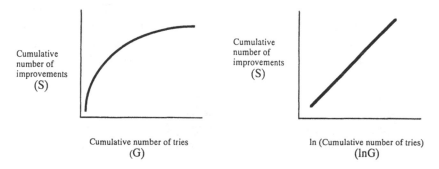

Cumulative number of improvements (S)

Cumulative number of tries (G)

Cumulative number of improvements (S)

ln (Cumulative number of tries) (lnG)

Figure 8.3. Theoretical prediction ("universal law") of long-jump adaptation in rugged fitness landscapes. The expectation is a logarithmic relationship [$S = a \ln G + b$] between cumulative number of improvements (S) and cumulative number of tries (G). *Left plot* untransformed. *Right plot* has x-axis turned into logarithmic scale.

nations as seen through the fossil record indeed reflect successful improvements in some way. Taxa at the ordinal level or above have been viewed as proxies for successful, relatively discrete, novel character combinations in morphospace (see Jablonski and Bottjer 1990a and references therein; Valentine et al. 1991a; Bambach and Sepkoski 1992). Also, the cumulative number of tries can be approximated by either cumulative time or the cumulative number of lower-level entities (e.g., genera), viewed as successive experiments in the generation of higher-level entitites (e.g., orders). Cumulative number of genera that have existed is preferred here because it is biologically more realistic (with number of tries assuming clear biological meaning, in contrast to time elapsed).

For phyla (figure 8.4), an arithmetic plot gives an impression of a logarithmic relationship, which is corroborated by inspection of the semilogarithmic plot. The pattern is clearly linear in such a plot (with no apparent serial correlation of the residuals), although the small numbers involved make it not very robust. Given that yet unformalized or undiscovered phylum-level taxa might potentially change the pattern (see Valentine et al. 1991a), one can accept such linearity provisionally as a regularity deserving additional testing but especially in need of detailed explanation and scrutiny. If the fit is not artifactual, the relationship should be maintained as more phyla are discovered, more taxa are formalized as new phyla, or both.

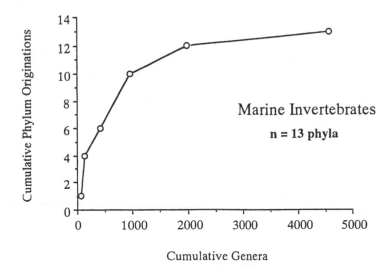

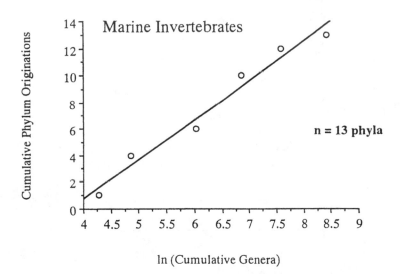

Figure 8.4. Plots of cumulative number of phylum originations versus cumulative number of genera through time for 13 phyla of marine invertebrates. Untransformed plot and semi-logarithmic plot. Note clear fit to linear prediction.

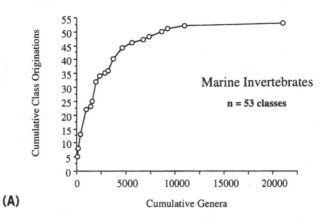

(A)

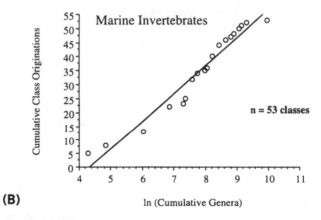

(B)

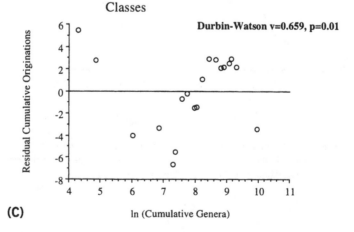

(C)

For classes (figure 8.5), although the general pattern in an untransformed plot again would seem to support a roughly logarithmic relationship, we see a clear deviation from the linear expectation in a semi-log plot. Because of an inherent autocorrelation encapsulated in cumulative curves, correlation coefficients are meaningless (McKinney 1990a). Of interest here is the *shape* of the cumulative function in a semi-log plot, and how nonlinear it might be. For classes, there is a somewhat sigmoidal profile for the whole curve. Residual analysis shows significant nonhomogeneity of the residuals ($P = .01$, Durbin- Watson test for positive serial correlation). A sigmoidal function fits the data significantly better ($P < .001$, F test).

For orders (figure 8.6), the logarithmic expectation is somewhat damped in a conventional plot, and the deviation from the rugged fitness landscape prediction is even more pronounced, with a stark curvilinear, quadratic pattern in a semi-log plot. Analysis of residuals again demonstrates that there is nonhomogeneity of the residuals ($P = .01$, Durbin-Watson test for positive serial correlation). A quadratic function fits the data significantly better ($P < .001$, F test).

The discrepancy across taxonomic levels is not consistent with a logarithmic expectation that should have been isomorphically applicable to phyla, classes, and orders. Different cumulative origination functions appear to hold at each level. Even if the patterns at the phylum level were discarded because of small numbers and if one would still not be confident in assignment of taxa to class level, the cumulative pattern of appearance of marine invertebrate orders by itself disproves the prediction of the rugged fitness landscape model. Orders have been used successfully in the past as proxies for morphological distinctness (see Jablonski and Bottjer 1990a; Valentine et al. 1991a), and there is clearly a break in patterns of diversity at the ordinal level: lower levels show self-similar patterns that differ from ordinal ones (Jablonski and Bottjer 1990a,b,c, 1991; Bambach and Sepkoski 1992). If anything, the larger sample size for orders ($N = 219$ in the present study, a very conservative estimate that tends to diminish biases arising from the quality of the fossil record and taxonomic practice) greatly reduces the

Figure 8.5. (Opposite) Untransformed plot **(A)** and semi-log plot **(B)** of cumulative number of class originations versus cumulative number of genera through time for 53 classes of marine invertebrates. Note the roughly sigmoidal shape of the underlying function. **(C)** Residual semi-logarithmic plot of cumulative class origination versus cumulative generic origination. *Open circles* represent classes. Results of Durbin-Watson test on residuals are $v = 0.659$, $P = .01$. Results of F test for increase in variance explained by sigmoidal fit over linear fit in (B) are $F = 20.08$, $P < .001$.

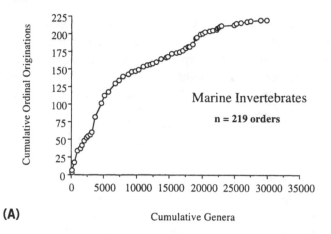

(A)

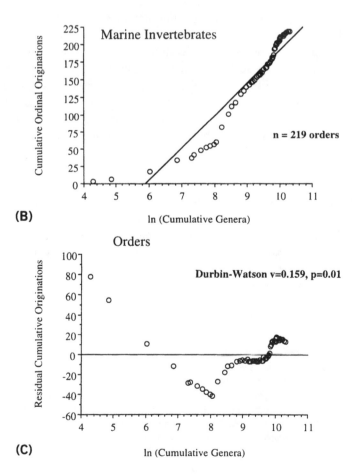

(B)

(C)

chance that a certain proportion of spurious orders would significantly affect the results.

Discussion

It must be noted that testing of the logarithmic prediction (yielding a linear prediction in a semi-log plot) is plagued by the cumulative nature of the variables. In a different context, one faces the same issues surrounding demonstration of linearity for survivorship curves (Van Valen 1973; Raup 1975b; Pearson 1995). Cumulative curves can look deceptively regular, and, in borderline cases, statistical testing (however difficult) should be attempted. In the preceding analyses, I concentrated on the pattern of the residuals for its visual appeal (with lack of homogeneity clear for classes and orders) and used the Durbin-Watson test statistic (with expected value of 2) to rigorously check positive serial correlation of the residuals: they should be independent and with constant variance for the linear regression model to be correct (this again was not the case for classes and orders). The rugged fitness landscapes model incorporates serial correlation of the *predicted values,* because of its reliance on cumulative variables, but it implies no expectation of strong serial correlation of the *residuals,* which should behave more or less randomly around the regression line (see Kauffman and Levin 1987). Apart from residual analysis, the strong curvilinearity of the semi-log pattern for orders, and to a lesser extent for classes, is sufficient to test the fit of the model prediction. A parabola fits the ordinal pattern significantly better than a straight line, the same being true of a sigmoidal curve for the class pattern. The meaning of such cumulative origination functions is not of concern here, but one might suspect that an unconsidered variable was involved, with historical effects being a likely factor.

I do not claim taxonomic data to be perfect. Measures of morphological disparity could, in principle, be used in the same way, with first appearances of character combinations being traced back in the fossil record, or inferred with the help of morphological and molecular phylogenies. In view of documented discordances between taxonomic and morphological diversity (Foote 1993b), there is nothing that should preclude morphological data from be-

Figure 8.6. **(Opposite)**   Untransformed plot **(A)** and semi-log plot **(B)** of cumulative number of ordinal originations versus cumulative number of genera through time for 219 orders of marine invertebrates. Note the roughly quadratic shape of the underlying function. **(C)** Residual semi-logarithmic plot of cumulative ordinal origination versus cumulative generic origination. *Open circles* represent orders. Results of Durbin-Watson test on residuals are $v = 0.159$, $P = .01$. Results of F test for increase in variance explained by quadratic fit over linear fit in (B) are $F = 483.42$, $P < .001$.

ing subjected to the same kind of analysis; perhaps the data will even support the rugged fitness landscapes model. A difficulty with the rugged fitness landscapes model is that one does not know a priori at what level the long-jump adaptation end of the spectrum ceases to be meaningful for analyses. If the same cumulative plots are constructed for families and genera, deviation from the model prediction again occurs. Rosenzweig's (1995, p. 68) analysis of a local assemblage of Ordovician invertebrate species over 5 my suggests again deviation from linearity in a semi-log plot (although he was testing species–area models). In view of the overly patchy and incomplete fossil record of species (Raup and Boyajian 1988; Valentine 1990a,b; Smith 1994), however, it remains to be seen whether species-level cumulative curves could be extended over longer time scales, thus allowing greater comparability with the present analysis (but see Rosenzweig 1997). Rosenzweig does provide a context in which Kauffman's model might tie in with more refined ecological arguments. His call for renewed attention to cumulative species–time curves should be expanded to all scales.

Another way of testing Kauffman's hypothesis would be to concentrate on more circumscribed clades or contrasting environmental contexts. Although additional analyses are beyond the purposes of this paper, it is worth mentioning that, for orders of mollusks, echinoderms, mammals, and insects (with cumulative number of genera being used as a proxy for cumulative number of tries for mollusks and echinoderms, and cumulative number of families as such a proxy for mammals and insects), the rugged fitness landscapes model also does not hold, at least as far as the assumption of a single landscape is concerned (Eble 1995a).

Deformation of landscapes with environmental change is a relevant issue here. It has been said that the ad hoc invocation of landscape deformation through time would render Kauffman's arguments essentially untestable (Levinton 1995). However, Kauffman (1993) implicitly assumes that deformation, even in the face of pronounced environmental perturbation, would be minor and not affect the general structure of the landscape at the higher taxonomic levels considered here. That the generation of morphological innovations in long-jump adaptation would be achieved by early ontogeny mutants lends support to claims for a relative stability of landscapes at higher levels, since the developmentally constrained set of possible changes at deeper levels of the hierarchy would be quite independent of environmental changes. Here is the irony of the rugged fitness landscapes model: by relying on ahistorical principles to explain history, historical effects must a priori be assumed to be largely ineffective; if history could change the nature of those ahistorical principles, then there would be nothing left to test. Landscapes in the present context would change only if fundamentally different genetic and morphogenetic controls were repeatedly appearing through time. This may well be the case in view of the present results, suggesting that history would need to be taken into account from the outset.

The whole approach to the rugged fitness landscapes model assumes that adaptation during radiation will occur by successively shorter long-jumps because of the correlational structure of the landscape—phyla are founded, followed by classes and then by orders (Simpson 1953). The argument breaks down if there is no obvious top-down pattern. Sepkoski's (1992c) newest analysis of the Vendian-Cambrian diversification shows that the top-down scenario may not apply any more to the metazoan radiation, because the patterns of diversification for genera, families, orders, and classes are all convergent, showing approximately the same pattern of diversification and the same timing of diversity increases. Although in other instances (e.g., the Ordovician radiations) a case could be made for a top-down pulse of filling (see Erwin et al. 1987), the generality of the pattern is greatly reduced by its absence early on. In the face of coeval diversification, the basic problem is reduced to explaining why that pattern of diversification (and accompanying disparity) has not recurred for higher taxa. While a hypothesis of differential developmental flexibility through time remains unscathed, ecology-dependent alternatives, including not only the rugged fitness landscapes model but also, for that matter, the empty ecospace model, may need to be framed in different terms—without reliance on lower rates of diversification at lower taxonomic levels—if earlier higher disparity is a problem to be explained.

It must be emphasized that I went on to test the model (within the limitations of paleontological data) *given* the assumptions and simplifications built into its structure. The model is indeed testable and has heuristic value that should not be underemphasized. There is certainly room for questioning of the assumptions of the rugged fitness landscapes model (Maynard Smith, personal communication, September, 1995). From a strictly theoretical point of view, thus, I would like to raise several points of contention pertaining to the argument as a whole (table 8.2). All the issues mentioned in table 8.2 are not problems with the model per se, but legitimate questions being debated by evolutionary biologists. Different views on each of them affect the reliability of the model, but its robustness can in principle always be maintained with appropriate modifications. What I attempted to demonstrate is that *as it stands,* the rugged fitness landscapes model is empirically questionable, at least as applied to the problem of the Cambro-Ordovician explosion and later quiescence. A possible explanation for this situation is Kauffman's use of a rather restrictive version of evolutionary theory to achieve a synthesis with self-organization. Self-organization may be very im-

TABLE 8.2.   Points of contention pertaining to the rugged fitness landscapes model[a]

1. Arbitrary fitness values; there is no discussion of how fitness should be appropriately measured.
2. It is asserted that the arguments do not rely on filling of ecospace, but success in the model is tacitly ecological.
3. Extant morphologies are fitter, early morphologies poorly fit; there is no definitive empirical evidence for that (but see Sepkoski 1984; Vermeij 1987, 1994).
4. The structure of the landscape should be sufficiently stable from the Cambrian to the Recent for cumulative improvement to occur; see text for discussion.
5. Reliance on mutation and selection only; drift and neutrality are considered, but not modeled.
6. Optimization is rampant; extinction is always constructive, there being no space for random or contingent extinction.
7. Fitness landscapes are claimed to extend to higher taxa; this is disputed on genetic (Provine 1986) and ecological (Eldredge 1989) grounds.

[a]These points pertain to the structure of the rugged fitness landscapes model as it is applied to the problem of the Cambrian explosion and later quiescence. Each issue implies different positions on the appropriateness of the assumptions of the model, and each relates to particular debates in evolutionary theory.

portant in evolution (Seilacher 1991; Goodwin 1994) and should receive more attention from paleobiologists and neontologists alike in its own right. But given the structure of the rugged fitness landscapes model, the wrong imagery may have been chosen for a synthesis.

Conclusions

Evolutionary radiations, as well as biodiversity change in general, must logically have a component of developmental causation in addition to that of ecology. Development provides the generative matrix on which origination depends, in terms of both constraints and opportunities. Ecology provides the usual arena for ultimate success or failure. We are now in position to test the role of development in many instances, directly or indirectly. The approaches outlined in this paper constitute an attempt at constructing a relevant agenda, but many other possibilities potentially await scrutiny. Focus on the description and analysis of developmental spaces, with different kinds of data and different degrees of approximation, seems central to the general task. Detailed analysis of entrenched characters, when they occur, should point to ways in which development affects structure, function, and rates of diversification. And the mechanistic study of stasis should illuminate the counterpart of radiations, failure to radiate.

Even so, attention should also be paid to the complementarity of development and ecology, both in origination and extinction. Kauffman's rugged fitness landscapes model has been an exciting first step in that direction. However, the model does not hold for the Cambro-Ordovician radiation and later quiescence in origination of phyla, classes, and orders. The analytical predictions do not apply uniformly across these levels, suggesting that modifications may be necessary, that the model is less general than originally implied, or that fossil classes and orders do not reflect morphological disparity to the extent traditionally perceived by paleontologists. Nevertheless, certain regularities do arise, and they deserve further study, indicating promise for models that attempt to describe an interaction of development and ecology. The virtual lack of such models may be indicative of the separate traditions and methodological agendas of evolutionary developmental biology and ecology and their sometimes overly ahistorical emphases. Paleontology has much to add here for its inherent focus on history over long-term scales. Where these disciplines meet, there is great potential for improved understanding, beyond the restrictive simplicity of "either-or" dichotomies.

Acknowledgments

I thank M. Foote, D. Jablonski, M. LaBarbera, M. L. McKinney, D. W. McShea, D. M. Raup, J. J. Sepkoski, Jr., L. M. Van Valen, and P. J. Wagner for discussion and reviews of different versions of the manuscript; C. R. Marshall, J. Maynard Smith, P. N. Pearson, and G. A. Wray for discussion; and J. J. Sepkoski, Jr. for kindly providing access to his unpublished generic database. This research benefited from CNPq grant 201542/91-9.

9

Evolutionary Turnover and Volatility in Higher Taxa

Norman L. Gilinsky

This chapter is principally about evolutionary turnover at a high level of taxonomic organization, specifically, turnover among families within orders and turnover among orders themselves. To the extent that characteristics of family-level turnover within orders are correlated with processes of species-level turnover, a view urged by Bambach and Sepkoski (1992) and others (Allison and Briggs 1993; Sepkoski et al. 1981), the chapter is also about turnover at the species level. And to the extent that processes of species-level turnover can be understood by reference to ecological properties of organisms, populations, and environments, the chapter is also about the links—just beginning to be articulated—between newly emerging ecological theories of metapopulation dynamics and long-term paleontological patterns. In this chapter, there are four main aims:

1. to discuss aspects of biological turnover that are common to all levels of the genealogical hierarchy;
2. to present a paleontological example that illustrates the importance of volatility in evolutionary turnover: the long-term decline of background familial origination and extinction probabilities;
3. to call for a systematic examination of possible relationships between the decline of background origination and extinction probabilities and the ecological processes that operate on shorter temporal scales; and
4. to argue that ecology and paleontology need to be studied together

for a fuller understanding of the processes that generate and maintain biological diversity.

Basic Concepts of Turnover

Turnover Generalized

Although paleontologists, ecologists, and population biologists usually focus on different kinds of biological entities when speaking of turnover, the concept can be generalized in the following way:

> Turnover is a change of composition of the Earth's biota at a specified taxonomic level that results from a sorting among entities at the same or another taxonomic level.

As a paleontologist, I usually think of the taxonomic composition of species, genera, or even higher taxa, as undergoing turnover in the sense that the taxonomic composition of the biota changes at those taxonomic levels [see, for instance, Larwood (1988); Sepkoski (1979, 1984, 1993); Signor (1990); Stanley (1979); Thayer (1983)]. But a population ecologist would often think of turnover in the sense of replacement of some populations (demes or avatars) by others of the same species, or, as in interspecific competition, in the sense of replacement of some populations of one species by some of another (Hanski and Gilpin 1991).

In the paleontological literature, the term *turnover* often implies that the taxonomic composition of the world has noticeably changed, brachiopod species of Paleozoic aspect having been replaced by those of Mesozoic aspect, for instance (e.g., Williams and Hurst 1977). A second example might be a brachiopod-dominated benthic community (measured in terms of species richness) giving way to a one dominated by bivalves (Gould and Calloway 1980). A third example, also typical in the paleontological literature, would be the concept of turnover as a continuous, but ecologically inconspicuous, "rollover" of species. Here species become extinct but are replaced by genealogically closely related ones with similar, or even identical, ecological roles, and only after the lapse of considerable time (tens of millions of years) does the basic complexion of the ecosystem show noticeable change (Bambach 1986; McGhee 1981). In the first and third examples, the turnover would be recognized as changes of species compositions within the same phyla. In the second example, the turnover would be recognized by a shift in the dominance relationships between two phyla. Other examples of turnover can also be identified up and down the taxonomic hierarchy.

In the turnover of the ecological literature, some populations might be replaced by others within the same species, but the species itself persists. This

TABLE 9.1   Genealogical and ecological hierarchies

Genealogical hierarchy	Ecological hierarchy
Monophyletic taxa	Regional biotas
Species	Communities
Demes	Populations
Organisms	Organisms
Chromosomes	Cells
Genes	Molecules

From Eldredge, N. 1985. *Unfinished Synthesis: Biological Hierarchies and Modern Evolutionary Thought.* New York: Oxford University Press.

type of turnover would be analogous to the first paleontological example of Paleozoic brachiopods giving way to Mesozoic ones, but with the phylum Brachiopoda continuing to exist. On the other hand, the ecological transfer of dominance from one species to another by competition within a local community might be analogous to the second paleontological example of the transfer of dominance from one phylum to another within the global ecosystem (brachiopods to clams, for example).

Paleontologists and ecologists might be tempted to regard their respective senses of turnover as fundamentally different because of the taxonomic and temporal scales involved. But the two kinds of turnover do seem to be formally the same, at least in the sense that both kinds of turnover describe the replacement of one kind of taxonomic entity by another. One might even go so far as to treat the sorting of organisms in organismal selection as a kind of turnover, here that turnover being represented as a change of the frequencies of different types of organisms (or genotypes).

It seems to me that any and all of the entities listed in Eldredge's genealogical hierarchy (table 9.1) can be targets of turnover (Eldredge 1985). This is not to say that turnover at one level is necessarily *caused* by processes acting specifically at that level. The "turnover" represented by the replacement of brachiopod species by bivalve species might be caused by organismal-level selection (Thayer 1983). (Brachiopod organisms are poor competitors against mussels, and also against the sediment-disturbing organisms that were becoming increasingly important in marine ecosystems. Thus loss of brachiopod organisms led to the demise of brachiopod species.) Or they might even have been caused by global-level processes. [Species of brachiopods were differentially extirpated by mass extinction (Gould and Calloway 1980)]. In each case, turnover is typically described at the species level, but it was not necessarily caused by processes operating at that level.

Vrba and Gould (1986), in defining *sorting*, as distinct from selection, also seem to imply the same basic concept I am proposing here, but using different terminology. Their *sorting* is responsible for "turnover," but the taxonomic level at which the turnover is observed may or may not be the locus of its biological cause. Nonetheless, the basic concept of turnover can be made hierarchical and quite general, with any of Eldredge's genealogical entities being participants in the process.

Another fundamental characteristic of turnover is that it involves the origination and/or extinction of genealogical entities at some level. At the high taxonomic levels typically studied by paleontologists, turnover is brought about by the origination and extinction of species, genera, etc., leading to changes in the global biotic composition. Within existing species, the ecologist focuses on the founding and extinction of populations, leading to change in frequencies of those populations. There would also seem to be a connection between the turnover among populations and Wright's interdemic selection (Wright 1931), but there is not enough space to explore that connection here.

If the concept of turnover is treated formally as a change of composition of the Earth's biota that results from a process of origination and extinction of genealogical entities at some specified focal level, as defined above, it would seem that the mathematics of turnover should also be quite general. Paleontologists have frequently turned to the mathematical theory of branching processes (in discrete and continuous time) to forward-model the processes of taxonomic turnover. This theory, best summarized by Raup (1985), predicts median and mean species diversities (richness), expected times to extinction for clades, and other parameters from given probabilities of species origination and extinction. Raup and other authors have also used the theory for the inverse problem of inferring the underlying probabilities from observed results of the evolutionary process (Foote 1988; Gilinsky 1991a, 1994; Gilinsky and Good 1991; Raup 1978).

Ecologists have at the same time elaborated their own mathematical theory of metapopulation dynamics, under which species are modeled as evanescent groups of quasi-independent populations [the Levins model; see Hanski and Gilpin (1991), and also Schoener (1991) for a discussion of the Boorman-Levitt metapopulation model]. Under that theory, the fates of species are determined by the propensities of their constituent populations to found descendant populations of like type and to resist becoming extinct. Turnover (within species) results from the processes of origination and extinction of populations in a manner that seems analogous to the paleontological case of turnover within clades resulting from the processes of speciation and extinction of constituent species. Although I am a newcomer to metapopulation concepts, it appears that the generalized, hierarchical concept of turnover introduced here—which seemingly applies to origination and extinction of entities at all levels of the taxonomic hierarchy—implies that a generalized mathematical theory of turnover must exist as well [see Leigh's (1981) highly mathematical treatment of some of the elements of the problem].

Turnover and Diversity (Richness)

Change of diversity (in terms of the richness of any of the focal entities in the genealogical hierarchy) implies turnover, but turnover does not imply change of diversity. Precisely what the relationship between turnover and diversity will be depends on the rates of origination and extinction of the genealogical entities involved. (I criticize the term *rate* later.) Very simply, if the rate of origination of entities within a given taxon (at any level of the genealogical hierarchy) exceeds the rate of extinction, not only will there be turnover (under the definition given at the outset), there will also be an increase in the diversity (the "richness") of entities within that taxon. Conversely, if the rate of extinction exceeds the rate of origination for entities within a taxon, not only will there be turnover, there will also be a decrease in the diversity within that taxon. And finally, if the rate of origination exactly equals the rate of extinction within a taxon, the diversity will not change, but there will nonetheless be turnover within the taxon, except under the special case where the rates are equal, but zero.

One might imagine the flow of water into and out of a bathtub as an analogy. Even when the total volume of water in the tub does not change (inflow equals outflow), there is nonetheless turnover of the water, except for the special case where the water supply is off and the drain is plugged. The bathtub analogy suggests a connection to yet another concept, that of flux. Under the flux model, origination serves as an input, extinction an output, mean taxonomic longevity is the residence time, diversity is the size of the reservoir, and the stage is set for thinking of turnover in the language (and mathematics) of systems analysis.

Rates and Frequencies

The paleontological literature is loaded with analyses and discussions of taxonomic evolutionary rates (Gilinsky 1991b; Stanley 1979; Stanley 1985). Historically, the use of the term *rates* for the origination and extinction of taxa can be traced at least as far back as Simpson (1944), who referred to character rates, organism rates, taxon rates, etc., as if they were all on a par.

(In fact, to Simpson, all of these rates *were* on a par, since all were indirect measures of genetic evolutionary rates. Linking these rates to genetic ones was a critical element of his attempted synthesis of genetics and paleontology.) Although the term does apply reasonably well to the temporal change in the dimensions of a biological character, just as it applies to the speed of a bullet, it does not seem to apply well to describing the originations and extinctions of taxa (at any level). The reason is that the originations and extinctions of species (or demes, etc.) are inherently discontinuous, not continuous, processes (Gilinsky 1986, 1991b, 1994; Mayr 1954, 1970). Furthermore, they are probabilistic. Thus, the founding and extinction of a species (or other genealogical entity) is more akin to obtaining a specified number in the roll of a die. For each species, there is a certain probability of splitting into two species, or of becoming extinct. But this is different from the case of a speeding bullet, because one cannot count on the expected outcome actually coming to pass. We have no problem talking about the speed of a bullet for the simple reason that, for all practical purposes, the motion of the bullet is a continuous, and predictable, process. But the same cannot be said of the founding and extinction of species or populations.

Although the term *rate* is commonly used when referring to species origination and extinction, organism birth and death, and population founding and extinction [see Hanski (1989) for a typical use of the term among metapopulation ecologists], what we are really talking about is frequencies, not rates. In each of these cases, we are referring to the frequency of the occurrence of a certain type of event or process (speciation, birth, etc.), not the amount of time it takes for a process to be completed. Although the distinction might seem to be nitpicking, I do not believe it is, because it is just this distinction that motivates the concept of volatility introduced later.

Revisiting the idea of the bathtub can be helpful in seeing why it is useful to think in terms of frequencies rather than rates. Suppose we fill a bathtub half full, but we allow a homunculus to control the tap in such a way as to allow new water to enter the tub in bursts at will. A second homunculus controls the drain plug. Let us specify that the flow of water will not be continuous, but that water must be let in and out in units of a certain volume, say one gallon. Now we have a process that offers a better analogy to the processes of origination and extinction, and their consequence, turnover. Here the events are the bursts of inflow and outflow, analogous to origination and extinction, and these events (however long each one takes) occur at some frequency that will be determined by probabilities. The level of the water at any one time is analogous to the diversity (or richness, the number of coexisting biological entities).

Let us now blindfold our homunculi and plug their ears (do they even have ears?) so that one is not aware of what the other is doing. And, finally, let us give them marching orders. Each is ordered to move gallon packets of water (in or out) at an average frequency of six packets per minute. But— and this is the key—the homunculi are allowed to move packets any time they want, so long as they maintain the long-term specified average. They are emphatically not required to move their packets at 10-second intervals. A homunculus might, for instance, choose to let 6 gallons out during a single second, and then rest for an indeterminate period before acting again. Another way of putting it is that the homunculi were instructed to abide by the restriction that the probability of adding or removing a packet of water is 0.1 per second.

The consequences for the level of water in the bathtub now become interesting. Both homunculi move water packages at the same frequency (averaged over a long time). But this does *not* mean that the water level will remain the same as when they started. In fact, it is quite possible, under the frequency model, for the tub to empty completely or even to overflow, since the water movements occur in packages at uncertain intervals. Under a frequency model, the discontinuous events have a specified probability of occurrence, but the events are not uniformly distributed in time. As with dice, speciation and extinction events have some probability of occurring per unit of time, but there is nothing to stop a glut of extinctions or originations from piling up. It is only over a very long interval that their characteristic frequencies emerge.

This is much different from the case where water flows in and out at a characteristic rate. Under the rate model, which assumes continuous flow, all that matters is the difference between input and output, *not their actual values.* The bath will remain at the same level, so long as the inflow and outflow are the same, for every flow value from infinitesimal to infinite. But in thinking of water movement in terms of frequencies of events, the actual flow rates do matter, as I will show.

Volatility

If originations and extinctions (of entities at any level in the genealogical hierarchy) are probabilistic, and if long-term collections of such events describe frequencies, then diversity fluctuations of various degrees naturally arise. As mentioned earlier, such fluctuations will arise even when probabilities of origination and extinction are exactly the same. These fluctuations can be measured in various ways, one of which I term volatility. In a recent paper (Gilinsky 1994), I defined volatility (V) for a taxonomic order i as

$$V_i = \frac{\sum\limits_{n=1}^{N} \dfrac{|v_{n+1} - v_n|}{v_{n(tot)}t_n}}{N} \qquad [1]$$

where v_{n+1} is the number of families entering stratigraphic stage $n + 1$, v_n is the number of families entering stage n, $v_{n(tot)}$ is the total number of families that left a record in stage n, t_n is the duration of stage n in millions of years, and N is the number of stages survived by the order. In effect, the volatility of an order was defined as the average net proportional change of diversity per million-year, estimated over the entire known history of the order. The higher the volatility, the greater the average proportional change of diversity (up or down).

Although volatility was originally defined for families within orders, the concept can be generalized to any taxonomic level. For the volatility of a species in terms of its constituent populations, one needs only to replace *family* by *population*, and *order* by *species*. Geological stages will need to be replaced by time intervals appropriate to the analysis, but the essential concept of volatility as a measure of fluctuation remains unchanged. Other measures of such fluctuation are also available (e.g., Schoener 1991). In general, for any of the genealogical entities given in table 9.1, the volatility for a specified entity at focal level F would typically be calculated in terms of the net proportional change of richness of its contained entities at hierarchical level F − 1.

Volatility describes the average change in number of entities (organisms, populations, species, etc.) per unit of time. It arises in part as a natural consequence of the probabilistic nature of the events that give rise to or extinguish those entities and, in that sense, volatility is a quantified sampling error. For organisms, volatility is in part a result of the probabilistic, discontinuous nature of birth and death processes (termed *demographic stochasticity* by Lande (1993)). For clades, it is in part a result of the probabilistic, and discontinuous, processes of speciation and extinction. Not all volatility results from sampling error, of course. Some results from deterministic processes as well, such as organismal selection on individual members of an ensemble (of populations, species, etc.). Nonetheless, I emphasize that fluctuations of numbers of entities in an ensemble (here measured as volatility) will occur even if no deterministic process operates on the ensemble. Deterministic processes at work on the individual members of the ensemble are not precluded, however. Furthermore, volatility will occur even if the probabilities of origination and extinction for members of an ensemble are equal.

Why does volatility matter? It matters because the expected degree of

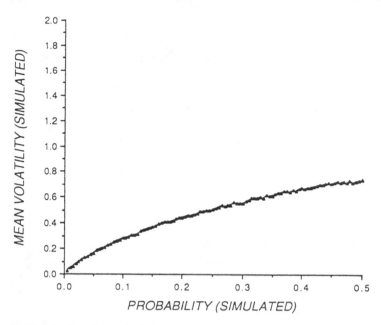

Figure 9.1.   Plot of mean (= statistically expected) volatility (see text definition) versus probability of origination and extinction constructed using data from 1 million simulated clades. Probabilities of origination (dichotomous branching) and extinction were set equal for simplicity. Mean volatilities were calculated from 10,000 simulations for each of 100 probability combinations. Volatility increases with probabilities with a concave-down curve-shape.

fluctuation, or volatility, depends on the *actual values* of the origination and extinction probabilities, not just on their relative values. Low probabilities of origination and extinction entail low volatility, and high probabilities of origination and extinction entail high volatility, even when the probabilities of origination and extinction are exactly equal to each other (see later). Consequently, natural fluctuations can lead to diversification or extinction simply by good or bad luck, respectively. The theory of metapopulation dynamics also recognizes the concept of volatility in various ways. For instance, the probability of extinction by "demographic accident" is known to be an approximate exponential function of population size (Lande 1993; see also Leigh 1981).

Figure 9.1 is a plot of the expected volatility of a hypothetical clade as a function of the probabilities of origination and extinction of its constituent members. The probabilities are assumed to be equal, for simplicity, but similar graphs could be generated for cases where one probability exceeds the other. The volatilities were obtained by simulating the process of branching evolution in the manner discussed in Gilinsky (1994). The simulation was meant to model the branching and extinction of families within orders of marine organisms, but it seems that the basic principles should apply to any process of branching and extinction, including the births and deaths of organisms in a population, the founding and extinction of populations within a species, the speciation and extinction of species within a clade, etc. The principal result is that the expected volatility is an increasing, concave-down function of the probabilities of origination and extinction. The higher the probabilities of origination and extinction for a clade, the larger the expected fluctuations in the standing diversity. The reason for the relationship can be seen intuitively by noting that a high probability of origination might push a clade to high diversity very rapidly, but a correlated high probability of extinction might equally rapidly cut the clade down to size. Low probabilities of origination and extinction will have much more limited capacity to modify diversity over short times.

Although figure 9.1 shows the expected, or average, volatility for specified probabilities of origination and extinction, higher and lower volatilities are possible (and were, in fact, observed). Volatility can even be sufficiently great as to cause a clade to falter to extinction by chance. Since high volatility implies highly fluctuating numbers, extinction will be more likely in clades experiencing high volatility than low.

High-volatility clades typically will not live as long as low-volatility clades. Figure 9.2 shows the expected relationship between volatility and duration for clades generated in the same simulation as that in figure 9.1. As anticipated, the expected duration is a decreasing function of the volatility. This is because high volatilities provide frequent opportunities for diversity to plummet to zero, as I just mentioned. Low volatilities minimize that danger.

Figure 9.3 shows the relationship between probabilities and duration. As probabilities of origination and extinction increase, the expected durations of clades go down. This relationship demonstrates that it is not just the net difference between origination and extinction that matters to the diversities of clades. Because volatility is an increasing function of the probabilities of origination and extinction, and because volatility poses a danger to clades, the actual values of the probabilities are critical as well.

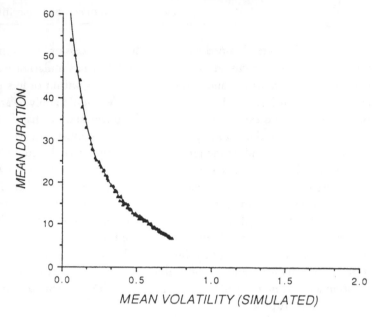

Figure 9.2. Plot of mean clade duration (in units of computer time steps) versus mean volatility for the same 1 million simulated clades used to construct figure 9.1. Mean (= statistically expected) duration declines as volatility increases.

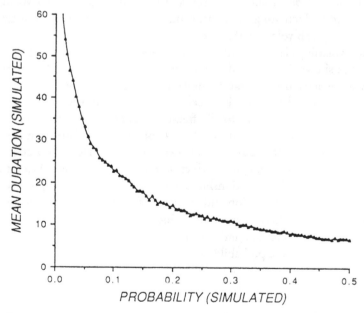

Figure 9.3. Plot of mean (= statistically expected) duration versus probability of origination and extinction for the same 1 million simulated clades used to construct figures 9.1 and 9. 2. Expected clade duration decreases, in concave-up fashion, as probabilities of origination and extinction increase. Higher volatilities are responsible for shorter life spans at higher probabilities of origination and extinction.

Volatility and the Decline of Global Background Origination and Extinction Probabilities

Gilinsky and Good (1991) and Gilinsky (Gilinsky 1991a, 1991b, 1994) used the mathematical theory of branching processes in discrete time and maximum likelihood to estimate the probabilities of familial branching, persistence, and extinction within some *137* orders of marine invertebrate organisms. Sepkoski's "Compendium of Fossil Marine Animal Families" (Sepkoski 1992a) provided the stratigraphic range data; these data were then manipulated to obtain the numerical paths of diversity actually experienced by the 137 orders. Then the diversity paths were subjected to analysis to estimate probabilities. We assumed in those analyses, for simplicity and as a first guess, that probabilities of familial origination and extinction were time-homogeneous for the families within orders. Probabilities might in fact not be time-homogeneous, but we wanted to see how much of the basic pattern in the taxonomic history of marine animal life could be explained by the simple time-homogeneous model before introducing further complexity.

Figure 9.4 gives the most recent set of results (Gilinsky 1994). Two observations are immediately obvious. The first is the extraordinary correlation between origination and extinction probabilities. For extinct orders (the pluses), some degree of correlation might be anticipated simply because orders that are now extinct ultimately experienced equal numbers of extinctions and originations. But for living orders (the squares), where the number of originations need not even be similar to the number of extinctions, the strong correlation remains nonetheless.

Some sort of correlation seems almost required at some level, because orders whose familial probabilities of origination grossly exceeded those of extinction would rapidly have taken over the world. Conversely, orders where probabilities of extinction grossly exceeded those of origination would so rapidly have become extinct we would never be aware they existed. So some pressure toward roughly equal probabilities would seem to be required for a diversity of life forms even to exist. The strength of the observed correlation is nonetheless surprising. Stanley has written extensively on the correlation at the species level (Stanley 1990c).

The second obvious feature of the graph is the disparity between living and extinct orders. Extinct orders span the entire range of origination and extinction probabilities, but living orders are almost entirely confined to the lower left corner of the graph, in the region of low probabilities of origination and extinction. One possible explanation for this might be that low-probability orders arise preferentially in the record as we approach the geological recent. But figure 9.5 shows that this is a minor effect, if any. The

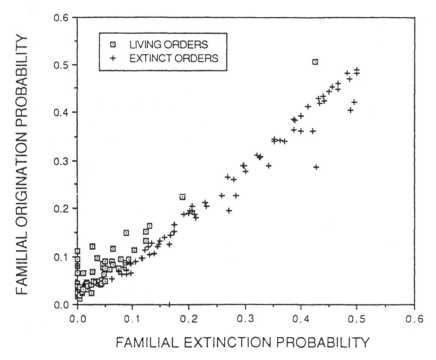

Figure 9.4. Plot of familial origination versus extinction probabilities for 137 orders of marine animals. *Squares,* living orders; *pluses,* extinct orders.

first-order explanation for the difference between living and extinct groups is the differential extinction of the high-probability groups. Orders whose families had high probabilities of origination and extinction (top of the diagram) had shorter life spans than those whose families had low probabilities. Orders with low probabilities tended to span the entire Phanerozoic from their origin to the Recent. Because high-probability orders tended to become extinct sooner than low-probability orders, today's biota is composed almost exclusively of the low-probability orders.

The disparity of familial probabilities between living and extinct orders implies that global background origination and extinction probabilities must have declined through time. Raup and Sepkoski, in their landmark analysis of mass extinction (Raup and Sepkoski 1982), also discovered and documented a decline of the global rate of familial extinction. Figure 9.6 illustrates that their observation just is one element of decline of probabilities of both origination and extinction. Each point on the graph is the average probability of origination (squares) or extinction (pluses) of all orders in existence at that point in geological time. As mentioned earlier, the analysis assumes that probabilities of familial origination and extinction have not changed during the histories of orders, at least to a first approximation. So the graph assumes that the global decline was not a result of any decrease of familial origination and extinction probabilities within orders. Direct examination of familial extinction rates within orders (Gilinsky and Bambach 1987; Van Valen 1985b) indeed shows little if any evidence for declines in extinction probabilities. There is some evidence that familial origination intensities have declined within suprafamilial taxa (Gilinsky and Bambach 1987; Van Valen and Maiorana 1985), but these declines need not be invoked to explain the global pattern.

In summary of the pattern, it appears that the observed global declines of background origination and extinction probabilities can be adequately explained by the preferential loss of orders whose constituent families experienced high probabilities of origination and extinction. The strong correlation between origination and extinction probabilities (see figure 9.4) strongly suggests that the decline of global background extinction intensity, for which

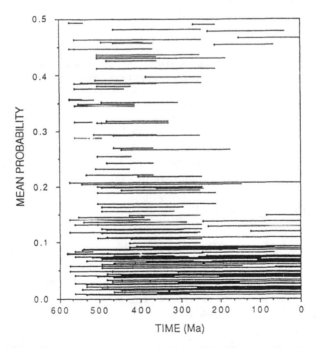

Figure 9.5. Plot of probabilities of origination and extinction versus life spans of orders (depicted by "time-lines"). Virtually all still-living orders had low probabilities of origination and extinction, and long life spans. High-probability orders had short longevities and became extinct before reaching the Recent.

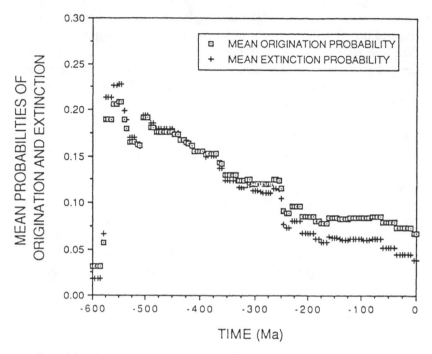

Figure 9.6. Plot of mean probabilities of origination (*squares*) and extinction (*pluses*) versus geological time. Each point is the mean familial probability (of origination or extinction) calculated from all orders in existence at the time indicated. Except for the first few Cambrian intervals, the graph depicts the Phanerozoic decline of global background origination and extinction intensities.

a number of independent explanations have been proposed, is actually just a part of this, more general, evolutionary process.

Volatility Enters the Mix

Why were the high-probability groups lost? Here we discover the critical role played by volatility. As shown earlier, a simulation model demonstrated an intricate set of relationships among probabilities, volatilities, and expected clade durations. Given that the model predicts shorter longevities for clades with high probabilities of origination and extinction (simply by virtue of higher induced volatility), and given the observed decline of global background origination and extinction probabilities (see figure 9.6), the question arises whether the observed decline might simply be explained by the loss of high-probability orders to volatility.

Figure 9.7 shows, using squares and pluses, the relationship of volatility (vertical axis) to probabilities of familial origination and extinction (horizontal axis) for 137 real orders of marine organisms. Although there is scatter in the plot, as one would expect for real-world data under a probabilistic process, the basic pattern of volatility increasing to the right is nonetheless clear. Superimposed on the plot is the curve for that relationship obtained by the simulation discussed earlier. Although the axes for two kinds of data are necessarily scaled differently (because computer time steps and geological stages are not in the same units), it nonetheless appears that the relationship between probabilities and volatilities for real data is just what would be expected if the only factor contributing to diversity fluctuation is the "sampling error" that is entailed by familial origination and extinction being a probabilistic process. That is, the shapes of the relationships suggest that the observed volatilities are entirely explained by each group's (assumed time-homogeneous) probabilities of origination and extinction.

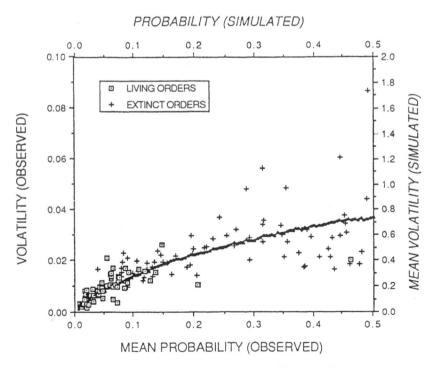

Figure 9.7. Plot of volatility versus probabilities of origination and extinction for the 137 orders superimposed on the simulated relationship depicted in figure 9.1. The volatility axis for real clades was scaled to fit, showing a strong correspondence between simulated and real data. *Squares,* living orders; *pluses,* extinct orders.

Figure 9.8 is a plot of the observed duration and volatility for each order, again with the solid curve showing the superimposed relationship from simulation. Here, the similarity between the solid curve and the scatter of points strongly suggests that the observed durations of taxa can be entirely accounted for by the measured volatilities. In other words, no explanation beyond the volatilities themselves is needed to explain the longevities of taxonomic groups.

Finally, figure 9.9 is a plot of the observed durations for orders as a function of their probabilities of origination and extinction, with the solid curve giving the relationship obtained by simulation. Again the fit is quite remarkable, suggesting that the observed durations of orders may be entirely explained by their (assumed fixed) probabilities of origination and extinction.

Because (1) the decline of global background origination and extinction probabilities (see figure 9.6) is accounted for by the sorting of orders (figure 9.5), and because (2) the loss of orders is accounted for by the perils of high

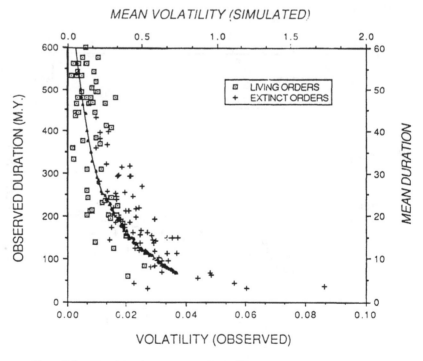

Figure 9.8. Plot of duration versus volatility for 137 orders superimposed on the simulated relationship depicted in figure 9.2. Duration axis for real data was scaled to fit that for simulated data and showed a strong correspondence between simulated and real data. *Squares,* living orders; *pluses,* extinct orders.

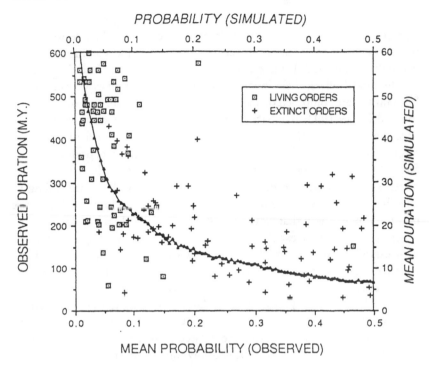

Figure 9.9.   Plot of duration versus probability for 137 orders superimposed on the simulated relationship depicted in figure 9.3. Graph shows a strong correspondence between simulated and real data. *Squares,* living orders; *pluses,* extinct orders.

volatility (see figure 9.8), and because (3) the observed volatilities were entailed by the probabilities of origination and extinction themselves (see figure 9.7), it follows that the decline of background origination and extinction intensities can be accounted for by the familial origination and extinction probabilities that are attached to the various orders. In sum, then, by treating origination and extinction as probabilistic processes, volatility emerges as a critical element in macroevolution. In the case discussed here, the probabilities of familial origination and extinction within orders determined the level of diversity fluctuation (volatility), and the likelihood of fluctuating to zero accounted for the observed change of global background origination and extinction intensities.

Causation

Although the model proposed here appears to explain the different life spans of the orders, and the decline of global background origination and extinction intensities, there nonetheless seems to be something missing in the explanatory account: organisms, populations, species, and their relationships to environment. The account offered here is abstract and statistical, but not particularly biological. Is there a way to bring biology back into the mix?

The model offered here involves the histories of families within orders. But families are composed of species. Sepkoski, Bambach, and others (Allison and Briggs 1993; Bambach and Sepkoski 1992; Sepkoski et al. 1981) have shown that a useful way to treat families, in the context of studies of taxonomic diversity and evolutionary rates, is as proxies for species. That is, familial diversity and rate patterns can be likened to species diversity and rate patterns for purposes of exploring biological causation. If we are willing to take this step, at least provisionally, we might be able to bring biology back into the mix in a way that would unify paleontological and ecological processes, at least to some extent. For if families are proxies for species, we can then examine ecological models of diversification and extinction and ask whether long-term paleontological patterns can be consequences of the extended operation of known shorter-term processes. Specifically, we can ask whether the measured probabilities of familial origination and extinction, from which the volatility pattern and the declines of global origination and extinction intensities follow, can be correlated with organismal or population properties known or thought to be causally important in speciation and extinction.

Elucidating in detail the organismal and population properties that promote or inhibit speciation and extinction would require an entire review paper in itself. I refer the reader to Stanley (1979, 1990c), Eldredge (1992), and Cracraft (1992) for recent discussions from a paleontological standpoint, and to Mayr (1970) for a more classical, but still current, treatment from a neontological standpoint. Maynard Smith (1989) attempts another such review. Boyce (1992) reviews the theory and procedures of population viability analysis, which aims to analyze propensity for species survival for conservation purposes. Parsons (1993), Pimm et al. (1988), and Tracy and George (1992) also review the subject, the last two largely from the standpoint of neontology. Of course, there is nothing wrong with neontological or paleontological approaches to the problem, but I make the distinction between paleontological and neontological approaches because it is not entirely clear that paleontologists and neontologists are really attempting to answer the same questions (see later).

Cracraft (1992) offers a useful pictorial model of the various factors that are causally efficacious in determining rates of speciation and extinction. In brief, both speciation and extinction rates are influenced by intrinsic and extrinsic factors. Intrinsic factors causally efficacious in speciation include biological (and genetic) propensity for generating novelties, propensity for fixing novelties in populations, and intrinsic rates of population growth. Extrinsic factors include properties of the geomorphic environment, such as amount of habitat fragmentation, topographic variety, climatic fluctuation, and others. Because most speciation is presumably vicariant, high speciation rates should be favored in species that combine certain aspects of both: species that tend to form novelties frequently [e.g., "stenotopes" (Eldredge 1992)], tend to fix novelties frequently (those with structured populations), and they tend to inhabit geomorphic regions that have high rates of fragmentation.

For extinction, causally efficacious intrinsic factors include "niche breadth" of avatars within species, and the capacity to disperse into new, favorable habitats. Extrinsic factors would include level of geomorphic instability (perhaps tectonic), and harshness of climate. Thus, highest extinction rates would be expected in stenotopes, with limited dispersal capacity, living in an unstable geomorphic region. "Fortuitously," to use Stanley's term (1990c), the same criteria apply to propensity for extinction as for origination, thus providing another lead in explaining the observed strong correlation between origination and extinction rates.

Eldredge (1992, pp. 10 to 12) outlines a causal model that relates "ecological niche width parameters" to speciation and extinction rates. Under his model, both eurytopes and stenotopes might have equal susceptibilities to disruption of their specific mate recognition systems. But stenotopes, by virtue of their higher potential for modifying their use of resources, might be able to produce descendant species more frequently. That is, stenotopes might have a greater capacity to evolve new economic adaptations, thus enhancing the probability of success for establishing new genealogical entities (species). Although nearly all of the work remains to be done, I believe that an effort to characterize groups of organisms known from the fossil record and from the Recent in terms of resource-use specialization and other ecological variables might lead to the emergence of generalizations that could explain the long-term decline of background origination and extinction probabilities and, perhaps, other paleontological patterns.

Vrba's studies of speciation rates in African mammals (Vrba 1987) exemplify the approaches just mentioned [i.e., those of Cracraft (1992), Eldredge (1992), and Stanley (1979)]. In African mammals, resource-use specialization (on certain resource-use variables) is correlated—Vrba believes causally—with high rates of speciation.

The difficulty with any such predictive scheme is, of course, that it is practically impossible to isolate all the possible factors that might be involved in determining speciation and extinction propensities in any particular case

(McKinney 1995). Nonetheless, it may be possible to tabulate the most critical controls and ask whether they correlate statistically with calculated origination and extinction frequencies, volatilities, and observed durations. Once tabulated, a systematic survey of the biological properties of the various taxonomic groups would be needed to test adequately whether the observed pattern of turnover can be convincingly attributed to particular properties or constellations of properties.

Does Metapopulation Dynamics Hold the Key to the *Global* Extinctions of Species?

Although I am optimistic that ecological theory and paleontological patterns can eventually be integrated, I believe that one of the challenges that will need to be met is the challenge of scale. When paleontologists write of extinction, they almost invariably mean the global and permanent extinction of a species or group of species. Ecologists, on the other hand, are typically referring to the *local* extinction of populations. (See Maurer and Nott, chapter 3, for a refreshing effort explicitly to address the problem of scale.) Paleontological and ecological concepts seem to converge most readily only when a species is already down to its last remaining population.

Excellent examples of recent studies by ecologists of the metapopulation dynamics of extinction include those by Bengtsson (1989), Schoener (1991), and Pimm et al. (1988). In all three studies, emphasis was on the risk of local extinction of relatively small populations. Bengtsson's work was on local extinction of *Daphnia* populations in natural and artificial rock pools. Pimm et al. studied the extinction of populations of British Island birds, where mean population sizes were typically 15 pairs or fewer. Schoener studied extinctions in Bahamian spiders and lizards, for population sizes typically of 150 individuals or fewer. Each study reached interesting and relevant conclusions, especially for conservation biology, where endangered species have small population sizes almost by definition. But as a paleontologist, I cannot help but wonder whether the conclusions drawn by these and other research workers can really be applied to the problem of the global extinction of a once widespread species. R. M. May raised similar questions recently (May 1993).

Not all species with a paleontological record were geographically widespread and abundant in life. But many, probably most, were (see also Raup 1992). Consequently, it was the global extinction of species such as these that forms most of the paleontological database on rates of speciation and extinction, and on the global decline of background origination and extinction probabilities discussed earlier. A typical example is the Cenozoic clam

Marvacrassatella undulata from the Atlantic Coastal Plain (Ward et al. 1991; also 1994, personal communication). This species had a geographical range that encompassed effectively the entire Atlantic Coastal Plain. It also had a 3-my temporal range, successfully negotiating some five glacially mediated transgressions and regressions of the Atlantic Ocean. (The species eventually did at last become extinct.) Under classical metapopulation theory (the Levin's model), *Marvacrassatella* must have consisted of a tremendous number of quasi-independent populations that probably spanned multiple communities. [See Eldredge (1985, 1992) on how species can only rarely be construed as playing a singular ecological role.]

With the preceding as background, let us now examine the process of extinction. First, *Marvacrassatella* must have had millions and millions of individuals, effectively ruling out extinction by demographic accident. Next, because the species covered such a wide geographical range, I speculate that competition or predation could not have been responsible either. For these factors to have been the proximate cause would require complete overlap of their geographical ranges—not impossible, but improbable. So how *are* widespread, successful species such as *M. undulata* brought to complete, global extinction?

Paleontologists have recently been favoring models involving wide-scale deterioration of the physical environment. For *Marvacrassatella,* and for many other Plio-Pleistocene Atlantic Coastal Plain species, temperature, with the other changes associated with glaciation, is a favored explanation (Allmon et al. 1993; Stanley 1986a, 1988; Ward and Gilinsky 1993). Indeed, at least one researcher has gone so far as to speculate that some 60 percent of all extinctions of fossil species might be caused by just one type of environmental deterioration, namely, meteorite impacts (Raup 1992)!

Although the idea that all global extinction results from physical environmental processes might seem radical at first, part of the motivation behind such claims seems to be a widespread feeling that both classical competition and predation, and the newer theory of metapopulation dynamics, have not yet provided a mechanism for extinction that is up to the task. Perhaps deterioration of the physical environment brings widespread species down to size, thus rendering them vulnerable to the factors that metapopulation dynamics has identified as causally efficacious in extinction. Or maybe the theory of metapopulation dynamics can indeed explain the total extinction of widespread species by effects that cascade beyond immediate local areas. [See Maurer and Nott (chapter 3), who question the capacity of current metapopulation theory to deal with entire geographic ranges, and Plotnick and McKinney (1993) on how disturbances might affect entire ranges via effects that percolate throughout ecosystems.] But maybe Raup is right after all, and the extinction of widespread paleontological species really is typically caused by processes that are extraordinary in ecological time. In my view, this is precisely the kind of problem that needs to be resolved if we are to bring metapopulation dynamics and extinction dynamics together.

Conclusion

Ecologists and paleontologists have long been interested in "turnover" of genealogical entities, but usually at different scales, ecologists focusing on turnover among populations, paleontologists on species and higher taxa within clades. Both have, rather independently, recently developed conceptual and mathematical theory to evaluate turnover processes. Ecologists have the theory of metapopulation dynamics, paleontologists the theory of branching processes. Although the two groups developed their respective approaches largely independently, the problems that both groups aim to address are, in a broad sense, the same, involving the origination and extinction of genealogical entities within larger entities. Originations and extinctions of populations determine the dynamics of species expansion and survival, while the originations and extinctions of species and higher taxa determine the dynamics of clade expansion and survival. These broad similarities strongly suggest a commonality of concepts and mathematics that needs to be more fully articulated.

Furthermore, it is not only in a formal sense that ecologists and paleontologists are on common ground. We should not lose sight of the fact that there has been only one history of life on Earth, and that all of the levels of the evolutionary hierarchies have been relevant and interconnected. Perhaps higher-level and lower-level patterns and processes can be integrated more fully in a causal sense as well. Maybe the loss of orders with high probabilities of origination and extinction, which explains the Phanerozoic decline of global background origination and extinction probabilities, can itself be accounted for by properties of organisms and populations in their interactions with their environments. Making such a link successfully will help both to legitimize the paleontological study of long-term patterns at higher taxonomic levels and to extend the scope of ecological theory beyond the local habitat in ecological time. My hope, ultimately, is that paleontologists and ecologists will eventually be able to converge—paleontologists from the global side, ecologists from the more local—to effect a more satisfactory understanding of the dynamics of biological diversity on earth.

Part Two

Community Turnover:
From Populations Through Global Diversity

10

Scaling the Ecosystem:
A Hierarchical View of Stasis and Change

Kenneth M. Schopf and Linda C. Ivany

The fauna characteristic of specific benthic environments,
and represented by recurrent faunal assemblages, remain
virtually locked or static for periods of millions of years
as constituent taxa merely track relatively persistent
environments. —BRETT ET AL. 1990, P. 200

Communities are thus shown to be merely temporary
assemblages of species brought together by the
environmental conditions prevailing at any particular
time. —HUNTLEY AND WEBB 1989, P. 5

It is a common practice in field geology to examine an outcrop in close detail only after getting the "big picture." One gains this broader perspective by backing off a suitable distance so that the whole (or most) of the exposed section is within the field of view. The reason? We get closer to understanding the whole by considering its parts at more than one scale.

Like an outcrop, an ecosystem is formed from a hierarchy of processes and must therefore be defined relative to time and space. Communicating ideas using these definitions is often made more difficult by the plasticity of terms such as *ecosystem,* whose meaning is partly dependent on context. Because many of the characteristics commonly used to define ecosystems (such as geographic limits, temporal duration, environmental factors, taxonomic composition) are themselves dependent on the temporal and spatial scale of observation, it is easy to see that each new research project may yield a different ecosystem concept. What we perceive as the behavior of ecosystems is therefore intimately tied to our scale of observation. What appears as flux up close may be stability from a few steps away.

As the introductory quotations suggest, polarized accounts of the "long-term" behavior of ecosystems are not difficult to find in the literature. Such conflicting views may arise because researchers assume consistency across scales (or have considered only one scale, which they see as long-term) in the hope of generalizable results. Recent research showing fluctuating, "open" benthic foraminiferan ecosystems over 55 million years (my) of the Cenozoic Western Atlantic Coastal Plain (Buzas and Culver 1994) has sounded to some workers like "the death knell of tightly knit benthic communities" (Jackson 1994a). These findings agree with classic studies of Holocene floras (Huntley and Webb 1989) that observe individual plant species responding to the last deglaciation (12,000 years BP to Recent) with independent migration patterns, rather than as interlocked groups. However, recently reported data from a temporal scale intermediate to these two cases suggest the opposite. Brett and Baird (1992, 1995; Brett 1995) report species-level morphological and ecological stasis in Silurian-Devonian benthic invertebrate assemblages on time scales of 3 to 7 my. Was the funeral pyre lit prematurely?

Herein, we examine the documented behavior of ecosystems across a spectrum of spatial and temporal scales. Of primary interest are the characterizations that may be drawn about ecosystem behavior (for our purposes, the stability versus instability of taxonomic composition) at each of these scales of observation. We ask whether the seeming inability to use theory derived from ecological observation and modeling to successfully predict the relative stability of ecosystems at various scales is pointing to our limitations, or perhaps to emergent system properties that reflect the operation of "different rules" (sensu Jablonski 1986a). In so doing, we are not seeking to gain insight into a generalized truth about whether or not ecosystems are stable entities [Clementsian versus Gleasonian extremes (Gleason 1926; Clement 1916)], but rather we are pointing out that scale-dependent questions require appropriately scaled sets of observations. Incorporating scale into our methodology and theory at their inception may allow two such differing points of view about the nature of ecosystems to be reconciled, and this may allow new insights from one scale to illuminate the conundrums of another.

Ecosystems and Nature's Continuum

Paramount in the consideration of ecosystems at any scale is the way in which they are defined. The mire surrounding the term *community* is an excellent example of the loss of a word's utility as a result of indiscriminate usage. Because it is likely that no two people will approach this term in the same manner (especially given the diverse group of scientists interested in communities, including neontologists and paleontologists), our ability to talk meaningfully about these groups of organisms is lost. Indeed, it has become routine in paleontological literature to include a brief definition of the author's personal take on the term *community* when it is going to be used (e.g., Brett et al. 1990; discussed by Järvinen et al. 1986). The same holds true for the less controversial time-averaged versions—*assemblage* and *paleo-*

community. Ecosystem may be a more contentious term, if anything, because it seems at once more inclusive and more technical. It is our premise, however, that as with *community,* such a term is useful if clearly defined in a given context. We use *ecosystem* loosely, in reference to the organismal systems tied to particular portions of time and space in figure 10.1. The mere fact that such words may be used to describe distinctly different phenomena points to more than linguistic confusion; it also stresses the unique ability of words like *ecosystem* to describe some underlying quality, independent of scale, that is shared by things as disparate as the ecosystem of human oral bacteria and that of a tropical coral reef. However, freedom from a particular frame of reference is a double-edged sword. If natural groups of organisms are observed on a variety of temporal and spatial scales, the single term *ecosystem* can be of little use when talking about their scale-dependent features without extensive elaboration. But this begs the question: *are* there scale-dependent features of ecosystems?

The world approached by the biological scientist is a continuum that runs along two dimensions: organisms existing in space through time. This continuum is broken into pieces by researchers in various ways, depending on their interests and the part of the continuum to which they actually have access. In many instances, the researcher—a grassland ecologist, a specialist in Ordovician brachiopods, or a student of extinct aquatic sloths—is bound to a particular region of time and space simply by the subject matter. In such instances, the fact that there exist additional levels of the continuum outside that of the research program may be irrelevant to the goals of the research. Yet, this is clearly not the case when studies have as their aim not merely the description of some localized part of the continuum, but rather a generalizable "truth," or set of rules, applicable to all scales at once. In the former case, data may be *dismissed* as outside the scope of a project; in the latter, there lurks the possibility that a global conclusion based on more local observations will result in *missed* data and thus an incomplete picture.

Much attention has been paid lately to aspects of scale in the natural world. The fractal revolution has penetrated thinking across many scientific disciplines—from the self-similarity of drainage networks (Turcotte 1994) and the extinction statistics of the fossil record (Solé et al. 1997), to the architecture of organismal design (Boyajian and Lutz 1992). It is becoming apparent that "scale is a matter that pertains to all ecological data" (Allen and Hoekstra 1991), and subsequent developments in ecological theory [e.g., hierarchical ecosystem theory (Allen and Star 1982; O'Neill et al. 1986)] actually have taken this issue as their foundation. Recent approaches to the assessment of fossil faunas (papers in Miller 1990c) stress the inclusion of scale as an integral part of describing and understanding natural systems of organisms, although the importance of this issue seems to have been slow to be considered in paleontology relative to ecology (Wiens et al. 1986). This broad attention to scale, at least as it originates from the paleontological end of the continuum, is no doubt partly derived from recent advances in areas such as stratigraphy, where the scaled division of the rock record into self-similar packages has made sequence stratigraphy the new context for organic evolution (e.g., Brett and Baird 1995; Holterhoff 1994, 1996; Holland 1995).

How has the continuum been parsed? With a pool of data ranging from the local population observed over a number of hours to global biodiversity tallied over hundreds of millions of years, it depends on the questions being pursued. It is not our purpose here to suggest that anything close to life's actual multiscale record is resolvable to the modern researcher. Instead, we would suggest that individual case studies be viewed as parts of this continuum, rather than as isolated proxies for underlying tendencies, and that scale always be considered when investigating ecosystem behavior. Relative stability and instability of organismal associations can be only that—*relative* measures with respect to other such measures at a *similar scale* of inquiry.

This issue comes to the forefront when considering whether the tendencies observed in one setting, over a particular temporal and spatial scale, may be extrapolated (either up or down) to predict or explain the behavior of an ecosystem at a different level. This might be taken at its most ridiculous—it is certainly true that any particular day in the life of a group of organisms might not be representative of its entire existence (a Boston Celtics fan would surely not want an average game of the '95–'96 season to be put forth as typical of franchise history). Yet there exists an intermediate scale of observation that seems to pass as a proxy for *the* behavior of ecosystems (see Valentine and Jablonski 1993). This portion of the continuum, which we later examine in detail as region 2 of figure 10.1, encompasses the broadest case studies (in terms of time—up to 1.5 million years, and space—approaching provincial) that comfortably conform with Recent observations of ecosystems characterized by flux.

Valentine and Jablonski's conclusion, "that change rather than stability is the normal lot of communities over ecological as well as evolutionary time" (1993, p. 341) not only is a statement about the basic congruence of modern and Pleistocene accounts of ecosystem behavior, which they document convincingly, but it also presents a world view that encompasses all scales of the continuum (emphasized in Jablonski and Sepkoski 1996). Yet, there exists another 95 percent or so of the Phanerozoic record that could be considered within the generous limits of their title: "Fossil communities: compositional variation at many scales." The authors address this silent majority of the fossil record largely by citing, and building on, the work of Sepkoski (see

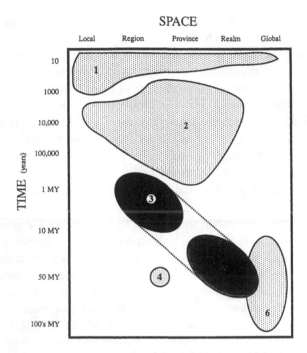

SPACE

Figure 10.1. A graphic representation of the available data pertaining to ecosystem behavior, presented in a matrix defined by time and space. Positions and boundaries of regions are approximate. *Stippled areas* are those in which ecosystems have been characterized as unstable. *Solid bodies* represent regions characterized by ecosystem stability, *darker areas* imply documented stability at the species level. See text for discussion of individual regions.

figure 10.1, region 6) whose outlook and data are decidedly global and at the taxonomic rank of family. Are there other pertinent data between the Pleistocene and the Precambrian? We have undertaken this paper because we believe the answer to this question to be yes, and because a synopsis of ecosystem stability and instability specifically couched in terms of scale may bring sense to seemingly irreconcilable accounts of ecosystem behavior.

The Time–Space Matrix

Perusing the literature, it is easy to observe that the continuum of naturally occurring organisms in space and time has traditionally been broken into a number of discrete steps along the spatial axis: the local population, region, province, and realm. It has likewise usually been studied over distinct slices of time. Figure 10.1 is a representative, though certainly not exhaustive, attempt to graphically display the range of ecosystem studies that have been undertaken within a matrix whose axes are time and space. We have concentrated on studies that are characteristic of whole programs of research that might occupy similar areas of the matrix where possible (our primary goal is not to summarize all available literature, but to characterize its conclusions), although in several instances it is the work of a sole or only a few researchers that characterizes a region. This has produced an unevenness in the time spent explicating the consensus views for several of the regions; it is easier to summarize the findings of a single paper (e.g., region 4) than the whole of ecological data and theory (e.g., region 1). This said, we identify six areas of the continuum that have been probed by researchers, recognizing that this division is somewhat arbitrary and based on only a small subset of the available literature.

It is immediately apparent from figure 10.1 that there is a general trend toward working at a larger spatial scale with increasing temporal magnitude. It is part of the unavoidable trade-off involved in the collection of data within such a continuum that, with all else equal, looking at a wider geographical area will usually entail sparser sampling, and looking for a longer period of time usually necessitates less-continuous sampling [the trade-off between "grain" of data versus its "extent" (Allen and Hoekstra 1991)]. In both instances, the greatest possible detail is sacrificed to encompass a bigger picture. This, coupled with the fact that the vagaries of preservation prevent any particular local setting from being preserved over very long periods in the fossil record, forces the tandem increase of temporal duration with spatial spread observed in the figure.

Region 1: The Modern World and Recent Past

The most immediate scale of observation is that which surrounds us in our day-to-day living. Almost all studies that may be called ecological (in the sense that they deal with living organisms) take place in this temporal domain, on the order of days to tens of years. Temporal limits are imposed by the life span of researchers—or more accurately their grants. Spatially, these studies run the gambit from localized populations and experimentally controlled plots (Naeem et al. 1994; Tillman and Downing 1994) to global assessments of biodiversity (May 1988b). Unlike cases of chance fossil preservation and exposure, a far greater expanse of geography is accessible to the researcher. In this section, we do not attempt to summarize the vast libraries of ecological data or theory; rather, we have tried to distill general research approaches and conclusions from this much larger (and rather overwhelming) sink.

Case Studies

Studies of modern ecosystem behavior take one of several forms. Most obvious is the tactic involving observation of the chosen fauna over some period of time in its natural state. Such studies usually involve detailed scrutiny of a relatively circumscribed area and so may sacrifice broad geographic coverage to obtain exhaustive information about a local area (e.g., Diamond 1975, for New Guinea bird communities). Because they take place over a relatively short time (usually several years), they may also fall victim to a particularly uncharacteristic or infrequent event—or they may miss one that *is* characteristic of a given habitat (e.g., hurricane, fire). A classic example of the importance of such rare events is the notable evolutionary change observed in Galápagos finches after severe El Niño events (Grant and Grant 1993). Recognizing the effects of such events requires data encompassing both "normal" and "perturbed" times, calling for a level of persistence (or serendipity) in ecological research that is difficult to achieve.

A related group of studies uses this approach but combines it with a devised (or, fortuitously, natural) experiment involving partial or complete destruction of the local community to observe how the area is repopulated. Such studies range from defaunating small mangrove islands to track reestablishment of arthropod communities (e.g., Wilson 1969; Wilson and Simberloff 1969; Simberloff and Wilson 1969), to recording the response of grassland systems to severe natural drought (Tillman and Downing 1994), to surveying the effects of hurricanes on benthic marine invertebrate communities (Miller et al. 1992). The massive Hubbard Brook deforestation experiment (e.g., Bormann and Likens 1979) also falls under this category.

Another approach is to investigate a broader geographic area (e.g., regional) at a single moment in time, to take note of various subsets of the area that seem to be in different stages of succession, and to extrapolate the resultant spatial patterns through time. This is an approach most commonly taken with relatively sessile communities, e.g., rain forest trees (Denslow 1987) or rocky intertidal organisms (Underwood et al. 1983). A necessary assumption involved with these "space-for-time substitution" studies is that subsets of the area studied differ because they are at different stages of successional development along an invariant trajectory toward a common endpoint. This assumption is derived directly from Clementsian views of ecosystems, and it implies that no alternate pathways or unique site-specific characteristics exist (Pickett 1989). Such studies may also be compromised by ignoring smaller-scale heterogeneity within the area of interest (see discussion in Pickett 1989; also Wiens 1989).

Last, there are the smaller-scale experimental designs that allow careful manipulation of species composition under controlled environmental conditions. These include various exclusion studies that seek to define the role of predators or grazers in a community by removing them from (or adding them to) an enclosed portion of the habitat and observing the response (e.g., Paine 1992, on intertidal-zone grazers), and laboratory plots set up to mimic real communities so as to determine their behavior under a variety of carefully constrained environmental conditions (Naeem et al. 1994). These have the advantage of allowing the possibility of replication and direct comparison of results under different conditions. However, there will always be the potential criticism (particularly for the latter studies) that experimental plots do not reflect what is occurring in the real world. These same limitations apply to mathematical models that have been developed to explain ecosystem behavior in nature.

Ecosystem Behavior

The debate about Clementsian versus Gleasonian ecosystem dynamics has colored the course of ecological research and theory for much of this century. As related by Roughgarden (1989), studies emphasizing the stability and internal structuring of ecosystems dominated the field until the last 10 to 20 years, when the publication of papers such as Connell (1978) on nonequilibrium dynamics initiated a swing toward more Gleasonian views of individualistic species behavior and open systems.

To the layman and scientist alike, the notion that Modern ecosystems display a stability of sorts is patently obvious. Each summer we return to the same beach or favorite camping spot to meet familiar aspects of the general flora and fauna. Indeed, such a notion of compositional stability in ecosystems implicitly underlies undertakings such as the massive cataloging enterprise of *Ecosystems of the World* (29 volumes for terrestrial and aquatic ecosystems, e.g., *Coral Reefs*, Dubinsky 1990). If this work is to hold any meaning for future researchers aside from providing them with many "snapshots" of a single instant in time, ecosystems must be assumed to have some temporal and spatial persistence.

Yet, studies rigorously demonstrating the compositional stability of ecosystems appear to be the exception rather than the rule in an ecological literature increasingly devoted to studying turnover (e.g., see Russell et al. 1995). Reported cases of stability appear limited but have been documented from wildly differing habitats. For example, McGowran and Walker (1979) have made observations in the marine planktonic realm showing that abundance and structure of copepod communities in ocean gyres is very consistent over time spans of as much as 4 years. This coherence persists despite fluctuations in nutrient influx (Hayward and McGowran 1979).

Similarly, Herbold (1984) was able to demonstrate that a seemingly random association of stream fishes recorded over 13 years was an artifact of the mixing of two distinct ecological associations, one of which displays considerable coherence through time (as reflected in significantly concordant relative ranks and times of fluctuation). Bengtsson's (1994) study of forest soil communities is a rare, but excellent, example of compositional stability evident over a somewhat longer time scale—circa 20 years. This author analyzed data largely culled from literature sources (67 studies/plots in total), concluding that such communities display a high degree of predictability (constancy) in terms of relative abundances of organisms. Between-year relative abundances of soil organisms were predicted at the species level in over 80 percent of the studies used in this analysis, and values close to 95 percent were attained by using higher taxonomic levels.

The counter to stable, persistent ecosystems is the situation where stochastic processes dominate, and co-occurrence of taxa is solely a result of the coincidence of shared environmental preferences. A number of examples can be cited in which continual flux in ecosystem composition is observed, and where it is therefore difficult to infer any kind of "structure" underlying the community. A typical example is the prelude to Wilson and Simberloff's classic mangrove island experiment, in which the arthropod fauna on several nearby islands was found to be constantly changing through time and different on each island at any given time, supposedly reflecting continual exchange with the "species pool" (Simberloff and Wilson 1969; Wilson and Simberloff 1969; Wilson 1969).

There is also evidence that disturbance on a variety of scales (Connell 1978) combines with chance processes (Drake 1990) and metapopulation dynamics (e.g., Hanski and Gilpin 1991) to play a huge role in determining the local composition of communities. Denslow (1987), for example, concludes that tree species composition in rainforests is largely a function of disturbance via formation of gaps in the canopy; rapid growth of propagules, by chance already germinated in the soil prior to the disturbance, determines the composition of the ensuing community. In a Venezuelan forest, such "advance regeneration" accounted for 95 percent of the tree species present. Extending this result into time suggests that community composition, at least in a local area, is not predictable or stable, but instead is dependent on the vagaries of disturbance, seed dispersal, and fortuitous timing of germination.

Marine ecosystems, too, are heterogeneous on a variety of temporal and spatial scales because of a high susceptibility to chance physical processes and disturbance, a theme echoed by numerous authors including Probert (1984), Dayton (1984), Barry and Dayton (1991), and Karlson and Hurd (1993). Underwood et al. (1983), for example, in a study of rocky intertidal communities off New South Wales, found that community composition is variable and strongly controlled by the crop of larvae that arrive and settle out of the water column in a given area by chance. This viewpoint can be summarized by Connell and Sousa (1983), who, after an extensive search for stability in ecosystems, concluded, "If a balance in nature exists, it has proven exceedingly difficult to demonstrate" (p. 808).

Because stability in species composition is not always apparent, researchers have tended to look instead for structure in ecosystems that attests to the importance of long-term interspecific interactions, and then, by inference, to long-term stability in composition. Diamond (1975), for example, outlines a set of rules of limited membership for the assembly of local faunas drawn from observations of bird communities in New Guinea that sets limits on which taxa from the species pool can coexist (see also Roughgarden 1989). Brown and Bowers (1984) identify character displacement and competitive exclusion in chipmunks, and strong interspecific competition and competitive exclusion in both desert rodents and hummingbirds. Demonstrations of structure such as niche partitioning, limited membership, character displacement, and coevolution strongly suggest that the taxa involved have been interacting more or less continually for a significant period of time. However, Brown (1995) also points out that most of the good examples of structured communities come from closed systems, such as islands, which house a limited number of species and hence are not directly extrapolatable to the large open systems (marine, continental) that are more commonly represented in the present-day and fossil record.

While ecosystem stability therefore does not appear to be generally manifested in stable compositions of communities throughout time, there is a body of literature that sees "stability" manifested in other ways. One approach is that of diversity-stability studies, in which a constant species *richness* is equated with stability. In the mangrove island experiments of Wilson and Simberloff, arthropod species diversity through time on any one island maintained constant levels, despite continual change in actual species composition (Wilson 1969). Predisturbance diversity levels were reached within months after the initiation of recolonization and were maintained over the next several years of observation. This and other studies support the species–area curves of MacArthur and Wilson (1967), showing a balance between immigration and extinction in local settings of a given size. Another school of research treats the productivity of communities as a proxy for stability and monitors the effects of increasing diversity on that stability. In both experimental plots (Naeem et al. 1994) and natural prairie communities (Tillman and Downing 1994), higher diversity was shown to correlate with higher productivity and faster recovery from disturbance (in the latter case, a drought). In each of these cases, however, "stability" is manifested in some ecosystem property *other* than composition.

Studies exploring ecosystem composition on a provincial level (the far right tail of region 1) are not generally reported for the modern world. Those that exist are compilations of species censuses from a number of different studies, many of which are one-time transects across a region that record everything encountered but give little information about spatial or temporal variability [e.g., the exploratory cruises of the U.S. Steamer *Blake* through the Caribbean by Alexander Agassiz (Agassiz 1888; Agassiz et al. 1878); see also examples in Hedgpeth 1957]. The degree to which ecosystem composition varies within a province is therefore difficult to assess. Valentine (1973a), however, comments that communities within a province generally share many of the same (eurytopic) taxa, and provincial boundaries can be quite abrupt, coinciding with some sort of physical barrier to most of the taxa (e.g., ocean currents, mountain range, climate zone). Faunal provinces therefore may be taken to support the notion of unique, persistent assemblages, and those types of communities that span provincial boundaries change their compositions significantly as they make this transition.

Brown (1995) points out that much of the seeming confusion surrounding the behavior of modern ecosystems can be resolved when one considers that well-documented examples of *both* stability and flux can be found in the literature, and that much of the apparent dichotomy can be attributed to differences in the scale of analyses. Rahel (1990) also emphasizes that the degree of persistence observed is often related not only to scale but also to the way stability is measured (relative versus rank abundance, or presence/absence). Notwithstanding these realizations, it seems to us that the prevailing view of ecosystem behavior today is still primarily one of flux and individualistic response.

Because this is the level at which such facets of ecosystem dynamics as competition, predation, and recruitment may be observed in action, it has been the cornerstone of ecological theory for the last century. Findings made here, primarily on terrestrial plant ecosystems, have been enshrined as *the way ecosystems behave*. Implications from this practice have extended as far afield as to the study of Paleozoic marine communities (e.g., the application of ecological succession by Nicol 1962 to Silurian reefs). Observations made at this level can give us information about the life habits of organisms and their communities that is simply not recoverable from any other level. But, as will be shown, a unique insight does not necessarily guarantee generality.

The question then becomes, is this most immediate scale of observation displaying characteristics that are consistent across all other levels?

Region 2: The Youngest Fossils

The upper edge of the place where "ecology and paleoecology begin to merge" (Davis 1994, p. 357) occurs on the time scale of thousands to hundreds of thousands of years. Such studies are exemplified by research done to retrace the pattern of ecosystem behavior in response to the last deglaciation (Huntley and Webb 1988, 1989; Rousseau 1992) and the longer record of climate accommodation of the Pleistocene (references in Huntley and Webb 1988; Valentine and Jablonski 1993). Spatially, these studies range from local to continental in scale. This scale of investigation allows a more expanded temporal view than that afforded by working solely in ecological time, while retaining much taxonomic and spatial affinity with the present. In many respects, it is the best of both worlds. It is worth remembering, nonetheless, that the interval of time accessible to these studies is relatively small in comparison to the history of life on earth, and that, at least in some cases, the resolution available does not allow the recovery of ecological data comparable to that accessible in region 1 (Roy et al. 1996).

Case Studies

There are in actuality two broad programs of research represented within this region, although they are close enough in our spatiotemporal matrix to be treated together. On the one hand, there are numerous accounts of the dynamics of organisms (primarily trees) over the last 18,000 years or so, centered on the period following the most recent glacial maximum (Huntley and Webb 1989). Such pollen-based studies can have very broad spatial dimensions, often on the scale of continents. Additionally, sampling strategies are often able to resolve changes occurring over very short time scales. Davis (1989), for example, cites work done on pollen assemblages from lake sediments up to 5600 BP with sampling resolution of 3 to 4 years, approaching that available to modern researchers. Precise dating techniques such as lead-210 and cesium-137, in combination with carbon-14, pollen stratigraphy, and paleomagnetic stratigraphy, have been especially useful in the careful calibration of sections (Davis 1989).

Second, there are researchers who have attempted to follow faunal changes over a number of glacial and interglacial cycles (see references in Huntley and Webb 1988). Prime examples are the recent articles by Valentine and Jablonski (1993) and Roy et al. (1995) that examine the Pleistocene mollusk record of the northeastern Pacific across seven coastal provinces. Latitudinal ranges were compiled for the living fauna and compared with fossil migration patterns for taxa in the region from 1.6 to 0.12 my.

Ecosystem Behavior

Both of these schools of inquiry into the fossil record have offered firm confirmation of the view of ecosystems as environmentally mediated groupings of species that may change their membership and structure unpredictably in response to climatic fluctuations [but see Pandolfi (1996) for a compelling counterexample documented in Pleistocene coral reef assemblages]. After their examination of mollusk assemblages, Valentine and Jablonski (1993) surmised that assemblages of taxa existed in the past that were "anomalous" by today's standards of comparison, and that seemed to have been generated by climatic changes [an assessment similar to the one that led to the notions of "community disequilibrium" by Davis 1986, and "intermingled biotas" in Graham's (1986) study of late Quaternary North American rodents]. As an example, they state that 17.4 percent of the species found only in Pleistocene deposits *outside* the Californian Province today call this area home. Valentine and Jablonski (1993, p. 344) concluded:

> The Pleistocene picture, then, is one of species shifting independently up and down the shelf as sea-level rises or falls, and northward or southward as temperatures warm or cool; the living biota is merely a "snapshot" of this process. So far as can be told, the shifts do not involve entire communities as such, but rather, species respond individually, some changing their ranges by hundreds of miles and some by tens.

Similar insights have arisen from inquiry into terrestrial floral distribution in the Holocene. When tree species are tracked over time and space "individualistic responses" are seen as species move along paths independent of any over-arching group dynamic (Huntley and Webb 1989). In at least one case, the implication that these "migrations" were in response to climatic cycling led to an explicit analogy being drawn between this phenomenon and the yearly trek made by birds in response to the seasons (Huntley and Webb 1989), an analogy that becomes loaded by its extrapolation of pattern and process across scales (the boundaries of regions 1 and 2 of figure 10.1).

It is duly noted by workers in this portion of the matrix that the Quaternary was atypical climatologically ["a time of unusually rapid environmental change, geologically speaking" (Valentine and Jablonski 1993, p. 343)]. Yet, the dynamics displayed at these levels of inquiry were, in the aforementioned cases, explicitly shown to be consistent with those displayed at lower levels (i.e., the Recent). Does this call into question the atypical environmental nature of region 2 settings, or should we instead question what consistency with the present really tells us? The agreement between two instants in life's history is surely useful in a quest to understand those particular instants, but in the quest to identify the typical behavior (if such a thing exists) of ecosystems over the broader history of life, two data points so closely spaced in time as regions 1 and 2 may be only enough to reassure us that the present is the key to the Pleistocene. The extent of our gained wisdom seems testable by exploring further along the axes of figure 10.1.

Region 3: The Level of the Biofacies

This is the level of temporal and spatial observation most traditionally allied to the paleontological pursuits of biostratigraphy and paleoenvironmental reconstruction. In figure 10.1, this region occupies a patch stretching from approximately 1 to 10 my, and it falls anywhere in space from local to near-provincial.

Case Studies

Although the literature surrounding the subject of biofacies is far too extensive to be summarized in this context, it is fair to point out that the concept of identifiable packages of fossil organisms in the rock record characteristic of particular paleoenvironments has received the most attention from those researchers using them for stratigraphy. Many of these practical accounts remain difficult to access, even for those with a fair grasp of geology and sedimentology, because they are often hidden within larger studies that did not have characterizing ecosystem behavior as one of their main goals. A second cache of biofacies accounts was amassed during the paleontological fascination with synecology beginning in the late 1960s (e.g., McKerrow 1978).

That biofacies may have an important part to play in our understanding of ecology and evolution was early noted by Boucot (1975, 1978, 1983), who has published extensively on the identification and hierarchical nature of biofacies (which he terms *communities*) in the rock record. His 1978 and 1990 papers contain extensive lists of examples of the relative fixity of these units throughout the Phanerozoic. Partly as a result of the particularities of the field of paleontology when he began putting these ideas forth (coinciding with the early blossoming of paleobiology and its emphasis on random processes), Boucot's work in region 3 is often overlooked today, except by students of biostratigraphy (but see Boucot, 1996).

More recently, Brett and Baird (1992, 1995) have sparked new interest in what biofacies may actually be telling us by presenting findings from nearly two decades of field work in the U.S. Appalachian Basin. Their work is fairly typical of the larger opus that encompasses this scale of inquiry in that it is marine and deals with Paleozoic strata. Although *biofacies* is also a term used in modern studies (e.g., Ivany et al. 1994a), we restrict it here to a characteristic faunal assemblage with temporal persistence as recorded in the rock record and hence beyond the scope of regions 1 and 2. Although the claim has been made that biofacies, at least in terms of recurrent assemblages of organisms, are identifiable in younger strata [Pleistocene (Valentine and Jablonski 1993; Valentine and Mallory 1965)], the necessarily shorter duration of these phenomena (coupled with typically higher resolution) leaves open the possibility that they record ecosystem dynamics distinct from those observable in older strata (see discussion in Schopf 1996).

Ecosystem Behavior

It is readily apparent that no such phenomenon as a recurrent biofacies would be recognizable in the fossil record if these characterizable groups of organisms did not persist for geologically significant periods of time (see Bambach and Bennington 1996). The mere fact of their widespread recognition and utility, therefore, is an indication of their apparent stability in time. No amount of gestalt, however, could have engendered the volume of discussion and interest that inquiry at this scale now enjoys [recent reviews by DiMichele 1994; Miller 1993; Brett 1995; Morris 1995a; Roy and Wagner 1995; as well as a GSA Theme Session in 1994 and a theme issue of *Palaeogeography, Palaeoclimatology, Palaeoecology* (1996, volume 127) devoted to this topic; and, more recently, Ivany 1997; Tang and Bottjer 1996, 1997; Schopf and Ivany 1997; Miller 1997; Boucot 1997]. This renewed interest was largely a result of the Brett and Baird (1992, 1995) study, which differed from past accounts in two ways: (1) The authors document an unexpected amount of ecological and evolutionary stability through time rather than describing synecological relationships for their own sake. (2) They were able to support their conclusions of stability with extensive species-level data. Brett and Baird identified 14 stable units [ecological-evolutionary subunits (Brett and Baird 1992, 1995; see Boucot 1996)] in the Silurian-Devonian, each lasting 3 to 7 my. Persistence of species throughout the stratigraphic extent of these geographically widespread units is high, on the order of 65 to 80 percent. Major turnover events occurring between stable units are relatively much briefer (on the scale of 100,000 years), with restructuring and extinction resulting in only 10 to 20 percent carry-over between stable units. Ecologically, whole gradients of biofacies track preferred environments over millions of years of sea-level fluctuation, preserving their ecological structure and relative abundances while resisting invasion and showing surprisingly few cases of in situ evolution. It is the ecological, as well as the evolutionary, stability that prompted Brett and Baird to describe this pattern as "coordinated stasis."

The Brett and Baird case does not seem to be unusual (see Brett et al. 1996), although it is exceptional in terms of the scope of the study, the reliability of the stratigraphic framework, and the taxonomic resolution of the analysis. Similar patterns of bipolarity (long-lasting stability and briefer intervals of flux) have long been recognized for Cambrian trilobite faunas (biomeres; Palmer 1984; Westrop 1994, 1996) and some vertebrate faunas (Olson's chronofaunas 1952; Vrba's Turnover-Pulse 1980; and by Barry et al. 1995a). DiMichele (1993; with Phillips 1992) offered a startling, similar example of coordinated stasis in terrestrial coal swamps concurrent with Brett and Baird's report. Fossil reefs over the last 3 my also seem to be characterized by stable species composition (Jackson 1992). Aside from the long lists compiled by Boucot (some 50 reports in 1978, 1990a), additional examples await the retooling of data sets originally used for purely stratigraphic purposes, and new studies undertaken in different settings. There are also indications that such stability may sometimes be manifested at the generic level, rather than at the species level (Westrop 1996; Patzkowsky 1994). Whether the Brett and Baird case is generalizable remains in question, but the existence of biofacies and the stability they imply does not (Ivany et al. 1994b; Ivany and Schopf 1996; see Morris et al. 1995).

Region 4: Across Biofacies Boundaries

This is the point at which the paleontological trade-off between temporal coverage and resolution begins to become even more evident. Factors such as geographic and stratigraphic accessibility are similar to that of region 3, but they are multiplied (as are research hours) if the same distribution of samples is sought. The limits of feasibility are very near in this region, as the shift from a temporal window of 5 to 50 my may represent the difference between a Ph.D. thesis and a lifetime of faunal documentation.

Case Studies

Recently, attention has been drawn (Jackson 1994a) to a study of benthic foraminifera conducted by Buzas and Culver (1994) that typifies region 4. These authors document the pattern of faunal turnover over 55 my in the mid-Atlantic Coastal Plain. Six transgressive units were sampled, varying in duration from 1.5 to 8 my in length. Significant intervals of time lay unsampled between several of these units, the longest being approximately 24 my. Lists of taxa (at the species level) were prepared for each formation as a whole, and these were used for interformational comparisons and tracing taxon movement between the onlap coastal embayments and the open sea. Much of the data that produce the pattern revealed in region 3 is therefore forfeited to a broader temporal perspective.

Ecosystem Behavior

The pattern Buzas and Culver document is one of little displayed unity and periodic interchange with the larger open ocean species pool. During the transgressive portions of each cycle, the entrants to the newly created shelf setting are composed of newly evolved forms and members of the neritic species pool (including some shelf emigrants that escaped the last regressive cycle by moving offshore). Little recurrence of anything called a community (or an ecosystem with characteristic taxonomic membership) is observed. The authors themselves conclude that their results relegate the groups of species occupying each formation to ephemeral associations, and (temporal differences noted) they point out that this result is in accord with those made of terrestrial Quaternary faunas (our region 2).

Region 5: The Ecological-Evolutionary Units of Boucot

At this level of the natural continuum, a global perspective is taken for the entire geologic record of the Phanerozoic. Time has been parsed into units on the order of tens to a hundred million years. It at once incorporates all the available data, in a sense, but must forsake the finer scales of taxonomy and spatiotemporal resolution to get at broad pattern. Accordingly, data are no longer at the species level but are stepped up in taxonomic rank to genera and families.

Case Studies

This region is largely defined by the work of Boucot (1975, 1978, 1983, 1990a), with subsequent modification by Sheehan (1991, 1996). Building on the conventional wisdom of the biostratigrapher, dating back to the Etages of d'Orbigny (1849–1852; 1850–1852), Boucot identified 12 stratigraphic units of ecological and evolutionary stability characterized by "an apparently fixed number of individual community groups" (Boucot 1983, p. 5). As such, these higher-order units are a hierarchical extension of his inference of community stability from region 3 (the dashed lines in figure 10.1 are an attempt to show this theoretical continuity). Because of this claim to bottom-up architecture, Boucot's broad scale EEUs are theoretically collapsible to the level of his Benthic Assemblage Zones (collections of biofacies) and to their constituent biofacies and communities; i.e., the ecological-evolutionary units (EEUs) are inherently ecological in the sense that they are ultimately based on organisms living within the same environmental settings and the same communities. In practice, EEUs nicely characterize the paleontologist's impression that "a Late Ordovician community looks different from one in the Cambrian" while identifying periods of earth history during which relative ecological stability reigned, in a very broad sense.

It is this same sense of confirming intuition that has caused the EEUs, and indeed the framework on which they are based, to be criticized as being less than rigorously documented. Interestingly enough, Alroy (chapter 12), working at a similar temporal scale and at a somewhat narrower spatial one, reports stability in generic richness that he implies may be "homeostatic" over 65 my of North American mammal history. Although the EEU that has been identified over this interval of time (Boucot's EEU XII encompasses the entire Cenozoic) has not been singled out for criticism, its relatively long duration makes it an interesting test case. Alroy's data set is not hierarchically collapsible to the community level in the same sense as Boucot's, because it is based on compiled stratigraphic ranges rather than documented co-occurrence of taxa. Alroy's analysis is thus more akin to the diversity-stability studies of region 1, because it examines how the appearance and disappearance of taxa affect diversity through time. It is interesting to note, however, that stable diversities are another aspect of each of Boucot's EEUs (1983, 1990a), and this expectation seems to be corroborated by Alroy's very different mode of analysis.

Ecosystem Behavior

By definition, Boucot's EEUs identify periods of quiescence in the broad history of life. They are also another instance of the bipolarity of stability and change in the fossil record. Like biofacies and the histories of their constituent taxa, Boucot's EEUs point to enormous expanses of time being characterized by the stability of families across the globe *in their ecological associations.* This is not to say that no change occurs within EEUs (or biofacies, for that matter). But such change is predicated on the relative abundance of the organisms (the more rare taxa turn over more quickly) and is restricted to phyletic changes within the already-established genera (see Boucot 1983; and figure 10.1). This is starkly contrasted with the bursts of quantum evolution (as well as ecological re-sorting and emigration/immigration) that Boucot invokes to account for the relatively abrupt boundaries of his stable units.

Evolutionary and ecological change, when it is recorded, occurs across entire stable units; from the reestablishment of entire biofacies (as discussed earlier) to the adaptive radiation and establishment of a new EEU. Change in this view is as hierarchical as stability, and it is the truly rare event that will provide the end of a stable EEU (and its constituent community groups). For this reason, Boucot has implicated long-term environmental stability and accompanying stabilizing selection to explain the nested stability manifested at the level of the biofacies and observed within the EEUs. Although this has met with positive responses from workers in regions 1 and 2 (see Valentine and Jablonski 1993), other interpretations are possible (Morris et al. 1995) and the stability documented by Alroy (chapter 12) in the environmentally turbulent Cenozoic certainly seems to contrast with this view [as does that documented by Prothero and Heaton (1996) across the early Oligocene climatic "crash"].

Necessarily, Boucot's EEUs are often bounded by mass extinctions, after which a period of recovery involving adaptive radiation and the development of a new stable ecological landscape is seen. Sheehan (1991) added to Boucot's interpretation of the biostratigrapher's common wisdom by pointing out that several of Boucot's shortest EEUs (his V, VI, and IX) followed mass extinctions and were interpretable as "recovery EEUs." Sheehan added two more of these postextinction units of reorganization to make a total of 14 recognizable units.

Boucot's vision is of particular interest in the context of this paper because it is an attempt to incorporate many scales of inquiry hierarchically. Like Valentine and Jablonski (1993), Boucot has explicitly compared ecosystem behavior at one level of the continuum with that at another. Although the nature of this comparison may be much more dependent on the "wisdom of the farmer" (Boucot 1981) in Boucot's case, the results are very similar— and surprisingly different. He sees consistency within the regions of the matrix with which he is concerned (the extent of regions 5 and 3), as do the other authors (regions 1 and 2). But instead of flux, he sees ecological (and to a degree evolutionary) stability as the norm, and change as a periodic but brief interruption.

Region 6: The Compendium and the Continuum

The compendium of fossil marine family ranges compiled by Sepkoski (1982 and continually updated, 1993) has a strictly global outlook on the entire Phanerozoic. It records the stratigraphic ranges of all fossil families as retrieved from literature searches. It is therefore a uniquely approachable data set from which to distill large-scale patterns in the history of life, for it has taken the raw data out of its sometimes pesky biological environs. Temporally, although the data itself is usually resolved to the level of stratigraphic stages [roughly 4 to 6 my; its "grain" sensu Allen and Hoekstra (1991)], the most significant patterns that have arisen from these data have been manifest on the order of 10s to 100s of millions of years.

Case Studies

The number of studies utilizing these data has exploded since they first became widely available (in hard copy as well as magnetic media courtesy of the author). The number of purported patterns has also exploded, perhaps outstretching the understanding of meaning *behind* those patterns (e.g., see cautionary notes by Koch and Morgan 1988; Boucot 1990a). Nevertheless these data have opened new areas of the continuum to fruitful paleontological scrutiny.

The seminal study, and that which is of primary concern in the present context, is Sepkoski's (1981) factor analysis of the numbers of families within metazoan classes and orders, from which he identified three "evolutionary faunas." Each of these three faunas (the Cambrian fauna dominated by trilobites, the Paleozoic fauna dominated by brachiopods, and a Modern fauna typified by mollusks) characterizes a particular portion of the Phanerozoic fossil record. Additionally, each fauna was shown to encompass a period of rapid diversification followed by an equilibrium (save the Modern fauna, in which the part of the equilibrium is played by an ongoing diversity maximum). The transitions between faunas appear to be periods of overlap in which an outgoing evolutionary fauna dissipates exponentially as the incoming fauna outstrips the equilibrium diversity of its predecessor.

Ecosystem Behavior

Whether one should call the three great evolutionary faunas of Sepkoski stable in the same sense as Boucot's EEUs seems debatable. Even the contention of a stable diversity with a changing cast of characters (sensu region 1, although in this case stability derives from a balance between extinction and origination, as opposed to emigration and immigration) seems difficult to interpret given the nature of these data. Although recognizable entities (and confirming the basic gestalt of what a Paleozoic fossil assemblage looks like, even better than Boucot's EEUs), they are certainly not ecologically coherent groups in the same sense as presented in other regions of figure 10.1.

Given that elements of all three faunas overlapped in time and space (and still do), it is apparent that ecological insights are difficult to recover from these data. These statistical groupings are aptly named *evolutionary* faunas, having been described based purely on the conjunctions of temporal ranges, rather than ecological co-occurrence in space.

The dynamics within these three faunas has at least in part been teased apart by adding an environmental parameter to the mix. This has taken the shape of a series of onshore–offshore studies designed to trace various members of the three faunas over time in relation to a depth gradient (Jablonski et al. 1983; Sepkoski and Miller 1985; Bottjer and Jablonski 1988). Rather than observing stability behind the faunas and the apparent distinctness that facilitated their identification, these studies have universally uncovered complex and independent movements of the clades of which they are composed. It remains unclear, however, what level of ecosystem behavior we are being informed about by such conclusions (i.e., where exactly do these studies at the ordinal level, over a quasi-global yet environmentally parsed spatial scale, belong in figure 10.1).

For those researchers primarily interested in ecosystems and organismal patterns within more restricted settings (than the globe itself), the fact that these data are largely ungrounded in ecological and environmental parameters may seem frustrating. Although there is not a direct way to glean the underlying pattern from these conclusions, they do imply a similar parsing of the fossil record into characterizable pieces much like EEUs (see Sheehan 1996). This similarity must, at least in part, be ecological (see discussion in Jablonski and Sepkoski 1966). Yet it is very difficult to propose that two families of organisms were not important contributors to the same stable community complexes based solely on discordant compendium turnover rates (see Valentine and Jablonski 1993 for an attempt to suggest local ecosystem behavior from these data). It is an understatement to stress that patterns at lower taxonomic levels and actual patterns of co-occurrence are much more important indicators.

Short of combining the numerical rigor of Sepkoski's approach and the emphasis on bottom-up documentation of Boucot, there seems little possibility of reaching a consensus as to the nature of ecosystem behavior in the lowest right corner of our figure 10.1. The two approaches are as rooted in different scales of observation as they are in philosophical outlook. It is interesting to note, however, that both schools of thought, while encompassing the most expansive views of the natural world possible, have looked elsewhere (to lower levels, be they taxonomic, spatial, or temporal) for explanations of their observed patterns. This approach is well illustrated by the recent work of Patterson and Fowler (1996): their analysis of the timing of originations and extinctions of planktonic foraminifera *species* on a *global* scale (Cretaceous to Recent) revealed a deterministic pattern of extinctions, leading them to propose causal explanations derived from hierarchically lower levels (specifically, the degree of interdependency within ecosystems). How far should this search go, and how do we know when to stop?

Comments and Conclusions

The "Slide" Rule of Ecosystem Behavior

The apparent clash in ecosystem behavior exhibited between regions 1, 2, and 4 characterized by flux, and those of 3 and 5 seen as stable is not a new observation. As a result, authors have often chosen sides, usually deciding that the data set (and thus the region of the matrix) with which they were in conflict was anomalous [e.g., Boucot's (1990a) "Pleistocene Paradox" and Valentine and Jablonski's (1993) emphasis that lesser-known cases of pre-Pleistocene ecosystem variability undercut the usual impressions of stability in region 3]. Viewing the record of ecosystem dynamics as a window on behavior that may slide along axes of time and space emphasizes the possibility that both observations/interpretations may be correct.

The apparent discord that can be resolved with a sliding concept of ecosystems is nowhere more evident than in comparing region 3 to region 4. Taken as an independent proxy for how ecosystems work, the patterns documented by Buzas and Culver (1994) in region 4 point to instability in ecosystems, and communities as happenstance associations. However, each of Buzas and Culver's transgressive units span 2 to 8 my, roughly the same duration as each of Brett and Baird's 14 stable EE subunits; only a single data point was recorded for each transgressive interval. Were Brett and Baird to sample the 65 to 70 my of the Siluro-Devonian encompassed by their study in a comparable way, sliding from their frame of reference to that of region 4, the distinct differences between each of their EE subunits would guarantee a pattern of flux. Yet, extensive sampling *within* the units reveals a pattern of coordinated stasis. Likewise, characterizing the behavior of ecosystems within each of Buzas and Culver's transgressive intervals would provide a truly comparable data set with which to test the assertion that coordinated stasis is general to region 3.

The fact that ecosystems appear stable (stable for a significant amount of geologic time) at some scales of observation and as ephemera at others should sound a warning for ecologists and paleoecologists alike (see also Pandolfi 1996). If this phenomenon reflects the fact that either (1) different processes *act* over different spatial and temporal scales, or (2) the results of different processes are only *observable* within particular portions of figure 10.1, it gives new importance to a comparative approach to ecology. Akin to backing off from the outcrop in our introduction, it might justifiably be said that to understand "ecology" as it has acted over the many years and acres of earth history, one must be able to explain the dynamics of "ecosystems" as seen in regions 1 to 6. Even if one accepts the upper-most levels of the continuum (region 6) as being indicative of fluctuating ecosystems, there seems little hope of identifying a single characteristic behavior that typifies the entire continuum. The stability observed in region 3, that of the biofacies, seems to present a truly different insight into ecosystem dynamics, even after we realize that scale of observation can color our comparisons (i.e. that ecosystems can slide).

This has deep implications for the role of ecology in the study of evolution (Brett 1995; Brett et al. 1996; Miller 1993, 1996; Morris 1995a, 1996; Morris et al. 1995) and presents us with the possibility that regions 1 and 2 in figure 10.1 may be decoupled from region 3 and subsequent levels of the continuum. A similar heterogeneity of pattern in the history of life led Jablonski (1986a) to conclude that there existed a barrier to extrapolation of pattern (and therefore underlying cause) between micro- and macroevolution. Is there a similar need to call for a "macroecology" (Plotnick et al. 1994), or is the proposed consistency between the lowest levels (regions 1 and 2) and the highest (region 6; as in Valentine and Jablonski 1993) evidence that ecosystems are messy, but it all works out in the end?

The means to addressing such issues lies in comparing the ecosystems that are displaying the disparate behaviors (in what ways are regions 1 and 2 really different from 3?) and attempting to fill in empty portions of figure 10.1 where possible. Are there instances of brief, high-resolution windows on the past outside of the Quaternary? What would the ecosystem dynamics from a 1-my interval in the Paleozoic (under fairly stable environmental conditions) look like under high resolution? The concepts of metapopulations (Hanski and Gilpin 1991) and the species pool as used in modern ecological studies may provide some resolution to the observed differences in behavior across scales (particularly between regions 1 and 3). The species pool in this sense is the set of taxa present in a broad area that are potentially available to colonize a particular location. The fact that species populations are patchy and discontinuous (the metapopulation) means that only a few of these species will be present at a given time, but averaged over longer time scales most taxa will make an appearance in a given area. The time-averaged nature of the fossil record suggests to us that perhaps what we as paleontologists are seeing preserved in the rock record (as biofacies, region 3) is more akin to the species pool of the modern ecologist, the group of taxa that are never all seen together on the time scale of a single field study but whose local populations have equal chances of appearing at a given spot at some time in the future.

Metapopulation studies tell us that two species with concurrent geographic ranges may in fact not be able to coexist at any one local site, yet they persist together regionally because of continually shifting populations. Observed on a short time scale, the interplay between two such taxa may appear unstable, yet when averaged over time their ranges merge into one (e.g., see Brown 1995). If anything, the time-averaged nature of the fossil record may act to smooth over the high-frequency fluctuations that influence our impressions on shorter time scales and afford a view of ecosystem behavior from a different, longer-term perspective. The fact that we often see stability from this different vantage point may suggest that species pools (as opposed to their local populations) do in fact have some degree of coherence in time and can track preferred environments as a unit. This also implies that the structure of metapopulations may not be directly reflected in the fossil record except under extraordinary circumstances (but see Harrison, chapter 2).

Ultimately, there are limits imposed by the nature of the data within any region of figure 10.1. Paleozoic and Quaternary students alike cannot go beyond the spatial and temporal extent of a bed, and workers in the Recent cannot transcend the frame of reference tied to their own observations (although information at other scales, such as that of the biofacies, may relieve the frustration noted in studies such as Stenseth and Maynard Smith (1984) at the lack of adequately long time series to test their model-derived hypotheses). The trick is not to see these boundaries and limitations as somehow indicating that an end, the truth, has been reached. Rather, they should be seen as borders to foreign countries: the language is different, you may have to go through customs, and there is a time change, but you almost always learn something on your way to getting a new perspective of the world.

In the Long Run

Still, there is the question, does all this really matter? In our attempt to understand the behavior of ecosystems at any scale of the natural continuum, it seems as if each region is fairly autonomous. Overpeck et al. (1991) were correct to use the conventional wisdom gleaned from the last 18,000 years to test their hypotheses about future ecosystem dynamics of terrestrial plants. Information about ecosystems at another scale (the Recent or Boucot's EEUs) would have done them little good. Ensuring that the scale of the question accords with that of the data being used to answer it concurs with Allen and Hoekstra's (1991) suggestion of converting observations to a common scale before comparison. Yet, even those of us who work at a single scale should realize that in a broader sense, the outlooks we have on our fields are colored by tangential glimpses of work done on other scales. The origin of species through natural selection is a process that is not often observed on the scale investigated by most biological scientists. Yet, Darwin's theory about the behavior of the natural world unites otherwise distinct and inexplicable aspects of biology (Dobzhansky 1973) and permeates our philosophical outlook profoundly (Dewey 1909). Ecosystems are out there at all scales, even if we don't notice them.

Summary

1. Ecosystems characterized by change are observable at a number of temporal and spatial scales, ranging from the Recent to hundreds of millions of years, local to global.

2. Stable ecosystems are now well documented at the species level on the scale of the biofacies (1 to 10 my), intermediate to the scales at which ecosystem flux has been reported.

3. Long-standing differences noted between the documented behavior of ecosystems studied may be reconciled as consequences of our scale of observation.

4. Novel ecosystem dynamics at the level of the biofacies appear to reflect a boundary to extrapolation along the continuum of ecosystem behaviors.

5. "Paleoecology, once regarded as a quaint but irrelevant specialty, is on the verge of joining mainstream ecology" (Davis 1994, p. 357), *not* because of any refinement of sedimentologic procedures alone, but because it offers unique insight into the nature of ecosystems through time.

Acknowledgments

We thank Mike McKinney for the invitation to contribute to this exciting volume, and for his patience. Two anonymous reviewers made insightful comments on an initial draft that improved the paper—particularly its approach to the ecological literature—as did correspondence with Gareth Russell. We both thank Dick Bambach for stretching our minds and allowing some of the ideas presented here to ferment in his course entitled Communities through Time. Ivany acknowledges the support of the American Association of University Women in the form of a doctoral dissertation fellowship. Schopf acknowledges the support of Rufina.

11

Nested Patterns of Species Distribution: Processes and Implications

Alan H. Cutler

Species are not distributed randomly across the face of the Earth. It is a fundamental observation in ecology that species occur in particular habitats, and that the occurrence of a species often correlates, positively or negatively, with the occurrence of other species. Describing and interpreting patterns of species occurrences—over different spatial scales and through time—are central concerns of ecology and paleoecology, as well as of their new cousin, macroecology (Brown 1995).

One simple and extremely common pattern of species distributions has been termed nested subsets by Patterson and Atmar (1986). Given a set of replicate habitats with biotas varying in species richness, less species-rich biotas tend to be subsets of more species-rich biotas. In the case of perfectly nested subsets, if a species is present in a given biota, it is also present in all larger (i.e., more species-rich) biotas. If a species is absent from a biota, it is also absent from all smaller (i.e., less species-rich) biotas. An example of a perfectly nested series of fish faunas in desert springs is shown in table 11.1. Species are ranked by number of occurrences among springs; springs are ranked by species richness. The wedge-shaped pattern of occurrences in the species presence–absence matrix is characteristic of nested biotas.

Nested patterns have been documented for a variety of vertebrate, invertebrate, and plant taxa [Patterson 1990; Wright and Reeves 1992; Mikkelson 1993; Wright et al. (in press)], and on spatial scales ranging from vacant lots

TABLE II.I   Native fish species in desert springs in Ash Meadows, Nevada

Species	Spring														
	1	*2*	*3*	*4*	*5*	*6*	*7*	*8*	*9*	*10*	*11*	*12*	*13*	*14*	*15*
Cyprinodon nevadensis	x	x	x	x	x	x	x	x	x	x	x	x	x	x	x
Rhinichthys osculus		x	x	x	x	x	x	x	x	x					
Empetricthys merriami	x	x	x	x	x										

Key to springs: 1, Big; 2, Jackrabbit; 3, Point of Rocks; 4, Forest; 5, Longstreet; 6, Crystal Pool; 7, Fairbanks; 8, Soda; 9, Tubbs; 10, Devil's Hole; 11, School; 12, North Indian; 13, South Indian; 14, Scruggs; 15, Mexican.

Data from Miller, R. R. 1948. The cyprinodont fishes of the Death Valley system of eastern California and southwestern Nevada. *University of Michigan Museum of Zoology Miscellaneous Publication* 42:1–80; cited in Brown, J. H. and A. C. Gibson. 1983. *Biogeography*. St. Louis: Mosby.

in urban Chicago to mountain ranges in the Great Basin of western North America and oceanic islands surrounding New Zealand (Patterson 1990). The development of a nested pattern among a set of sites requires that (1) the sites vary in species richness, (2) all sites share the same potential species pool, and (3) the sites comprise comparable habitats (Patterson and Brown 1991; Cutler 1994). In terms of Brown's (1984, 1995) niche-breadth model of species distribution and abundance, a nested pattern implies that species vary in the breath of their niches, and that there exists a kind of hierarchy of niches. How nested patterns may be generated and what they imply will be discussed later.

Measuring Nestedness

Perfectly nested patterns such as that in table 11.1 are unusual. More commonly, the pattern is like that in table 11.2A and 11.2B; the wedge-shaped pattern is there, but there is a certain amount of noise. Some otherwise widespread species are absent from some of the larger biotas (e.g., in table 11.2A, species C is absent from the biota at site 2) creating "holes" in the pattern. Some depauperate biotas harbor species that are absent from larger biotas (e.g., in table 11.2B, species G in the biota of site 7) creating outliers on the pattern.

Several metrics have been devised to measure the fit of a given presence–absence matrix to the nested subset model, most notably N (Patterson and Atmar 1986), U (Cutler 1991), C (Wright and Reeves 1992), and T (Atmar and Patterson 1993) (for a discussion of other metrics, see Cutler 1994). N,

TABLE 11.2 Comparison of nestedness matrices

A.

Species	1	2	3	4	5	6	7	8	9	10	Total occupancy	U_a	U_p
A	X	X	X	X	X	X		X	X	X	9	1	0
B	X	X	X	X	X	X	X	X			8	0	0
C	X		X	X	X		X	X			6	1.5	0.5
D	X	X	X		X	X	X				6	1	0
E	X		X	X		X					4	1	0
F	X	X	X	X							4	0	0
G	X	X									2	0	0
H	X	X									2	0	0
I		X									1	0.5	0.5
J	X										1	0	0
Richness		9	7	6	5	4	4	4	2	1	1	5	1

$$6 = U_t$$

B.

Species	1	2	3	4	5	6	7	8	9	10	Total occupancy	U_a	U_p
A	X	X	X	X	X	X	X	X	X		9	0	0
B	X	X	X	X	X	X	X			X	8	0	1
C	X	X	X	X	X	X					6	0	0
D	X	X	X	X	X				X		6	0	1
E	X	X		X		X					4	1	0
F	X	X	X					X			4	0	1
G	X						X				2	0	1
H	X		X								2	0.5	0.5
I		X									1	0.5	0.5
J	X										1	0	0
Richness		9	7	6	5	4	4	3	2	2	1	2	5

$$7 = U_t$$

T, and U assess nestedness by counting holes and outliers, although they use different algorithms for this. T weights holes and outliers by the size of the biotas and the incidences of the species involved. Statistical significance of nested patterns is determined by comparing the index of the pattern with random matrices generated through Monte Carlo simulation (Patterson and Atmar's programs RANDOM0 and RANDOM1; see later for more description). C is based on conditional probability theory: if a species occurs in a given biota, what is the probability that it occurs in richer biotas? Unlike N and U, the value of C is not dependent on the size of the matrix. Also, the statistical significance of a given value of C can be determined analytically, without the need for computer simulations. Each of these metrics has merits, and the results are highly correlated [Wright and Reeves 1992; Cutler 1994; Wright et al. (in press)]. In the following discussion and analyses, I will use the U metric, which I will now describe in more detail.

The U metric counts unexpected absences (holes) and unexpected presences (outliers) in a given presence–absence matrix (see table 11.2A and 11.2B). These are tallied under the subindices U_a and U_p, respectively. U_a and U_p are summed to provide U_t, the total number of deviations from perfect nestedness. The algorithm for identifying holes and outliers is as follows: for each species, all absences from sites more species rich than a focal site are counted as holes. All presences in sites less rich than the focal site are counted as outliers. The focal site is chosen to minimize the sum of the holes and outliers for that species. Where more than one combination of holes and outliers yields the same total, fractional values are assigned to U_a and Up. For example, in table 11.2A, species I could be interpreted as being anomalously absent from site 1 (a hole) or anomalously present in site 2 (an outlier). The ambiguity is resolved by splitting the difference and assigning the value of 0.5 to U_a and U_p. U_t is the sum of Ua and U_p for all species; it can be thought of as the minimum number of steps (filling in holes and erasing outliers) required to transform a given pattern into a perfect nested series.

Naturally, the raw U indices of matrices of different sizes are not directly comparable—a larger matrix may have a higher U than a smaller matrix simply because it has more sites and species. Also, some degree of nestedness is to be expected by chance—a species-rich biota is more likely to include a given species than is a depauperate biota, and an abundant species is more likely to occur in a depauperate biota than is a scarce species. To compare matrices of different sizes and to assess the statistical significance of nested patterns, a population of random matrices is generated to serve as a null distribution of U values. Two programs for this purpose were written by Patterson and Atmar (1986) and modified by Cutler (1991) for use with U. In both programs, the number of sites, the size of the species pool, and the

TABLE 11.3 Boreal mammals on Great Basin mountain ranges

Species	Range																			U_a	U_p
	1	2	3	4	5	6	7	8	9	10	11	12	13	14	15	16	17	18	19		
Neotoma cinerea	X	X	X	X	X	X	X	X	X	X	X	X	X	X	X	X	X	X	X	1	0
Tamias umbrinus	X	X	X	X	X	X	X	X	X	X	X	X	X	X	X	X	X	X		0	0
Tamias dorsalis	X		X	X	X	X	X	X	X	X	X	X	X	X	X	X	X	X	X	2	0
Spermophilus lateralis	X	X	X	X	X	X	X	X	X	X	X	X	X	X	X	X				2	0
Sylvilagus nuttallii	X	X	X	X	X	X	X	X	X		X	X	X						X	2.5	0.5
Marmota flaviventris	X	X	X	X	X	X	X	X	X	X	X	X	X	X						0	0
Microtus longicaudus	X	X	X	X	X	X	X	X	X	X	X	X			X					0	1
Sorex vagrans	X	X	X	X	X	X	X	X	X					X						0	1
Sorex palustris	X	X	X	X	X	X	X													1	0
Mustela erminea	X		X	X	X	X														2	0
Ochotona princeps	X	X	X	X						X										0	1
Zapus princeps	X	X									X									0.5	1.5
Spermophilus beldingi	X	X																		0	0
Lepus townsendii		X								X										0.5	1.5
Richness	13	12	11	11	10	10	9	8	8	8	8	7	6	6	5	4	3	3	3	12.5	6.5
																				$U_t = 19$	

$P < .01$, RANDOM1 assumptions.

Key to mountain ranges: 1, Toiyabe-Shoshone; 2, Ruby; 3, Toquima-Monitor; 4, White-Inyo; 5, Snake; 6, Oquirrh; 7, Deep Creek; 8, Schell Creek-Egan; 9, Stansbury; 10, Desatoya; 11, Roberts; 12, White Pine; 13, Diamond; 14, Spring; 15, Grant-Quinn Canyon; 16, Spruce-South Pequop; 17, Pilot; 18, Sheep; 19, Panamint.

Data from Grayson, D. K. and S. D. Livingston. 1993. Missing mammals on Great Basin Mountains: Holocene extinctions and inadequate knowledge. *Conservation Biology* 7:527–532.

distribution of species richnesses match those of the real matrix. In RAN-DOM0, all species have an equal probability of occurring at a given site. In RANDOM1, species occurrence probabilities are weighted according to the actual number of occurrences in the real matrix.

In addition to assessing the overall nestedness of a matrix, U can be used to determine whether a pattern is biased toward holes or outliers. The matrices in table 11.2 are about equally well nested (for 11.2A, $U_t = 6$; for 11.2B, $U_t = 7$), but 11.2A is hole rich ($U_a = 5$, $U_p = 1$) and 11.2B is outlier rich ($U_a = 2$, $U_p = 5$). Hole-rich and outlier-rich patterns occur in real matrices as well. Table 11.3 shows the hole-rich pattern of boreal mammals on mountain ranges in the Great Basin. Table 11.4 shows the outlier-rich pattern of resident birds on mountain ranges in the Great Basin.

The Dynamics of Nested Patterns

What produces nested subsets? Darlington (1957) observed nested species distributions along dispersal routes and dubbed it the immigrant pattern. Patterson and Atmar (1986) attributed the strongly nested patterns of montane mammals in the American southwest to selective extinctions accompanying post-Pleistocene faunal "relaxation." Cutler (1994) recognized three general kinds of mechanisms that can produce nestedness: (1) passive sampling, (2) nested habitat distributions, and (3) colonization/extinction dynamics. These mechanisms are not mutually exclusive, and all may play a role in generating a given nested pattern.

Passive Sampling

Nested patterns can result simply from differences in abundance among species in the source pool, especially if the differences are large. Indeed, in many biotas the species-abundance curve is lognormal, and differences in abundance can span several orders of magnitude (Preston 1962a). Cutler (1994) tested the efficacy of passive sampling in producing nested patterns through a computer simulation. A hypothetical source biota of 100 species was generated, using the "canonical" lognormal species-abundance curve of Preston (1962a). Individuals were then sampled at random (with replacement) until an archipelago of ten islands was filled. The islands ranged in size from 2 to 1024 individuals. One hundred simulated archipelagos were produced and all were significantly nested at the 1 percent level (RANDOM1 assumptions). Ninety-eight percent of the simulated archipelagos were outlier rich.

In principle, then, passive sampling can be a potent mechanism for generating nested patterns. The salient features of the passive sampling model are

TABLE 11.4   Boreal birds on Great Basin mountain ranges

Species	Range 1	2	3	4	5	6	7	8	9	10	11	12	13	U_a	U_p
Dendrocopus villosus	X	X	X	X	X	X	X	X	X	X	X	X	X	0	0
Parus gambeli	X	X	X	X	X	X	X	X	X	X	X	X	X	0	0
Sitta carolinensis	X	X	X	X	X	X			X	X	X	X	X	2	0
Dendragapus obscurus	X	X	X	X	X		X	X			X		X	0.5	1.5
Cyanocitta stelleri	X	X	X				X	X	X		X	X		0.5	1.5
Cinclus mexicanus	X	X	X	X	X	X	X	X						0	0
Dendrocopus pubescens	X		X		X		X	X						1	0
Sitta pygmaea	X	X				X				X				0.5	1.5
Oreortyx picta		X		X		X								1	2
Glaucidium gnoma									X	X				0	2
Picoides tridactylus	X													0	0
Richness	9	8	7	6	6	6	6	6	5	5	5	4	4	5.5	8.5

$$U_t = 14$$

$P < .01$, RANDOM1 assumptions.

Key to mountain ranges: 1, Snake; 2, White-Inyo; 3, Deep Creek; 4, Toiyabe-Shoshone; 5, Ruby; 6, Spring; 7, Stansbury; 8, Oquirrh; 9, Panamint; 10, Sheep; 11, Grant-Quinn Canyon; 12, Desatoya; 13, Spruce-South Pequop.

Data from Brown, J. H. 1978. The theory of island biogeography and the distribution of boreal birds and mammals. *Great Basin Naturalist Memoirs* 2:209–227.

(1) all individuals of all species have the same chance of being included in a biota—and the differences that emerge among species are the result of sheer numbers, and (2) all sites are equally accessible and homogeneous, although they vary in capacity. Few real situations will completely satisfy these conditions, but, for highly vagile species, repeated colonization and recolonization of sites may create conditions that approximate the passive sampling model. Cutler (1994) suggested that because the resident bird biotas of montane forests in the Great Basin show no isolation effects or evidence of faunal

relaxation (Brown 1978), their highly nested pattern may be the result of passive sampling. The bird pattern, like the simulated archipelagos produced by passive sampling, is outlier rich ($U_a = 4.5$, $U_p = 9.5$). Andrén (1994) has similarly interpreted significantly nested bird communities in Finnish old-growth forest fragments as being the result of random sampling.

Nested Habitat Distribution

Nested species distributions may simply reflect the distribution of habitats among sites and in that sense be an epiphenomenon. Cody (1983) observed that bird diversity on islands in the Gulf of California increased in a stepwise fashion as island area increased. He suggested that this was controlled by the geomorphology of drainage basins on different-sized islands: larger islands could support higher-order (larger) stream channels and therefore a greater range of riparian habitats than smaller islands. Bird distributions are highly nested (Patterson 1990), and this could be a reflection of this hierarchy of habitats. Other geologic or geomorphic hierarchies, such as altitude gradients on mountain ranges, could also produce nested habitats and therefore nested species distributions. Unfortunately, no study of nestedness in a particular archipelago has directly addressed this question, so it is unknown to what degree this accounts for nestedness among biotas in general.

Colonization/Extinction Dynamics

Nested patterns can be produced if species vary significantly and consistently in their abilities to colonize sites and/or resist extinction once they are established. This may or may not involve interactions among species.

One can easily imagine a biota with "assembly rules" (Diamond 1975) in which species A could not successfully colonize a site unless species B was present, B required species C, and so on. A hierarchy of dependence would also affect patterns of extinction at a site; extinction of a species at any given level in the hierarchy would induce a cascade of extinction at lower levels (Plotnick and McKinney 1993). Either of these scenarios would produce nestedness among biotas of differing size [although assembly and disassembly of communities may not necessarily proceed according to the same rules (see Mikkelson 1993)]. Presumably, the greater the strength and number of linkages within the system, the more perfectly the nested pattern would be expressed.

Alternatively, success in colonization or resistance to extinction could be intrinsic and "individualistic" (sensu Gleason 1926) properties of species. Differences in vagility among species might control ability to colonize sites regardless of the species composition of the site. Highly vagile species would be able to colonize sites a greater distance from the source area than could less vagile species. Accordingly, proximal biotas would harbor species with a range of dispersal abilities, remote biotas would be restricted to superior dispersers (figure 11.1). A similar colonization-induced pattern might be apparent in a set of sites that varied in their age (i.e., the amount of time they have been available for colonization). Relatively young sites would harbor only species that could disperse rapidly, and older sites would have had more time to accumulate species over a range of dispersal ability (figure 11.2).

Species might differ in their resistance to extinction because of differences in population size, variability of their resource base, or nested tolerance to

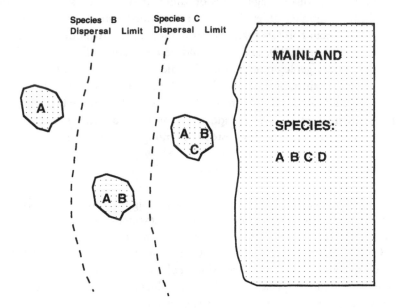

Figure 11.1. Colonization-produced nested biotas on islands or patches varying in isolation from mainland source of species but of identical age. Species differ in their dispersal range: Species D is unable to colonize even the closest island, whereas A can colonize the most distant island. Species B and C are intermediate in dispersal ability. Richness of islands correlates with proximity to mainland.

Figure 11.2. (Opposite) Colonization-produced nested biotas on islands or patches varying in age, but not in isolation from mainland. Species A rapidly colonizes islands as they are formed and is followed by species B and C in sequence. By time 3, the first island formed has all three species, whereas younger islands have incomplete biotas. Richness of biotas correlates with age of island.

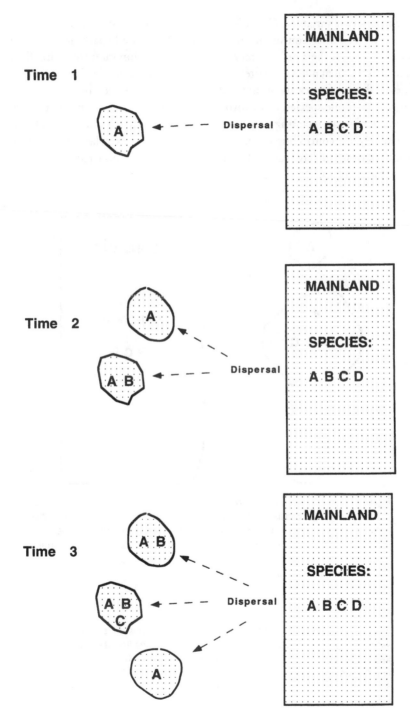

environmental variables. Consider a group of sites that represent isolated fragments of a once-continuous habitat, and that initially harbored the same set of species. If there are differences in local extinction rates (resulting from area effects), then a nested pattern could result through selective extinction. Sites suffering few extinctions would retain larger biotas, including some extinction-prone species. Sites with high extinction rates would be reduced to smaller biotas, retaining only extinction-resistant species (figure 11.3). Alternatively, the controlling variable could be time—sites isolated for a greater length of time will have suffered more extinctions than more recently isolated sites (figure 11.4).

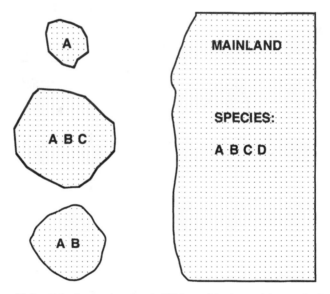

Figure 11.3. Extinction-produced nested biotas on islands or patches varying in size but of identical age. Smaller islands have lower population sizes and hence higher extinction rates. Species A is the most extinction-resistant species and survives on all islands. Species D is the most extinction-prone species and can survive only on the mainland. Species B and C are intermediate in their resistance to extinction. Richness of islands correlates with size.

Figure 11.4. (Opposite) Extinction-produced nested biotas on islands or patches varying in age but of similar size. As islands become isolated from the mainland, they lose species in sequence through faunal "relaxation." Species D is the most vulnerable to extinction and disappears from biotas shortly after isolation. By time 3, only species A survives on the first-formed island. Richness of islands correlates inversely with island age.

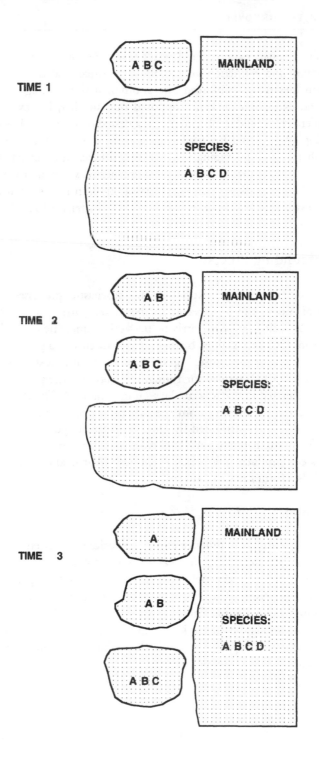

Finally, a nested pattern could result from the combined effects of dispersal ability and extinction vulnerability, provided that the dispersal and extinction hierarchies reinforced one another. If the rankings of dispersal ability and extinction resistance opposed one another, however, then the overall hierarchy would be obscured, and a nested pattern would be less likely to develop (Wright and Reeves 1992). It is for this reason that it has been argued that biotas undergoing relaxation (and hence dominated by extinction) will tend to show a more pronounced nested structure than will biotas in a MacArthur-Wilson (1967) type of equilibrium between colonization and extinction (Patterson and Atmar 1986; Patterson 1990; Wright and Reeves 1992).

Confounding Processes

Because colonization and extinction are stochastic processes, a certain amount of "noise" can be expected in the nested patterns observed in real data sets. Also, sites will only rarely be perfectly homogeneous, so holes and outliers can often be explained by the unusual absence of a particular habitat from a large site, or the unusual presence of a habitat in a small site. Atmar and Patterson (1993) termed such situations idiosyncratic sites. An example of a hole generated in this way can be seen in table 11.3. The cliff chipmunk *Eutamias dorsalis* is missing from the otherwise species-rich Ruby Mountains. Brown (1978) attributed this to the unusually poor development of *E. dorsalis*'s prime habitat(juniper-pinyon woodland) on this range. Other processes can produce "idiosyncratic species" (Atmar and Patterson 1993) that will also disrupt nested patterns.

Speciation

Allopatric speciation can produce endemic species on isolated sites that will appear as outliers on the pattern. In an extreme case, each site would have its own unique biota, totally obliterating the nested subset pattern. Given enough time for dispersal between sites, however, a nested pattern can persist despite rampant speciation. Table 11.5 shows the distribution of Darwin's Finch species among the various islands of the Galápagos. All of these species were derived through speciation within the archipelago (1986), yet the presence–absence matrix is significantly nested.

TABLE 11.5 Darwin's Finches on the Galápagos Islands

Species	1	2	3	4	5	6	7	8	9	10	11	12	13	14	15	16	17	U_a	U_p
Certhidea olivacea	X	X	X	X	X	X	X	X	X	X	X	X	X	X	X	X	X	0	0
Geospiza fulginosa	X	X	X	X	X	X	X	X	X	X	X	X	X		X			0.5	0.5
Geospiza fortis	X	X	X	X	X	X	X	X	X	X	X	X	X					0	0
Geospiza magnirostris	X	X	X	X		X		X	X	X				X		X	X	1.5	2.5
Geospiza scandens	X		X	X	X	X	X	X	X	X		X	X					0	0
Camarhynchus parvulus	X	X	X	X	X	X	X	X	X		X							0.5	0.5
Camarhynchus psittacula	X	X	X	X	X	X	X		X	X								0	0
Platyspiza crassirostris	X	X	X	X	X	X	X			X	X							0	0
Cactospiza pallida	X	X	X	X				X			X							0	2
Geospiza difficilis		X	X		X									X	X	X	X	1	3
Geospiza conirostris														X	X			0	2
Cactospiza heliobates	X	X																0	0
Camarhynchus pauper							X											0	1
Richness	10	10	10	9	8	8	8	7	7	7	6	4	4	4	3	3	3	3.5	11.5

$U_t = 15$

$P < .01$, RANDOM1 assumptions.

Key to islands: 1, Isabela; 2, Fernandina; 3, Santiago; 4, Santa Cruz; 5, Pinta; 6, Rábida; 7, Floreana; 8, Pinzón; 9, Santa Fe; 10, Marchena; 11, San Cristóbal; 12, Seymour; 13, Baltra; 14, Genovesa; 15, Española; 16, Darwin; 17, Wolf.

Data from Grant, P. R. 1986a. *Ecology and Evolution of Darwin's Finches*. Princeton: Princeton University Press.

Competition

Species may be missing from sites as a result of competitive exclusion, occurring only at sites where their competitors are absent. "Supertramp" species (Diamond 1975), highly vagile generalist species confined to species-poor islands because they are outcompeted on species-rich islands, will appear as outliers on a presence–absence matrix. Checkerboard patterns of species distribution, attributed to competitive exclusion (Diamond 1975), are necessarily non-nested. In table 11.5, the two cactus finches *Geospiza scandens* and *G. conirostris* do not coexist on any island; *G. conirostris* occurs on only two of the smaller islands. This is presumably a result of competition between the two species (Grant 1986a). A near-checkerboard pattern is also seen in two other inferred competitors, *G. fulginosa* and *G. difficilis*. These species only co-occur on larger islands with enough relief for the two species to segregate by altitude (Grant 1986a). As a result, three of the occurrences of *G. difficilis* appear as outliers.

Hole/Outlier Asymmetry

Presence–absence matrices usually show some degree of asymmetry between holes and outliers. That is, some matrices are hole rich and others are outlier rich. Montane mammals of the Great Basin (table 11.3) have a hole-rich matrix ($U_a = 12.5$, $U_p = 6.5$); birds of the Great Basin (see table 11.4) and Galápagos Finches (see table 11.5) have outlier-rich matrices ($U_a = 5.5$, $U_p = 8.5$, and $U_a = 3.5$, $U_p = 11.5$, respectively). A survey of 170 presence–absence matrices of a range of taxa has revealed that outlier-rich matrices are more common than hole-rich matrices (Wright et al., in press). This is at least partly an artifact of the relative "fullness" of matrices, that is, the total number of species occurrences relative to the number of potential occurrences. Outliers, by definition, occur at species-poor sites, and a matrix with many species-poor sites will be more likely to have outliers than a "fuller" matrix with few species-poor sites (Cutler 1991). By this argument, matrices with less than 50 percent fill would tend to be outlier rich, and, conversely, matrices with greater than 50 percent fill would tend to be hole rich. Not surprisingly, then, Wright et al. (in press) have found that outlier-rich matrices tend to have a lower fill than hole-rich matrices. Still, a slight outlier bias persists even when only matrices with at least 50 percent fill are considered— out of 55 such matrices from Patterson et al.'s compilation, 18 are hole rich, 23 are outlier rich, and 16 have equal numbers of holes and outliers. These are the matrices in which there would be expected to be a bias in favor of *holes*, so hole–outlier asymmetry is clearly more than an artifact of fill.

Cutler (1991, 1994) suggested that in matrices where fullness could be accounted for, hole–outlier asymmetry might provide a clue regarding the mechanism responsible for producing the nested pattern; hole-rich patterns would indicate patterns shaped by selective extinctions, and outlier-rich patterns would indicate patterns produced by colonization events. This prediction was partially corroborated by computer simulation of passive sampling, which showed a strong bias toward generating outlier-rich patterns—98 out of 100 simulated archipelagos were outlier rich (Cutler 1994). The efficacy of selective extinction in producing patterns that are relatively hole rich has yet to be tested rigorously, but it is consistent with analyses of inferred extinction-produced patterns (Cutler 1994).

In effect, these explanations for hole–outlier asymmetry make statements about the relative probabilities of different rare events. Is it more likely that a widespread (or extinction-resistant) species will get unlucky (i.e., fail to colonize a suitable site or fail to survive in a large biota) or that a rare (or extinction-vulnerable) species will get lucky (i.e., unexpectedly colonize a remote site or survive where more successful species have perished). In the case of the passive sampling simulations described, in which outliers consistently outnumbered holes, it is clear that it is easier for a rare species to get lucky and colonize a small biota, than it is for an abundant species to get unlucky and fail to be represented in a large biota. This is because, in the simulations, to be present in a biota a species needs to get lucky just once, but to be absent it has to be unlucky many times (for the largest site, it has to be missed in all 1024 draws from the source pool). Is there an analogous asymmetry in the extinction process? Given the current interest in the role of "bad luck" (Raup 1991b) and "contingency" (Gould 1989a) in shaping patterns of extinction and evolution, the roots of hole–outlier asymmetry are worth exploring further.

Nested Subsets and the Species–Area Relationship

The variation of species richness among sites is often attributable to the area encompassed by the sites (Patterson and Atmar 1986; and others). This, of course, is the venerable species–area relationship (Preston 1962a; Connor and McCoy 1979; Rosenzweig, chapter 14). The species–area relationship and the nested subset pattern are similar in a number of ways. They share the same ground rules: the sites must consist of ecologically similar habitats and must share a common species pool. A data set consisting of ecologically or biogeographically incomparable sites with distinctly different sets of potential colonists would be unlikely to show either pattern. Also, similar kinds of mechanisms have been proposed to explain both patterns (Connor and McCoy 1979; Cutler 1994).

Despite their similar preconditions and mechanisms, the species–area effect and the nested subset pattern do not necessarily co-occur. Species richness can vary among sites because of distance from the source of colonizing species, or some other factor unrelated to area, yet the biotas might still be strongly nested. A weakly nested biota, on the other hand, may show a perfect fit to a species–area curve.

Cutler (1991) used the Great Basin montane mammal data of Brown (1971) and Brown and Gibson (1983) to compare the respective fits of the species–area and nested subset models (as measured by U). In both cases, deviations from the model predictions can be expressed in terms of extra or missing species. Cutler found that the nested subset pattern was the more robust of the two. Subsequent work has added to the species lists of many of the mountain ranges (Grayson and Livingston 1993; Kodrick-Brown and Brown 1993; Lawlor, personal communication, 1994). This additional data has strengthened the nested subset pattern but weakened the species-area correlation (Grayson and Livingston 1993; Lawlor, personal communication, 1994).

It would appear, then, that at least in some cases species composition is a more predictable property of biotas than is species richness. This runs counter to the common practice in island biogeography of treating species as nameless, ecology-less "particles" (see comments in Brown 1986). The nested pattern of montane mammals in the Great Basin and elsewhere has been attributed to nonrandom local extinctions over the past 10,000 years (Patterson and Atmar 1986; Cutler 1991; McDonald and Brown 1992; Atmar and Patterson 1993). Whether this can be extrapolated to other ecosystems and longer time scales is uncertain, but it does suggest the value of looking at the deterministic, rather than the stochastic, aspects of extinction (see Atmar and Patterson 1993; Van Valen 1994). McKinney's (1995) study of nonrandom species extinctions within Cenomanian–Late Miocene echinoid genera is a case in point.

Scale Dependence

As with many ecological phenomena, the nested subset pattern emerges only when viewed at the appropriate spatial scale (May 1994). Data sets encompassing too large or too small a scale will violate one or more of the preconditions of nestedness, such as (1) ecological comparability of sites and (2) a shared pool of potential species for all sites. Data sets that include widely separated sites are unlikely to satisfy condition 1, because they may straddle environmental or latitudinal gradients. Data sets on too large a spatial scale may also run afoul of biogeographic boundaries and thus violate condition 2—a shared species pool. Such problems of scale account for the failure of Glen (1990) to find significant nestedness in the mammalian faunas of Canadian parks—her data set included sites from the Cordilleran, Laurentian, and Hudsonian regions, each of which have distinctive faunal elements (Wright et al., in press).

Data sets that are at too small a scale may also fail to show the nested pattern. For example, Reice (1994) found that stream invertebrates colonized disturbance patches as random subsets of the total available species pool. He attributed this to chance effects in recruitment. Also, if sites are small relative to patchiness in the environment, or if time scales of colonization are short relative to temporal variability in the environment, then sites may fail to meet the condition of ecological comparability—different sites may be dominated by different types of habitat or be at different successional stages.

A more interesting (but, to my knowledge, thus far unstudied) aspect of scale dependence is the extent to which nested relationships within a system can or cannot be extrapolated across temporal and spatial scales. Put another way, are nested relationships spatially and temporally self-similar? In table 11.3, depicting the nested relationships of boreal mammals on Great Basin mountaintops, the Toiyabe-Shoshone range fauna includes nearly all of the mammals occurring in the montane archipelago. If sites *within* this mountain range were censused, would they also show a nested structure? Would the species rankings be the same as for the entire archipelago? If the nestedness of these mammalian faunas is the result of selective local extinction, can this be used to predict species' vulnerabilities to global extinction, and hence their relative longevities over geologic time?

The answers to these questions are not known. It is likely that the self-similarity of nested patterns will depend on the mechanisms producing them. Nested patterns produced by passive sampling would likely be self-similar, because of the general correlation between abundance and distribution among species (see Brown 1995 for discussion). Nested patterns produced by nested habitat distribution would be vulnerable to disruption by habitat graininess at small spatial scales, and therefore it would frequently not be self-similar. A similar lack of self-similarity resulting from environmental grain was observed for species-richness patterns in forest plots by Palmer and White (1994).

The potential self-similarity of nested patterns produced by extinction–colonization dynamics is more difficult to predict, especially since different processes might prevail at different scales. Cutler (1991) argued that for a nested pattern to have developed at all in the Great Basin ranges implies at least some degree of similarity across scales: species evidently became extinct in roughly the same sequence on small ranges (where extinctions occurred at a high rate) as on large ranges (where extinctions occurred at a low rate). If extinctions occurred in a different sequence depending on the size of the range, ranges of different sizes would harbor unique sets of survivors, producing a non-nested pattern. But this example covers only a limited temporal range (10,000 years) and addresses only local extinctions within an archipelago. All else being equal, one might expect that vulnerability to local extinction would render a species vulnerable to global extinction. After all, the fate of a species is in a sense the summation of the fates of all its populations. But if all else is *not* equal, then the relationship between local and global extinction vulnerability becomes less clear. If high dispersal ability compensated for high extinction vulnerability, as with so-called fugitive species, then locally ephemeral populations could be globally long-lived.

One can imagine an archipelago in which distribution of species *among* islands is determined by their relative powers of dispersal, but the distribution of species *within* islands is determined by their relative abilities to survive in a range of habitats. In this case, in the nested pattern for the entire archipelago, the species would be ranked according to vagility, but in the nested pattern for each island, the species would be ranked according to niche breadth.

Summary

1. The biotas of archipelagos and other replicate habitats are frequently not random subsets of the available species pool. Rather, they may form a nested series in which smaller biotas tend to be subsets of larger biotas. The nested subset pattern is usually not perfect. The degree of nestedness of a given set of biotas can be measured by any of several metrics.

2. Nested subsets can arise in a number of ways: passive sampling of species differing in abundance, underlying nestedness of habitats, and hierarchies of dispersal ability and extinction resistance among species.

3. The determinism implied by highly nested biotas is consistent with the notion of communities as highly interconnected entities (c.f. Plotnick and McKinney 1993), but it does not necessarily imply linkages between species—individualistic models such as passive sampling can easily account for the degree of nestedness observed in real systems.

4. Hierarchies of dispersal abilities and extinction resistance among species can result in nested subsets. In systems where species richness of sites is

a function of isolation of sites or time available for colonization, nested patterns can plausibly be attributed to selective colonization. In systems where species richness is a function of site size or time since separation from the mainland, nested patterns can plausibly be attributed to selective local extinction.

5. Nested patterns usually show asymmetry with regard to holes and outliers. Most often, outliers dominate. Passive sampling, as modeled by computer, tends to produce outlier-rich patterns. The signature of patterns generated by selective colonization and extinction is not as well established, although there is some evidence that selective extinction favors hole-rich patterns.

6. The nested pattern can be disrupted if (1) selective colonization and extinction act in opposition, and thus no clear hierarchy of occurrence exists among species, (2) allopatric speciation introduces unique species in smaller biotas, (3) competition exclusion restricts the occurrence of some species, and (4) the sites are nonhomogeneous, because they encompass either too large a geographic scale (making biogeographic boundaries and environmental gradients a factor) or too small a geographic scale (making patchiness a factor).

7. Self-similarity of nested patterns is a topic that has received little attention, but it may depend on the mechanism generating nestedness. Non-self-similar patterns may result if different mechanisms are operating at different scales.

Acknowledgments

Thanks to Mike McKinney for inviting me to participate in this volume. Many of the ideas in this chapter are products of discussions and other interactions with Wirt Atmar, Greg Mikkelson, Bruce Patterson, and David Wright, especially during workshops funded by NSF Grant BSR-9106981 to Bruce Patterson. Tim Lawlor provided updated information on Great Basin montane mammal distributions. Bruce Patterson, Wirt Atmar, and two anonymous reviewers improved the manuscript with their helpful criticisms. This work has been supported by the Office of the Director, National Museum of Natural History, through the Smithsonian Institution Office of Fellowships and Grants.

12

Equilibrial Diversity Dynamics in North American Mammals

John Alroy

The study of diversity patterns has long been one of paleobiology's principal preoccupations. Simpson (1944) was only one of many early workers to quantify diversity trends using paleontological data. Despite this, "taxon counting" has been plagued by enough practical difficulties to seriously undermine our confidence in statistical results. These difficulties are so profound that Benton (1995) ignored two decades of advances going back to MacArthur (1969) and boldly declared that "there is no need to assume equilibrium models of global diversity, nor to apply logistic models to the investigation of past diversification patterns." Is this true? Can ecologists safely assume that the fossil record is so plagued by sampling bias and masked by unpredictable extinction and diversification episodes that no coherent interpretation is possible?

The goal of this paper is to urge ecologists and paleontologists alike to reconsider the possibility of mathematically predictable diversity dynamics on evolutionary time scales. I will present an analysis of diversity dynamics in a model group—the Cenozoic fossil mammals of North America—that will control for four different sources of sampling bias at once. The analysis will show that, at least for this group, diversity is equilibrial; equilibrium is maintained by the suppression of origination rates, perhaps through competition; there is no long-term trend in the equilibrium diversity level; origination rates are kept to about one quarter of the theoretical maximum by diversity dependence; there are many origination pulses but very few extinction pulses; and origination and extinction rates are uncorrelated.

Despite being foreshadowed by the work of MacArthur (1969), Sepkoski (1978), Walker and Valentine (1984), Maurer (1989), Raup (1991a), Nee et al. (1992), Wagner (1995a), and others, these macroevolutionary conclusions are sufficiently radical that they raise serious questions about the quality of the data. Therefore, I will begin by outlining the four major statistical biases and hinting at solutions to these problems. First, taxonomic data often fail to maintain consistency in taxonomic usage and may rely on spurious identifications. This led some workers (Patterson and Smith 1987; Doyle and Donoghue 1993) to question an entire research program carried out by Raup and Sepkoski (1984, 1986) and collaborators (but see Sepkoski 1987b; Sepkoski and Kendrick 1993; Wagner 1995a). Taxonomic standardization is discussed in the following section.

Second, almost all previous diversity analyses have used a traditional, qualitative time scale (but see Stanley et al. 1988; Barry et al. 1990). Such time scales consist of unevenly sized time intervals to which faunas and taxonomic age-ranges are assigned subjectively, or at best by interpreting such quantitative guides as ordination plots (e.g., Ausich et al. 1994). The boundaries of these intervals often, but not always, correspond to intervals of rapid taxonomic turnover. Because such correspondences are subjectively determined, it is difficult to standardize turnover data to take their effects into account. These problems can seriously distort our picture of diversity dynamics (Foote 1994). The only clear-cut solution is to discard the traditional time scale, define a new one quantitatively, and sample from it using regularly spaced time intervals. This paper expands on previous ones (Alroy 1992, 1994a; Wing et al. 1995) that have employed new multivariate ordination methods to construct time scales, and it introduces a new method for defining the optimal sampling interval.

Third, previous analyses of continental and global biotas have not controlled for variation through time in the geographic clumping of fossil localities. For example, the earliest land-mammal age in North America (Aquilan: early Campanian) is known almost entirely from a single cluster of localities in Alberta. The only previously utilized means of avoiding this problem is to restrict analyses to a limited region, such as a single intermontane basin (e.g., Wing et al. 1995). In this chapter, I discuss a quantitative algorithm for equalizing geographic clumping among time intervals by discarding selected geographic regions. This is made possible by tying the data set directly to locality-by-locality faunal lists, instead of relying on generalized age-range data such as those employed by Stucky (1990) and many others.

Finally, previous studies have documented that variation in sampling intensity can have a major impact on apparent diversity patterns (Raup 1976b; Koch and Morgan 1988; but see Sepkoski et al. 1981). However, parametric "corrections" based on regression analysis such as those employed by Raymond and Metz (1995) can be misleading because they may impose an overly simplistic, linearized relationship between sampling and diversity that fails to capture fine-scale biases. Here, I instead advocate subsampling of the data using the nonparametric rarefaction method (Sanders 1968; Simberloff 1972; Tipper 1979). Previous workers have not attempted to rarefy an entire well-resolved diversity curve at once, as is done here, and again this is probably because most studies have not made direct use of faunal lists (but see Alroy 1996; Miller and Foote 1996). For example, Foote (1992) was constrained to rarefy taxa, not localities, and his data were lumped within only a half-dozen epochs representing two thirds of the entire Paleozoic.

Those who are interested in using the fossil record to test ecological models will find little encouragement here. I will make almost no mention of conditional incidence functions, source-sink dynamics, habitat selection, or metapopulation dynamics (reviewed by Holt 1993), all of which could have a major impact on paleoecological patterns. This, however, is merely because the current data set concerns only continent-wide taxonomic diversity patterns. Eventually, it will be possible to relate these patterns to recent advances in mesoscale ecology by adding ecomorphological data, subdividing the data set into major taxonomic and/or ecological groupings, and focusing explicitly on geographic patterns that are here treated only as a potential source of bias. Furthermore, with the further development of mesoscale models, it may soon be possible to make explicit predictions about the continental-scale patterns demonstrated in this chapter.

The next section briefly describes the raw data used in this analysis. Because a long series of statistical manipulations is necessary, the discussion of methods and results is divided into four sections, the first on preparation of the data and the remainder on analyses of diversity dynamics per se. The section on data preparation is intended for paleobiologists who may wish to carry out similar analyses; it may be skipped by readers with more strictly evolutionary and ecological interests. The diversity analyses in the next three sections have direct ramifications for a series of major macroevolutionary hypotheses that have been published by other authors, so I will combine the presentation of each result with a discussion of its implications. These results broadly focus on constrained diversification models, extinction models, and alternative, nonequilibrium diversification models.

Data

It is no coincidence that students of macroevolution often have focused their studies on the mammalian fossil record (e.g., Simpson 1944; KurtYn 1959; Stanley 1979). Many other frequently fossilized groups, such as brachiopods, bivalves, gastropods, trilobites, and crinoids, are more abundant and diverse. Mammals, however, have a well-studied fossil record that makes up for these deficiencies in several ways. Mammalian alpha taxonomy is well understood; mammals are intensively studied by paleoecologists and phylogeneticists; mammalian fossil localities are frequently reported, despite their rarity in comparison to micropaleontological or marine macroinvertebrate localities; North American mammals have a relatively continuous record stretching back to the middle Cretaceous; mammals are found in terrestrial sediments that frequently present opportunities for geochronological study; and, perhaps most importantly, mammals are evolutionarily volatile. For example, I have used the half-life method of KurtYn (1959: see also Van Valen 1973, Raup 1975b, Foote and Raup 1996) to determine that the median duration of North American mammals is only 3.4 my at the genus level and 1.6 my at the species level. Comparable figures for most other major groups are much higher (Van Valen 1973; Stanley 1979; Raup 1991a). Therefore, mammals are unusually amenable to studies of mass extinction, pulsed turnover, equilibrial diversity dynamics, and so on. If there are any truly general laws of diversity dynamics in the first place, they should be found in a 65-my mammalian fossil record.

The data set used here includes 4015 species-level mammalian faunal lists and is based on a review of 2415 publications. A previous study used these data to revise the North American mammalian time scale (Alroy 1994b). The faunal lists pertain to individual fossil localities that range in age from the Campanian to the last interglacial (Aquilan to Sangamonian: about 85 to 0.01 Ma). Most of the localities correspond to small stratigraphic intervals and limited geographic areas (e.g., individual quarries), but a small minority encompass member-level stratigraphic units and/or basin-scale geographic areas. Because volant and marine mammals are rarely preserved with terrestrial mammals and may have fully independent diversity dynamics, taxonomic records for the orders Chiroptera, Cetacea, Sirenia, and Desmostylia and the carnivoran families Phocidae, Otariidae, and Odobaenidae were excluded from the data set.

A supplementary database documenting 451 invalid genus names, 2692 invalid species names, and 1197 invalid genus–species combinations was used to standardize the faunal lists, and individual identifications were removed or reassigned to a higher taxonomic level when published descriptions were inadequate. The taxonomically corrected faunal lists include 1196 valid genera and 3181 valid species. Thus, the genus-level taxonomy is far more reliable, with 72.6 percent of all genus names, but only 45.0 percent of all species-level combinations, currently being considered valid. The faunal lists include 27,241 separate generically or specifically determinate records, of which 8355 (30.7 percent) are specifically indeterminate. Although the average faunal list includes only 6.8 taxonomically distinct records, the lists provide enough information to establish both a time scale and a realistic diversity curve, as will be demonstrated.

Stratigraphic and geochronologic data also were recorded. A total of 2499 lists (62.2 percent) could be placed in 186 separate stratigraphic sections. "Absolute" dates included ^{40}K-^{39}Ar, ^{40}Ar/^{39}Ar, fission-track, and paleomagnetic age estimates. Fission track dates on glass were excluded because of their extreme unreliability (Naeser et al. 1980). This left 162 age estimates, of which 10 were excluded because they pertained to faunas that were indistinguishable from Wisconsinan (latest Pleistocene) or Recent faunas.

The data set encompassed Campanian and Maastrichtian localities that were included in the initial time scale analysis but were excluded from later analyses of diversity dynamics. There are several reasons for this. To begin with, the Cretaceous–Tertiary boundary is marked not only by a catastrophic mass extinction, but by a fundamental change in underlying diversity dynamics that results in an explosive adaptive radiation immediately after the event (Van Valen 1985c). This diversification affects both local and continental diversity. More importantly, the 144 Late Cretaceous faunal lists are spread far more thinly through time than the remaining Cenozoic lists, making it impossible to balance sampling intensity on both sides of the boundary.

Data Preparation

Appearance Event Ordination

The initial step of the analysis was to arrange the faunal lists in a relatively parsimonious temporal sequence using appearance event ordination [AEO (Alroy 1994a)]. This method incorporates information regarding faunal association ["conjunction" (Alroy 1992)] and stratigraphic superposition by recording F/L statements, which are records of individual first appearances that are known to predate last appearances of other taxa. F/L statements are important because they capture the only unambiguous temporal information presented by both types of data. Other types of relationships between events can be disproved by additional lists or sections; only F/L statements are necessarily true once demonstrated, barring misidentifications or tapho-nomic problems such as reworking.

AEO produces an appearance event sequence, i.e., a hypothetical relative age ordering of first appearance events (FAEs) and last appearance events (LAEs). The sequence is subject to three constraints: the FAE of any one taxon I (FAE_i) must predate its own LAE (LAE_i); FAE_i must predate the LAE of a second taxon J if this relationship is matched by an F/L statement demonstrated by the raw data; and otherwise, FAE_i should postdate LAE_j whenever possible. The last constraint is a parsimony criterion, namely, that the event sequence should imply the smallest possible number of F/L state-ments. To do otherwise is to contradict the raw data with ad hoc hypotheses of temporal overlap between age-ranges.

The ordination algorithm used here is largely the same as that described by Alroy (1994a) and uses the same C language program (CONJUNCT), but it has been augmented in five ways. First, since the publication of that paper I have developed a new localized optimization routine. Once an initial appearance event sequence has been generated, neighboring appearance events are swapped whenever this would reduce the total number of implied F/L statements. This is similar to a standard "bubble float" algorithm for arranging a list of numbers in ascending order. A more sophisticated swap-ping algorithm allowing swaps between non-neighboring events would be desirable but computationally demanding.

Second, AEO in its original form makes no use of the fact that certain taxa, such as those occurring in the Recent, may be known to survive past the time interval represented by any of the faunal lists. Therefore, their LAEs must come last in the event sequence. An easy and sensible way to implement this constraint is to reset the LAE scores for all "surviving" taxa to 1.0 during each iteration of the score computing procedure, 1.0 being the highest pos-sible value. The list of "surviving" genera and species used here is based on that of KurtYn and Anderson (1980) for the Wisconsinan (last glacial), aug-mented in a few cases when previously Wisconsinan and/or Recent taxa oc-curred in pre-Wisconsinan faunal lists unknown to those authors.

Third, in an effort to use as much biochronological information as pos-sible, both genus- and species-level F/L statements were employed in the same data matrix. This marks a significant departure from standard practice in multivariate analyses of biotic presence–absence data (e.g., Gauch 1982). It is justified by the fact that genus-level age-ranges not only represent the sum of species-level age-ranges but may extend beyond them whenever there are specimens assigned to a genus but not to an individual species.

Fourth, taxa confined to a single faunal list (singletons) must have an age-range of zero years in any hypothetical temporal scheme. Therefore, they are always biochronologically uninformative apart from the fact that they may demonstrate differences in overall diversity, which generally is of unclear relevance. In light of this, the 1156 singleton genera and species (26.4 percent of all taxa) were discarded for the purpose of generating an event sequence, although they were considered in most of the diversity analyses.

Fifth, I have developed a new algorithm to compensate for the influence of biogeographic effects on F/L data. These effects are unique because no amount of sampling can remove their overprint, unlike paleoecological, taphonomic, monographic, or sample size biases, which can be eliminated from a raw F/L data set as long as the number of faunal lists is very large (Alroy 1992). If two taxa I and J never occurred in the same basin even though their age-ranges overlapped, the F/L statements $FAE_i < LAE_j$ and $FAE_j < LAE_i$ will never be demonstrated by taxonomic lists and/or stratigraphic superposition. Fortunately, there is a graph-theoretic solution to the problem. It was originally described as the square graph method by Alroy (1989).

Suppose there are four taxa: α, β, γ, and δ (figure 12.1). Taxon α is confined to a western region, and δ to an eastern region, whereas β and γ are cosmopolitan. All of the taxa except β and γ have overlapping age-ranges. After sufficient sampling, four overlaps (indicated by $\wedge$, and by solid lines in figure 12.1C) should be demonstrated in the form of conjunctions or complementary F/L statements: $\alpha\wedge\beta$, $\alpha\wedge\gamma$, $\beta\wedge\delta$, and $\gamma\wedge\delta$. A simple proof not given here shows that at least one of the two apparent nonoverlaps ($\alpha{:}\delta$ and $\beta{:}\gamma$) must represent either a sampling artifact or a biogeographic disjunction, rather than a temporal disjunction; this can be visualized by inspecting a graph in which taxa are represented by vertices, and known overlaps by edges (a square graph). In this particular example, one apparent disjunction is an artifact of geography ($\alpha{:}\delta$) and the other is temporal in origin ($\beta{:}\gamma$). If it can be shown that α and δ never occur in the same geographic area, but β and γ occur together in at least one area (perhaps even superposed in a stratigraphic section), then the square graph dilemma can be resolved by declaring α and δ to be conjunct, implying FAE $\alpha <$ LAE δ and FAE $\delta <$ LAE α. This is called a virtual conjunction.

Similar interactions among sets of more than four taxa could be demonstrated, but this would be computationally very demanding, and it appears that the current algorithm is sufficient. The version of the CONJUNCT program now available makes use of geographic information to generate virtual conjunctions following the above argument. In the current data set, I defined geographic areas as states, provinces, or Latin American countries (see following discussion of geographic bias) and identified a total of 61,596

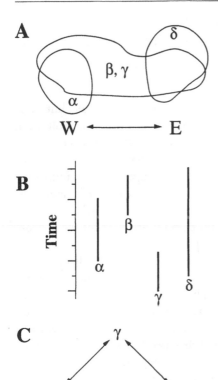

Figure 12.1. The "square graph" dilemma. **(A)** Geographic ranges of four hypothetical taxa labelled α, β}, γ, and δ; W and E indicate western and eastern regions. **(B)** Age-ranges of the four taxa. **(C)** Known conjunctional relationships of the four taxa, with conjunct pairs indicated by lines; relationships form a square graph. *Dashed line* indicates virtual conjunction.

virtual conjunctions using the square-graph algorithm. This increased the number of available conjunctions by 30 percent, from 205,655 to 267,251.

The raw data that resulted from these manipulations totaled 796,505 F/L statements for the 3238 nonsingleton genera and species. Of these, 262,003 (32.9 percent) were demonstrated strictly by stratigraphic superposition: if fossils of some taxon I are found lower in a section than those of another taxon J, FAE_i predates LAE_j. Another 534,502 statements (67.1 percent) were demonstrated by faunal conjunctions: whenever fossils of I and J are found in the same faunal list or are declared conjunct by the square graph algorithm, FAE of I predates LAE of J and vice versa. The optimized AEO routine generated an appearance event sequence of 6196 events (3221 FAEs, 2975 LAEs, 246 taxa surviving to the Wisconsinan) that implied 5,663,636 F/L statements and 477,826 conjunctions, with the conjunction index (Alroy 1992) equaling 0.430 (excluding virtual conjunctions). This is comparable to results obtained in earlier studies (Alroy 1994a; Wing et al. 1995).

Calibration

The event sequence by itself constitutes a valid relative time scale, but to be usable in diversity studies it must be calibrated to absolute time. Following previous studies (Alroy 1992, 1994a,b; Wing et al. 1955), this was accomplished by numbering the appearance events and then regressing 152 published absolute age estimates against concurrent range-zones for individual faunal lists. The concurrent range-zone is defined by the latest FAE and earliest LAE of the taxa in a faunal list, which sometimes pertain to different taxa. The average of the two event numbers for each list was used for the purpose of regression.

In 87 cases (51.9 percent), age-range zones were defined by adjacent FAEs and LAEs, i.e., they spanned an interval of zero events. The dates for these maximally "precise" lists were augmented by 65 others (40.1 percent for lists having zones that spanned multiple events, half of which had a span of less than or equal to 110 events. A 55-event "error" spans only 0.9 percent of a 6196-event-long sequence. An additional six dated lists were redundant because they had age-range zones so broad that they encompassed those of other dated lists, and four Pleistocene dates (2.5 percent) pertained to lists of taxa that are all known from the Wisconsinan and/or Recent.

A plot of the 152 age estimates against range-zones (figure 12.2) shows that the relationship is complexly curvilinear. One could use a polynomial regression to characterize the relationship, and, in fact, such a regression would generate entirely significant coefficients for a fifth-order polynomial function. However, the scatter indicates that there is a very abrupt change in the relationship at the K-T boundary, calling for a less smooth fitting function. The method used here to find such a function, called hinge regression, works by iteratively defining a series of linear interpolation segments. The segments are separated by "hinge points" corresponding to individual calibration points. Because the interpolation segments are constrained to meet each other cleanly at the hinge points, defining the hinges automatically defines the slopes and intercepts. The two most extreme data points are defined as hinges by default, and additional, intermediate hinges are then picked in a way that minimizes the unexplained sum of squares, as in least-squares regression. All possible hinge points are evaluated during each pass. A standard F-test is used to see if the improvement in fit provided by each hinge is significant; using a $P < .01$ cutoff, five hinges that define seven segments are accepted; one further point is allowed so as to break up a very

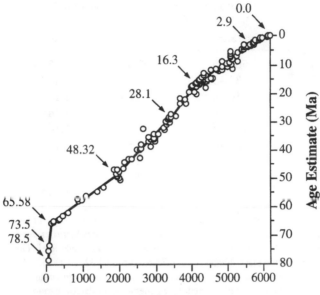

Figure 12.2. Calibration of the North American mammalian appearance event sequence using 152 radioisotopic and paleomagnetic age estimates. A seven-segment hinge regression is shown (see table 12.1); the six internal break points and two end points are indicated by *arrows* and the appropriate age estimates, in Ma. The appearance events are for all genera and species appearing in more than one faunal list, and they are numbered from the Campanian through the Recent. Event numbers for dated localities (*circles*) are based on concurrent age-range zones.

long segment in the Paleogene. The F-tests and resulting hinge points are described in table 12.1.

Raw Diversity Curve

Once the appearance event sequence is calibrated, it can be used to compute a diversity curve with no further manipulations other than to add back the singleton taxa omitted in previous steps. This can be accomplished by computing the concurrent range-zones of all the individual faunal lists and then equating the range-zone of each singleton taxon with the concurrent range-zone of the list that includes it. A secondary issue concerns the taxonomic level to be used in the analysis. As mentioned previously, genus-level names are far more taxonomically stable than species-level names. Another problem is that species often are defined as temporal segments of long-lived, phyletically evolving lineages, creating pseudoextinctions within the lineages (Stan-

TABLE I2.I   Hinge regression analysis[a]

SSnew[b]	SSun	F	df	r^2
6633.45	569.96	1757.40	1,151	0.9904
334.53	235.43	213.14	1,150	0.9961
24.99	210.44	17.69	1,149	0.9965
15.31	195.14	11.61	1,148	0.9967
15.59	179.54	12.77	1,147	0.9970
3.45	176.10	2.86	1,146	0.9970
7.86	168.23	6.78	1,145	0.9972
3.00	165.23	2.61	1,144	0.9972
1.29	163.94	1.13	1,143	0.9973

[a]Regression calibrates the North American mammalian appearance event sequence to absolute time using 152 radioisotopic and paleomagnetic age estimates. Statistics are given for the first nine hinge points. The first five hinges are significant at the $P < .001$ level. The preferred six-interval hinge solution places hinges at 78.5, 73.5, 65.58, 48.32, 28.1, 16.3, 2.9, and 0.0 Ma.

[b]SSnew, sum of squares explained by new hinge point; SSun, remaining unexplained sum of squares; F, F ratio; df, degrees of freedom; r^2, coefficient of determination.

ley 1979). The prevalence of such a bias is suggested by the data set's low species-to-genus ratio of 2.66:1. As it happens, there is a tendency for pseudoextinctions and matching pseudo-originations to be concentrated at the boundaries between traditional land-mammal "age" boundaries, creating largely spurious "extinction" and "origination" episodes. This too seems to be much less a problem at the genus level.

Despite all these problems, it still would be desirable to make use of the information provided by species-level occurrences. Therefore, the major analyses reported here also were performed for "minimal lineage" curves in which the species-level data were "corrected" using the genus-level data. This involves making two straightforward assumptions: (1) genera are present at all times within their age-ranges, even if no named species have been sampled; and (2) separate species at opposite ends of such "gaps" represent mere segments of single lineages, and thus the extinction of one species at the beginning of a gap and the origination of another species at the end of a gap must be discounted. About 18.8 percent of all species-level turnover events were eliminated using this second rule, implying that at least this many events represent pseudoextinctions and pseudo-originations; detailed phylogenetic analyses imply that the true figure is not much higher (see later). Genus- and lineage-level analyses both yielded essentially the same results. All of this argues against the suggestion that because genera, as polytypic taxa, may include multiple independent evolutionary lineages, they may reflect a misleading macroevolutionary signal. In fact, traditional, higher-level taxa such as genera may be a *better* indication of diversity patterns when species-level sampling is poor, as shown by Sepkoski (1978), Sepkoski and Kendrick (1993), and Wagner (1995a).

Optimal Sampling Interval

One could simply stop at this point and compute a diversity curve from the appearance event sequence. However, such a curve would have 702 distinct peaks, or about one peak every 0.1 my. Analyzing the data at such an extremely fine level of temporal resolution would be highly unrealistic: other studies of diversity dynamics in fossil mammals either have used arbitrary temporal intervals of 0.5 my (Barry et al. 1990) or 1.0 my (Prothero 1985), or more commonly have used highly variable mammal zones and "ages" spanning anywhere from 0.5 to 10 my (Barnosky 1989; Cifelli 1981; Jaeger 1994; Van Valkenburgh and Janis 1993; Lillegraven 1972; Savage and Russell 1983; Simpson 1944; Stucky 1990). This presents a dilemma. There is no established, objective method for defining the sampling interval, but a careful choice must be made because using shorter intervals will introduce noise into the analysis by putting more and more events in the wrong intervals, whereas using longer intervals will reduce statistical power and make it difficult to identify medium-term turnover events.

One possible approach is to examine the tradeoff in terms of the apparent versus "true" ranks of events (figure 12.3). The basic idea is to estimate the age of the "true" events using the 50 percent confidence intervals around the ages of the apparent events (Marshall 1994). The apparent and "real" events are each rank-ordered, and the inaccuracy implied by a particular sampling interval is then defined by giving all events in a given interval a tied rank, and then summing the absolute values of the differences in ranks between the observed and "true" event sequences. This definition of inaccuracy has two strengths. First, it generally allows that apparent events falling in the "true" interval are fully accurate, because the difference in ranks should be close to zero in these cases. Second, it generally allows that larger interval sizes reduce the accuracy of apparent events falling in the *wrong* interval, because both the apparent and "true" events are given more and more vague rank-orderings as more and more events are lumped in the same intervals. As a result of these two features, the degree of inaccuracy should illustrate a trade-off, with very high inaccuracy at very small sampling intervals because almost all events are in the wrong interval; a rapid fall-off as more apparent and "real" events are lumped together; and then a gradual increase as the larger interval sizes blur the event ranks.

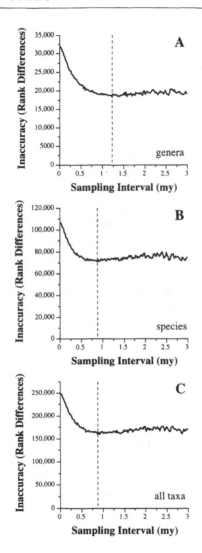

Figure 12.3. Trade-off between sampling interval, in millions of years, and perceived inaccuracy in age ranges. Inaccuracy is measured as the sum of the differences in age-range ranks between the observed appearance event sequence and a "true" sequence, which was estimated by adding 50 percent confidence intervals to the age ranges. Taxa assigned to the same interval were given tied ranks. **(A)** Species-level trade-off; optimum is about 1.2 my. **(B)** Genus-level trade-off; optimum is about 0.9 my. **(C)** Combined genus- and species-level results; optimum is about 0.9 my.

One minor computational problem is that the nonparametric "median gap size" method of Marshall (1994) makes a crucial assumption that is not borne out by the data, namely, that the gaps in temporal distributions of occurrences are randomly distributed within any given age-range. Instead, the data show clearly that the first and last gaps in an age-range tend to be much longer: for species, the median median gap is 0.13 my, but the median first gap is 0.23 my, and the median last gap 0.20 my. It is not even clear whether the source of this bias is methodological or biological. In any event, substituting the first or last gap length for the median gap length as an estimate of the 50 percent confidence interval should suffice to compensate for the bias regardless of its source.

The results for 982 nonsingleton genera, 2120 nonsingleton genera, and all taxa lumped together seem to show a consistent pattern (see figure 12.3). The absolute minimum of the inaccuracy curve varies slightly, falling at 1.2 my for genera and 0.9 my otherwise, but the shallowness of the curve makes this difference appear not to be important. Because most of the taxa are species to start with, a compromise figure of 1.0 my seems conservative.

Geographic Bias

North American fossil mammal localities are neither widespread nor randomly distributed across the continent. There are virtually no Tertiary localities east of the Mississippi in the Paleogene, apart from a few scattered finds in coastal deposits; and with the major exception of Florida, this pattern persists through the Neogene. Within the West, Paleogene localities are concentrated in the Rocky Mountains, whereas Neogene localities are most common in the Great Plains and the Great Basin. At the start of some intervals such as the Uintan to Duchesnean [46 to 38 Ma (Alroy 1994b, 1996)], the bulk of the record suddenly shifts from one part of the continent (here, the northern Rocky Mountains) to another (southern California) and eventually back again.

This pattern of bias is most easily depicted by counting the number of geographic sampling areas, defined as provinces (in Canada), states (in the United States), or countries (in Mexico/Central America), that are represented in each 1.0-my sampling interval (figure 12.4A). Mexican states were lumped together because of the small total number of Mexican localities. A variable but broadly consistent increase in geographic representation through time is apparent. Some of this trend is related an increase in the overall number of reported fossil localities, as discussed in the next section, but most of it has simply to do with the underlying amount of available fossil material. Simply put, processes such as erosion, diagenesis, and burial make less and less sediment available as one goes back farther into the record. A better way to quantify this trend is to use a measure, such as the Shannon-Wiener evenness index H (May 1975), that is relatively insensitive to sample size, which in this case is the overall number of reported localities. Here I have computed the index separately for each time interval by examining the number of localities in each sampling area. H is defined as -1 times the sum of $p_i \ln p_i$ for N areas, where pi equals the proportion of all localities that are present in the ith sampling area. Hence, if in one interval there are eight localities in Wyoming and two in California, $H = -1 \times [(8/10) \ln(8/10) + (2/10)$

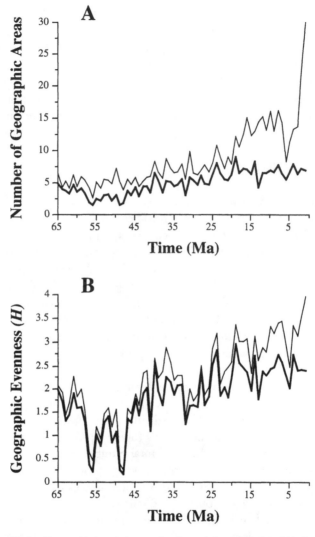

Figure 12.4. Geographic trends in sampling through time. Raw data (*thin line*) and sampling-standardized data (*thick line*) are shown. **(A)** Number of geographic regions sampled, defined as provinces (Canada), states (United States), or countries (Mexico/ Central America). The trend results from increases in both the geographic availability and reporting of fossil localities. **(B)** Geographic evenness in sampling, i.e., the evenness in the frequency distribution of taxonomic records across geographic areas, quantified with the Shannon-Wiener index *H*. Because this index controls for sample size, the trend is interpreted as an artifact of the availability of fossil localities, not of a reporting bias. Truncation of the curve in the late Neogene is caused by restricting the standardized data to western North America.

ln(2/10)] $= 1.833$. The resulting plot (figure 12.4B) shows that despite high fine-scale variation, the long-term increase in the geographic distribution of localities now appears to be even more steady. This is because the Neogene bias in sampling intensity is largely removed by the Shannon-Wiener index.

Because leaving the geographic bias in the data would have a serious impact on the results, some correction must be made. I explored objective algorithms for excluding geographic areas one at a time in a way that minimizes the temporal variation of H, with the algorithm coming to a halt when no further reductions can be made. The problem with this and similar methods is twofold. First, the remaining, included areas typically form an irregular patchwork, with large gaps created by the excluded areas. This is biologically unrealistic and has unclear effects on perceived beta diversity. Second, because a maximal number of areas per interval is imposed by the ability of the algorithm to flatten the temporal trend, so many areas may be excluded that there may be almost no lists left in certain time intervals.

Although I consider this a topic worthy of further investigation, I here take the simpler approach of simply excluding entire geographic regions on a priori grounds. Clearly, the eastern part of North America is far too poorly known throughout most of the Cenozoic to provide a reliable long-term record. Excluding it would seem to be a minimal step. The Central American record is similarly biased, and it is quite poor to start with. This leaves the western North American region. Within that region, some states such as California, Nevada, and Oregon present a temporally biased record, but experiments with removing those states show that they provide indispensable information on certain parts of the Neogene record. Therefore, the current analysis focuses on Mexico and the western region of Canada and the United States. Although it may seem that the retained sampling areas form a distinct minority, they actually include 95.5 percent of the faunal lists and 97.3 percent of the genera (table 12.2).

Rarefaction

Temporal biases in the data set are even greater than geographic biases. The most notable features of a plot of the number of faunal lists per interval (figure 12.5A) are (1) a very high peak in the early Eocene, and (2) a trend toward increase through the Neogene following an Oligocene trough, which, unlike the Paleogene pattern, mirrors the world-wide trend through the entire Phanerozoic in the availability and preservational potential of sediment (Raup 1976b). The Eocene peak relates to a large amount of paleontological interest in that interval, as reflected by work in the Bighorn Basin (Gingerich

TABLE 12.2 Size of the full North American data set and various subsamples

Data subset	Lists	Pct	Records	Pct	Genera	Pct	Lineages	Pct
Full	4015	100.0	27,265	100.0	1196	100.0	2583	100.0
Cretaceous	144	3.6	878	3.2	52	4.3	903.5	35.0
Cenozoic	3871	96.4	26,387	96.8	1156	96.7	2510	97.2
Paleocene	504	12.6	3754	13.7	244	20.4	436	16.9
Eo-Pleist	3367	83.9	22,642	83.0	993	83.0	2169	84.0
Western	3834	95.5	5,470	93.4	1164	97.3	2470	95.6
Standardized	987.8	24.6	7068.1	25.9	963.1	80.5	1579.0	61.1

Lists, number of faunal lists; Pct, number in preceding column as a percentage of that found in the full data set; Records, number of records, i.e., sum of number of identifications in each faunal list; Genera, number of genera present in data subset; Lineages, number of species-lineages present in data subset; Eo-Pleist, Eocene-Pleistocene data; Western, western North America data, including Mexico; Standardized, data rarefied from the geographic subset to the level of 100 records per 1.0-my interval.

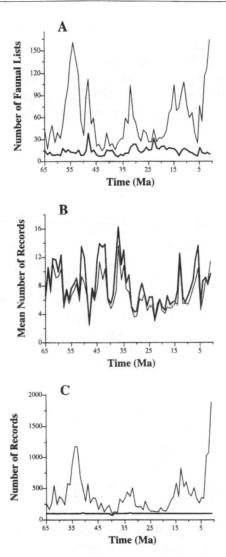

Figure 12.5. Sampling intensity through time. Raw data (*thin line*) and sampling-standardized data (*thick line*) are shown. The high Early Eocene sampling peak in the raw data is attributed to paleontological interest based on the restricted geographic availability of fossil localities (see figure 12.4); the increase through the Neogene reflects both availability and interest. **(A)** Number of faunal lists per interval. **(B)** Mean number of taxonomic records (i.e., alpha diversity) per list per interval, believed to represent reporting biases. **(C)** Number of taxonomic records per interval, i.e., the product of the preceding two curves.

1980) and other parts of the northern Rocky Mountains such as Alberta (Fox 1990), Colorado (Kihm 1984), and other parts of Wyoming (Gazin 1962, 1976; Stucky 1984). This level of interest may be related to the first appearance in North America of true primates, as well as perissodactyls, artiodactyls, and hyaenodont creodonts, at the Paleocene–Eocene boundary (Gingerich 1989). Because the Eocene "primate peak" does not appear in the geographic sampling curves (see figure 12.4), I believe it only reflects strong paleontological interest and not the greater availability of fossil localities in the field, despite the fact that this interval is well known for the Laramide orogeny.

Order of magnitude variation in sampling intensity surely will have a major impact on diversity patterns, necessitating some kind of standardization. It would not be appropriate to employ a parametric method such as the Jolly-Seber capture–recapture technique recommended by Nichols and Pollock (1983), which makes several substantive methodological assumptions that are violated by the current data set. Although other approaches eventually could prove superior, for the moment I have chosen to employ the nonparametric rarefaction method first developed by ecologists (Sanders 1968; Simberloff 1972; Tipper 1979). This involves subsampling an equal number of faunal lists from within each sampling interval. Sampling is without replacement, and age-ranges are recomputed from the remaining faunal lists (Alroy 1996; contra Miller and Foote 1996). There are three complications with this approach.

First, variation in apparent alpha diversity (i.e., mean within-faunal list richness) appears to be artifactual (see figure 12.5B). This claim is based on the observation that the "richest" intervals are the ones characterized by reporting that is biased towards generalized faunal localities including numerous specimens (e.g., the early Oligocene and late Neogene). Therefore, rarefaction should take both alpha diversity and the number of faunal lists into account. One way to do this is to rarefy on the basis of the number of records, i.e., the sum of alpha diversity for all lists in each interval. The curve showing the number of records per interval (see figure 12.5C) is in any event very similar to the previous lists-per-interval curve (see figure 12.5A). Note that if alpha diversity trends are artifactual, it is impossible to discuss beta (among-region) diversity patterns sensu Whittaker (1960). This is unfortunate because beta diversity has been discussed in several earlier analyses of the North American mammalian record (Flynn 1986; Storer 1989; Stucky 1990; Van Valkenburgh and Janis 1993), it has been shown to be important in the Paleozoic marine record (Sepkoski 1988), and it could provide a critical test of the idea that metapopulation dynamics influence long-term, large-scale diversity patterns (Harrison, chapter 2).

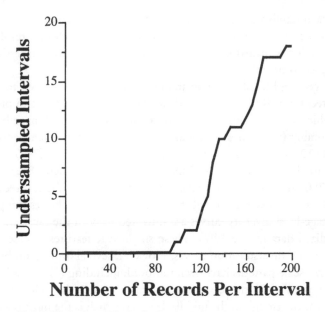

Figure 12.6.   Effect of possible rarefaction levels (i.e., the number of taxonomic records per time interval) on the number of undersampled intervals (i.e., the number of intervals failing to include that number of records). The abrupt change in slope at 100 records per interval indicates the optimal rarefaction level across the entire time series, which allows almost all of the intervals to be sampled fully.

A second, more difficult problem is that the standard sampling level for all intervals must be set arbitrarily. I have attempted to make the procedure more objective by basing it on a curve relating the number of subsampled faunal records per interval to the number of intervals that fail to supply that number of records (figure 12.6). The figure indicates an initially very shallow climb, with all of the intervals meeting the cut-off at low rarefaction levels. The slope suddenly increases when intervals start to be "lost" at an inflection point of 100 records per interval, which therefore will be used as the standard sampling level for the remainder of this discussion. Beneath this level, the curve climbs at a rate of 0.00 intervals per record, and above it at 0.17 intervals per record.

Finally, single rarefaction trials can introduce unwanted variation to an already noisy pattern. Turnover rates and diversity magnitudes may swing up and down wildly just because an individual rarefaction trial has, or hasn't, subsampled a few particularly rich localities. To compensate for this, I performed 100 separate rarefaction trials and then computed all diversity statistics as averages across all the trials. This made some of the statistical tests much more algorithmically complex, as it required performing separate analyses of each rarefied diversity curve and then averaging some of the resulting statistics such as variances in turnover rates. However, cleaner results made the extra effort worthwhile, and I recommend that future workers follow a similar procedure.

After geographic subsampling and rarefaction of each interval to the level of 100 records 100 times, the trends in the number of states and provinces, geographic evenness, faunal lists, and taxonomic records all have been flattened notably (see figures 12.4 and 12.5). The sampling-standardized data (see table 12.2) now include an average of 987.8 faunal lists (24.6 percent of the raw total), 7068.1 records (25.9 percent), 963.1 genera (80.5 percent), and 1579.0 lineages (61.1 percent). Despite the fact that about three quarters of the data have been discarded, fine-scale patterns in the resulting genus- and lineage-level diversity curves are mirrored in both the raw data and the standardized data (figure 12.7). These small-scale features represent minor extinction and diversification events. However, several important, but spurious, coarse-scale patterns have been removed, including (1) the early Eocene "primate peak"; (2) the downward trend through the middle and late Eocene; (3) the trough in the middle Tertiary, which is the most poorly sampled interval in the record; (4) the apparent middle Miocene diversification associated with Webb's "Clarendonian chronofauna" (Webb 1983; Webb and Opdyke 1995); and (5) the extraordinarily steep diversification in the Pleistocene. The Pleistocene spike is caused by a massive increase in sampling intensity (see figure 12.5); it should not be confused with the "pull of the Recent" effect in other data sets, whereby living taxa inflate apparent diversity (Raup 1978).

In summary, standardizing the data for variation in sampling intensity dramatically alters long-term diversity patterns. Instead of wild fluctuations from one epoch to the next, we now see a remarkably flattened trajectory at both the genus and lineage levels. In fact, the two standardized curves are far more similar to each other than to their respective raw-data cousins. This is surprising in light of earlier studies (e.g., Sepkoski 1993) showing that large taxonomic databases tend to converge quickly on a stable diversity pattern. For example, the diversity pattern of Phanerozoic marine animal families outlined by Valentine (1969) has been reproduced by all later studies (Sepkoski et al. 1981; Sepkoski 1993; Benton 1995). This should come as a warning to future workers that no amount of additional, but undirected, sampling can correct for an underlying bias in the distribution of fossils through time.

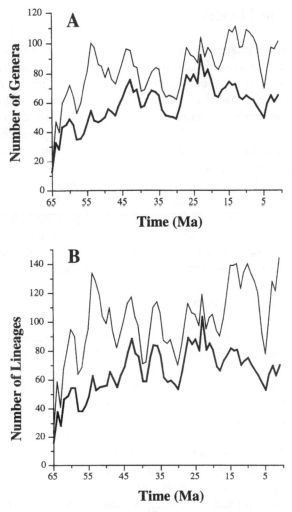

Figure 12.7. Genus-level **(A)** and lineage-level **(B)** diversity histories of North American mammals in the Cenozoic, based on appearance event ordination of 4015 faunal lists. The sampling interval is set at 1.0 my, and 65 data points are shown. Raw data (*upper, thin lines*) are contrasted with sampling-standardized data (*lower, thick lines*). The two plots scaled identically to emphasize the great similarity between the standardized genus- and lineage-level curves, despite the great disparity of the raw data. Note also that the standardized curves duplicate the fine-scale patterns in the raw data and remove artifactual, coarser-scale variation related to sampling effects (see figures 12.4 and 12.5).

Constrained Diversification Models

In this section, I present several analyses that outline a remarkably simple model of diversity dynamics. To summarize the results, extinction rates are stochastic but fluctuate around a nearly invariant background rate. Meanwhile, the absolute magnitude of diversity is governed by a subtle negative dependence of origination rates on standing diversity, which results in the maintenance of a dynamic equilibrium. Variation through time in extinction and origination rates seems to have no apparent long-term effect on the underlying equilibrium point.

This model of diversification is complex, so the argument is broken up into several parts. First, I outline the logistic model itself and present evidence for it. Two alternative models of diversity dependence are then considered. After presenting evidence that the underlying equilibrium point is stable throughout most of the Cenozoic, I apply the niche occupation formulas of Walker and Valentine (1984) to the data and discuss their implications.

The Logistic Growth Model

MacArthur (1969) and Rosenzweig (1975) argued from ecological principles that continental diversification should be regulated by competition. In their models, both extinction and origination rates were dependent on standing diversity: at moderate to high diversity levels, per-taxon extinction rates should rise and per-taxon origination rates should fall as diversity increases. When the two are equal, an equilibrium is attained. These ideas were rooted in the equilibrial island biogeography model of MacArthur and Wilson (1963, 1967) but differed importantly because the total speciation rate rises as a function of diversity, starts from zero, and may never level off. In the island biogeography model, however, total immigration rate begins at a maximum and falls continually with rising diversity. Sepkoski (1978) formalized these ideas as a logistic growth model in which the per-taxon diversification rate ΔD_t per fixed time unit was equal to the difference of two quadratic functions,

$$O_t D_t = O_0 D_t - s_o D_t 2 \qquad [1]$$

and

$$E_t D_t = E_0 D_t + s_e D_t 2, \qquad [2]$$

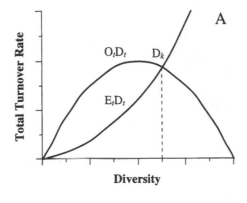

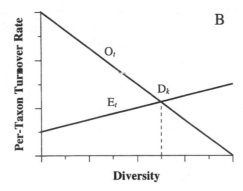

Figure 12.8. Hypothetical dependency of total **(A)** or per-taxon **(B)** origination and extinction rates on standing diversity, based on MacArthur (1969) and Sepkoski (1978). The linear functions in (B) imply an overall logistic growth function. E_t, per-taxon extinction rate; O_t, per-taxon origination rate; D_t, diversity; D_k, equilibrium diversity level.

where D_t = diversity at time t, O_t = per-taxon origination rate at time t, s_o = a slope constant, O_0 = per-taxon origination rate when diversity approaches zero, E_t = per-taxon extinction rate at time t, s_e = a second slope constant, and E_0 = per-taxon origination rate when diversity approaches zero (figure 12.8A). Note that I have formulated the equations around per-taxon rates, although Sepkoski redefined them as total rates. The preceding equations can be reformulated as two linear functions (see figure 12.8B):

$$O_t = O_0 - s_o D_t \qquad\qquad [3]$$

and

$$E_t = E_0 + S_e D_t, \tag{4}$$

which imply

$$\Delta D_t - O_0 - E_0 + s_o D_t - s_e D_t, \tag{5}$$

a version of the well-known logistic growth function assuming evenly spaced sampling intervals.

Sepkoski (1978, 1979, 1984) used logistic growth functions to model family-level diversification in marine animals throughout the Phanerozoic, with a better fit being obtained when the model is augmented by a few mass extinctions and by coupled dynamical interactions among three major taxonomic subgroupings ("evolutionary faunas") identified on the basis of a factor analysis (Sepkoski 1981: see also Flessa and Imbrie 1973). Sepkoski (1991a) also successfully showed that widely recognized temporal onshore-offshore trends in marine taxonomic diversity could be reproduced using a simple, independent model that also predicted logistic growth in overall diversity. However, many workers have continued to interpret diversity patterns in different parts of the marine fossil record as if logistic growth dynamics did not exist (e.g., Stanley et al. 1988; Oliver and Pedder 1994; Benton 1995; but see Miller and Sepkoski 1988; Wagner 1995a), and previous studies have not applied the model to the terrestrial fossil record (e.g., Stucky 1990; Van Valkenburgh and Janis 1993).

Maurer (1989) recognized that the logistic growth model could be tested most easily by plotting per-taxon rates against diversity, allowing linear regression methods to be used. The logistic model predicts negative linear correlations of net per-taxon diversification and per-taxon origination rates with diversity, but a positive correlation of per-taxon extinction rate with diversity (see figure 12.8B). The success of these predictions for the current data set is illustrated by table 12.3, with correlations given for all three rates and all three data subsets (sampling-standardized genus level, raw data genus level, and raw data species level). Figure 12.9 shows the relationships for per-taxon changes in diversity, origination rates, and extinction rates in the standardized data. Because (1) diversity levels are far lower in the early Paleocene than at any later time, regardless of how diversity is measured; (2) turnover rates appear to be more rapid and volatile in the Paleocene; and (3) including an apparent exponential growth phase would predispose the analysis to confirm the logistic growth model, I excluded the first ten data points for the Ceno-

TABLE 12.3   Diversity dependence regression analyses

NET DIVERSIFICATION RATES[a]

Data set	r^b	t	P	$s_{\Delta D}$	ΔD_0
Genus-level, S/NS	−0.356	2.769	<.01	−0.00338	0.227
Lineage-level, S/NS	−0.416	3.334	<.01	−0.00440	0.326

ORIGINATION RATES

Data set	r	t	P	s_o	O_0
Genus-level, S	−0.559	4.907	<.001	−0.00482	0.526
Genus-level, NS	−0.529	4.542	<.001	−0.00398	0.420
Lineage-level, S	−0.563	4.963	<.001	−0.00519	0.693
Lineage-level, NS	−0.615	5.683	<.001	−0.00486	0.560

EXTINCTION RATES

Data set	r	t	P	s_e	E_0
Genus-level, S	−0.218	1.622	n.s.	−0.00144	0.298
Genus-level, NS	−0.111	0.813	n.s.	−0.00059	0.193
Lineage-level, S	−0.114	0.836	n.s.	−0.00079	0.367
Lineage-level, NS	−0.088	0.640	n.s.	−0.00046	0.234

[a]Data are for the 55 Eocene—Pleistocene intervals. The independent variable is standing diversity; the dependent variables are per-taxon net diversification, origination, or extinction rates. Separate results are reported for genera and for lineages and have been standardized for geographic and sampling intensity effects by excluding some sampling areas and rarefying the remaining data (see text). Results are also given for analyses that include (S) or exclude (NS) single-interval taxa; note that these results are of necessity identical for net diversification.

[b]All coefficients are for data that have been averaged over 100 rarefaction trials. r, correlation coefficient; t = Student's t test; p, two-tailed p-value; $s_{\Delta D}$, slope of diversification rate against diversity; ΔD_0, = intercept of same, the intrinsic rate of diversification; s_o, slope of origination rate against diversity; O_0, intercept of same, the intrinsic rate of origination; s_e, slope of extinction rate against diversity; E_0, intercept of same, the intrinsic rate of extinction.

zoic, the 11th (55 to 54 Ma) being the first that falls entirely within the Eocene.

The predictions of the logistic model concerning diversification and origination are borne out for all three data sets, but interestingly there is no consistent relationship at all between per-taxon rates of extinction and standing diversity. Similar results were found by Sepkoski (1978) for Phanerozoic marine invertebrates. In this case, extinction even falls slightly with increasing diversity, the opposite of what any model would predict. Clearly, then, diversity is regulated by the dampening of origination rates and not by an increase

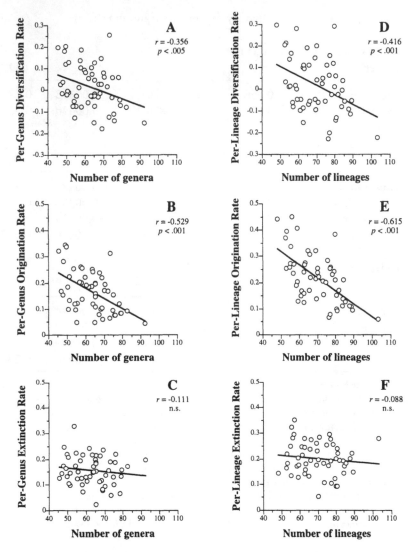

Figure 12.9. Dependency of per-taxon change in diversity **(A, D),** per-taxon origination rate **(B, E),** and per- taxon extinction rate **(C, F)** on standing diversity. Genus-level data are shown in A, B, and C, and lineage-level data in D, E, and F. The analyses are based on singletons-excluded sampling-standardized data for the Eocene–Pleistocene interval, and least-squares regression lines are plotted. The relationships show that diversification is nonrandom and can be modeled as a logistic growth function.

in extinction rates. Instead, extinction acts as a stochastic mechanism and presumably is related to abiotic or biotic factors that are independent of mammalian ecology and evolution, such as global climate change, sea level fluctuation, volcanism, bolide impacts, and evolutionary events in other terrestrial organisms. Although these patterns are fully compatible with the logistic growth model, as noted by Sepkoski (1978), they are more specifically in accord with the niche incumbency model described by Walker and Valentine (1984) that is discussed below.

There are several reasons to be cautious about these results. Most obviously, Foote (1994) showed that per-taxon, per-million-year rates of this kind are biased by variation in interval lengths, in addition to errors in interval-length estimates. Hence, if there is wide variability in the actual durations of time intervals, or if a time scale is poorly calibrated, per-taxon rates can be highly biased. In the current analysis, these problems are not a concern because the time intervals are of uniform duration (1.0 my) and the calibration is objective and relatively precise.

However, two additional concerns are not so easily dismissed. One is that per-taxon rates are proportional estimates, which means that their precision will scale to the denominator—in this case diversity. Because this kind of binomial error scales to the square root of the denominator, all regressions involving per-taxon rates are computed here by using the square root of diversity as a weighting factor. For example, a rate based on a standing diversity of 10 is given half as much weight as a rate based on standing diversity of 100. This correction makes it very unlikely that the negative slopes seen in the preceding analyses are an artifact of increased variance at low diversity levels due to binomial error.

Finally, and perhaps more importantly, Sepkoski (1978, 1991c) was aware of the linear per-taxon rate–diversity relationships outlined, but he chose not to perform direct regressions. Instead, he outlined an iterative curve-fitting method that employed plots of diversity against time. Because this method is computationally burdensome, vulnerable to small errors in diversity estimates for the initial growth phase, and hard to apply to curves that lack such a phase, it was abandoned by Wagner (1995a) and replaced with quadratic regressions of total rates against diversity. However, the details of Sepkoski and Wagner's methods are independent of Sepkoski's argument against per-taxon rates: if a turnover rate R is a completely random variate and one plots R/D_t against D_t, one always expects to see a spurious, curvilinear relationship of the form $1/D_t$ versus D_t— even in the absence of a true relationship between the variables.

There are three practical reasons not to be concerned with this problem in the current analysis. First, no model of diversification has ever been proposed that ignores standing diversity and per-taxon rates. Models as diverse as the "distance structured" (Walker 1985), "hierarchical" (Maurer 1989), and inverse exponential (Nee et al. 1992) all presume that whatever governs turnover, it acts on a per-taxon basis. Second, if such a bias really were significant, it would cause not just a negative correlation between per-taxon origination and diversity, but between per-taxon extinction and diversity. Thus, either a positive correlation for extinction, as predicted by the logistic, or no correlation at all, as predicted by the niche incumbency model of Walker and Valentine (1984) and seen here, would be a conservative result relative to this hypothesis.

Finally, the curvilinear relationship that could be introduced by regressing R/D_t against D_t could be recast as a linear relationship by logging both variables, yielding $\ln(R) - \ln(D_t) = \ln(D_t)$. Note that $\ln(R)$ is hypothesized to be a random variable and therefore can be treated as a "noisy" constant; and, more importantly, the slope of this log-log relationship is predicted to be -1. Analyses not presented here show that rather than improving the fit of a linear regression, log-transforming the variables has no effect; and that the true slope of the relationship is probably much steeper than -1 for origination rates, and shallower than -1 for extinction rates.

Hierarchical Diversification Model

Maurer (1989) perhaps was the first to publish an analysis comparing per-taxon origination and extinction rates with diversity levels, based on the earlier ideas of Rosenzweig (1975) and Sepkoski (1978). Unlike earlier workers, he attempted to predict the shape of the relationship by formal extrapolation from ecological principles. Because Maurer's argument is oblique and dependent on several nontrivial assumptions, I will focus on his prediction: per-taxon turnover rates should be a linear function not of diversity, but of the logarithm of diversity.

Maurer (1989) performed some analyses on the relatively small data set of MacFadden and Hulbert (1988) and was unable to find strong support for either formulation. Using the same F-ratio test, I compared the fit of the standard logistic (linear) and hierarchical (semi-log) models to the three data sets (table 12.4). The standard and hierarchical regressions differ only very marginally, and these differences are not even remotely significant. This is not a surprise for the extinction rate analyses, which only provide further evidence that such rates are independent of diversity. However, it is disappointing to find that the two models give the same fit to both the diversification and origination rates. Perhaps this is because the range in diversity is too narrow in the first place to distinguish models whose predictions should differ most greatly when diversity is very high or very low. Not surprisingly, then, the logistic and hierarchical models generate very similar functions.

There are additional reasons to disregard Maurer's model. First, similarly plausible ecological scenarios have been given for deriving the logistic equation (Sepkoski 1978, 1991a,c; Walker 1985). Second, Maurer's semi-log equations for origination and extinction rates cannot be summed to generate a simple, two-parameter net diversification function. This stands in stark contrast to the logistic model, whose parameters can be determined via linear regression for all three types of rates. For all these reasons, I will employ the standard logistic functions in the rest of this chapter even though the issue is not yet resolved.

Inverse Exponential Model

Yet another model has been offered by Nee et al. (1992), who were concerned with back-estimating net diversification rates from molecular phylogenies. As published, their method is vulnerable to violations of molecular clock assumptions, the possibility of mass extinctions in a group's history, and an intractably unrealistic assumption that diversity has always shown a net increase through time. In theory, however, all of these problems could be avoided by extending the model to origination and extinction rates instead of net diversification rates, and then applying it to paleontological data. This would yield a model that differs from the preceding ones by predicting turnover rates to be an inverse function of standing diversity raised to an exponent. A linearized function could be obtained by log transforming both turnover rates and diversity (as opposed to just diversity, as in Maurer's model), allowing the parameters to again be derived by linear regression.

Taking this approach yields the r^2 values reported in table 12.4. Surprisingly, F-tests show that the inverse exponential model is just as bad as the logistic or the hierarchical. I consider this unfortunate in light of the fact that a log-log model avoids the linear logistic model's most unrealistic assumption—that origination (or extinction) rates can take on negative values at very high (or low) diversity levels (Sepkoski 1991c). It does, however, share with Maurer's semi-log model the inability to add origination and extinction functions to produce a tractable, two-parameter diversification function.

Coupled Logistic Model

Sepkoski (1979, 1984) and Miller and Sepkoski (1988) demonstrated apparent long-term trends in the underlying equilibrium diversity levels of several Phanerozoic marine animal groups. Their explanation invoked coupled lo-

TABLE I2.4 Comparison of diversity dependence analyses using the standard logistic model (L), the hierarchical model of Maurer (1989) (M), and the inverse exponential model of Nee et al. (1992) (N). The differences among them are in regard to the scaling of the variables: the second and third models log-transform diversity, and the third also log-transforms the turnover rate.

NET DIVERSIFICATION RATES

Data set	r_L	r_M	r_N	F_M	F_N
Genus-level, S/NS	−0.356	−0.336	N/A	0.984	N/A
Lineage-level, S/NS	−0.416	−0.406	N/A	0.990	N/A

ORIGINATION RATES

Data set	r_L	r_M	r_N	F_M	F_N
Genus-level, S	−0.559	−0.543	−0.536	0.975	0.965
Genus-level, NS	−0.529	−0.515	−0.511	0.980	0.975
Lineage-level, S	−0.563	−0.569	−0.586	1.010	1.040
Lineage-level, NS	−0.615	−0.616	−0.630	1.002	1.031

EXTINCTION RATES

Data set	r_L	r_M	r_N	F_M	F_N
Genus-level, S	−0.218	−0.237	−0.223	1.010	1.003
Genus-level, NS	−0.111	−0.137	−0.136	1.007	1.006
Lineage-level, S	−0.114	−0.145	−0.145	1.008	1.008
Lineage-level, NS	−0.088	−0.117	+0.107	1.006	1.004

Data sets as in table 12.3. r, correlation coefficient for appropriate model; F, F statistic, ratio of transformed correlation $1 − r^2$ coefficients contrasting alternate models with the standard logistic model. F-ratios normally would involve residual mean square values, but equal sample sizes allow correlations to be used directly. All remaining F statistics are insignificant even at the $P < .10$ level.

gistic growth dynamics, i.e., interactions between pairs of diversity curves that could be modeled using Lotka-Volterra equations. However, it also is logically possible that interactions have no effect and that the equilibrium point evolves due to either abiotic environmental trends with potentially differential effects (temperature, CO_2 levels, habitable geographic area, etc.), or the evolution of the fauna itself, opening up new adaptive zones. I will not discuss these alternatives here because a more fundamental question—whether there is any statistical evidence for a change in the underlying equilibrium point—needs to be answered first.

There is, in fact, a significant correlation of diversity and time during the Eocene–Pleistocene interval (sampling-standardized genus-level data from figure 12.7A: $N = 55$; Spearman's $r = -0.338$; $t = 2.612$; $P < .01$). However, the correlation disappears in the lineage-level data (figure 12.7B: $N = 55$; Spearman's $r = -0.259$; $t = 1.949$; n.s.). Unless the equilibrium point really is stable, this result is surprising: time series are notorious for showing strong but fully artifactual correlations with other time series—such as time itself, as in this case (McKinney 1990a). Such problems should occur whenever the time series are Markovian, i.e., the values at any one time are direct functions of the values preceding them. This is certainly true of diversity in the standardized data set, which shows a very strong serial correlation of diversity at time t with diversity at time $t + 1$ (genus level: $N = 54$; Spearman's $r = +0.830$; $t = 10.743$; $P << .001$; lineage level: $N = 54$; Spearman's $r = +0.759$; $t = 8.417$; $P << .001$).

Despite these suggestive results, I will discuss a more robust approach: examining the residual values of per-taxon turnover rates that have been regressed against standing diversity levels (McKinney 1990a). If the underlying equilibrium point is changing, there should be a temporal trend in the residuals because the predicted changes are based on an assumed static equilibrium. If the real equilibrium point is low, the residuals should be negative; if it is high, they should be positive. Therefore, if the equilibrium point increases through time, the residuals should as well. Because it already has been shown that there is no relationship between standing diversity and extinction rate, I will focus on the relationship between diversity and origination rate and assume that the diversity/change-in-diversity relationship is a side effect of the latter.

In fact, the residuals show no correlation with time itself (figure 12.10) (genus level: $N = 54$; Spearman's $r = +0.111$; $t = 0.813$; n.s.; lineage level: $N = 54$; Spearman's $r = +0.039$; $t = 0.284$; n.s.). Therefore, the underlying equilibrium point does not change as a linear function of time. It also is important to show that the residuals have no serial correlation, because even if there is no long-term trend in the equilibrium point, there still could be long stretches of time during which the equilibrium level is suppressed or inflated. This would surely produce a strong serial correlation, but in fact there is almost none (genus level: $N = 54$; Spearman's $r = +0.240$; $t = 1.782$; n.s.; lineage level: $N = 54$; Spearman's $r = +0.272$; $t = 2.038$; $P < .05$). Based on these results, there seems to be little need to invoke either abiotic environmental trends, progressive evolution within the Mammalia, or coupled diversity interactions with other groups to explain the remaining variation in diversity levels and origination rates. However, this conclusion rests on excluding the Paleocene data; preliminary analyses show that there

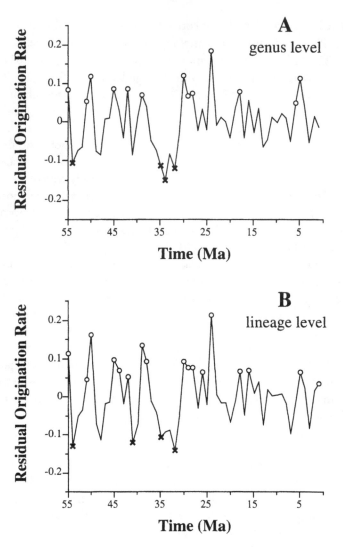

Figure 12.10.   Trends through time in residuals of predicted origination rates at the genus level **(A)** and lineage level **(B)**. Predictions are based on the relationships shown in figure 12.9B,E. There are no significant correlations between time and the residuals, or between neighboring residual values, showing that the underlying equilibrium point does not evolve through time. Origination pulses and shutdowns were identified by an iterative rate simulation test (see text). *Open circles,* origination pulses; *crosses,* origination shutdowns.

is some possibility of coupled logistic dynamics contrasting the dominant Paleocene orders with their surviving relatives.

Niche Incumbency

Walker and Valentine (1984) hypothesized that diversity may approach an equilibrium simply because origination rates are suppressed at high diversity levels, i.e., because niches are increasingly preoccupied by "incumbent" taxa. Based on ecological studies showing, for example, that competition only rarely leads to extinction (e.g., Simberloff 1981), Walker and Valentine claimed that extinction could be modeled as a constant, diversity-independent process. This already had been seen in an empirical analysis of the Phanerozoic marine fossil record by Sepkoski (1978), and a similar claim was made by Rosenzweig and McCord (1991) in their discussion of "incumbent replacement." In general, the niche incumbency model's predictions for diversity dynamics are corroborated by the results reported earlier, even though the fine-scale ecological interpretation of the model is debatable.

It is important to note that the island biogeography model of MacArthur and Wilson (1967) makes completely different predictions: they claim that equilibrium results because per-taxon extinction rates track diversity, and they do not address origination because of the short time scales they consider. Furthermore, they make an additional prediction that migration rates should fall with rising diversity. This opens the door for a final test of Walker and Valentine's model. Because newly evolved taxa are thought to derive from ecologically similar ancestors, native taxa should form small clusters in niche space, and evolutionarily distinct migrants should differentially occupy new, sparsely occupied parts of niche space. Therefore, diversity dependence should be weaker for migration rates than in situ origination rates. It also follows that if extinction is truly stochastic, then abiotically facilitated pulses of immigration should temporarily raise diversity above its normal dynamic equilibrium, and the rate of decay back to this level may be predictable from general diversity dependence of origination rates (see Marshall et al. 1982). These exciting predictions have clear ramifications for conservation biology and will be tested once immigrant and autochthonous genera in the data set have been distinguished.

Niche Saturation and the Cost of Equilibrium

Walker and Valentine (1984) went on to argue that the relative occupation of niche space (i.e., niche saturation) could be estimated from the extinction rate and intrinsic rate of increase. This was based on equating the size of the

niche space with the maximal number of taxa that could evolve in the absence of extinction. The maximum is set by the diversity dependence relationship and is equal to the number of taxa that correspond to a zero origination rate. This definition of niche space differs from that of many ecologists,
who equate the number of niches with observed diversity levels. Furthermore, it suffers from a logical flaw that seems fatal: the very notion of a "zero"
origination rate in the absence of extinction. Such a value is mathematically
permissible under the standard logistic growth model, in which per-taxon
origination rates are a linear function of diversity. However, this heuristic
over-simplification requires the biologically unrealistic possibility of negative
turnover rates (Sepkoski 1991c). Instead of reifying such an assumption, it
seems much more reasonable to admit that origination probabilities never
reach absolute zero under any circumstances, rendering the concept of "maximal" diversity useless.

Despite this, the niche saturation theory does include the kernel of a more
satisfactory replacement. This revised theory will rest on the concept of the
"cost of equilibrium" to origination rates, and its derivation depends on an
algebraic argument.

To begin with, note that Walker and Valentine (1984) admitted that their
estimate of the intrinsic rate of increase was based on an indirect and presumably unreliable method. They were unable to make direct niche saturation
estimates because no data set at the time had equally spaced temporal intervals, statistical control of sampling effects, and adequate sample sizes. Because these features are available in the current analysis, the parameters computed previously can be used to generalize Walker and Valentine's method to
allow for slight dependence of extinction rates on diversity (figure 12.11).
The equilibrium diversity level D_k can be computed by examining the two
linear diversity dependence functions that were introduced previously (equations 3 and 4). At equilibrium,

$$O_k = E_k,   [6]$$

where O_k is the equilibrial origination rate, and E_k is the equilibrial origination rate. Hence,

$$O_0 - s_o D_k = E_0 + S_e D_k,   [7]$$

leading to

$$D_k = (O_0 - E_0)/(s_e + s_o).   [8]$$

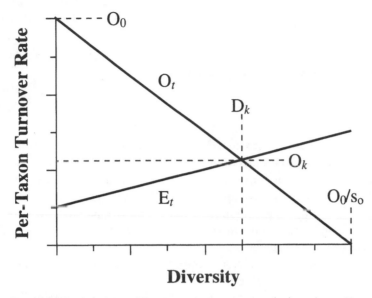

Figure 12.11. Calculation of niche saturation based on diversity dependence of turnover rates. Symbols as in figure 12.8, except the following: O_0, per-taxon origination rate when diversity is zero; O_k, per-taxon origination rate at equilibrium; s_o, slope of the origination rate versus diversity regression line.

Note that if the data are robust, the definition of ΔD_t as $O - E$ means that a regression of ΔD_t on D_t should yield an intercept equal to the numerator of equation 8, and a slope equal to its denominator. Within rounding error, this in fact is the case for analyses of all three data sets (see table 12.3). We now define the niche saturation ratio S_k as the ratio

$$S_k = D_k/([O_0/s_o]),$$ [9]

where O_0/s_o is simply the x-intercept of the origination dependence function, i.e., the "maximal" diversity. One then could compute S_k from the terms of the appropriate regression equation using

$$S_k = ([O_0 - E_0]/[s_e + s_o])/(O_0/s_o)$$ [10]

This computational exercise would of course be useless, because it depends on reifying the biologically unrealistic denominator, as argued previously. However, let us return to equation 9. Moving the s_o term and adding an arbitrary $O_0 - O_e$ term to the denominator leads to the following:

$$S_k = (s_o D_k + [O_0 - O_o])/O_0, \qquad [11]$$

which then, by using the definition of O_k as $O_0 - s_o D_k$ to justify a substitution, leads to a completely surprising result:

$$S_k = (O_0 - O_k)/O_0. \qquad [12]$$

Hence, we now have redefined niche saturation as the proportional decrease in the origination rate from its intrinsic value that is attained at equilibrium. In plain English, whenever niches are highly saturated, intrinsic origination rates are largely shut down, and when niche saturation is low, there is little impact on origination rates. Thus, equation 12 defines the "cost" of equilibrium to origination rates resulting from the preoccupation of niches by competing species.

The importance of this "cost of equilibrium" index is four-fold. First, by mathematically sidestepping the "maximal" diversity term, it allows us to describe niche saturation in biological terms that avoid the "maximal" diversity concept. Second, like the slope of the diversity dependence function, the new index can be combined with the intrinsic rate of increase and the equilibrium diversity level to specify the extinction rate, if that rate is constant. Third, because the index is dimensionless, its value is independent of the time scale used in a study, or of the fraction of true diversity that is represented in the analysis. This stands in stark contrast to the slope of the diversity dependence function. Finally, because it refers only to intrinsic and equilibrial origination rates, the index can be used regardless of which equilibrial diversification model one assumes. It makes no difference if these models force origination rates to always be positive, or if a model's slope coefficients are in units that differ from those of the linear model used here. There is only one difficulty with extending the index in this way: if a model manipulates the log of diversity instead of the raw values, as in the equations of Maurer (1989) and Nee et al. (1992), the "intrinsic" rate has to be redefined as the predicted value when $D = 1$, not when $D = 0$.

Applying the cost-of-equilibrium equation to the sampling-standardized data for the Eocene–Pleistocene interval leads to the niche saturation estimates given in table 12.5. Different results are given for all eight combinations of the following parameters: genus-level versus lineage-level, singletons included versus singletons excluded, and constant extinction assumed versus diversity-dependent extinction assumed. Note that the equilibrium diversity level seems robust to these manipulations, although the equilibrium is consistently about 10 percent higher at the lineage level (about 74 versus 67). A more striking result is that excluding singletons makes the estimated intrinsic

TABLE 12.5   Equilibrium and niche saturation estimates for the Eocene—
Pleistocene mammals of North America, based on the dependence of
origination and extinction rates on diversity levels (see text).

Data set	Singletons	Extinction	D_k	O_k	O_0	S_k
Genus-level	Included	Dependent	67.5	0.201	0.526	0.618
Genus-level	Included	Constant	66.4	0.206	0.526	0.608
Genus-level	Excluded	Dependent	67.0	0.153	0.420	0.636
Genus-level	Excluded	Constant	66.6	0.155	0.420	0.631
Lineage-level	Included	Dependent	74.1	0.308	0.693	0.556
Lineage-level	Included	Constant	73.6	0.311	0.693	0.551
Lineage-level	Excluded	Dependent	74.1	0.200	0.560	0.643
Lineage-level	Excluded	Constant	73.9	0.201	0.560	0.641

Level, taxonomic level of analysis; singletons, single-interval taxa included or excluded in analysis; extinction, model of extinction; D_k, equilibrium diversity level; O_k, rate of origination at equilibrium; O_0, intrinsic rate of origination; S_k, niche saturation, the difference of O_0 and O_k divided by O_0.

rates of origination, and therefore the cost of equilibrium, consistently approach the value of 0.64.

The problem here may be that the singleton taxa inflate the extinction figures in an unpredictable way, being more vulnerable to residual sampling effects that the rarefaction procedure failed to remove. Another possibility is that because lower niche saturation at the lineage level is to be expected, the singletons may contain useful information after all. The argument here is that genus-level niches are little more than the sum of species- (and thus lineage-) level niches. If this is true, then the niche of a genus is occupied whenever at least one of its species is extant; all of the genus-level niches could be occupied even when many of the species-level niches are empty; and a random distribution of species across genera *should* produce a lower degree of species-level saturation.

All of this creates a minor dilemma: are the patterns an artifact of poor data quality, or do they show that species are nonrandomly clumped within genus-level "mega-niches"? Testing this will require a combination of more sophisticated randomization tests and acquisition of additional data such as an outline of genus-level phylogenetic relationships.

Effect of Pseudoextinction on Niche Saturation

There are two additional empirical problems with the species-level results. One is that the extant terrestrial mammalian biota of the Nearctic Realm includes 509 species and 126 genera (Cole et al. 1994), which for species is far more than the supposed equilibrium diversity levels derived from this analysis. This, however, is a deceptive comparison because the diversity dependence statistics pertain only to taxa that have been sampled. All that is expected of these statistics is that they are proportional to total diversity. Therefore, if the extant fauna is close to equilibrium, then, for example, its current per-taxon turnover rates should have been predicted accurately by the preceding analyses.

A more serious problem is that niche space (and therefore niche saturation) estimates can be biased severely by pseudoextinction, i.e., anagenesis leading to strictly taxonomic changes that falsely indicate extinctions and originations within a continuing lineage (Stanley 1979; Walker and Valentine 1984). Whenever pseudoextinction is common, niche space estimates will be too high and saturation estimates will be too low.

In addition to the 19 percent pseudoextinction rate that was derived by the "minimal lineage" computation, there are several phylogenetic studies that allow direct estimation of such rates in North American fossil mammals. Two of them provide large sample sizes. Archibald (1993) used taxonomically wide-ranging phylogenetic analyses to estimate that early and middle Paleocene mammals in North America had pseudoextinction rates of only 19 percent at the species level (7 out of 37 disappearances). In a study of Neogene North American equids, Hulbert (1993) gives data to support a rate of 16 percent (6/38) in the extinct tribe Hipparioni, 33 percent in the surviving tribe Equini (11/33), and 24 percent overall (17/71).

These figures are all very similar. Supposing that, at worst, one third of the species-level extinctions are pseudoextinctions, then the niche saturation proportion can be recomputed as follows. The revised intrinsic per-lineage, singletons-excluded origination rate $O_{0'}$ decreases to

$$O_{0'} = O_0 - (1/3 \, E_0) \qquad [13]$$

or $0.560 - (\tfrac{1}{3} \times 0.234) = 0.482$. The revised equilibrial origination rate $O_{k'}$ falls to

$$O_{k'} = 2/3 \, O_k \qquad [14]$$

or $\tfrac{2}{3} \times 0.200 = 0.133$. The revised niche saturation is thus $(0.482 - 0.133)/0.482 = 0.724$, only slightly higher than the original estimate of 0.641.

If the estimated pseudoextinction rate is reasonable, the impact of genus-level pseudoextinction on the saturation statistics is probably similar. This is because lineage-level anagenesis rates should be about the same as genus-level rates; the extinction of a genus is defined by the extinction of the last species within the genus, and if the last species' extinction is actually a pseudoextinction, so is that of the genus. Note here that pseudoextinctions or real extinctions of lineages other than the latest one have no direct bearing on the fate of the genus per se. Similar recomputations for the singletons-excluded genus-level data yield revised intrinsic and equilibrial origination rates of 0.356 and 0.102, and a niche saturation estimate of 0.713. Based on these modified saturation figures, it appears that when diversity is at equilibrium, an impressive five sevenths of all potential origination is shut off because of the preoccupation of niches by competing species.

Extinction Models

This section discusses three models of extinction rates: Van Valen's law, the Phanerozoic decline of extinction rates, and the kill curve. The data support the first, disagree with the second, and are ambiguous with respect to the third. Of course, it should be remembered that these results apply only to terrestrial mammals.

Van Valen's Law and the Red Queen's Hypothesis

Van Valen (1973) proposed as a new "evolutionary law" the notion that the environment of "any homogeneous group of organisms deteriorates at a stochastically constant rate." This essentially was to say that for any ecologically uniform group, extinction probabilities are independent of time. After demonstrating his law using dynamic survivorship data for many different fossil groups (Van Valen 1973), Van Valen (1985b) showed it in a separate way using stage-level diversity data on Phanerozoic marine families taken from Sepkoski (1982). Groups ranging from sponges to bony fishes all showed insignificant correlations of time and extinction rate (see also Gilinsky and Bambach 1987). In other words, extinction probabilities are largely independent of either the age since origin of individual taxa or of time itself.

The current results bear Van Valen out. Using the singletons-excluded sampling-standardized data for the entire Cenozoic (figure 12.12), there is a correlation of time and singletons-excluded, per-taxon extinction rate (two-tailed test: $N = 65$; genus-level: $r = +0.343$; $t = 2.902$; $P < .01$; lineage-level: $r = +0.376$; $t = 3.225$; $P < .01$), but this disappears after excluding the ten Paleocene data points, with their high binomial errors and below-equilibrium diversity levels ($N = 55$; genus-level: $r = +0.016$; $t = 0.113$;

n.s.; lineage-level: $r = +0.081$; $t = 0.594$; n.s.). Interestingly, per-taxon origination rates show similar patterns (figure 12.13): a significant overall trend ($N = 65$; genus-level: $r = +0.352$; $t = 2.984$; $P < .01$; lineage-level: $r = +0.379$; $t = 3.252$; $P < .01$), but no significant trend using the Eocene—Pleistocene data only ($N = 55$; genus-level: $r = +0.078$; $t = 0.566$; n.s.; lineage-level: $r = +0.126$; $t = 0.921$; n.s.). Note that the correlations including the Paleocene are positive because turnover is higher in that interval and the time scale, in Ma, increases into the past.

Van Valen (1973) went on to hypothesize that the biotic cause for the pattern was competition and predator–prey interactions leading to constant evolution within each lineage. As the lineages adjust to each other, some fail to keep up and go extinct. Lineages, in other words, are like Lewis Carroll's Red Queen: They run continually but never make any progress. The Red Queen's hypothesis is fully compatible with the general concept of equilibrial diversity dynamics. However, it makes differential predictions with respect to the relative importance of extinction and origination in regulating that process. If it is true that extinction is an epiphenomenon of ecological interactions, then in an equilibrial system extinction should scale to diversity. As diversity increases, the Red Queen forces the competitors to run ever faster. However, the current results show that it is origination and not extinction that is dependent on standing diversity.

I see no way to reconcile this result with Van Valen's original hypothesis. However, Van Valen (1985a) himself recognized that the theory in its raw form would have to be modified; perhaps competition within major groups is more important in regulating origination than in regulating extinction, at least once an initial exponential diversification has been surpassed. This generally is concordant with the opinions of Walker and Valentine (1984) and Sepkoski (1991a), and with the empirical results of Sepkoski (1979), Gilinsky and Bambach (1987), Maurer (1989), and Gilinsky and Good (1991). Given these results, the important task for ecologists and geneticists is to explain how microevolution and short-term population dynamics could lead to the regulation of origination rates but not of extinction rates by diversity levels. These results also are relevant to conservation biology.

The Phanerozoic Decline of Extinction Rates

Raup and Sepkoski (1982) argued that the overall extinction rate for marine animals has declined through the Phanerozoic (see also Van Valen 1984; Gilinsky 1994, and papers cited therein). It might seem that Van Valen's law is in conflict with their claim. There are several reasons to think that it is not. To begin with, Van Valen's law originally was based on regressions of survivorship against the time since origination for each taxon, not against time per se. Perhaps survivorship is independent of taxonomic age, but overall extinction rates still fall through time. However, the regressions used here, like those of Van Valen (1985b), explicitly demonstrate that extinction rates are static during the interval of concern.

Van Valen (1985a,b) was well aware of this problem, and in fact his data were used previously to demonstrate the overall decline pattern that creates the dilemma (Raup and Sepkoski 1982; Van Valen 1984). He explained the discrepancy by arguing that the overall decline is a result of sorting among groups; as time goes on, groups with higher intrinsic extinction rates become depauperate or disappear entirely, even though the intrinsic rates are invariant for each group. Similar arguments were made by Sepkoski (1984) and Gilinsky (1994). Thus, it may be the case that mammals are just one of many terrestrial groups that are being sorted by their intrinsic diversity dynamics through time; within mammals, the dynamics are invariant, but the terrestrial biosphere as a whole may show varying dynamics.

Another possibility is that the results of Raup and Sepkoski (1982) point to a fundamentally different process operating in a fundamentally different system. After all, their data pertain to a much longer time interval (the Phanerozoic), a much higher operating taxonomic level (families versus narrowly defined mammalian genera), and a very different group of organisms (marine animals versus terrestrial mammals). Finally, Pease (1992) may be correct in interpreting the marine pattern as an artifact of the increasing availability of fossils through time, as conceded by Gilinsky (1994). The current sampling-standardized genus-level data set suffers from no such bias.

The Kill Curve Hypothesis and the Myth of Pulsed Extinction

Raup (1991a) suggested that there may be no clear difference between "background" extinction, "pulsed" extinction, and true mass extinctions such as the Cretaceous–Tertiary boundary event. Perhaps all observed extinction rates instead form an unbroken continuum, or kill curve, whose distribution follows a simple mathematical equation that also explains natural disasters such as earthquakes. His argument also could be extended to origination rates.

Raup's hypothesis may seem intuitive because the paleontological literature is rife with purported "extinction" events. Most relevant here, Webb (1984) has claimed that no less than six pulsed extinctions have affected North American mammals during the last 10 Ma. This widely-cited conclusion is based on simply tabulating the number of last appearances and the total number of extant taxa within each land-mammal "age" or sub-"age," and then dividing the former figure by the latter. By this methodology, the high intrinsic turnover rates of mammals guarantee that virtually every "age" boundary will appear to be marked by an "extinction event," even if the actual underlying rates are completely invariant.

As it happens, both the kill curve theory and Webb's pulsed extinction model make a simple prediction: If there are few or no pulsed extinctions to start with, then the variation in rates should be indistinguishable from a binomial distribution—or, in the worst case, removing just a few apparently pulsed events from the time series should produce such a pattern. Here I will test the hypothesis using an iterative turnover rate simulation algorithm (Alroy 1996). The method is inspired by Raup and Marshall (1980) and Barry et al. (1990), who tried to identify peaks in turnover rates by computing chi-square values for time series of extinction and origination counts. Those intervals contributing most greatly to the overall chi-square were treated by Barry et al. (1990) as if they were extrinsically produced turnover pulses.

The chi-square approach suffers from a fatal problem, also noted by Wagner (1995a): It requires a minimal count of turnover events for each time interval, but real data sets frequently include few or no events in many intervals. Furthermore, the fact that the diversity curves are averages across 100 rarefaction trials would require that the test be performed separately on each of the rarefied curves. Because some intervals are poorly sampled to start with, and therefore vary little across the rarefaction trials, the curves are too complexly interdependent to be analyzed with standard parametric techniques.

I therefore developed an iterative rate simulation test as a substitute (Alroy 1996). The algorithm is as follows: Compute the variance in turnover rates for the time series; repeatedly simulate turnover events using appropriate underlying probabilities and observed standing diversity levels; compute the variance across each simulation; remove the interval contributing most greatly to the observed variance if it differs from the simulated variance at the $P < .01$ level; and continue to remove intervals in this way until the remaining ones are homogeneous. Variability in origination rates caused by their dependence on standing diversity levels is taken into account by predicting underlying origination probabilities on the basis of diversity dependence equations; for extinction, these probabilities are simply equated with the mean rate across all the intervals. Single-interval taxa are excluded: Because they are not present at the beginning of the interval in which they appear, their turnover probabilities should be lower than those of other taxa, but making such an adjustment requires unwanted ancillary assumptions.

There are several reasons to think that if the test finds significant variation in rates, this could result solely from sampling biases that artifactually clump originations and extinctions. However, the major types of sampling bias appear to be inconsequential in the current analysis. For example, the problem of not having enough taxonomic lists relative to the number of sampling intervals is unlikely to be important, because the average rarefied data set includes 987.8 lists and 55 intervals. That of taxonomic records tending to be concentrated in a small minority of unusually rich taxonomic lists can be ignored for the same reason. Finally, the general problem of lists and/or records being concentrated in certain time intervals was dealt with earlier by rarefying the data set.

Another concern is that Wagner (1995a) presented a different approach that involves computing the binomial probability that an observed number of extinction (or origination) events could be generated by a uniform underlying turnover probability across all time intervals. Why, then, revert to a simulation approach? The most important reason is that Wagner's test statistic is the number of turnover events, but the statistic of concern here is the variance among turnover rates. Finding the exact probability of generating the observed variance across an entire time series would require prohibitive computation. Furthermore, Wagner's test makes the additional simplifying assumption that a single underlying turnover probability can be used as a null hypothesis for all the time intervals, despite the fact that finding any one "significant" interval renders this probability meaningless in testing all the others. Iteratively removing intervals and recomputing turnover probabilities avoids this difficulty.

Despite all of these precautionary measures, the iterative test identifies a surprisingly large number of pulses and shutdowns in the sampling-standardized data. Each of the 100 curves was simulated 100 times. For extinction, only five (genus-level) to ten (lineage-level) of the 55 intervals need to be removed to make the variance among the remaining intervals insignificant. Note that extinction in the 6- to 5-Ma interval is pulsed at both taxonomic levels. For origination, however, the number of outlying intervals inflates to 17 or 21, with respectively 13 and 17 of these intervals being pulses (see figure 12.10). This clearly goes far beyond the predictions of simple equilibrial dynamics.

Another way to express the extreme variability in rates is to compare the observed standard deviations for the time series (figures 12.12 and 12.13) with values predicted by the homogeneous underlying rate model. For the Eocene–Pleistocene, observed extinction rates average 0.161 ± 0.070 (genus-level) or 0.207 ± 0.078 (lineage-level). The 10,000 simulations produce rates with average standard deviations of 0.048 or 0.050. Observed origination rates, on the other hand, averaged 0.177 ± 0.102 (genus-level)

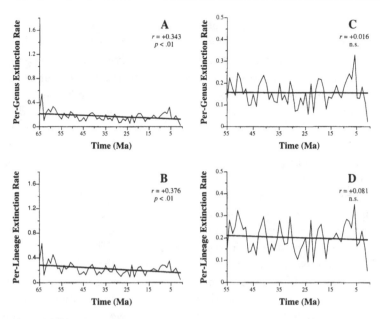

Figure 12.12. Trends through time in singletons-excluded per-taxon extinction rate. Least-squares regression lines are plotted; scaling is intended to facilitate comparisons with figure 12.13. The significantly negative trends disappear when the Paleocene (65 to 55 Ma) points are excluded. **(A)** Genus-level data including the Paleocene. **(B)** Lineage-level data including the Paleocene. **(C)** Genus-level data excluding the Paleocene. **(D)** Lineage-level data excluding the Paleocene.

or 0.289 ± 0.121 (lineage-level), whereas the simulated standard deviations were 0.050 or 0.052. Therefore, observed extinction rates are less than half again more variable than one would expect, but origination rates are more than twice as variable.

These highly variable origination rates may be caused by major immigration episodes mediated by the short-term emergence of land bridges (e.g., Gingerich 1989; Webb and Opdyke 1995; Woodburne and Swisher 1995), or more likely by major adaptive radiations (e.g., MacFadden and Hulbert 1988). Interestingly, Foote (1994) found the opposite pattern in Phanerozoic marine animals, whose extinction rates are more variable than origination rates. Together with the simulated rate test results, the variability ratios show that for mammals at least, origination rates are far too heterogeneous to be attributed to steady "background" turnover rates punctuated with a few qualitatively distinct turnover pulses, whereas the notion of "pulsed" extinction rates is a myth.

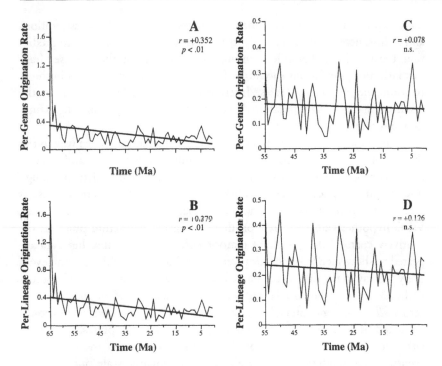

Figure 12.13. Trends through time in singletons-excluded per-taxon origination rate. As in figure 12.12, the negative trends are created by high rates in the Paleocene. Note that origination is much more variable than extinction. **(A)** Genus-level data including the Paleocene. **(B)** Lineage-level data including the Paleocene. **(C)** Genus-level data excluding the Paleocene. **(D)** Lineage-level data excluding the Paleocene.

Alternative Diversification Models

This final set of analyses presents evidence against three generalized, non-equilibrium models of turnover: the turnover pulse hypothesis of Vrba, the correlated extinction and diversification hypothesis of Stanley, and the coordinated stasis hypothesis of Brett and Baird and other authors. Although the first and last of these appear to be tested fairly, my discussion of Stanley is essentially a warning against applying his ideas to patterns within individual groups, as opposed to comparisons among groups.

The Turnover Pulse Hypothesis

Vrba (1985) has constructed an elaborate model of population dynamics, ecological specialization, and macroevolutionary process, called the turnover pulse hypothesis. Although the theory involves many subsidiary assumptions and claims, here I will discuss only its most clear-cut and easily tested prediction: During times of abiotic environmental disturbance, the increased fragmentation of populations should lead to simultaneous increases in both extinction and origination rates. Interestingly, this prediction rests on the assumption that metapopulation dynamics are crucial to determining extinction and origination rates.

As recognized by Barry et al. (1990), the most straightforward way to test the model is to see if statistically significant pulses in extinction and origination occur together. Of course, this depends on being able to identify enough discrete pulses to see a correspondence, and the simulated rate test results given earlier pick out only a handful, especially for extinction. Fortunately, Vrba's hypothesis does not rely on there being *any* extinction pulses at the relatively coarse 1.0-my scale of temporal resolution, because her data appeared to show such frequent turnover pulses that one would expect an event every 1.0 my in the first place. Therefore, one really should test the hypothesis by examining the correlation between extinction and origination rates across *all* of the time intervals (Alroy 1996). A strong positive correlation is predicted because intervals encompassing many fine-scale pulses, or particularly strong fine-scale pulses, should show many extinction and origination events, whereas intervals with weak or infrequent fine-scale pulses should show little overall turnover.

At first glance, there does seem to be such a pattern. For the sampling-standardized Eocene–Pleistocene data, per-genus extinction and origination rates are weakly correlated (figure 12.14A, two-tailed test: $N = 55$; $r = +0.283$; $t = 2.144$; $P < .05$), as are per-lineage rates (figure 12.14B: $r = +0.181$; $t = 1.337$; n.s.) Unfortunately, these correlations are a statistical mirage. An algebraic proof not presented here shows that the observed correlation would be the same if taxa restricted to one interval were subjected to the same underlying extinction and origination probabilities as any other taxa. However, when turnover rates are recomputed excluding these single-interval taxa, the correlations disappear both at the genus level (figure 12.14C: $r = +0.026$; $t = 0.188$; n.s.) and at the lineage level (figure 12.14D: $r = -0.226$; $t = 1.687$; n.s.). Because there is no plausible biological mechanism that might create such a difference, I must conclude that the original correlations largely reflect a sampling bias. The correlations are strong because counts of originations and extinctions both are inflated artificially whenever large numbers of rare, temporally restricted taxa are sampled.

The fact that the sampling-standardized data still show this subtle bias is cause for concern, but it did not affect the other analyses discussed in this

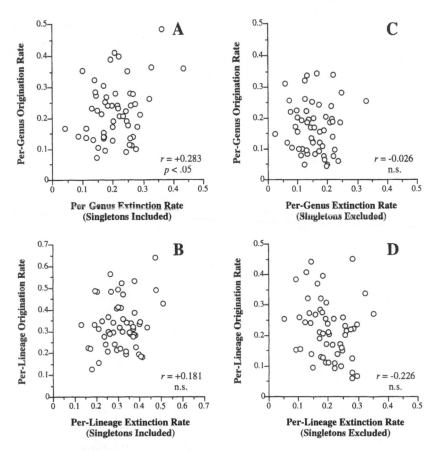

Figure 12.14. Correlation of extinction and origination rates. Data are for the Eocene—Pleistocene (55 to 0 Ma) and are sampling-standardized. The significant correlations disappear after excluding taxa that are restricted to a single time interval, indicating that apparently correlated turnover pulses are merely a statistical artifact. **(A)** Genus-level data including single-interval taxa. **(B)** Lineage-level data including single-interval taxa. **(C)** Genus-level data excluding single- interval taxa. **(D)** Lineage-level data excluding single-interval taxa.

chapter. This was true, for example, of excluding single-interval genera from the diversity dependence analyses: The correlation of per-genus origination rate with diversity was about the same ($r = -0.559$ versus -0.529), whereas that of extinction rate with diversity grew even weaker ($r = -0.218$ versus -0.111). In summary, I conclude that Barry et al. (1990) were right to reject the turnover pulse hypothesis: Long-term mammalian diversity histories disconfirm it despite the fact that the theory was developed in the first place using the fossil record of mammals (Vrba 1985). Although Flessa and Levinton (1975) found no origination–extinction correlation for raw Phanerozoic marine invertebrate families, Mark and Flessa (1977) did find a positive correlation for Phanerozoic brachiopod genera. I suspect that the latter result was influenced by including single-interval taxa in the data set.

Causally Linked Extinction and Diversification

Stanley (1979, 1990) has shown that there is a positive correlation among groups between overall rates of extinction and diversification ("Stanley's law"). Stanley has argued that this is the result of a "fortuitous" biological cause called the fission effect: Ecologically, geographically, and behaviorally restricted and/or variable populations are both more likely to go extinct and more likely to undergo cladogenesis. Many of these factors depend on metapopulation substructuring, but not all of them. Because the illustrated relationship is between extinction and net diversification (i.e., origination minus extinction), not origination alone, Stanley's claim is quite strong: everincreasing "fission" should lead to more extinction but still more origination. This prediction also serves to distinguish Stanley's law from the turnover pulse hypothesis of Vrba (1985) and from the among-group correlation of extinction and origination rates described by Gilinsky (1994).

The present data set does not provide among-group data such as those used by Stanley (1979), because only one group is available for study. Of course, the data could be subdivided into lower taxonomic categories, such as orders, but that is not the purpose of the present chapter. My goal here is merely to warn that Stanley's model is broadly incompatible with constrained diversification, and in particular logistic growth; when applied to a single, logistically diversifying group, it yields very limited predictions that fail to be corroborated. Furthermore, the analysis Stanley used to support his model crucially depends on logistic growth not being present.

The first point is illustrated as follows. To begin with, suppose that Stanley's fission mechanism applies within each group, not just among groups. This seems plausible because most of the causal factors Stanley (1979, 1990) invokes to explain fission—niche breadth, population size, dispersal ability, and habitat fragmentation—are influenced by abiotic environmental circumstances, and therefore they should vary not just among groups, but through time within any particular group. The only likely exception is behavioral complexity.

The logistic model, of which the niche preoccupation model is a special case, predicts that per-taxon origination rates will fall with rising diversity, whereas per-taxon extinction rates will either rise or remain constant. This has a clear implication: Origination and extinction should be either negatively correlated or uncorrelated. The results in the last section showed that the latter is true. Because diversification is equal to origination minus extinction, it now follows that extinction should be *negatively* correlated with diversification in either case. This can be visualized by imagining the falling linear function "−1 times E" being subtracted from the flat or falling origination–extinction correlation line to yield an even more steeply negative relationship.

In fact, an analysis of the Eocene-to-Pleistocene interval does demonstrate a significantly negative relationship between per-genus extinction and diversification rates (figure 12.15A, two-tailed test: $N = 55$; $r = -0.444$; $t = 3.609$; $P < .001$). The same pattern is seen in the lineage-level data (figure 12.15B: $r = -0.497$; $t = 4.165$; $P < .001$). This holds despite each analysis including taxa that are restricted to one interval, which previously created a strong, artifactual correlation between extinction and origination rates (see earlier). When single-interval genera are excluded, the correlation grows substantially more negative at the genus level (figure 12.15C: $r = -0.590$; $t = 5.322$; $P < .001$), and it plunges even further at the lineage level (figure 12.15D: $r = -0.675$; $t = 6.663$; $P < .001$).

In theory, at least, there is a way to resolve the conflict between constrained growth and Stanley's model. Perhaps both per-taxon origination and extinction rates fall with diversity, but extinction rates start out being lower at very low diversity levels and then fall off more slowly. This would allow positive diversification rates at low diversity levels and negative rates at very high levels, resulting in an equilibrium. Of course, the diversity dependence regressions shown earlier do not strongly confirm these predictions, which comes as no surprise because the restrictions on the predicted slopes and intercepts are so narrow in this scenario.

My second warning, which concerns Stanley's data, follows that of Van Valen (1980). Stanley (1979, 1990) computed the intrinsic rate of diversification as a simple function of time and diversity in the Recent. This is appropriate for exponential growth, as his equations show; but if each group has separately attained an equilibrium after growing logistically, and the amount of time since reaching equilibrium is unknown, his method provides no information on the initial, intrinsic rate of diversification. Instead, this rate would have to be computed by performing a diversity dependence analysis for each group of the kind presented in this chapter, and then using the resulting equations to predict the diversification rate at very low standing diversity levels.

This point is important because Stanley (1979, 1990) used mammals as one of the key animal groups supporting his extinction–diversification relationship. Because mammals reached their equilibrium point no later than in

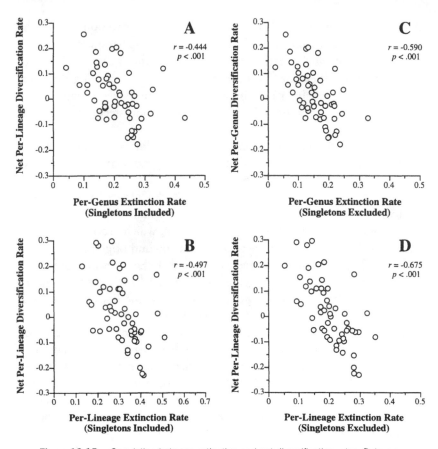

Figure 12.15. Correlation between extinction and net diversification rates. Data are for the Eocene—Pleistocene (55 to 0 Ma) and are sampling-standardized. The significantly negative relationships show that Stanley's hypothesis linking extinction and diversification rates does not apply within a single clade. The result is not affected by excluding single- interval taxa. **(A)** Genus-level data including single-interval taxa. **(B)** Lineage-level data including single-interval taxa. **(C)** Genus-level data excluding single-interval taxa. **(D)** Lineage-level data excluding single-interval taxa.

the Early Eocene, Stanley's exponential growth-based coefficients surely are not accurate. His other data points included marine groups such as gastropods, shown by Wagner (1995a) to have experienced logistic growth as far back as the middle Paleozoic, and terrestrial groups such as insects and birds, which again have been shown to diversify logistically by Labandeira and Sepkoski (1993) and Nee et al. (1992). Of course, logistic growth was demonstrated among Phanerozoic marine groups in general by Sepkoski (1978) before Stanley (1979) even published his initial hypothesis.

All of this argues that whatever Stanley's "intrinsic rates" represent, they are not a direct function of the typical diversification rates for any of the animal groups he studied. Instead, they must be some function of those groups' equilibrial diversity levels. Perhaps, then, his analysis indicates that groups with high extinction rates have high equilibrium levels. I hope to test this interestingly counter-intuitive hypothesis in the future.

Coordinated Stasis

Because the current results show that a form of macroevolutionary stasis is at work, they could be construed as an endorsement of the coordinated stasis hypothesis (Brett 1995; Brett and Baird 1995). On the contrary, the coordinated stasis scenario has little to do with equilibrium in taxonomic diversity levels. It instead posits that morphological evolution, ecological abundance, and taxonomic diversity are almost completely static across entire basins for millions of years. Then, during brief "reorganization" intervals, communities across a basin are completely disrupted, resulting in widespread extinction, rapid speciation among the survivors, and the establishment of previously unsuccessful immigrants. This generally is compatible with either equilibrial or nonequilibrial diversity dynamics, because the important idea is merely that rates of morphological change and taxonomic turnover should be highly variable.

Coordinated stasis shares many features of the turnover pulse hypothesis (Vrba 1985) and the much earlier chronofauna concept (Olson 1952, 1983), but it differs from them quite importantly in having ideas about morphological stasis that derive from punctuated equilibrium theory (Eldredge and Gould 1972; Gould and Eldredge 1977), in also postulating stasis in relative abundance, and in making some specific numerical estimates of turnover rates. However, the connection to punctuated equilibrium was elaborated decades earlier by Beerbower (1953) in a hypothesis linking the quantum evolution theory of Simpson (1944) to the breakdown of chronofaunas sensu Olson (1952). Therefore, the most clearly new idea associated with the coordinated stasis theory is a hypothetical mechanism for creating the chronofaunal pattern: "ecological locking, . . . whereby incumbent species resist invasion and maintain nearly constant proportions" (Morris 1995a).

Although these ideas are foreshadowed by Olson (1952), for example, and imply nothing about community structure that ecologists have not already debated for decades (see review by Pimm 1994), they at least provide a concrete means of testing the hypothesis. Six predictions seem to be germane in this context. Because another paper discusses the first three of these tests in detail (Alroy 1996) and the second two cannot yet be addressed in detail, I will summarize the argument briefly.

First, the turnover pulse hypothesis (Vrba 1985) must be correct and extinction and origination rates across different time intervals therefore must be strongly correlated. A correlation of this kind in the raw data already has been shown to be an artifact related to the sampling of rare, temporally restricted taxa.

Second, variation in turnover rates should be so extreme that a bimodal distribution should be observed, with most intervals evidencing zero or near-zero turnover rates and a few others showing extremely high rates. Specific claims about such rates were made by Brett and Baird (1995), who state that "<10%" of species should go extinct during 3- to 7-my intervals of stasis, interrupted by 0.01- to 0.1-my intervals in which 80 to 90 percent of species are replaced. Although turnover rates in this data set are highly variable, they cannot be shoehorned into such a simplistic, two-mode model. For origination, the large number of "pulses" simply makes it impossible to distinguish major turnover episodes and intervals of "background" turnover, as shown earlier, whereas for extinction, the number of "pulses" is miniscule. Therefore, there is no long-term stasis in diversity, origination and extinction bear no consistent relationship, and there are no sweeping "reorganizations."

Third, there should be positive interactions among the diversity histories of major ecological and/or taxonomic groups (e.g., carnivores, insectivores, small herbivores, ungulates) because constant relative abundances should scale up to constant relative diversity levels if diversity patterns have as strong an ecological foundation as proponents of coordinated stasis believe. In other words, when herbivore abundance and therefore diversity increase, carnivore abundance and therefore diversity should increase as well. This conjecture is demonstrably false for North American Cenozoic mammals (Alroy 1996).

Fourth, the Red Queen's hypothesis (Van Valen 1973) must be false because it posits continual competition-driven extinction, which should be absent if relative population abundances are "nearly constant" on geological time scales. The failure of the Red Queen's hypothesis in these terms already has been demonstrated, which could lend some support to the coordinated stasis theory.

Fifth, migration events should be frequent in the aftermath of extinction pulses and almost absent when turnover rates are otherwise low. Recently published data that identify intercontinental migrants in the data set appear to contest this claim (Woodburne and Swisher 1995), but pending a full analysis I offer no further comment.

Finally, most major turnover events should correspond to major climate changes, although some events should have no impact on turnover. Again, at present I have not been able to fully analyze the relevant data, although climate proxies such as oxygen isotope ratios and gradients in floral physiognomy are available. However, it does seem relevant that mammalian diversification patterns in the face of the three most important global climate changes of the Cenozoic do not form a coherent picture.

The first of these is the Paleocene–Eocene boundary isotope excursion at 55 Ma (Rea et al. 1990; Koch et al. 1992). This severe, short-lasting warming pulse was followed by a long-term greenhouse climate episode that persisted through the Early Eocene. It is marked by clear-cut origination pulses in both the genus- and lineage-level data, but no such pulse is seen in extinction rates. The second is the earliest Oligocene Antarctic glaciation episode [34 Ma (Zachos et al. 1992)], roughly coincident with the Grand Coupure in Europe (Legendre and Hartenberger 1992), which similarly was followed by an icehouse interval that persisted for several million years. Here the mammals show a shutdown in origination rates and again no clear pattern in the extinction data. The third is the Messinian salinity crisis that ends at the Miocene–Pliocene boundary [5 Ma (Möller and Mueller 1991)]. This event is accompanied by severe global cooling and seems to result both in the most severe extinction pulse of the Cenozoic and in a slightly later origination pulse. One should note that there are several other Cenozoic climate events that nearly rank in this company, but they have no obvious tie to turnover rates. For example, the Great American Interchange [2.5 Ma (Stehli and Webb 1985)], which is primarily a tectonic and biogeographic event but is coincident with substantial climate change, seems to have no clear impact on diversity patterns in North American mammals, contra Marshall et al. (1982).

The important point is that climate events rarely have a clear-cut relationship to turnover patterns, and when they do, they can have unpredictable effects: cooling accompanies an origination slowdown in the earliest Oligocene, but an extinction burst followed by a possibly compensatory origination burst in the end-Miocene. Although the fact that diversity patterns are independent of some climate changes generally confirms the coordinated stasis model, the contradictory effects seen in these two cases do not. Webb and Opdyke (1995) and Woodburne and Swisher (1995) presented similar evidence that oxygen isotope excursions bear only a complex relationship to North American mammalian turnover episodes.

On balance, these results are strongly discordant with the coordinated stasis model. Perhaps, then, the original pattern of Brett and Baird (1995) is a large-scale taphonomic artifact of the kind that can be created by major sea level changes (Holland 1995). If not, one might argue instead that the present data set does not allow a fair test of coordinated stasis, because that hypothesis was developed using basin-scale, species-level data on middle Paleozoic marine macroinvertebrates. However, the correspondence between the mammalian data set and that of Brett and Baird (1995) is actually quite close. First, each of Brett and Baird's coordinated stasis intervals corresponds to between three and seven of the standardized 1.0-my intervals used in this study, which means that the static interludes between such faunal replacement events would have been observable if they had existed in the mammalian data set. Second, their study and this one pertain to the same taxonomic level: all of the genus-level analyses discussed here also are supported by lineage-level results. Third, the difference in scale between a marine basin and the continent of North America may seem substantial, but it should be remembered that the mammalian data pertain almost entirely to the western United States and that mammals have relatively large geographic ranges, with most genera today being found throughout the region if anywhere within it.

Finally, it could be argued that the coordinated stasis pattern is somehow unique to either marine communities, the Paleozoic era, or both. This would be unsatisfying: Paleobiologists should be able to offer general models that have relevance for every group and every time interval. If we fail to do so, ecologists and other biologists are right to ignore us. In summary, then, the coordinated stasis theory takes an unnecessary leap from the unavoidable fact of variation in turnover rates through time to some very strong, but dubious, ideas about the integration of paleocommunities on very long time scales. Further studies may bear out some of its predictions, but the present results show that at least for mammals, the major pattern in long-term, large-scale diversity patterns is instead a dynamic numerical equilibrium, a result that coordinated stasis does not even address.

Conclusion

Paleontologists were inspired to look for equilibrial diversity dynamics on global and continental scales three decades ago by MacArthur and Wilson's (1967) model of equilibrial diversity dynamics on ecological temporal and spatial scales. For example, Webb (1969) attempted to demonstrate equilibrial dynamics in the late Miocene through Pleistocene mammal record of North America. These initial efforts failed because the wrong tests were employed: Correlations between origination and extinction rates were not corrected for variation in standing diversity levels and in the lengths of sampling intervals (e.g., Mark and Flessa 1977; see also Cowen and Stockton 1978), and in any case, they did not directly address the issue of whether per-taxon origination and extinction rates depend on standing diversity levels. When the correct approach was outlined by Sepkoski (1978), the advance was obscured by questions about his raw data (Signor 1985) and by a proliferation of alternative models that lacked distinct hypothesis tests (Carr and Kitchell 1980; Walker 1985).

Fortunately, equilibrial diversity models have been revived over the last few years (Maurer 1989; Nee et al. 1992; Wagner 1995a), despite the persistence of profound methodological confusion in some camps (e.g., Benton 1995). Thus, the present contribution is only one of several to show that diversity equilibria have been attained quickly in major taxonomic groups. The key difference is that this study benefits from recent advances in quantitative time-scale definition (Alroy 1992, 1994a), combined with the new methods described here that objectify sampling interval definition, eliminate geographic bias, and equalize sampling intensity. I hope that the greater confidence imparted by these refinements, in addition to the apparently interesting contrasts between origination and extinction patterns, will encourage future workers to discuss equilibrial diversity dynamics. This eventually may help us to reformulate our views of macroevolution by revealing its foundation in ecological patterns and processes.

Acknowledgments

I thank M. McKinney for his invitation to participate in the 1994 ESA biodiversity dynamics symposium, and D. Jablonski for refusing said offer. P. Holterhoff, K. Flessa, and M. Rosenzweig encouraged me as I worked through various aspects of the statistical analysis. I thank M. Foote, B. Maurer, J. Sepkoski, P. Wagner, and several anonymous reviewers for no-holds-barred comments on the manuscript, although the many remaining intemperate insinuations and implausible leaps of logic are entirely my responsibility. This contribution is based on dissertation research at the University of Chicago and postdoctoral research at the University of Arizona, the latter funded by the Research Training Group in the Analysis of Biological Diversification.

13

Scales of Diversification and the Ordovician Radiation

Arnold I. Miller and Shuguang Mao

SWEET ARE THE USES OF DIVERSITY: For the study of paleodiversity, the final goals involve synthesis of the processes connecting the ecological structure of the environment, the taxonomic structure of the biota over all categories, and the history of the individual clades at each taxonomic level. We are far from such goals, which are tantamount to understanding macroevolution, but nevertheless they appear to be within eventual reach. Paleontologists have been piling up data on biotic history for many decades now at an accelerating pace. Perhaps within some of these messy piles of data there are elegant principles struggling to get out? If so, it may well be diversity studies that eventually set them free.

—VALENTINE 1985

Three decades of intense paleobiological research have yielded a substantially improved calibration of global biodiversity trends throughout the Phanerozoic. The overall pattern among marine biotas, evaluated in numerous studies (e.g., Newell 1959; Valentine 1970, 1973b; Raup 1972, 1976a,b; Bambach 1977; Sepkoski 1976, 1978, 1979, 1981, 1984, 1993; Sepkoski et al. 1981; Signor 1990), is now understood with some confidence (but see Miller and Foote 1996). Among the early highlights of Phanerozoic marine diversification were (1) an initial radiation in the Cambrian Period (the *Cambrian explosion*), when most phyla originated (Erwin et al. 1987) and the trilobite-dominated Cambrian evolutionary fauna flourished (Sepkoski 1981); and (2) a subsequent diversification through the Ordovician Period (the *Ordovician radiation*), during which the articulate brachiopod- and bryozoan-dominated Paleozoic fauna rose to a position of diversity dominance and the mollusc-dominated Modern fauna underwent its first major radiation (Sepkoski 1981).

While perhaps not as profound as the Cambrian explosion with respect to the origination of major body plans, the Ordovician radiation was nevertheless a highly significant, and unique, interval in the history of life. During the Ordovician Period, standing marine familial diversity, as depicted directly in the fossil record, increased more than threefold, from about 160 to about 530, and genus-level diversity increased from about 700 to about 2100; these increases far surpassed, in numbers, the familial and genus-level increases exhibited during the Cambrian [family data are from Sepkoski 1993, and genus data are from an unpublished global scale compendium of fossil marine genera compiled by Sepkoski; the genus-level numbers differ from those reported by Miller and Mao (1995) because a later version of Sepkoski's genus compendium is used herein]. This is not to suggest, however, that the Ordovician radiation was discernible only at lower taxonomic levels. The number of classes and orders that originated during the Ordovician rivaled Cambrian levels of origination (Erwin et al. 1987), suggesting a continuation of the extensive morphological innovation that characterized the Cambrian explosion.

Despite the extent of the Ordovician radiation, explanations for this and other major Phanerozoic diversity increases remain enigmatic (but see Miller and Mao 1995; Vermeij 1995). In part, this is a consequence of our general lack of knowledge concerning the ways in which these phenomena, detectable as major transitions on a coarse, global scale, are actually played out in more local, geographic, and environmental contexts. For example, in the case of the Ordovician radiation, it would be useful to know, among other things, whether the diversity increases occurred randomly around the world or, alternatively, were concentrated in particular regions (e.g., see Babin 1993 for an initial evaluation of Ordovician bivalves in this context). Recent research on geographic selectivity of mass extinctions (e.g., Raup and Jablonski 1993; Jablonski and Raup 1995) has demonstrated the importance of these kinds of data in narrowing the search for causal mechanisms. Clearly, similar kinds of analyses should be considered essential in evaluations of radiations.

Moreover, a new generation of research on the local distributions and dynamics of species (e.g., the metapopulation approach described by Levins 1969 and Hanski and Gilpin 1991, among others) has led to the suggestion that temporal changes in biodiversity, at hierarchical levels ranging from local to global, are manifestations of the same kind of general mechanism operating at different scales (McKinney and Allmon 1995). In a related vein, Aronson (1994) has argued that biological processes affecting distributions of organisms on a local level are commonly scale independent and, thus, might well be manifested at the global scale (see Miller 1990a for a related discussion of hierarchies in the context of bivalve diversification). With respect to global radiations, the implication is that these diversity increases may be linked mechanistically to changes taking place on a local, community level. This kind of linkage is also suggested in the recognition by some researchers of comparable temporal transitions in Phanerozoic species richness at local and global scales (see Bambach 1977; Sepkoski et al. 1981). However, as Sepkoski (1988) demonstrated, global diversity trends cannot be accounted for entirely by transitions recognized locally, and processes promoting diversification at different levels (e.g., the local, community, and "intermediate" provincial levels) may impart their own, unique signatures that combine to produce diversity trends detected in global-scale compilations.

Thus, in an effort to better understand the kinetics of global-scale radiations, the purpose of this paper is to describe a preliminary exploration of the Ordovician radiation at several geographic/spatial scales, ranging from global to local. Most of the patterns described here involve utilization of a database that depicts documented occurrences of Ordovician genera around the world. Below, we first discuss the global pattern of diversification, as delineated with Sepkoski's unpublished genus-level compendium. Then, after briefly describing the database on worldwide genus occurrences and its relevance to the topics addressed in this paper, we focus on provincial diversification patterns exhibited in marine settings associated with two major Ordovician paleocontinents: Laurentia (much of present-day North America) and South China. This comparison permits the exploration of province-scale effects on diversification, as well as the further dissection of each province to assess local diversity trends. Finally, we consider the significance of the observed patterns in the context of possible relationships among macroevolutionary processes operating at different scales.

Ordovician Diversity Trends on a Global Scale

The global-scale Ordovician diversity trajectory has been documented and discussed in detail by Sepkoski (e.g., 1979, 1981, 1988, 1993, 1995b). As noted earlier, standing familial and generic diversities increased through the Ordovician to levels far exceeding those achieved earlier, during the Cambrian explosion. Subsequent to the Ordovician radiation, diversity levels remained fairly constant through the remainder of the Paleozoic era, although they were reduced temporarily during mass extinctions in the Late Ordovician and Late Devonian.

The global, genus-level trend is illustrated in figure 13.1, which was constructed at the series level of stratigraphic resolution using Sepkoski's unpublished compendium (the data source for all global patterns presented in this paper). This relatively coarse stratigraphic treatment appears to inflate the level of standing diversity during the lengthy Caradocian series, relative to its depiction at the finer, subseries level (Sepkoski 1995b). Nevertheless, the

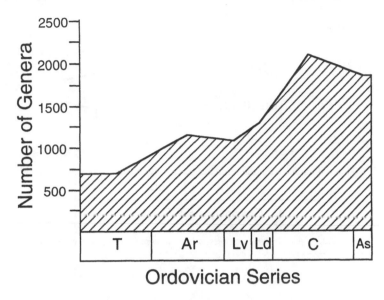

Figure 13.1. Series-by-series representations of genus-level, global marine biodiversity through the Ordovician. Data are from an unpublished compendium compiled by J. J. Sepkoski, Jr. The following abbreviations apply to this and subsequent Ordovician time-series graphs: T, Tremadocian; Ar, Arenigian; Lv, Llanvirnian; Ld, Llandeilian; C, Caradocian; As, Ashgillian.

general Ordovician trend, including the threefold increase in genus richness, is consistent with other published depictions (but see Miller and Foote 1996 for a new assessment of the Ordovician radiation that suggests a somewhat different global diversification trajectory).

Most of the analyses that will be described focus on a group of higher taxa that comprise a cross section of major faunal elements from each of Sepkoski's (1981) three evolutionary faunas: (1) trilobites, (2) articulate and inarticulate brachiopods, and (3) bivalve, gastropod, and monoplacophoran molluscs (referred to collectively herein as benthic molluscs). The Ordovician global diversity trajectories of most of these taxa are illustrated in figure 13.2. Whereas trilobite diversity was fairly static throughout the period, the global genus richness of articulate brachiopods and benthic molluscs increased. When these numbers are recalibrated as percentages of the aggregate diversity exhibited by trilobites, brachiopods, and benthic molluscs (figure 13.3), it is evident that trilobite dominance was waning through the period, whereas that of articulate brachiopods and benthic molluscs was increasing; this reflects the global transition from the Cambrian evolutionary fauna to the Paleozoic and Modern faunas that characterized the Ordovician radiation (Sepkoski 1981).

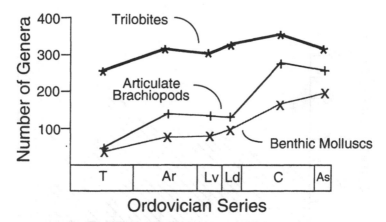

Figure 13.2. Series-by-series representations of genus-level, global biodiversity for the select groups discussed in the text.

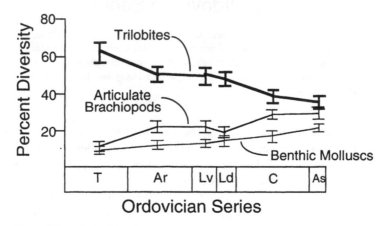

Figure 13.3. Series-by-series representations of genus-level global biodiversity for the select groups discussed in the text, depicted as percentages of total genus diversity for these groups (in this and subsequent graphs that depict percent diversity, error bars are 95 percent confidence intervals around the percentages, calculated using an algorithm described in Raup 1991c).

The Worldwide Database of Ordovician Genus Occurrences

To analyze possible patterns of environmental and geographic selectivity associated with the Ordovician radiation and the Late Ordovician mass extinction, we have been compiling, from the worldwide literature, a database that documents individual occurrences of genera in Ordovician strata around the world. [Throughout this paper, an individual occurrence of a genus in the rock record is referred to as a genus occurrence. This term is synonymous with genus appearance, used by Miller and Mao (1995). It was changed because of concern that usage of the word *appearance* necessarily implies *first* appearance, which is not its intended meaning. Rather, a genus occurrence is a single, documented occurrence of a fossil genus in an Ordovician horizon recognized at an individual locality or in a limited region.] Redundancy of occurrences of individual genera was common, in that many genera recurred in numerous horizons or localities around the world. The database includes, for each genus occurrence, a variety of stratigraphic, geographic, paleogeographic, and paleoenvironmental information. To date, we have catalogued some 15,000 genus occurrences from the Ordovician paleocontinents of Laurentia (much of present-day North America); several paleocontinents that today make up most of China; East Avalonia (most of present-day England and Wales); Baltoscandia (present-day Scandinavia and the Baltic countries); Bohemia; and, to a limited degree, Australia. Of these occurrences, 6576 belong to the higher taxa that are the main foci of the analyses presented herein. These taxa were highlighted for two primary reasons. First, as suggested earlier, they comprise a cross section of major elements from Sepkoski's three evolutionary faunas. Second, they tend to be among the biotic elements that most paleontologists identify down to genus level in published faunal lists. Thus, to a first approximation, they can be viewed as providing a reasonable proxy of broader taxonomic attributes that characterize the Ordovician radiation.

Although data have not yet been collected, or are only now being collected, from several other paleocontinents with appreciable Ordovician strata and fossils (e.g., Siberia, Kazakhstan, and South America), the database is already global in scope (see Miller and Mao 1995). This is demonstrated in figure 13.4, which compares the Ordovician global diversity trend for the aggregate group of taxa highlighted here (figure 13.4A) to a similarly styled figure for the same taxa generated independently from the database (figure 13.4B). The two diversity trajectories are highly comparable, and, on a series-by-series basis, standing diversities in figure 13.4B are roughly half to two thirds of those in figure 13.4A, except during the Ashgillian. Thus, it is evident that, for the taxa in question, the database already contains an appreciable sample of Ordovician genera worldwide. [The database of genus occurrences contains no data from the Silurian or later, in contrast to Sepkoski's global compendium. Because estimated global diversity in the Ashgillian depends in part on post-Ordovician data, relative levels of Ashgillian global diversity (see figure 13.4A) are higher than our estimates (see figure 13.4B), likely reflecting an effect directly analogous to "pull of the Recent" (Raup 1979)].

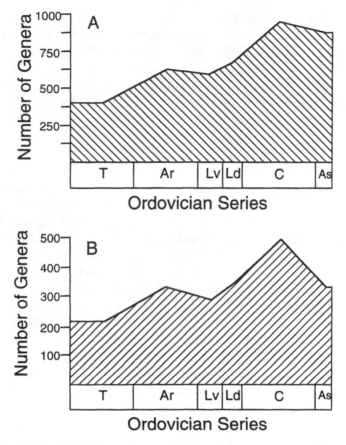

Figure 13.4. **(A)** Series-by-series representations of genus-level, global marine biodiversity through the Ordovician for the aggregate pool of genera belonging to the select groups discussed in the text. Data are from Sepkoski's unpublished compendium. **(B)** The same representations for the same select groups seen in 13.4A, except that this graph was tabulated using the database of genus occurrences described in the text.

To investigate diversification patterns at, or below, the provincial level, the analyses presented here focus on a comparison of two paleocontinents: South China and Laurentia. The data from South China are particularly well suited for an investigation of patterns down to the alpha level (see later) because most of the South Chinese data were presented in the literature on a bed-by-bed basis; these delineations were maintained when the database was compiled. Thus, for South China, it is possible to tabulate genus richness at the level of individual collections from single beds; this degree of reduction is not possible with most of the data from other continents. Laurentia was chosen for comparison with South China because, in virtually all Ordovician paleogeographic reconstructions, the two paleocontinents are thought to have been isolated from one another, although they occupied similar latitudinal regimes in the southern tropics (see Scotese and McKerrow 1990).

Aggregate Diversity Trends in South China and Laurentia

The Ordovician diversity trajectory exhibited by the group of highlighted taxa in South China is illustrated in figure 13.5A. It shows a substantial diversity increase from the Tremadocian into the Arenigian: in fact, peak Ordovician diversity in South China may have occurred at that time. After a

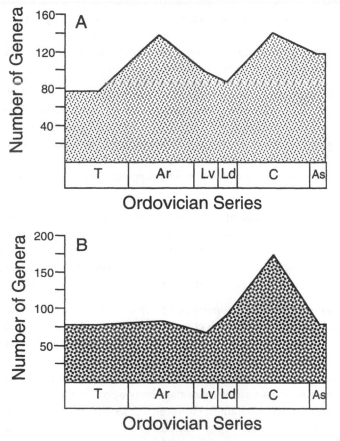

Figure 13.5. Series-by-series representations of genus-level biodiversity trends for the select groups based on the database of genus occurrences described in the text. **(A)** South China; **(B)** Laurentia.

decline through the Llanvirnian and Llandeilian, diversity increased in the Caradocian to a level comparable to the Arenigian, before apparently declining somewhat into the Ashgillian. By contrast, in Laurentia (figure 13.5B), diversity remained nearly unchanged from the Tremadocian into the Arenigian, it declined slightly in the Llanvirnian, it increased through the Llandeilian to an Ordovician peak in the Caradocian, and then it apparently declined precipitously into the Ashgillian. Interestingly, the global time series (see figure 13.4B) appears to exhibit an amalgamation of attributes of the Laurentia and South China time series. As with South China, the global curve shows a diversity increase from the Tremadocian into the Arenigian. However, as with Laurentia, it exhibits a much greater diversity peak during the Caradocian. Finally, the global rate of diversity decline from the Caradocian into the Ashgillian was more comparable to that of South China than it was to that of Laurentia (but see earlier caveat about the global Ashgillian value).

However, there are two potential problems with this kind of assessment of raw diversity trends within and among paleocontinents:

1. As suggested earlier, the relative durations of series may well affect their perceived relative diversities. For example, all else being equal, the Caradocian would be expected to yield a greater number of genus occurrences and, thus, a greater genus richness, than the preceding Llandeilian, simply because it was of much longer duration (assuming a rough relationship between duration and sedimentary rock volume per aggregate volume of available fossil material).

2. Each paleocontinent may have its own individual peculiarities with respect to its preserved Ordovician record. For example, the relatively large diversity exhibited in the Arenigian of South China might well be associated with an Arenigian stratigraphic package that is unusually extensive relative to Laurentia and other parts of the world.

For intercontinental comparisons of the kind conducted here, the first of these problems is not of much consequence. To the degree that series durations affect the Ordovician time series, this artifact should be similar all over the world. In other words, the substantial difference in the Tremadocian-to-Arenigian transitions exhibited by South China and Laurentia, as well as other series-to-series differences among the two continents, were not affected by variations in series duration. However, the second problem might be quite significant: the unusual genus richness of South China relative to North America for the Arenigian might well have been associated with intercontinental differences in the extents of the preserved Arenigian records.

To further explore this issue, we conducted rarefaction analyses, on a se-

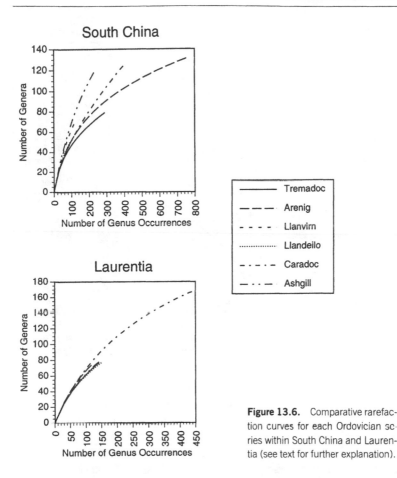

Figure 13.6. Comparative rarefaction curves for each Ordovician series within South China and Laurentia (see text for further explanation).

ries-by-series basis, for each paleocontinent (see Hurlbert 1971; Raup 1975a). In this context, rarefaction involved numerical estimations of genus richness at sample sizes (i.e., the number of genus occurrences) *smaller* than those actually catalogued for each series on each paleocontinent (rarefaction *cannot* be used, however, to estimate diversity at sample sizes *larger* than those of the original samples). Thus, it permitted exploration of the degree to which series-to-series changes in genus diversity may have been consequences of fluctuating sample sizes (tied to likely differences in the amount of rock volume available to sample) that overprinted, and perhaps masked, the actual biological signal. For example, in the case of the Arenigian diversity-peak in South China (see figure 13.5A), rarefaction (figure 13.6) indicated that, indeed, the Arenigian peak was inflated artifactually by the unusually large number of genus occurrences tabulated for that series. In figure 13.6, the Arenigian rarefaction curve for South China resided above

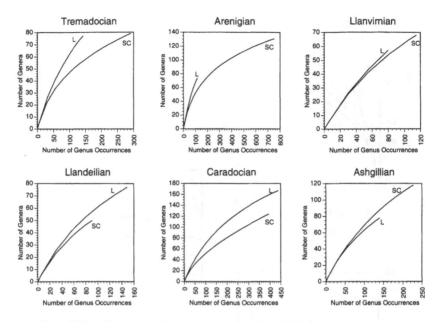

Figure 13.7. The same rarefaction curves seen in figure 13.6, except that the curves for South China (SC) and Laurentia (L) are compared directly on a series-by-series basis.

only that for the Tremadocian. This, in turn, indicated that, if the number of genus occurrences (i.e., the sample size) from the Arenigian were limited to the levels of other Ordovician series, the measured genus richness for the Arenigian would exceed only the Tremadocian level. In fact, this comparison of rarefaction curves suggested that, when sample size differences among the stages were accounted for, genus richness in the Middle and Late Ordovician probably exceeded that of the Early Ordovician (Tremadocian and Arenigian).

For Laurentia, rarefaction suggested a diversity history (see figure 13.6) that closely approximated the path exhibited by the raw diversity trajectory in figure 13.5B. The raw trajectory indicated a fairly static diversity history for Laurentia, except for a large diversity increase in the Caradocian. For the most part, this was corroborated by the close grouping of the Laurentia rarefaction curves in figure 13.6. However, the rarefaction analysis also attested that the Caradocian diversity peak (see figure 13.5B) was inflated by the unusually large sample for the Caradocian of Laurentia, relative to the sample sizes of other series.

From the standpoint of intercontinental comparison, additional insight was gained by comparing directly the rarefaction curves from South China and Laurentia on a series-by-series basis (figure 13.7). For all but one series (the Ashgillian), the rarefaction curves suggested that, after accounting for sampling differences, the genus richness of Laurentia exceeded that of South China (although differences among the two paleocontinents for the Llanvirnian, Llandeilian, and Ashgillian were only slight and might not be statistically distinguishable). Thus, for the group of higher taxa evaluated here, the Ordovician radiation may have been more extensive in Laurentia than it was in South China.

Furthermore, a comparison of their taxonomic compositions revealed that faunal attributes of the radiation were rather different between the two paleocontinents (figure 13.8). When calibrated as percentages of the aggregate diversity exhibited by trilobites, brachiopods, and benthic molluscs, articulate brachiopod diversity (figure 13.8A) in South China significantly exceeded that in Laurentia during the Arenigian and the Ashgillian. Differences during the other series were not significant, although the calculated

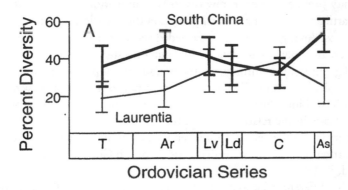

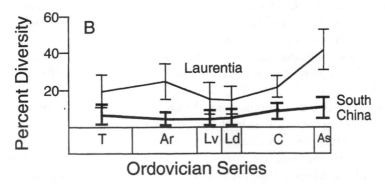

Figure 13.8. Series-by-series representations of percent diversity for selected taxa through the Ordovician in South China and Laurentia. **(A)** Articulate brachiopods; **(B)** Benthic molluscs.

articulate brachiopod percentages for South China tended to exceed those in Laurentia throughout the period (except in the Caradocian). Moreover, the South Chinese articulate brachiopod percentages generally exceeded global values (see figure 13.3) on a series-by-series basis, whereas those for Laurentia did not.

Clearly, the limited percentage diversity of articulate brachiopods in Laurentia, relative to those in South China, was associated with the rather extensive percentage diversity of benthic molluscs in Laurentia, particularly late in the period. The percent diversity of benthic molluscs in Laurentia significantly exceeded that in South China in three of the six series (and nearly so in the other three), and it increased significantly within Laurentia late in the period.

Thus, after accounting for possible sampling artifacts, the Ordovician radiation appears to have transpired somewhat differently in South China than in Laurentia. In South China, rarefaction analyses suggested that the Arenigian diversity peak, apparent in raw diversity compilations, was almost certainly an artifact of sampling; diversity levels in the Middle and Late Ordovician (i.e., post-Arenigian) probably exceeded those of the Early Ordovician. By contrast, the analyses of data from Laurentia suggested a rather static pattern of genus richness throughout the period, except perhaps for a slightly higher diversity level in the Caradocian. This is not to suggest, however, that the Ordovician biotic history of Laurentia was static. Clearly, there was a notable increase in the relative contribution of benthic molluscs to Laurentian diversity late in the period. Moreover, rarefaction indicated that diversity levels in Laurentia generally exceeded those in South China through most of the period.

Why the differences between the two paleocontinents? Miller and Mao (1995) suggested a general linkage between the Ordovician radiation and orogenic activity (i.e., mountain building); in large part, the differences between South China and Laurentia may well have been associated with substantial differences in levels of orogeny. Much of what is now eastern North America was affected extensively by the Taconic Orogeny, which was manifested by a variety of tectonic events through much of the Middle and Late Ordovician (see Rodgers 1971). Among the consequences of the Taconic Orogeny was a dramatic increase, throughout much of Laurentia, in the amount of terrigenous (i.e., land-derived) sediment. This, in turn, may have helped to fuel the diversification of benthic molluscs, many of which were most diverse and abundant in terrigenous sediments at that time (see Bretsky 1969, 1970; Miller 1988b, 1989; Babin 1993). By contrast, orogenic activity in South China was rather limited during the Ordovician, when carbonate sedimentation predominated on a platform that extended over much of the

continent. Substantial uplift that changed the sedimentation pattern occurred later (Hsü et al. 1990). Clearly, South China during the Ordovician
was much less rich in terrigenous sediments than Laurentia. To the degree
that these differences resulted in different Ordovician diversification patterns
between the two paleocontinents, they lend credence to the suggestion that
continent-scale differences in geological phenomena may lead to profound
differences in component biodiversity, both quantitatively and qualitatively
(see Westoby 1993).

Alpha Diversity Trends in South China and Laurentia

To understand the potential linkages among diversification at different
scales, the aggregate paleocontinental patterns must be dissected further. In
part, this involves exploration of diversity within individual communities or
habitats [i.e., alpha diversity (see, e.g., Whittaker 1975)], evaluation of
trends in mean alpha diversity through geologic time, and a comparison of
these patterns with trends exhibited at broader scales (Sepkoski 1988). Although alpha diversity has been calculated in numerous ways, sometimes
with indices that take into account the relative abundances of constituent
taxa, the simplest method involves direct tabulation of taxonomic richness.
This approach has been used previously, with compelling results, in paleobiological investigations of changes in alpha diversity through geologic time
(e.g., Bambach 1977; Sepkoski 1988).

For South China, determination of alpha diversity for individual assemblages is relatively straightforward because most of the data were reported in
the literature as lists of genera present in individual beds; the integrity of
these lists was maintained in the database. Thus, the number of genera in a
single list depicts the known genus richness of a single fossil assemblage and,
by extrapolation, the alpha diversity of the preservable elements of the living
assemblage (albeit time-averaged) from which it was derived. This asserted
relationship rests on the assumption that postmortem transport and other
taphonomic processes have not obscured the biological character of the assemblage. Although taphonomic effects are doubtless of importance in some
instances, there is a growing body of evidence suggesting that, even at fine
spatial scales (10 m or less in some instances), fossil assemblages should provide reasonable renditions of (time-averaged) genus richness and other community attributes [see Kidwell and Bosence 1991; Miller 1988a, 1996 (in
press); Miller et al. 1992].

The mean alpha diversity for a particular Ordovician series can be determined simply by averaging the values for all faunal lists tabulated for the
series. However, the analyses conducted here were again limited to the group

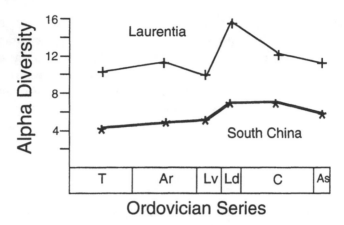

Figure 13.9. Series-by-series representations of mean alpha diversity for South China and Laurentia (see text for further discussion).

of higher taxa evaluated earlier. Thus, calculated values underestimate "true" alpha diversity, at least as depicted by the fossil record. Nevertheless, they permit exploration of alpha diversity *trends*, in comparison with trends at other scales, which are the primary topics of interest herein.

The mean alpha diversities for the Ordovician series of South China are depicted in figure 13.9. The time series suggests that mean alpha diversity was fairly stable in the range of four to five genera from the Tremadocian through the Llanvirnian, but increased to approximately seven genera during the subsequent Llandeilian. This higher level was maintained through the Caradocian but declined slightly in the Ashgillian. In a broad sense, this increase in the Middle and Late Ordovician parallels that exhibited by the aggregate signal for South China, as depicted in the rarefaction analysis (see figure 13.6). However, when compared in detail, there are notable differences. Rarefaction suggests that, after adjusting for sampling disparities among series (see earlier discussion), aggregate diversity for South China was greatest in the Ashgillian and, perhaps, the Llanvirnian. Alpha diversity, on the other hand, apparently peaked in the Llandeilian and Caradocian.

In contrast to those for South China, the data files that make up the Laurentia database do not record the faunal compositions of individual beds. Instead, they generally depict the broader compositions of "paleocommunities," typically consisting of aggregate information from several beds and, perhaps, localities, as reported in the literature (e.g., see Sepkoski and Miller 1985; Miller 1988b; Sepkoski 1988). Thus, mean alpha diversities tabulated using the individual data files for Laurentia are not directly comparable to their South Chinese counterparts; because they are aggregates of several beds, the mean genus-richness values reported for Laurentia should generally be greater than those tabulated for South China. Indeed, this is evident when comparing the respective values for the two paleocontinents, as depicted in figure 13.9. Nevertheless, because the Laurentia database, like that of South China, was tabulated with an internally consistent methodology, it should be reasonable to compare alpha diversity trends among Laurentia and South China. This comparison (see figure 13.9) suggests that the time series for Laurentia was broadly comparable throughout the Ordovician to that of South China. Like South China, Laurentia exhibited its greatest mean alpha diversities in the Llandeilian and Caradocian.

The Ordovician alpha diversity trend for Laurentia is difficult to compare with the aggregate, continent-wide trend (see figures 13.5B, 13.6) because of the general *lack* of a trend in the aggregate pattern (see earlier discussion). However, it is interesting that alpha diversity trends for South China and Laurentia are roughly comparable, despite the differences described earlier in their aggregate diversity trends and methods of data collection. Both paleo-continents exhibited alpha diversity increases in the Llandeilian to levels not recognized in earlier series. The discovery of this similarity must be viewed as preliminary; the pattern could be documented further by exploring alpha diversity trends for other paleocontinents, and by expanding the analysis to include additional taxa (these analyses are ongoing). Nevertheless, the similarity hints at the possibility of a global trend in alpha diversity that transcended the peculiarities of individual continents and regions, in contrast to the aggregate signals for individual paleocontinents, described earlier. Although we cannot yet identify the mechanisms that might have produced this pattern, its (potential) global manifestation might suggest biological, rather than local geological, mediation. Perhaps this is the level at which ecological processes associated with metapopulation or metacommunity dynamics would be most definitively expressed.

Beta Diversity Trends in South China and Laurentia

For a full assessment of diversification patterns throughout the Ordovician, it is important to consider an additional intracontinental measure beyond that provided by an evaluation of alpha diversity trends. Given the probable organization of organisms on the sea floor into "communities," be they bounded discretely or diffusely along environmental gradients (see Springer and Miller 1990; we make no claims here about the functional significance of communities), it is reasonable to ask whether the degree of dissimilarity among communities, known commonly as beta diversity, changed through the Ordovician. This kind of transition might be expected to occur in association with increasing within-community (alpha) diversity (see discussions in Whittaker 1975), and Sepkoski (1988) suggested that the Ordovician radiation was characterized by an increase in beta diversity.

Beta diversity has been calibrated previously using a variety of methodologies. Here, we followed, to some degree, the procedure utilized by Sepkoski (1988), which involved numerical assessments of similarity among pairs of assemblages as determined with the Jaccard coefficient:

$$S_j = T_c / (T_1 + T_2 - T_c),$$

where T_1 is the number of genera in one assemblage, T_2 is the number of genera in a second assemblage, and T_c is the number of genera in common among the two. However, in his analyses of beta diversity trends through the Paleozoic era, Sepkoski pooled together all the assemblages contained within each of six environmental zones in several stratigraphic intervals, and he then compared these pooled assemblages to one another within and among the intervals in question. In contrast, no pooling of assemblages was conducted in the present study, nor was an environmental classification utilized for this tabulation. Instead, within a series, each of the assemblages for which alpha diversity had been determined previously (see earlier) was maintained for calibration of beta diversity. For each series, the Jaccard coefficient was utilized to assess pairwise similarities of every assemblage with every other assemblage, and a mean value for the series was determined based on the individual values.

Using this methodology, mean Jaccard coefficients were calculated (with 95 percent confidence intervals) for each Ordovician Series within each of the two paleocontinents (figure 13.10; in this graph, the Jaccard coefficient increased in a downward, rather than upward, direction to reflect the understanding that an increase in the coefficient reflected a *decrease* in beta diversity; i.e., the greater the similarity among assemblages, the less the beta diversity). Given the differences in the structures of the South China and Laurentia databases (see earlier discussion), there were more (smaller) assemblages, and within-series pairwise comparisons, for each South Chinese series than there were for Laurentia. Thus, it is not surprising that the 95 percent error bars were substantially narrower for South China than they were for Laurentia. Nevertheless, despite these differences among the two databases in the numbers of assemblages and their averages sizes, there was no clear-cut tendency for the mean beta diversity to be higher for one paleocontinent than for the other.

The beta diversity time series for South China suggests a surprising pattern: Throughout the Ordovician Period, as alpha diversity increased, beta

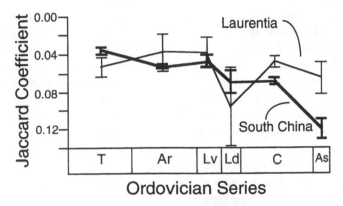

Figure 13.10. Series-by-series representations of mean beta diversity for South China and Laurentia, as depicted with the Jaccard coefficient (see text). Error bars are 95 percent confidence intervals about the means.

diversity appears to have *decreased* significantly. Beta diversity was significantly less (i.e., the average Jaccard coefficient was significantly greater) during the Llandeilian, Caradocian, and Ashgillian than it had been earlier. At a finer temporal scale, several significant series-to-series decreases can be recognized: from the Tremadocian to the Arenigian, Llanvirnian to Llandeilian, and Caradocian to Ashgillian. In fact, the mean Jaccard coefficient determined for the Ashgillian was approximately three times that exhibited by Tremadocian assemblages.

For Laurentia, the time series was more ambiguous, given the smaller number of samples and, thus, the greater uncertainty in the calculated mean values of the Jaccard coefficient. Although the mean beta diversity appeared generally to be less (greater Jaccard coefficient values) in the latter three series of the Ordovician than it was previously, these differences among the series were not statistically significant (at $\alpha = 0.05$), except perhaps for the Llandeilian. Nevertheless, there is no indication that beta diversity increased in Laurentia through the period; if anything, there is limited evidence of a decrease.

Because the method of calibrating beta diversity in this study was somewhat different from, and the spectrum of taxa more limited than, those utilized by Sepkoski (1988), a direct comparison of Sepkoski's results with the findings presented herein should be viewed as only preliminary. For a more definitive comparison, an analysis that more directly parallels that of Sepkoski must be conducted; this will be accomplished in the future as the database continues to grow.

Sepkoski's (1988) study, which focused on a database limited almost exclusively to Laurentia, suggested that increased alpha diversity during the

Paleozoic era was accompanied to some degree by increased beta diversity. In his period-by-period analysis, alpha and beta diversity increases were both most appreciable from the Cambrian to the Ordovician. In this regard, Sepkoski's findings do not necessarily conflict with those of the present study, which suggested that in South China, and perhaps Laurentia, beta diversity decreased in the face of increasing alpha diversity. Although Sepkoski utilized stratigraphic, and temporal, subdivisions finer than the period scale in his calculations, his suggestion that the Ordovician radiation was accompanied by an increase in beta diversity was based on the transition recognized from the Cambrian to the Ordovician, rather than a documented increase through the Ordovician. Beta diversity transitions throughout the Ordovician were not highlighted in his presentation.

Given the discussions of other workers, including Whittaker (1975, 1977) and Sepkoski (1988), the relationship between alpha and beta diversity recognized here seems counterintuitive. Although it is possible to envision a lack of correlation between these two levels of diversity, it is more difficult to understand why there might be an inverse correlation. Why might beta diversity decrease while alpha diversity was increasing? Because the results of the present study are preliminary, any suggested explanation can be viewed only as speculation. Nevertheless, we propose the following explanation, which can be tested in the future with an augmented database.

Given that this study was conducted at the genus level, there is no direct sense of possible changes in the species richness of individual genera that contributed to the Ordovician radiation. It may well be that the radiation was characterized not only by an increased number of genera, but also by an increase in the number of species per genus. One possible offshoot would likely have been increased environmental and geographic ranges for genera that were experiencing increases in species richness. By this logic, the Ordovician radiation would have been accompanied by increases in the environmental and geographic ranges of "participating" genera. This, in turn, would have served to increase similarities among Ordovician assemblages at the genus level, which would have decreased genus-level beta diversity. Using the database of genus occurrences described earlier, which includes geographic and paleoenvironmental information for each occurrence, it will be possible to calibrate the geographic and environmental ranges of Ordovician genera, and to test directly the hypothesis that these ranges increased, on average, through the period. This test will be conducted in the future.

Discussion and Conclusions

> *I feel that I am beginning to understand how the different*
> *levels of biological organization and the different spatial*
> *and temporal scales of pattern and process are related to*
> *each other.*   —BROWN 1995

Because the analyses presented herein provide only preliminary glimpses at a still-developing database, the results must be interpreted cautiously; the inclusion of additional taxa and paleocontinents should provide further insight with respect to the central issues of this paper. Nevertheless, these analyses suggest a clear methodological pathway for the dissection of global diversification, and they provide a series of tentative conclusions concerning the Ordovician radiation that can be verified and, perhaps, clarified further with additional data:

1. As illustrated by continent-scale differences between South China and Laurentia in diversity trends through the Ordovician, it is evident that the synoptic, global diversity trend was not divisible into roughly comparable diversity histories exhibited in different parts of the world.

2. Although the raw diversity trajectories for South China and Laurentia were affected by differences in sample size from series to series within each continent, there were clear differences between the two paleocontinents that transcended sampling effects. Most notably, it appeared that, after accounting for these sampling differences, genus diversity during most Ordovician series was greater in Laurentia than it was in South China. Moreover, from the standpoint of biotic composition, Laurentia exhibited a substantial Late Ordovician radiation of benthic molluscs that was not shared by South China. These kinds of intercontinental differences almost certainly reflect geographic differences in physical attributes, such as degree of orogenic activity (see earlier discussion), and they serve as a reminder that the world, at any given time, is patchy on even this large, continent-to-continent scale. Thus, at least in retrospect, there is no reason to have expected that marine settings associated with South China and Laurentia, or any other Ordovician paleocontinent, should have exhibited Ordovician diversity trajectories that were roughly comparable to one another.

3. To the degree that they could be compared, South China and Laurentia both appeared to exhibit alpha diversity increases in the Middle Ordovician (Llandeilian). If, indeed, these paleocontinents (and perhaps others) exhibited similar alpha diversity trends, it would suggest that some portion of

the Ordovician radiation had a local basis that was manifested worldwide and transcended the peculiarities of individual paleocontinents.

4. Perhaps surprisingly, South China provided a signal of decreasing beta diversity through the Ordovician. Although the pattern was less clear for Laurentia because of limited data, there was some indication of decreasing beta diversity there as well; at the very least, it can be said that the data for Laurentia showed no indication of a beta diversity increase through the period. A testable hypothesis was presented earlier to account for a decrease in beta diversity in the face of increasing alpha and global diversity.

These results have significant ramifications not only for our understanding of the Ordovician radiation, but also for the ways in which we analyze and interpret linkages among diversity at different scales throughout earth history. In evaluating the relationship between global and local diversification trajectories during the Paleozoic era, Sepkoski (1988) noted that increasing alpha and beta diversity fell far short, by about a factor of two, of accounting for the Ordovician diversity increase calibrated on a global scale. Sepkoski considered the possibility that this "missing" diversity could be accounted for by increased provinciality, but he noted that published analyses of levels of provinciality through the early and middle Paleozoic did not support such an interpretation. Thus, the missing diversity remained as an enigma.

However, as noted earlier, Sepkoski's study was limited to Laurentia; he apparently assumed implicitly that diversification trends for other paleocontinents would be comparable to Laurentia. Given the results of the present study, which illustrated differences between South China and Laurentia, it is clear that this assumption needs rethinking. Perhaps driven by differences in tectonic activity or other paleogeographic/paleoenvironmental attributes, the paleocontinents of the Ordovician world may have exhibited a variety of different diversity trajectories. If at least some paleocontinents exhibited radiations that exceeded the rate of diversification recognized in Laurentia, then the enigma suggested by Sepkoski might well vanish. This can be tested by evaluating the diversity trends for all the paleocontinents that comprised the Ordovician world and then evaluating explicitly whether, and how, their diversities "summed up" to yield the global signal. This is one of the end-objectives of our investigations of the Ordovician radiation for which this chapter serves as a progress report.

Another important conclusion of this study is that diversification trajectories at different spatial/geographic scales may well yield unique signatures that are perhaps best accounted for by different processes appropriate to each scale. For example, as suggested earlier, differences in overall diversity trajectories among paleocontinents might well be associated with large-scale physical processes that affect some paleocontinents and not others. On the other hand, possible similarities among paleocontinents at the alpha level might be manifestations of the kinetics of ecological processes tied to metapopulations and metacommunities (see McKinney and Allmon 1995). Finally, the surprising beta diversity trend exhibited for South China, to the degree that is ultimately found to be shared by other paleocontinents around the world, might be symptomatic of temporal changes in the geographic ranges of genera, not accounted for directly by the mechanisms responsible for diversification at other scales.

Thus, we offer the caution that our recognition of several meaningful scales of diversification is not intended to suggest that these patterns can be linked mechanistically to a single kind of process that transcends, and operates at, all (or several) scales (see Aronson 1994; McKinney and Allmon 1995; Aronson and Plotnick, chapter 18). Although much of the data to test the importance of scale-transcendent processes in global biodiversity have yet to be collected, we have found no manifestation of this during the Ordovician radiation.

We conclude that the Ordovician radiation was characterized in different ways at different scales; clearly, the unique attributes of various scales could not have been detected, let alone predicted, from analyses of the synoptic, global pattern. Conversely, the global signal was not simply the summed result of community-level diversification processes that took place (uniformly or varying randomly) worldwide. Diversification at different scales appears to have been associated with emergent properties perhaps unique to each scale. To the degree that some of these properties may have been linked to emergent characteristics of genera or other taxa at various levels of the taxonomic hierarchy (e.g., see earlier discussion of beta diversity), they testify to the reality of macroevolution (see discussion of hierarchy and emergent properties in Valentine 1990b, pp. 128–130). The Ordovician radiation cannot be accounted for by the global summation of microevolutionary processes.

Among paleobiologists studying temporal trends in diversity, there has been some tension with respect to determining the relative significance of biotic and abiotic factors, to the degree that they are distinguishable, in governing the global histories of clades (see Valentine 1990b, pp. 144–146). Some have argued for the overriding importance of biotic interactions as determinants of clade diversification (e.g., the escalation hypothesis championed by Vermeij 1977, 1987a), whereas others recognize what amounts to an interplay between biotic and abiotic factors in this regard (e.g., the biological bulldozing hypothesis of Thayer 1979, 1983, which involves the manipulation of a physical medium, the sediment, by newly diversified biotic elements). Still other studies have suggested a primary role for abiotic factors, including geochemical changes to the earth, as determinants of diversification (e.g., Martin 1995). An interesting outcome of the analyses presented here is the possibility that, during the Ordovician radiation, biotic factors may have been of primary significance at one scale (e.g., the local, community level), whereas abiotic factors, such as orogeny (Miller and Mao 1995) may have been of importance at another scale (e.g., the paleocontinental level).

Finally, we believe that this study serves as strong testimonial to the importance of dissecting global biodiversity trends, even though acquisition of data for this kind of task can be cumbersome and the initial exploration is largely inductive. These empirical data are essential for ultimately hypothesizing and testing the roles of processes to account for biodiversity trends throughout the history of life. This kind of agenda parallels the approach advocated recently by Brown (1995) for the detection of ecological patterns of large-scale significance. As Brown suggested (1995, p. 155), even local patterns cannot be understood without a global perspective: present-day species compositions and distributional patterns on a local level are the consequences not only of ongoing local ecological processes, but they also carry the signatures of larger-scale processes played out over geologic time (i.e., macroevolution). Similarly, only by investigating large-scale patterns of variation (e.g., comparisons among paleocontinents) is it possible to distinguish attributes and processes that are truly global in scope from those unique to particular regions at various scales.

Acknowledgments

This research was supported by a grant from the National Aeronautics and Space Administration, Program in Exobiology, grant NAGW-3307, to Miller. We thank the anonymous reviewers for several comments that helped us to improve the presentation and discussion of results.

14

Preston's Ergodic Conjecture: The Accumulation of Species in Space and Time

Michael L. Rosenzweig

Area is a pervasive and powerful influence on species diversity (Rosenzweig 1995). If we are to understand diversity, we must therefore understand its relationship to area. By the mid nineteenth century, Watson (1835) had described an important aspect of this relationship in terms that we would later understand as a power equation:

$$S = kA^z \tag{1a}$$

where A is area and S is the number of species. (S, species richness, is the only measure of species diversity I will use in this chapter.) Equation 1a is called the species–area curve, or SPAR.

To simplify their analyses, ecologists transformed equation 1a to logarithmic space:

$$\log S = c + z \log A \tag{1b}$$

where $c = \log k$. In equation 1b, as log-area grows, log-diversity grows linearly. Therefore, equation 1b allows ecologists to study SPARs using linear regression methods. Nevertheless, we should always remember that the true slope of a SPAR is the slope of 1a, not 1b (Lomolino 1989). That slope varies with A unless $z = 1$. In other words, z, the slope of 1b, is actually the curvature of 1a. If $z < 1$, the SPAR is convex upward. If $z = 1$, the SPAR is a straight line. If $z > 1$, the SPAR is concave upward.

Figure 14.1 presents two SPARs as typical examples. Both concern the plants of Great Britain. One begins with a small area in the county of Surrey; the other is in the county of Hertfordshire (Dony 1963). The former actually constitutes the first known SPAR (Watson 1835), and it is thus one of the oldest known patterns in ecology. These two SPARs are not identical. Because botanical knowledge advanced in the century between Watson and Dony, the older SPAR—Surrey's—reports fewer total species in Britain. And because Surrey has Britain's richest county flora, its diversities lie above those of Hertfordshire on the left side of the figure, but they head for the same point as Hertfordshire's on the right—i.e., all of the British flora. Hence, Surrey's SPAR rises a bit more slowly than Hertfordshire's ($z = 0.104$ versus $z = 0.192$). But the two curves are more similar than different. And they share those similarities—i.e., their linearity in log-log space and their approximate curvature—with floras and faunas all over the world (Rosenzweig 1995). (In contrast, c values vary over many orders of magnitude. They are taxon- and biome-dependent. They even depend on the exact biogeographical province in which they are measured.)

SPARs inspired island biogeography theory (MacArthur and Wilson 1967). They are used in conservation surveys to assess the value of reserves (e.g., Kitchener et al. 1980a,b, 1982). They have fueled the so-called SLOSS controversy (i.e., Is it better to establish a single large diversity preserve or several small ones?). And SPARs have been used to tease out the effect of area so that other variables influencing diversity can be studied. (For example, Wilcox 1978, did that to measure reptile extinction rates, and Rahbek (1997) did it to look at the effect of altitude on diversity.)

A SPAR is an example of a collector's curve. The "collector" keeps track of the number of categories in an ever-growing collection of observations. Naturally, as no observation is ever removed from the collection, the number of categories must continually increase until the sampling universe is completely known. In the case of SPARs, the categories are "species" and the new observations arise because the collector searches for them in a larger and larger area.

Another biological collector's curve extends time instead of space: The collector stays in one area and keeps track of the cumulative number of species recorded as time goes by. Let us call such a series of observations a species-time curve, or SPTI. Do SPTIs follow patterns? If they do, are those patterns like those of SPARs? Frank Preston (1960) conjectured that the answer to both those questions is Yes.

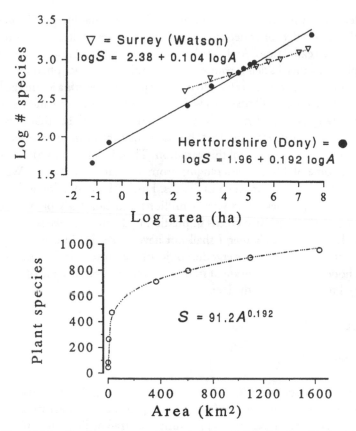

Figure 14.1. Species–area curves for plants in Great Britain. **Top:** Logarithmic plots: The shorter curve comes from Watson (1835) and is the oldest known pattern in ecology. The steeper curve comes from data of Dony (1963). The slope of such curves is termed their *z* value. **Bottom:** An arithmetic plot of most of Dony's data. The *z* reflects the curvature in arithmetic axes.

If Preston was correct, the SPAR-SPTI combination would form a powerful and unique ecological rule. It would say, in some deep sense, that time and space are interchangeable in their effect on diversity. This recalls the mathematical property called ergodicity. "In the presence of ergodicity, . . . time averages equal space averages" (Veech 1992). Perhaps it is a stretch, but I suggest that we call Preston's conjecture ergodic to evoke its proposal that time and space may have the same quantitative effects on diversity.

Having and understanding an ergodic rule in ecology would be a multifaceted coup. It would be fascinating in and of itself. It would give us a potent tool to use in our investigations of diversification: We could compare areas to investigate and predict temporal data. (Think how much boredom will be avoided by not having to wait around for 10 my while life proliferates!) It would lead to a deeper understanding of the processes of diversification (and perhaps even extinction). And—let's face it—it would add luster and prestige to ecology, a crucial science whose crude abilities to predict phenomena deprive it of the respect it must have to help the planet. There is a lot on the line.

In this chapter, I will first review SPARs, both their theory and examples at scales ranging from 1 m^2 to the whole Earth. We shall see that they vary much more than Preston thought, but that they conform to an even larger set of rules, rules that control this variation. Then I will turn to SPTIs. I will describe some SPTIs at scales ranging from 10 years to 10^8 years. We shall see that they vary just as much as SPARs. Finally, I will develop a theory of SPTIs at the grandest time scale—millions and hundreds of millions of years—and show that it succeeds in predicting the patterns we see.

By the end of the chapter, I shall not have succeeded in convincing you (or me either) that Preston's ergodic conjecture is correct or that it is wrong. But I hope to make you aware of it, of its potential, and of some of the steps we need to take to settle the issue.

Species–Area Curves

Lognormal Theory

As ecologists described SPARs from different taxa and different places, they noticed that SPARs usually conform to two quantitative regularities. First, as equation 1b summarizes, they approximate a straight line when plotted in log-log space (i.e., log diversity versus log area). Second, the slope of this line—its z value—is most often a fraction in the approximate range of 0.15 to 0.30. Ecology would like a theory to predict these two properties.

Ever since Fisher et al. (1943), ecologists have believed that the number of species with various abundances is a key variable in the study of diversity. If only there existed a standard rule for the number of species at each abundance, we could predict diversity from properties of that rule. Such a rule would possibly also predict the two basic properties of SPARs.

Preston (1948) proposed that lognormality constitutes the needed rule. He maintained that the number of species having various abundances follows a lognormal distribution. Preston's suggestion depended on his awareness of lognormality's ubiquity in all sorts of interactive processes. It is important to emphasize that point: He did not *prove* that abundances of species fit a lognormal distribution; he *suggested* it. Although he also looked at a few data sets to demonstrate that they might indeed fit a lognormal, he did not really test the idea. Today we know that it is far from an established rule

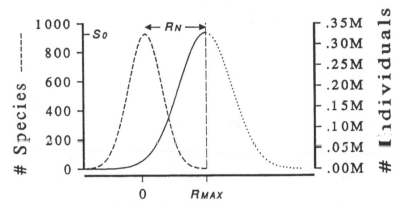

Truncated Canonical
Lognormal Distribution

Log$_2$ abundance (deviation from mode)

Figure 14.2. Preston's canonical lognormal abundance distribution. Values on the x-axis are standardized deviations from the modal abundance. The indivduals curve has two parts. The *solid line* is realized; the *dotted line* is not, requiring species to have more individuals than the environment can support. Notice that the peak of the individuals' curve lies over the maximum actual species abundance.

(Hughes 1986; Gaston 1994), but, for the time being, let us recall Preston's work. It was an elegant attempt; it has provoked much good research, and it retains its place in most basic ecology textbooks.

Preston (1962a,b) added a grace note to the lognormal distribution and used it to predict both the linearity and the slope of SPARs. That grace note is the individuals' curve. The species' abundance distribution shows us the number of species in each interval of abundance. Multiply that number by the abundance and you have the number of individuals that are members of species within that interval. The individuals' curve displays that number of individuals against the abundance. Preston showed that if the species' abundance distribution is lognormal, so is the individuals' abundance distribution.

Next, Preston turned his attention to the location of the peak of the individuals' curve. If this lies directly above the maximum (real) abundance, Preston called the combination of the two curves "canonical" (figure 14.2). Then he developed a mathematical argument which, he claimed, showed that canonical lognormal distributions produce SPARs that are approximately straight in log-log space and have z values of about 0.26. May (1975) expanded the allowable positions of the individuals' peak and discovered that it made little difference: Nearly canonical lognormal distributions also lead to nearly straight SPARs with a z value near 0.25. He suggested a range of 0.18 to 0.39 for z. We ecologists have taken this association between lognormality and SPARs so much for granted that Connor and McCoy (1979) reversed the sequence of logic, and began testing for lognormality by looking at SPARs!

Preston's and May's works have undergirded the use of SPARs by ecologists. Bolstered with the confidence that we understood them, we used them freely. In addition to the several examples mentioned earlier, we drew yet another application of SPARs directly from their apparent connection to lognormal abundance distributions. The minimal sustainable abundance of the lognormal—which Preston called its veil line—has become the crucial concept of the minimal viable population in conservation biology (Shaffer 1987).

Yet we shall now see that we have been too smug. Lognormal abundance distributions may have nothing at all to do with SPARs.

Real SPARs

After ecologists spent two or three decades of hard labor amassing SPARs, Connor and McCoy (1979) pointed out that we were not seeing the results that Preston's result predicted. Instead of SPARs neatly clustered near 0.25, they were spread out over the unit interval. We recognized other discrepancies, too. For a long time, we recognized that SPARs taken from nested mainland samples (as in figure 14.1) had consistently lower z values then those from sets of islands. Also, MacArthur and Wilson (1967) had predicted that island z values should not be the same for all islands but should be directly proportional to an island's isolation. That prediction too has been confirmed (Diamond 1972; Rosenzweig 1995). Then I noticed that SPARs from areas with nearly independent evolutionary histories always have z values greater than 0.6 (Rosenzweig 1992, 1995). Thus, although z values do not nestle snugly in the 0.25 region, they also do not scatter at random on the unit interval. They show pattern in their values, pattern unnoticed by Connor and McCoy.

The pattern points to a scaling rule: The longer the time scale of the dynamic processes governing diversities, the higher z (figure 14.3). Let us imagine a series of areas, *each the same size,* but located at different distances from a large continent. One area is on the continent (i.e., distance equals zero). Several areas are islands at different distances from the large continent. And one is so far away that—although it is very small—it has an independent

Three Scales of Species-Area Curve

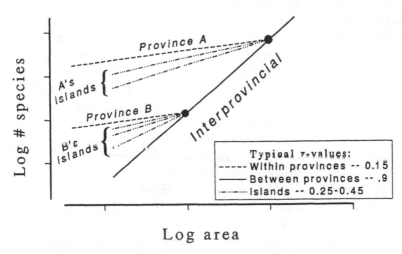

Figure 14.3.   Variation in SPARs depends on the scale at which we measure them. Nested SPARs taken from within one biogeographical province have the shallowest slopes. Interprovincial SPARs have the steepest slopes. Island SPARs lie in between. Their actual position depends on the degree of isolation of the island. After Rosenzweig (1995).

evolutionary history. Each area has a sample of habitats. In the piece of the continent, we find all the species whose habitats it contains (Williams 1943; Tonn and Magnuson 1982; Fox 1983) plus some sink species (i.e., those whose presence depends on generation-by-generation immigration from habitats not in the piece, because their replacement rate per generation falls short of maintenance in the piece) (Holt 1993; Rosenzweig 1995). Islands are too isolated for such regular immigration (by definition) and so have fewer species (MacArthur and Wilson 1963, 1967). More isolated islands have lower immigration rates and, so, fewer species than nearby islands. Small but separate provinces cannot even depend on irregular immigration, and so have mostly the species that have evolved endemically (Rosenzweig 1995). Because z is the slope of the line joining an area's point to that of the total large continent, z will rise as diversity falls. Thus areas that are regularly replenished with sink species have the lowest z's; islands— without sink species—have higher z's, and their z's rise gradually as their isolation increases (and thus as their immigration rates decline); and separate provinces, governed by endemic speciation rates, have the highest z's. Given all that we now know about diversity dynamics, it would greatly surprise me if all

SPARs had the same z value. Palmer and White (1994) also emphasize that SPARs are a collection of relationships at different scales.

In addition to predicting z values of about 0.25, lognormal theory predicted the near-linearity of SPARs in log-log space. In contrast to the z value prediction, the prediction of near-linearity has been fulfilled again and again by real SPARs. The exceptions come from very small sample sizes. Plant ecologists and sometimes animal ecologists find that semilog plots outperform log-log plots in fitting species–area relationships. But when they do, it is because their samples have penetrated deep into the left side of the area axis. In such areas, the linearity prediction itself actually breaks down. Instead, the theory predicted that SPARs should be convex upward over very small areas (Preston 1962a,b; May 1975). They are. Moreover, even if this prediction had failed completely over small areas, we could take refuge in the small sample sizes that inevitably characterize small areas. Small sample sizes tend to underestimate diversity. That also can cause the convexity of a SPAR in log-log space (Rosenzweig 1995). Investigators rarely correct such small samples for their bias (Rosenzweig 1995). Altogether, lognormal theory's prediction of the shape of SPARs has fared rather well. The dominant signal is that SPARs are linear in log-log space.

Yet we cannot ignore the failure of Preston's theory to predict z values accurately in all cases except nearby islands. What is amiss?

Other Theories and Simulations

Preston's theory makes quite the wrong predictions of z for all but one scale, islands fairly close to mainlands. Perhaps only such islands have lognormal species abundances? That in itself would be of interest. But, by performing some simulations, Leitner and Rosenzweig (1997) have discovered that it is not the explanation of the discrepancy between the data and the theory.

Figure 14.4 shows three sets of simulations. In each, Leitner and Rosenzweig (1997) assume that there is an assemblage of species whose abundances are lognormally distributed. The computer assigns each species a geographical range proportional to its abundance. (Thus, the population density of all species in the simulation is kept constant.) Then that range is randomly assigned coordinates on a virtual map. The computer constructs a SPAR by sampling larger and larger areas (with much replication). According to the data, if the theory is adequate, the simulation should result in z's of 0.1 to 0.2. Preston's prediction was a z close to 0.25. However, the simulation with a canonical lognormal did not achieve either of these results. Instead, its z averaged 0.77.

Canonicity depends on the distance between the peaks of the individuals'

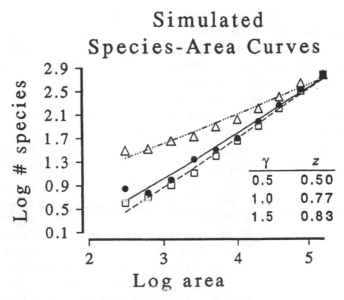

Simulated Species-Area Curves

γ	z
0.5	0.50
1.0	0.77
1.5	0.83

Figure 14.4. Simulations of nested SPARs. The simulated z values exceed those predicted by Preston (1962a,b) and by May (1975). The value γ is the ratio R_N/R_{MAX} in figure 14.2, and thus it measures canonicity. Redrawn from Leitner and Rosenzweig (1997).

curve and the species curve. May defined a useful statistic for this distance, which he called γ. When γ equals 1, the species curve is canonical. If γ is greater than 1.0, the peaks are closer than canonical; if it is less than 1.0, they are farther apart. May showed that as γ grows, z declines. The two other simulation sets in figure 14.4 did *not* use a canonical lognormal (although both maintained the assumption of lognormality). In one, Leitner and Rosenzweig (1997) set γ to 0.5; in the other, they set it to 1.5.

The result confirmed May's prediction of the trend; smaller γ led to larger z. However, the actual value of z did not agree with either theory or data. Both noncanonical simulation sets produced exceedingly high average z values (0.50 and 0.83, respectively).

So now we have lost the theoretical underpinnings of SPARs. The theory does not accurately predict either the data or the results of simulations. What is wrong? Leitner and Rosenzweig (1997) showed that Preston's math makes a hidden and invalid assumption: it assumes that if an abundance distribution is lognormal at one spatial scale, it is lognormal at all spatial scales.

The literature reports several recent attempts to construct a general theory of SPARs. In one, Caswell and Cohen (1993) produced a pair of models that join the well-known diversity-disturbance pattern to SPARs. The models depend on a fixed species pool and are thus not meant to handle evolutionary time and space. The model world is composed of a large number of patches, all of the same habitat. Species colonize patches, and patches suffer disturbances, thus losing all their species. One model allows no competition. The other assumes that all but one species will go extinct as a result of competition if disturbance is long enough delayed.

With or without strong competition, the models predict near-linearity and tolerably realistic z values for SPARs. But the results of the models must not be compared to the species–area curves we already have. Elements crucial to known SPARs do not appear in these models. In particular, the models lack both habitat variability and evolutionary processes. So, even if they match real data, we can be sure that known SPARs get their shape from other processes. Moreover, predictions of the models differ from real data in several substantive properties. The models predict convex-upward SPARs that glide to an asymptote (the species pool). The data show curves that are straight or even concave upward. The models also predict considerable variation (about an order of magnitude) in z values depending on disturbance frequency: the higher the disturbance frequency, the steeper the slope. Real z values, taken at scales for which disturbance can possibly govern diversity, cluster tightly in the range of 0.1 to 0.2. In addition, real trends in z values from several scales show the opposite relationship between scale and z—the slower the dynamics, the steeper is z.

Wissel and Maier (1992), in their attempt to construct a theory of SPARs, build and simulate a different model, designed to fit island conditions. Sometimes the model produces a linear SPAR with a reasonable z value and a species abundance distribution very like a lognormal. In these runs, the model contains both zero-sum interspecific competition and considerable stochasticity.

In this model, all species on the island have the same carrying capacity, the same immigration rate, and the same life history coefficients. The authors do not tell us how differences among species would affect the outcome, although they do point out the requirement for "sufficient similarity" among the species.

Wissel and Maier's model may represent useful progress toward the goal of a theory for island SPARs, but in my opinion it will be limited. Like the Caswell and Cohen model, it does not allow for evolutionary variation in diversity. It also assumes both stochastic and logistic population growth. But most important, it assumes that rates of birth and death are similar to rates of immigration. That is a serious shortcoming. Not only do we know that island immigration works much more slowly than birth–death, we also know this to be the crucial difference that gives island SPARs their characteristic z

values (Rosenzweig 1995). In contrast, the Wissel-Maier model shows z varying with parameters such as how strongly island area affects immigration rate. This parameter appeared relatively unimportant to MacArthur and Wilson (1967) and there is no evidence to contradict their intuition.

A third model (Seagle and Shugart 1985) predicts reasonable z's (0.14 to 0.31) but finds that they vary with the size of the species pool. But the data do not support that correlation at all. This model also finds that SPAR does not depend on whether there is competition, a conclusion diametrically opposed to that of Wissel and Maier (1992) and to any a priori model that assumes species' ranges are influenced by competitors.

Williams (1995) developed a model based on sampling theory, but its purpose is to supply a sigmoid function more reliable than the power function of equation 1. This it does for a fixed pool of species. But it is not meant to fit a growing pool, nor to predict the actual values of its parameters or the approximate value of z. In contrast, the Leitner and Rosenzweig (1997) function does allow for the actual species pool to grow with area, and it does predict z. Unfortunately, their prediction depends on the unknown parameter value of a second function, and on the shape and parameters of the abundance distribution. Thus it raises more questions than it answers.

In sum, we do not yet have a satisfactory model for known species–area relationships. None runs through the variety of spatial scales that require explanation. And none shows appropriate sensitivity to the suite of variables that data tell us must be important.

Species–Time Curves

As mentioned previously, Preston (1960) conjectured that space and time (in particular, evolutionary time) have equivalent effects on the accumulation of diversity. Thus, he guessed that species–time curves (SPTIs) should fit straight lines in log-log coordinates and have slopes (analogous to z values) resembling those from sets of islands. But we have far fewer SPTIs than SPARs. Let us examine a few to see if there is even a slight chance that Preston was right.

Short-term SPTIs

The first SPTIs were collected over very short time intervals, during which there was no change in habitat. That is, all their samples came from the same part of the day and during the same season. In such cases, we expect the samples to come from the same set of species. Nevertheless, we also expect a significant SPTI because more time means a larger sample of individuals.

Williams (1964) illustrates such an SPTI with Lepidoptera caught in six light traps at Rothamsted Experimental Station, England, from 30 June 1949 to 10 July 1949. As data accumulated during this time, the number of species increased steadily, fitting a straight line in log-log space very well (figure 14.5). I distinguish the slope of an SPTI from z by calling it ψ. The ψ-value of figure 14.5 is 0.25.

Preston also illustrates an SPTI from a constant set of species. Figure 14.6 shows the linear pattern of increase of bird diversity for one 69-acre (31-ha) area (called Neotoma). Data represent ten censuses conducted during an 18-year interval. The ψ-value is 0.12.

Finally, Arthur Shapiro has shown me a set of data for butterflies taken from 1977 to 1993 at the same place at Willow Slough in northern California every 4 July. Shapiro personally netted the animals at about the same time of day for about the same length of time each year. Figure 14.7 shows the SPTI of these data. It also fits a straight line. Its ψ-value is 0.13.

Several investigators have developed methods to reduce the bias of sample size from estimates of diversity (Colwell and Coddington 1994). These include Simpson's index and Fisher's α. If the SPTIs we have just seen depend

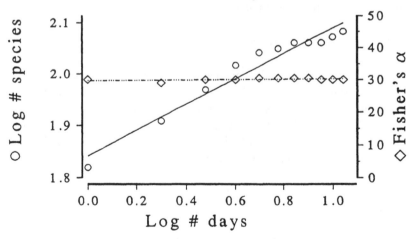

Figure 14.5. The species–time curve for a brief period in early summer, 1949. The SPTI is quite shallow and the reanalysis with Fisher's unbiased index of species diversity, α, shows that there is no real increase in the diversity of the collection during this period. Data from Williams (1964).

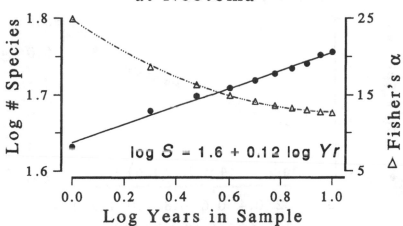

Figure 14.6.   SPTI for ten spring breeding-bird censuses at an estate in western Ohio. As in figure 14.5, converting S to Fisher's α shows that the slope, ψ, is greater than zero only because the sample size of the collection increases. In fact, the slope of the α's is strongly and significantly negative. This indicates both deviation from the log-series distribution and exhaustive, careful censusing. Such concave, declining α-curves occur when the sample size increases after the diversity has already been well determined (Rosenzweig 1995). Data from Preston (1960).

entirely on increasing numbers of individuals as time goes by, then these methods ought to rotate them to $\psi = 0$. I shall use the statistic called Fisher's α to test this idea. It is the oldest (Fisher et al. 1943) and still one of the best (Rosenzweig 1995).

R. A. Fisher deduced α, his index of diversity, based on the assumption that the abundances of species fit a log-series distribution (Fisher et al. 1943). That is, if the total number of individuals is N, and p is a constant proportion, then the most common species has pN individuals; the next most common, $p(1 - p)N$; the next, $p(1 - p)^2 N$; etc. If abundances fit that distribution and samples are taken from one fixed set of species, then the number of species actually occurring in the samples will obey this equation:

$$S = -\alpha \ln(1 - x) \qquad [2]$$

where α is a constant that depends on diversity alone and therefore may be used as a surrogate variable for unbiased species diversity. On the other hand,

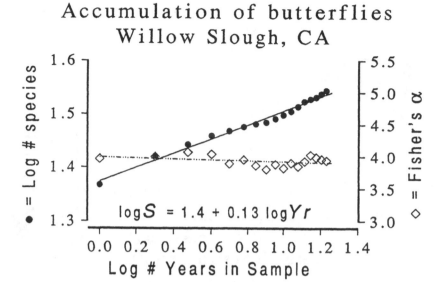

Figure 14.7. SPTI for 17 consecutive 4 July censuses at a single location in California. As in figure 14.5, converting S to Fisher's α shows that the slope, ψ, is greater than zero only because the sample size of the collection increases. In this case, there may even have been a decline in diversity—the negative slope is significant. The bias of increasing sample size hid the possibility of this decline. Data collected by Arthur Shapiro, University of California at Davis.

x is a variable that depends on sample size. The variable x satisfies the equation

$$S/N = [(x - 1)/x]\ln(1 - x) \tag{3}$$

Fisher's α may be used even though the abundances of the species being sampled do not fit a log-series distribution. Small, incomplete samples of other distributions also fit the log-series (Boswell and Patil 1971). Mitchell (p. 194 in Rosenzweig 1995) wrote a simple Q-Basic program to use in calculating α. (Note: the first printing of this book had a typographical error in this program. Consult subsequent printings or contact the author for the correct version.)

We have the individuals' data for the Lepidoptera at Rothamsted, for Shapiro's California butterflies, and for the birds of Neotoma. So, we can calculate the values of Fisher's α for the data in figures 14.5, 14.6, and 14.7. I included these values in the figures on a second y-axis. In all three cases, using Fisher's α removed the significant positive slope of the SPTI as expected.

SPTIs in Ecological Time

Let us now extend the sampling period so that increases in time reflect increases in the number of temporal habitats. Now we no longer expect Fisher's α, or any other technique for removing sampling bias, to be able to rotate the SPTI to $\psi = 0$. Instead, Fisher's α should also grow with time because the actual number of species in the pool from which we sample is growing. It may grow because of seasonal changes. It may grow because of differences between years. Perhaps both will be involved.

We do have a few SPTIs from longer periods during which we can suspect that some habitat changes may have occurred. For example, Preston (1948) produced an elegant SPTI from the record of moths collected by Seamans at a single light trap in Lethbridge, Alberta, for 22 years (1921 to 1943). Seamans identified 291 species among 303,251 individuals. But the number of species collected per year also increased, so that the raw record would have contained an increase in diversity even if the pool of moth species had remained constant. Preston ingeniously removed the trend by running the record both forward and backward, then summing the separate results. The SPTI fits a straight line with a ψ value of 0.257 (figure 14.8). I took the

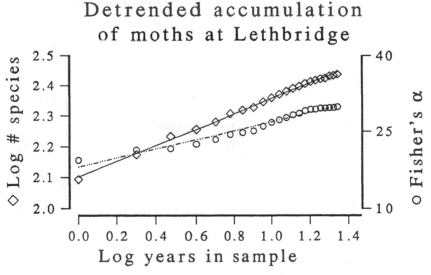

Figure 14.8. SPTI for a 22-year record. In this case, Fisher's α also shows an increase, indicating that the positive slope of the collector's curve, SPTI, is significant. During this interval, actual diversity grew (not plotted on the graph). So, Preston (1948) detrended the data by running them both forward and backward, then taking the average.

number of individuals for this data set from Preston's figure 14.13, calculated its values of Fisher's α, and added them to figure 14.8. Now α climbs significantly with time ($P = 10^{-15}$). Evidently, species were truly being added to this curve year after year. (Preston did the same for another 22-year record of Canadian Lepidoptera, but, although he reports similar results, I could not use the data because there is an error in his published data table.)

We also have the individuals' data for the Lepidoptera at Rothamsted Experimental Station. In fact, we have them for the 3 weeks immediately following the samples of figure 14.5. Thus we can construct a 32-day SPTI. We also have them for the next 8 years and can construct that SPTI as well.

Figure 14.9 plots the results for the first 32 days (including the 11 days of figure 14.5). As those first 11 days saw no real increase in diversity, I analyzed days 12 through 32 separately. A straight line still fits the data. But its ψ increases to 0.45. (The ψ for days 1 to 11 is 0.25.) Moreover, now Fisher's α also rises significantly with time ($P = 5 \times 10^{-10}$). In the first 11 days, there had been no significant rise ($P = .177$).

The 8-year record produces similar results (figure 14.10). Its ψ value is 0.30, and its Fisher's α rises significantly ($P = 4 \times 10^{-6}$).

Rothamsted Lepidoptera

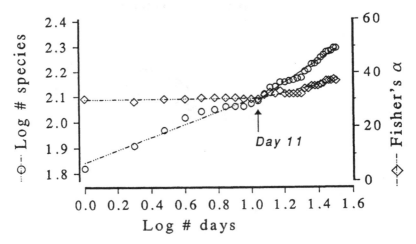

Figure 14.9. SPTI for 32 days in early summer, 1949. The first 11 days are the same as the data of figure 14.5. The rest tell us that after 11 days, the true diversity in the collector's curve began to increase. We know this because Fisher's α also grew from day 11 to day 32. Data from Williams (1964).

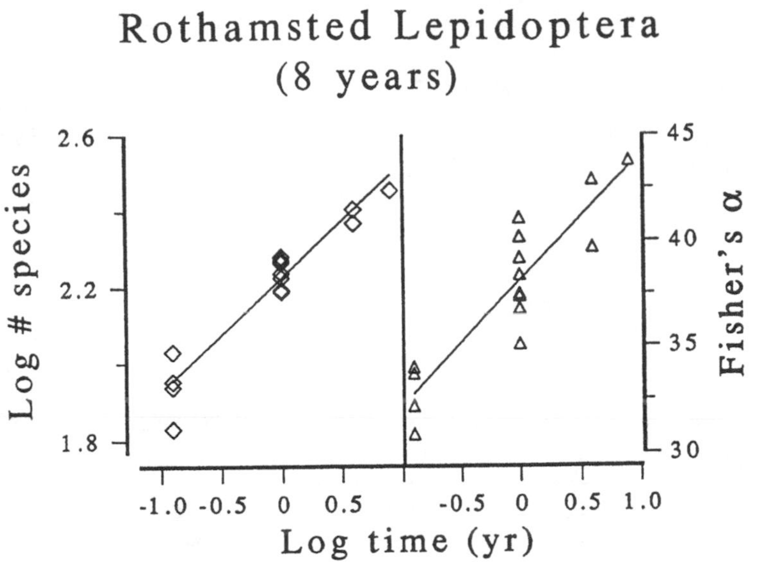

Rothamsted Lepidoptera
(8 years)

Figure 14.10.   SPTI for 8 years at light traps. As in figures 14.8 and 14.9, the increase proves real because Fisher's α also increases. Data from Williams (1964).

SPTIs in Evolutionary Time

From Peter and Sarah Bretsky (1976, and personal communication), I obtained a detailed, species-by-species and stratum-by-stratum fossil record from the late Ordovician muddy benthos of the Nicolet River Valley, Quebec. I measured its SPTI during the roughly 4-my period after the assemblage reached steady-state diversity. It does conform to Preston's ergodic conjecture, fitting the equation $S = 24t^{0.167}$ ($R^2 = 0.852$). So its ψ value is 0.167, similar to those of modern SPARs taken within a single continent, and only a bit lower than those Preston predicted for canonical, lognormally driven SPARs.

Raup (1976a) compiled a list of all the fossil invertebrate species published in the *Zoological Record* from 1900 to 1970. His compilation allows me to estimate the ψ value of the implicit SPTI from the entire (about 530-my) record of Phanerozoic marine invertebrates. For this purpose, I focus at the grandest (or, if you prefer, the crudest) level of resolution—that of the ten major time divisions, Cambrian to Tertiary.

Raup's data were collected by many different paleobiologists using a variety of techniques with variable intensity of group effort in a variety of epochs and from a variety of rock types (table 14.1). Lacking the uniformity of con-

TABLE 14.1   Data set for the Phanerozoic SPTI studies.

Period	Base[a]	Area[b]	Spp[c]
Tertiary	−65.0	20.87	39,561
Cretaceous	−145.6	16.73	21,622
Jurassic	−208.0	3.11	12,213
Triassic	−245.0	3.66	5,667
Permian	−290.0	2.87	6,712
Carboniferous	−362.5	3.57	14,709
Devonian	−408.5	3.37	12,137
Silurian	−438.0	1.96	5,411
Ordovician	−500.0	1.94	10,701
Cambrian	−530.0	2.48	8,102

[a]Base in my from Harland et al. 1990, with the generous advice of J. J. Sepkoski, Jr. to bring estimates up to date, especially the estimated base of the part of the Cambrian from which Raup's data were taken.

[b]Area in Mkm² digitized from Raup (1976b).

[c]Species of marine invertebrate fossils assigned to one and only one period (from Raup 1976a).

ditions associated with the Bretsky assemblage, the data must be addressed and analyzed with great caution.

In what follows, I deal with what I believe to be the two most important sources of variation in the data:

- The periods they come from have been sampled to a variable extent.
- The periods had a variety of durations.

Dealing with the first problem requires well-known methods. And we will see that the second can actually be turned to our advantage.

Most of the sample-size variation comes from variation in the area of remaining rock containing fossils of a particular period. This variation is substantial, covering an entire order of magnitude (Raup 1976b). It reflects the age of the period—recent periods having much more extensive remaining deposits (see table 14.1). Undoubtedly, it also reflects the great variation in the area of epicontinental seas from time to time. Sheehan (1977) suggested that all this variation was best reflected in the amount of attention that systematists give to different periods. The more effort systematists have expended on a period, the larger its sample size. He devised a measure called PIUs (paleontological interest units) to represent this.

PIUs are even more successful than estimates of remaining rock deposits at single-factor explanation of variation in diversity (Signor 1985; Sepkoski 1994). Yet Raup (1977b) pointed out that PIUs and rock availability reflect the same underlying variable. He argued logically that large areas of rock will generate large samples for systematists to take advantage of. Moreover, PIUs are a discrete, ad hoc, integer-based metric; we really do not know how to make a theory using them. On the other hand, as we have seen, area is a well-researched variable involved in many known diversity patterns that we can compare to whatever we find in the Phanerozoic data. Thus, I chose to use "remaining rock area" as the first independent variable in this study. I will assume that at the very large time scale I am examining, other sources of sample-size variation matter only as second-order phenomena. Table 14.1 shows the actual data values used in this study.

Just as regression analysis related area to diversity among different areas at a single time (i.e., our own), it can correct for the area variation among major periods (Flessa and Imbrie 1973; Flessa 1975; Raup 1976a,b; Sepkoski 1976; Flessa and Sepkoski 1978). Figure 14.11 shows the approximate quantitative influence of area on the Phanerozoic data. The signal is not quite significant ($P = .055$), and it leaves a great deal of variance to be explained ($R^2 = 0.39$). A bit of this residual variation reflects the variable proportion of deposits that come from marine habitats of any sort—for example, a relatively high

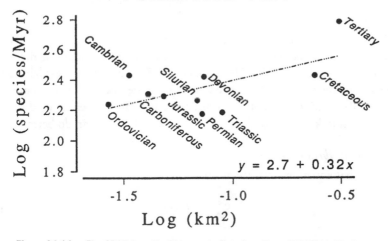

Figure 14.11.   The SPAR from the Phanerzoic. Data from Raup (1976a,b). Nonlinear regression showed that the univariate analysis (shown here) somewhat underestimates both the significance and the value of the slope. The true z value is close to 0.383 (see text).

proportion of Cretaceous deposits bear no fossils because they were laid down on continents as intermontane accumulations (Flessa and Sepkoski 1978). However, as we shall now see, the unequal durations of the different periods account for most of the residual variance.

Figure 14.11 already incorporates time, but as a linear influence. It does so by simply dividing time into diversity to obtain the dependent variable. However, that assumes our conclusion in advance; it tacitly assumes that new species appear in the fossil record at a constant average rate (Boucot 1975; Raup 1976a). We want to discover the value of the curvature of SPTIs (ψ) at evolutionary time scales. We already know that $\psi < 1$ for ecological time scales and also at the scale of a few million years in the Upper Ordovician. We must not assume it is 1 (i.e., linear) at the scale of tens of millions of years. If, for example, $\psi = 0.4$, then we should have divided the ordinate, not by my, but by $my^{0.4}$. We need an estimate of z that does not presuppose the pattern of time's influence on diversity.

We first approach the problem mechanically by expanding equation 1b to include the effects of duration. Using a bivariate linear regression, I fit the data to an equation of the form

$$\log S_{tA} = c + z \log A + \psi \log t \tag{4}$$

and obtained the following result:

$$\log S_{tA} = 3.1 + 0.437 \log A + 0.743 \log t \tag{5}$$

where A has the dimensionless units, fraction of Tertiary rock area. (Note: the values of the slopes and their significances are independent of the units of area used—Rosenzweig 1995.) So, $\psi = 0.743$. Equation 5 explains 82.3 percent of the variance, but only one of its slope coefficients is significant ($P_z = .019$ and $P_\psi = 0.061$). Perhaps the insignificant effect of duration came from lack of enough data points. Nevertheless, we can do better.

An Equilibrium Theory
Notice that equation 4 has the form

$$S_{tA} = kA^z t^\psi \tag{6}$$

As in the case of the SPARs, we have no theory to predict such a form. But we can make a few assumptions and predict a form fairly similar to equations

4 and 6. I will make these assumptions in the spirit of investigation. That is, I know that they are not true, but I wish to see if their untruth makes any difference. So, I will try them out on the data. Here they are:

1. The number of species has hovered around a single steady state value, C, for 530 my.

2. The species turnover rate has also hovered around a single steady state value, Ψ, for 530 my.

3. The area of the Earth in which marine invertebrates have been deposited as fossils has remained constant for 530 my.

If I may be permitted these outrages, we will see that they have much to teach us. They create the presumption of a single SPTI, which life has tracked for 530 my. Each of the major periods lasted a different amount of time and so would have tracked this SPTI to its position over a different value of t. This holds out the hope that by assembling all the data, we should be able to reconstruct the underlying SPTI. Let's try. Our success or our failure may reveal something useful about the assumptions.

According to our assumptions, each major period begins its interval with the same number of species, C. Then it experiences evolutionary turnover at the rate Ψ. The turnover allows new species to accumulate in the "collection." Thus, after t units of time, the collection will have $S(t)$ species, where

$$S(t) = C + \Psi t \qquad [7]$$

But most of the species do not survive even as fossils. Instead, we get a sample that depends on the area of rock remaining from each period. In particular, the number of species we expect to see is

$$S_{tA} = S(t)A^z \qquad [8]$$

providing we express A as the proportion of the rock area remaining from a period. In other words, A^z can be viewed as a classical logarithmic decay function.

Now I combine equations 7 and 8:

$$S_{tA} = (C + \Psi t)A^z \qquad [9a]$$

Equation 9a can be transformed into a quasi-logarithmic version:

$$\log S_{tA} = \log(C + \Psi t) + z \log A \qquad [9b]$$

But equation 9b merely reveals that, in contrast to equation 4, equation 9a is fundamentally nonlinear.

Nonlinear regression of the data using equation 9b produced

$$\log S_{tA} = \log(1708 + 383t) + 0.383 \log A \qquad [10]$$

Equation 10 has the following summary statistics: $R^2 = 0.834$; $p_C = .80$; $P_\Psi = .012$; $P_z = .028$. The high probability associated with C tells us that we must not take it very seriously, but the values for the coefficients are quite significant and encouraging.

Now let us compare this result to the previous one (equation 5), which came from a mechanical application of log-log relationships. First, notice that although both equations have three parameters, equation 10 has a slightly better R^2 (0.834 versus 0.823). It would appear that the outrageous assumptions leading to equation 10 have not done too badly! Second, notice that the two analyses produced very similar estimates of z: 0.437 versus 0.383. That strongly suggests that the effect of time is linear: We got about the same z value despite assuming linearity (i.e., a constant turnover rate) to obtain equation 9 but not equation 4.

How can we tease a test of linearity out of a fundamentally nonlinear equation? Equation 7 has an implicit coefficient. It is the power w. In equation 7, $w = 1$. That is what makes the arithmetic form linear. Let us therefore rewrite equation 7 to allow for the possibility of curvature:

$$S(t) = C + \Psi t^w \qquad [11]$$

And equation 9b, the equation used in the previous nonlinear regression, becomes

$$\log S_{tA} = \log(C + \Psi t^w) + z \log A \qquad [12]$$

The result of fitting the data to equation 12 is

$$\log S_{tA} = \log(140 + 594t^{0.91}) + 0.386 \log A \qquad [13]$$

Equation 13 ($R^2 = 0.8352$) confirms both the z value and the w value of the Phanerozoic data. Since its w value approximates 1.0, its SPTI is quite close to linear. In fact, the improvement in R^2 compared to equation 10 lacks significance. That could be a problem of too few data to estimate the four parameters of equation 13. (None of its parameters is significant.) But in any case, we have no statistical basis for changing w from 1.00 to 0.91. So, I am tempted to regard the value 0.91 as tolerably close to 1.00.

But comparing equations 10 and 13 reveals an even more useful value. Whether w is assumed to be 1.00 (as in equation 9) or not (as in equation 12), z lies between 0.383 and 0.386. That is not a very large range. So the value $z = 0.383$ will give a very good idea of the independent effect of area on each period's cumulative diversity. We get it by calculating how many species would have been found in its record (from 1900 to 1970) if it had the same area as the Tertiary does have.

First, I manipulate equation 8 into equation 14:

$$S(t) = S_{tA}/A^z \qquad [14]$$

and use it to obtain the vector of $S(t)$ values; these estimate S free of the effect of area. To produce this vector, I inserted into equation 14 the coefficient $z = 0.383$ from equation 10, and I used the vectors of observed diversities and proportional areas. Let us call $S(t)$ "detrended S" as it is statistically free of the area trend.

In compiling his data set, Raup (1976a) sequestered species found in more than one major period from those found in only one. I have not used those sequestered species in the data set of this chapter. Thus, I can combine the separate values of $S(t)$ into a monumental SPTI running for the entire Phanerozoic. I simply string together all the available data by accumulating the species period by period in order of occurrence (figure 14.12).

Figure 14.12 shows dramatically the tendency for speciation rates (at the million-year scale) to remain in a very narrow range during the entire Phanerozoic. If there were any tendency for them to decline over the past 500 my, figure 14.12 would be convex upward. Instead, the taxa of the Paleozoic, the Mesozoic, and the Tertiary all produced marine invertebrate species at about the same average rate.

Rosenzweig [1997 (in press)] produced a sort of poor man's bootstrap of the curvature intervals that might characterize these data. Maximal curvatures result if we reorder the data according to diversification slope rather than age. Maximal convexity comes from counting the time of fastest accumulation first. The results were $w = 0.866$ for the maximum and $w = 1.204$ for the minimum. I included these curves in figure 14.12. Neither lies very far from the actual linearity observed.

You may gain more confidence in the method if I repeat the analysis on the late Ordovician data of the Bretsky assemblage. We know these data do come from a time of equilibrial diversity, but their turnover rate was definitely not constant (figure 14.13). It fell from 0.105 new species per vertical meter of rock to about 0.034 during the several million years after equilibrium was reached (Rosenzweig 1995). Thus, this record ought to yield a w value less than 1.0.

It did. I used the actual diversity data in this case because no adjustments for unequal area were required. I defined a sample point as a stratum at which Bretsky's detected the evolution of one or more new species during the equilibrial period. From the data, I determined how many species in the assemblage were found in strata older than the period of equilibrium diversity and yet survived at its beginning (i.e., 33 species). I used that value as C in equation 11. Then I performed the nonlinear regression. The result was

$$S(t) = 33 + 0.554t^{0.735}, \qquad [15]$$

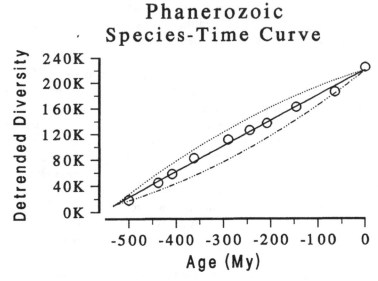

Figure 14.12. The Phanerozoic SPTI. The effect of different areas has first been removed from the accumulated diversities of the standard long geological periods. Then they are added together in order of their occurrence. The regression equation is $S = 220,962 + 397 \times Age$, with $R^2 = 0.995$. If turnover rate is at steady state, theory predicts a value of $w = 1$. By their remarkably close fit to a straight line, these data show that they agree with the theory. The curved lines are the most extreme deviations from linearity that these data might produce (see text). Redrawn from Rosenzweig (1997).

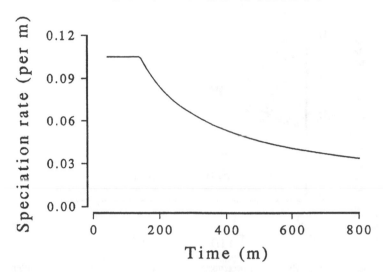

Figure 14.13.   Actual speciation rates of the Bretsky marine invertebrate assemblage during its period of steady-state diversity. Rates declined, at first sharply, then much more gently. Redrawn from Rosenzweig (1995).

for which $R^2 = 0.956$. We can see that $w = 0.735$. This turns out to be a very tight value with its $P = 4 \times 10^{-16}$. So this w is not all that close to 1.00. Instead, it reflects the considerable change of turnover rates during this record. A look at the graph itself (figure 14.14) confirms the change. The first part of the record grows linearly and so does the second. Each of these segments of the record has a w value near 1.00. But the turnover rate sharply increases about 1.2 my after the start of the steady-state period and then even more sharply decreases about 0.2 my later.

Notice also that 0.735 differs considerably from 0.167, the value of ψ that fit the SPTI analogue of equation 1 for the Bretsky assemblage! Such discrepancies are to be expected. In the example that covered the whole Phanerozoic, ψ and w were 0.743 and 0.91 (or, better, 0.743 and 1.00), respectively. The parameter ψ will often mislead us (see Appendix to this chapter). Thus warned, we ought to reexamine the other SPTIs—those measured in ecological time—to determine more reliable estimates of their curvatures.

The w Value of SPTIs in Ecological Time

Just as the steady-state diversity constitutes the initial value of an evolutionary SPTI, the diversity of one temporal habitat constitutes the initial value

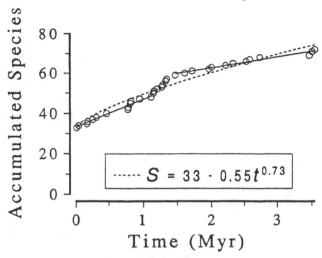

Ordovician Muddy Benthos

Figure 14.14. The Bretsky assemblage's SPTI in arithmetic space: An SPTI at an evolutionary time scale. The w value (0.735) reveals a considerable convex curvature. However, during most of this record, steady-state turnover seems to have prevailed. That is indicated by the linear SPTIs during both the first 1.2 my and the terminal, approximately 2 my periods. The convexity stems from the decrease in rates between these two periods. Data come from the late Ordovician just after a major increase in standing diversity. During the 4 my in which these fossils were laid down, diversity wobbled around a steady state. Data from Peter and Sara Bretsky (1976, and personal communication).

of an ecological SPTI. Using equation 11 as a guide, we therefore might expect ecological SPTIs to follow a power function plus a constant. As before, w shows the curvature of the SPTI. C is the number of species actually present in a fixed area at one time.

I reanalyzed the three ecological SPTIs with equation 11. In each case, I found an improved fit to the data (usually slight, as the fit was already quite good). As before, I also found that w exceeded ψ considerably. Equations 16, 17, and 18 show the following results:

$$\text{Lethbridge moths, years 1 to 22: } \psi = 0.26,$$
$$\text{but } S(t) = 124 + 34.36t^{0.50}, \text{ so } w = 0.50. \qquad [16]$$

$$\text{Lepidoptera, days 11 to 32: } \psi = 0.45,$$
$$\text{but } S(t) = 121 + 5.06^{0.90}, \text{ so } w = 0.90. \qquad [17]$$

$$\text{Lepidoptera, years 1 to 8: } \psi = 0.30,$$
$$\text{but } S(t) = 87.6 + 93.78t^{0.37}, \text{ so } w = 0.37. \qquad [18]$$

Discussion

Species–time curves (SPTIs) constitute the temporal analogue of species–area curves (SPARs). A SPAR consists of the accumulation record of species at a fixed time as we extend the space in which we make observations. An SPTI consists of the accumulation record of species in a fixed space while observation time increases.

Preston (1960) suggested that SPTIs would parallel SPARs in two ways. He conjectured that both patterns would be straight when plotted on logarithmic axes and that they would have the same values of curvature. He offered no theory in support of his conjecture. But because his conjecture suggests an ergodic property in ecology and, therefore, would be extremely interesting and useful if true, I have begun to test it.

Comparing SPARs and SPTIs

The empirical properties of SPARs are well known. SPARs are indeed straight when plotted on logarithmic axes. Their curvature varies from severe (exponent near 0) to very little (exponent near 1.00), depending on their scale. At the smallest scale, species accumulate merely because sample sizes grow (Rosenzweig 1995). There is no real change in diversity. If we do not adjust such samples for their size, we see collector's curves with exponents (z values) from about 0.08 to 0.15. At the next larger scale, the habitat diversity grows. Here we see exponents ranging from about 0.15 to 0.20. The next SPAR scale occurs on islands. As they become increasingly isolated, their z values change from about 0.25 to 0.55. Finally, SPARs have an exponent of 0.6 to 1.2 among different biogeographical provinces. At this scale, SPARs are governed by speciation and extinction rates.

Clearly, we have far fewer sets of SPTI data than SPAR data. Thus it is still premature to say whether there is empirical agreement of SPTIs with SPARs. Nevertheless, I am already encouraged that Preston's idea deserves our serious attention.

The shortest time scales accumulate species merely because their sample sizes grow. The mechanism—increase in the number of individuals counted—is clear, and SPARs share it. So, it is no surprise that the effects of time duplicate those of space. The three SPTI exponents I obtained were 0.12 (Preston's birds), 0.13 (Shapiro's butterflies), and 0.25 (Williams' Lepidoptera). But because underlying diversity itself is not changing, the resemblance of space and time curves depends on common statistical, rather than common biological, principles. So it fails to astonish or matter much to a biologist.

The next larger scale is that at which the habitat diversity grows—the ecological scale. Does species diversity grow with time as with increasing area? Certainly, at this scale true underlying diversity grows, so the scale poses an interesting biological question. Moreover, SPTIs and SPARs both seem approximately linear in log-log space. But, in the three examples I have found, SPTIs carry slopes (w values) that are quite high compared to those of SPARs. They varied from 0.37 to 0.90. None was in the neighborhood of 0.25.

However, SPTIs and SPARs that come from evolutionary time scales are remarkably similar. Different provinces—i.e., areas with virtually separate evolutionary histories—have z values from about 0.6 to 1.2 (Rosenzweig 1995). Evolutionary SPTIs—accumulated as new species evolve—also have w values near 1.00. Of course, we have only two such SPTIs. Yet one comes from the entire record of marine invertebrates during the Phanerozoic! And the other comes from a very special record for which we have a lot of detailed corroborating information.

Data will not teach us much more until we have considerably more SPTIs to examine. But we can turn to theory for some support. In this chapter, I presented a new mathematical theory that predicts $w = 1.00$ for evolutionary SPTIs (equation 7). Hence, the Phanerozoic and the Ordovician SPTIs are not just two data points. They are data that agree with the prediction of a theory.

I wish we had as reasonable a theory for evolutionary-scale SPARs as we do for evolutionary SPTIs. If we did, we would be in a position to say whether the agreement between SPARs and SPTIs at this scale is more than coincidental.

Theory at Intermediate Scales

Here, the news is discouraging. We have theory neither for SPARs nor for SPTIs at scales between sample-size effects and evolutionary time.

Until recently, we ecologists thought the work of Preston (1962a,b) and its extension by May (1975) gave us a solid mathematical foundation for all SPARs. Preston tried to demonstrate that both the general equation for SPARs and the actual value of its exponent were straightforward results of lognormal abundance distributions, particularly the one termed canonical. May relaxed the assumption of canonicity and showed that it mattered, but not very much—not at all to the form of the equation and only a little to the exponent.

But new patterns and simulations suggest that the papers of Preston and May only opened the door (Rosenzweig 1995). As Connor and McCoy

(1979) long ago noticed, the logarithmic slope (z value) of real SPARs varies over the interval from zero to one, and it does not fall narrowly in a range centered on 0.25 or 0.28 as suggested by Preston. (But they did not notice the regularity in this variation. As mentioned before, the z value grows monotonically with the time scale of the processes controlling diversity. The brief time scale of reproductive cycles yields low z values, whereas the long time scale of speciation and extinction yields the high z values.)

The basic scientist part of me wants to have a complete theoretical understanding of SPARs and SPTIs out of sheer curiosity. But there is a most pressing reason to discover a theory for intermediate scale SPARs, a reason that transcends the merely academic. SPARs provide a baseline for species conservation. They tell us how many species we can expect our nature preserves to hold. They predict overall losses to extinction. Their study also suggest ways to improve the conservation value of preserves.

In a stable, buffered world, we could simply rely on empirical determinations of SPARs. We would measure them for each taxon in each major biome and extrapolate the results to the preserves. But, in a dynamic world, pressed by an expanding human presence and faced with extensive global change, we ought not rely on such extrapolations. Instead, we should try to understand the reasons for SPARs. That means finding theories that work. Then we can better adapt our predictions to a variety of conditions now unforeseen.

Sink Species

One reason w exceeds z at the ecological time scale could be that SPTIs may have few sink species. Sink species are those that persist in a place despite having local population replacement rates below unity. Their persistence depends on regular immigration from habitats in which the species do better (Rosenzweig 1995).

Do temporal habitats have sink species, too? I see no physiological reason why a species could not persist past its temporal niche, so the answer may be Yes. However, considerations of optimal temporal habitat selection may require individuals to retreat from temporal habitats that profit them little (Rosenzweig 1995). Spatial habitat selection may sometimes cause such habitat restrictiveness, but often it does not. Often, some individuals are forced by optimization considerations to utilize sink habitats in space (Shmida and Ellner 1984; Pulliam 1988; Pulliam and Danielson 1991). Under certain circumstances, individuals are better off if they exploit sink habitats in space instead of continuing to search for a source habitat. I should like to see an extension of this spatial work to the dimension of time. Are there circumstances that would similarly favor an individual that used a moderately suboptimal habitat in time? If not, then we should not expect many temporal sink species compared to spatial ones.

Time's arrow (Blum 1951)—a reason even more daunting than physiology—may also reduce the presence of temporal sink species in SPTIs. Time moves in one direction only. As we add time to collect data for an SPTI, we may expect to add only species that can use the added temporal habitat. Why should a species appear earlier or emerge from diapause earlier than a time that would profit it? Yes, some of those we see during the new time may be holdovers from the older time, but during some part of that older time, they too were probably source species. So, the collection of temporal habitats should include ones in which they were source species. And species that will profitably appear tomorrow or next year should wait for tomorrow or next year. On the other hand, when we add space, we pick up not only the habitats of the new area, but also individuals of source species in surrounding habitats. When adding a slice of time, we have no way to sample the individuals that will also be present tomorrow, and they have no way to enter the world of today without jeopardizing the future of their descendants. Perhaps, then, there are very few sink species in SPTIs. If so, SPTIs should resemble not nested, mainland SPARs, but island SPARs. The difference between island and mainland z values comes from the absence of sink species on islands (Rosenzweig 1995). But the three w values in equations 16 to 18 exceed most island z values, too. So, although sink species may indeed play less of a part in SPTIs than in SPARs, that difference cannot entirely account for the difference between z's and w's. Ecological SPTIs differ from ecological SPARs. Preston seems to have guessed wrong about that.

The Phanerozoic SPTI Agrees with Steady-State Theory—A Surprising Conclusion

Why does the theory of evolutionary SPTIs (equation 7) work fairly well? After all, to apply it to the Phanerozoic data I had to make three preposterous assumptions: a constant sea-bed area for the evolution and deposition of species, a long-term steady state in diversity, and a long-term steady state in the speciation rate. Could these assumptions be close to the truth? Perhaps they are not but the theory is robust enough so that it does not much matter? Let us examine some possible sources of error and see how robust the theory may be.

Suppose the sea-bed area for the evolution and deposition of species varied from period to period? Much of this variation would be accounted for by the correction for unequal area in equation 8. But not all of it. Equation 8 corrects only for the effect of sampling a smaller proportion of a province.

The SPAR among biogeographical provinces remains, and it has a much higher z value. If a period had consistently less epicontinental sea floor than average, its internal SPAR would lie on a lower line than average. Sampling any proportion of it would yield a smaller diversity than would sampling the same proportion of a larger, richer period. Hence, violating this assumption would add variance to the estimate of z. But, unless there is a correlation between the proportion of a period's area in our sample and its true area, the variance would only reduce the precision of our z estimate; it would not change z's value. Although mathematically necessary, the assumption of constant area in the epicontinental sea floor is not very important to the estimate of z.

In essence, variation from period to period in the area of epicontinental sea floor would add great variance to our estimate of C, steady-state diversity. Perhaps this accounts for the very shaky C value I got from the Phanerozoic. On the other hand, an uncertain C probably did not have much effect on the estimates of w and ψ.

We can nullify the second assumption—that of a single steady-state diversity—in three distinct ways. Perhaps each period had a steady state, but it varied from period to period? Or, perhaps some life has spent most of each period far from its steady state? Or, perhaps it never even reached steady state during some periods? The first possibility is covered in the previous two paragraphs; variation in steady state is variation in C. There is much evidence for such variation, although little of it pertains to comparisons of steady-state species diversities among the major Phanerozoic periods. [Bambach (1977) is a notable exception.] More important for present purposes, variation in C mattered little to the other estimates.

With respect to the other possible deficiencies in the second assumption, one school of paleobiology contends, implicitly or explicitly, that life has rarely if ever experienced steady-state diversities (e.g., Knoll 1986; Signor 1990; and especially Benton 1995). But much evidence suggests otherwise (e.g., Bretsky and Bretsky 1976; Bambach 1977; Watkins and Boucot 1978; Benson 1979; Rosenzweig and Taylor 1980; Nichols and Pollock 1983; Walker 1985; Hansen 1988; Allmon et al. 1993; Van Valkenburgh and Janis 1993; Rosenzweig 1995; Alroy, chapter 12). Sepkoski and his associates have produced an entire corpus of work based on, and supporting, the likelihood that diversity grows sigmoidally when it does change (e.g., Sepkoski 1984; Miller and Sepkoski 1988). Webb (1989), a seminal mind in this area, goes so far as to suggest (p. 199) that the term *punctuated faunal equilibrium* should be considered (at least to describe the Neogene mammal condition), i.e., faunally stable periods of about 10 my interrupted by relatively brief times of lower diversity and rapid origination. Miller and Foote (1996) confirm the episodic nature of diversity changes during the Ordovician radiations of marine invertebrates. For present purposes, all this evidence is most fortunate because, if life has spent most of the Phanerozoic clambering to repair mass extinctions or climbing to fulfill a steady state so exalted that it has never been approached, then the theory of equation 7 would be in serious trouble. Rescuing it would require the absurd assumption that standing diversity matters little to total speciation rate. I do not mind an outrageous assumption now and again, but I draw the line at absurd ones.

In many ways, the third assumption—a long-term steady state in the speciation rate—is the most interesting and important. Given a long-term steady state in the global speciation rate among marine invertebrates, there would be a single SPTI tracked by each period for a variable amount of time. This, then, is the very assumption that allows us to combine the data of the ten periods in order to describe that long-term SPTI.

We need our assumption even if diversity itself achieves and mostly maintains a single steady state (a laughable hypothesis in which even I have no faith). Even with such a single diversity steady state, species turnover rates might vary. In fact, during the late Ordovician, that is exactly what we see in the muddy benthos: diversity stayed about the same while turnover rates declined. But turnover-rate variation on a smaller temporal scale does not contradict the assumption of a steady turnover rate at larger scales if—when we take the longer-term view of about 50 my, and the broadest view of space (all marine habitats)—such variations tend to cancel each other out and yield the same average speciation rate for each of the major periods.

So the data analysis that culminated in figure 14.12 really did test the assumption of a long-term, grand-scale steady state in speciation rates among marine invertebrates. Its linearity and its remarkably high R^2 argue that the assumption passed the test. Thus, I propose to accept—at least as a working hypothesis—the conclusion that speciation rates have stayed within a fairly narrow range during most of Phanerozoic time.

My conclusion disagrees from the traditional one. In this very volume, for example, Gilinsky (chapter 9) advances several interesting ideas to explain the decline in turnover rates during the Phanerozoic. And here I am, proclaiming that this decline is most probably illusory! Here I am, suggesting that the appropriate question is, What keeps turnover rates approximately constant for eons?

What analytical intricacy had previously tricked us into believing that turnover declines? It seems to be as simple as the need to correct fossil data for sample area (i.e., sampling intensity). Nevertheless, most investigators will not consider the issue fully resolved until they are sure of the problem with previous analyses or until they have discovered a flaw in the present one.

Will the Simple Power Equation Continue to Be Useful in Fitting SPARs?

In all work on collector's curves, we face the temptation to transform the variables by taking their logarithms, and then simply to regress them linearly. One lesson of the evolutionary SPTI is the danger in such a procedure if it is undertaken thoughtlessly. In the Appendix to this chapter, we see that the value of ψ, the slope in pure log-log space after such a mechanical transformation, can vary from 0 to 1 despite a true curvature of zero (i.e., $w = 1.00$). We did need a coefficient of curvature like z.

We got it (w) from equation 11 and it worked. But that very success, coupled with our lack of a quantitative model for SPARs, makes me anxious. Is there something simple we have overlooked in thinking about SPARs? We do know that point diversities (MacArthur 1964) exist (also called α diversities). They are the limit diversities toward which an area collapses as it shrinks toward zero m^2. Are these α-diversities mathematically like the steady-state diversities that characterize most eras? If so, shouldn't our spatial model, in its arithmetic form, contain a constant, just as the temporal model does? Perhaps it should look like the following:

$$S = q + kA^z \qquad [19a]$$

$$\log S = \log(q + kA^z) \qquad [19b]$$

And, if it does contain a constant, q, won't estimates of curvature based on equation 1 (i.e., z) be useless? Do we *need* the spatial equivalent of w?

The answers to the preceding questions come from data. The data say we are on firm ground. Yes, there are point diversities. But they are rapidly swamped by the increases in diversity that come from bigger areas. Thus, the term $\log(q + kA^z)$ rapidly approaches $\log(kA^z)$. The result of a simple logarithmic regression (equation 1) therefore converges very quickly on the true value of the curvature. Furthermore, data show that some very small provinces actually do have zero species. So, SPARs at the interprovincial level may have $q = 0$. For example, Hawaii is so small that its bracken ferns have no species of insect specialized on consuming them (Lawton et al. 1993). And data from all spatial scales do show curvature when plotted or analyzed with arithmetic variables (e.g., Wright 1981), so z is not obscuring absolute linearity.

As we continue to compile SPARs, we can generally rely on logarithmic regression to estimate their curvatures. Although nonlinear regression is certainly acceptable, we actually need it only when we are trying to estimate—

not z—but the α diversities themselves. SPTIs are a different matter. They cannot be safely analyzed with logarithmic regression. The evolutionary SPTI demands nonlinear regression according to the recipe of equations 11 or 12. And—unless the time series being studied finishes with at least one or two orders of magnitude more species than it began—SPTIs in ecological time also require nonlinear regression.

Paleobiology and the Problem of Sampling Biases

Paleobiologists at least since Raup (1976b) have grappled with the sampling artifacts that could influence our perception of diversity trends. Many addressed the sampling issue using rarefaction analysis. For instance, Benson (1979) compared modern and ancient rarefaction curves of marine ostracodes; he concluded that because there is no difference, there has been no change in their diversity. And Miller and Foote (1996) used rarefaction elegantly to remove the sampling inequities from their study of Ordovician invertebrate radiations.

Paleobiologists have also used the rarefaction method implicitly by trying to make their comparisons among similar sized samples of fossils. Using this approach by focusing on single fossil communities, Bambach (1977) nailed down the increases in marine invertebrate diversity at the end of the Ordovician and during the Tertiary. Simultaneously, he showed that other apparent increases were not supported by data from which he removed the sample-size bias. And Crane and Lidgard (1989), by studying diversity in a series of similar-sized collections, conclusively demonstrated the increase in flowering plant diversity during the Cretaceous.

Time limits also cause bias in detecting true diversities. Estimated diversities tend to rise without any other cause at the beginning of a sampling period as we learn about the species that were there initially. And they fall without any other cause at the end of a such a period (Rosenzweig and Duek 1979; Signor and Lipps 1982).

A few nonpaleobiologists studying fossil diversity trends have harnessed the machinery of capture–recapture censuses to remove much of the bias from estimates of species diversity (Rosenzweig and Taylor 1980; Nichols and Pollock 1983; Rosenzweig 1995). Capture–recapture censuses have much to recommend them. They eliminate the time-limit artifacts mentioned in the preceding paragraph. And they also eliminate the noise of unequal fossilizability among the species that are—at least theoretically—recoverable in the fossil record.

Finally, some paleobiologists have tried to identify the sources of bias and remove them directly. Raup (1976b) ascribed much potential bias to unequal preservation of rock. He obtained global estimates of remaining rock surface and volume, thus to adjust the raw diversities of different periods. Sheehan (1977) points out that another bias—unequal investigator interest in different periods—is an alternative bias that may account for Raup's results. But we need not take an all-or-none approach. Alroy (chapter 12) successfully combines corrections for the biases of different amounts of available rock and differing investigator interest. In this chapter, I followed the same philosophy as Alroy and tried to adjust the fossil data both for rock area and for duration of period. The combination seems to be justified.

SPTIs and Time Averaging

Deposits of fossils accumulate over a period of time. This results in the problem of time averaging: How can the paleobiologist tell what conditions or set of species resided at a particular time and place when "time averaging modifies the death assemblage by mixing organisms from a range of conditions broader than those present during any relatively short span of time" (Staff et al. 1986, p. 441)? In particular, "time averaging increases species richness" (p. 443).

Studies of time averaging (e.g., Kidwell and Bosence 1991; Miller and Cummins 1993) are therefore studies of the consequences of SPTIs (if the taxonomic units of the averages are species). One way to investigate them would be to treat a time average as one point in an SPTI, then get other points, fit the appropriate equation, and use it to compare different assemblages. Of course, that means taking data not just at the maximum time duration, but in a sequence of small steps leading up to it.

Clearly, Staff et al. (1986) amassed enough data to allow them to describe the SPTIs of their habitats, although as far as I know no one has yet applied SPTI methods to the time-averaging problem. Some investigators have come close to this by taking the ratio of diversity to the length of time during which it was amassed. But that implies linearity of the relationship. We may now suspect that not all SPTIs are linear, particularly those that reflect the shorter-term processes of ecology. If we can discover the degree of nonlinearity (i.e., the w value), we may resolve the different lengths of time by dividing diversity through by t^w.

Final Comments

SPARs and SPTIs clearly have relevance for each other. For example, we can apply what we now know about SPARs to some important issues of fossil history:

- Using a SPAR to detrend the enumeration of species is crucial to revealing the underlying signal of time in diversification.
- Signor (1978, 1985) brilliantly used the lognormal theory of SPARs to help him estimate the species diversities of the ten major periods. Yet, now we know that the lognormal distribution is an inadequate theory for SPARs (Leitner and Rosenzweig 1997). Sepkoski (1994) depended on the demonstration (Nee et al. 1991) of the over-representation of rare species compared to a lognormal, and recalculated Signor's estimates. He also tried other distributions such as the log-series. But Signor's estimates proved to be rather robust. (The estimate of diversity's growth during the Tertiary decreased substantially, however, if Sepkoski assumed that paleobiologists oversample rarer species.)
- Finally, Valentine (1973) and Valentine et al. (1972, 1978) hypothesize that different degrees of provincialization may account for the variation in global diversity (see also Schopf 1979; Wise and Schopf 1981; Signor 1985). Now that we know that the true value of interprovincial curvature approximates $z = 1.0$, we can see that this hypothesis can play but a small part in that variation (Rosenzweig 1995). (By the way, this finding also points to the importance of studying *species* diversity patterns using *species* diversity estimates—a view also held by Signor, 1985. Why? Flessa, 1975, shows that the value of z for mammalian genera is considerably less than 1.0. So, provincialization must play a large role in setting generic diversities.)

We can use the machinery of SPAR and SPTI analysis to refine our understanding of ecological communities. MacArthur (1964) long ago pointed out that the slope of SPARs measures the horizontal diversity of habitats. (Most ecologists now follow Whittaker's (1970) terms and call this a form of β diversity.) Similarly, α diversity measures the vertical habitat diversity plus the resource diversity. Now we can add analyses of the temporal habitat diversity by measuring SPTIs. These will tell us which communities emphasize temporal niche differences and which ones rely mostly on spatial differences. I could probably speculate for an entire chapter on the questions we will then be able to ask and on their importance to evolutionary ecology. Suffice it to say that I believe such investigations will be intriguing.

As a theorem to cover all scales of space and time, Frank Preston's conjecture that space and time are interchangeable in the study of species diversity seems to have failed its empirical test. Naturally, when small sample sizes are causing the collector's curve, SPARs and SPTIs are the same; there is no reason a quantity that depends on sample size alone should change if the differences in sample size come from collecting more individuals in time or more in space. But, once we look far enough and wide enough and long enough to allow for multiple habitats in space and time, SPTIs and SPARs do differ quantitatively. As we have seen, the curvatures of ecological SPARs seem to exceed those of SPTIs.

Yet, evolutionary SPARs and SPTIs do appear to agree: Both have little or no curvature. So, at the grandest scales, Preston's ergodic conjecture looks correct. Assuming that further analyses continue to support it, we will want to know why. Why does the z among separate provinces resemble the w created by species turnover in evolutionary time?

On the other hand, perhaps further analyses will not support Preston's conjecture. Then, we will at least be able to determine whether there is any general temporal pattern and, if there is, describe it. SPTIs are mostly unexplored territory. I suspect that Preston will turn out to have been right also in trying to awaken us to their possibilities.

Acknowledgments

This chapter is dedicated to the memory of Frank Preston, a friend and a sometime ceramics engineer, who was devoted to birds and their ecology and who generally managed to keep two generations ahead of the rest of us. Thanks to Michael McKinney for inviting me to participate. He, John Alroy, Arnie Miller, and Jack Sepkoski provided me with some of the best hints and suggestions I have ever received from manuscript reviewers. My participation in the symposium that led to this volume was supported by NSF grant DEB8905728.

Appendix

The sensitivity of ψ

Neither the z value nor the w value varies with either the scale of measurement used (e.g., square centimeters will produce the same z value as square miles) or the spatio–temporal range over which we sample the data. [Rosenzweig (1995) shows this for space; the proof for time is similar.] That is to say, either of these values may change in space and in time, but not because we alter our sampling effort. On the other hand, given the validity of a model such as equation 7 or 11, ψ is quite sensitive both to the size of the sampling regime and to the sizes of the underlying process coefficients it estimates. That is why w makes a better statistical analogue to z than does ψ.

TABLE 14.2   Effects of system values and sampling regime on ψ.

C	$\Psi = 500$		$\Psi = 3$		$\Psi = 0.15$	
	ψ	ψ_y	ψ	ψ_y	ψ	ψ_y
20	0.9965	0.9992	0.7829	0.8835	0.3032	0.2848
907	0.9007	0.9651	0.1819	0.1508	0.0130	0.0089
3567	0.7751	0.8764	0.0510	0.0435	0.0034	0.0023
6227	0.7034	0.8034	0.0361	0.0255	0.0019	0.0013
8000	0.6672	0.7614	0.0285	0.0199	0.0015	0.0010

C is the equilibrium in species diversity. The truncated range of time samples extends from 28 to 98, and its ψ is ψ_y. Full range is 1 to 355.

Let us briefly examine the sensitivity of ψ to the range and coefficients of the evolutionary process and even to the range of times over which we sample it. Its variation will emphasize the lesson that it should be avoided.

Assuming equation 7, I simulated some data at steady state in the range of the real data. I gave t 52 evenly spaced values from 1 to 355. I let C (the equilibrium number of species) vary from 20 to 8000. Then I set Ψ at 500 or 3 or 0.15. The first approximates the highest turnover rate [i.e., 594, that of the Phanerozoic (equation 13)] and the third falls below the late Ordovician result [i.e., 0.554 (equation 15)]. To display the effect of the time window during which we estimate ψ, I also estimated ψ using both the entire simulated period and a truncated fraction of it.

Table 14.2 shows that the turnover rate has the greatest influence on ψ. The larger Ψ, the larger ψ. If we recall that ψ is meant to measure the curvature of an SPTI, we will realize how misleading this dependence is. Equilibrium diversity does not influence ψ quite so much, but that influence is still considerable: ψ does fall as C rises, especially at lower values of turnover. Comparing the ψ columns with the ψ_y columns shows that the third variable in the table, the range of t over which we measure, seems hardly to matter at all.

15

An Intermediate Disturbance Hypothesis
of Maximal Speciation

Warren D. Allmon, Paul J. Morris,
and Michael L. McKinney

Diversity—by which we mean here the multiplicity of taxonomic entities we call species—is one of the most conspicuous aspects of living systems. It is therefore remarkable that we know so little about its causes. Maybe this is because it is genuinely one of the most difficult problems in biology; maybe it is because we have not looked at it the right way; maybe it is some combination of these two.

It is not for lack of trying; there is no shortage of literature on diversity. Yet some of the most sweeping attempts at explanation of the diversity we see in the living world have in fact skirted the issue. It has been noted by more than one observer, for example, that *On the Origin of Species* (Darwin 1859) did not really address the subject of its title (Mayr 1959; Coyne 1994). Rather, it concentrated on factors affecting the transformation of individual lineages. Exactly a century later, Evelyn Hutchinson's landmark paper (Hutchinson 1959) sought to address the question, Why are there so many kinds of animals? by discussing not mechanisms for the *origin* of species, but for the *maintenance* of different numbers of species in communities.

The same is true of the majority of ecological discussions of diversity since 1959, in which the analytical approaches and conclusions have been remarkably consistent: more species exist in a given habitat because more species *can*, not because more species arose there (e.g., Connell and Orias 1964; Mayr 1969; Janzen 1977; Sale 1977; Stenseth 1984; Hubbell and Foster

1986; Gentry 1989; Huston 1979, 1994). Even recent attempts at a more "historical" ecology (e.g., Ricklefs 1987; Ricklefs and Schluter 1993a), which include consideration of the effects of regional biotic and environmental history, focus more on what factors led to the co-occurrence of a particular number of species (often under nonequilibrium conditions) than on the factors that produced that number of species in the first place.

If we then set as our task the explication of the environmental and ecological context for species origin, our attention is quickly drawn to processes of environmental change or fluctuation. This is at least in part because species origin is a process of change, and we have a general expectation that particular environmental changes may elicit responses from organisms that may make conditions more or less favorable for speciation. Short-term environmental changes, which we will call *perturbations* (cf. Huston 1994, p. 216), are part of the everyday lives of all organisms. On occasion, environmental parameters may vary outside the range shown, for example, during the average lifetime of the organisms under study. Some of these perturbations may be severe enough to kill off or otherwise remove the organisms of a particular taxon from some geographic area, thus creating a patch of space from which that taxon is absent. Such perturbations we will refer to as *disturbances*. Disturbance is defined in this paper as an environmental perturbation that, at the temporal and spatial scale under consideration, removes all of the organisms under consideration from a unit area. Thus defined, disturbance may be very nearly ubiquitous in natural communities (e.g., Grubb 1977; Connell 1978; White 1979; Pickett 1980; Raup 1981; West et al. 1981). Virtually all ecosystems are mosaics of environmental conditions—patches—and a majority of this heterogeneity is a consequence of disturbance (Karr and Freemark 1985).

The temporal and spatial magnitude of change that qualifies as a disturbance will vary with the taxon and with the hierarchical level in the ecosystem (McKinney and Allmon 1995). What disturbs a population of crows, for example, may not disturb a co-occurring population of pine trees; what disturbs a single population of rabbits may not disturb the entire species. Disturbance is thus both scale and taxon dependent (Denslow 1985; Huston 1994; see also Levin 1992) (figure 15.1). Size (i.e., magnitude or intensity) and frequency may also interact to create disturbances of larger cumulative effect (Miller 1982).

Relationships between disturbance and the maintenance of diversity have been discussed by many authors. As discussed later, the relationship between disturbance and speciation has, in contrast, been scarcely considered. In this paper, we are interested primarily in the interface between processes that occur on ecological time scales with processes that occur on evolutionary time

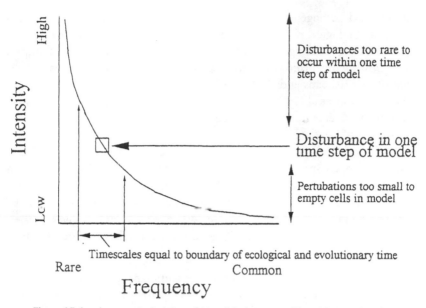

Figure 15.1.   A conceptual picture of how disturbance on different temporal scales is modeled in this paper. Disturbances (*curved line*) are seen as following a continuum from rare high-intensity events to frequent low-intensity events. The model isolates a single intensity of disturbance. The model assumes that disturbances more frequent than the chosen intensity are perturbations too small to entirely empty cells in the model. It also assumes that a single model run occurs over time scales too short to observe rare high-intensity disturbances. Setting disturbance intensity thus in effect determines the temporal scale that is being modeled by each time step. Appropriate levels of disturbance will thus allow us to model the time scales of interest on the boundary between ecological and evolutionary time.

scales. We therefore focus on events—disturbances—that occur on the long end of ecological time and on the short end of evolutionary time (i.e., about 10^4 to 10^5 years). We propose a hypothesis that connects disturbances at these time scales to the origin of species, and we test it with a simple computer simulation study.

Disturbance and the Maintenance of Diversity

Disturbances of natural communities can be conveniently divided into physical and biological. Physical disturbances in the terrestrial realm include wind storms, floods, lightning, fire, landslides, and volcanism (see Runkle 1985; Denslow 1987). In the marine realm, storms, freshwater floods, and sedimentation events are all common physical disturbances (see Sousa 1985). Biological disturbances are commonly divided into predation (including grazing and pathogens) and nonpredation. Nonpredation disturbance includes all biological phenomena that inadvertently kill or displace other organisms, from "hauling out" behavior of pinnipeds, to bulldozing of limpets or burrowing organisms, to senescence of trees or corals which leads to their falling over (Rhoads and Young 1970; Connell 1978; Sousa 1979; Probert 1984; Dayton et al. 1992). Both physical and biological disturbance may vary in areal extent, intensity, frequency, and seasonality (Connell and Keough 1985).

Biological and physical disturbances affect communities by creating patches of open space or habitat. This is species specific; what disturbs a population of one species may have little or no effect on others (Pickett and Thompson 1978; White 1979; Denslow 1985). Disturbance may thus be said to be, at least potentially, selective. When it happens to a given species, however, disturbance has at least three separate but related effects that may influence community diversity:

1. Disturbance may reduce numbers of abundant species, opening space and sometimes making other resources available that can be exploited by less common species or invaders; this may lead to increased community diversity (Paine 1966; Janzen 1970; Connell 1971; Dayton 1971; Dayton and Hessler 1972; Lubchenco 1978). Disturbance resets local successional trajectories, and it may suppress the abundance of successful local competitors, thereby allowing poorer competitors to expand (White and Pickett 1985). This may depend, however, on the structure of competitive hierarchies in the community (Connell and Keough 1985; Petraitis et al. 1989). If competitive relationships are more symmetrical than asymmetrical, the outcome of competitive interactions may be less predictable and thus the effects of disturbance may be difficult to predict. Connell and Keough (1985, p. 151) go so far as to conclude that "there is no uniform way in which disturbances influence communities of subtidal sessile organisms."

2. Disturbance may reduce numbers of rare species, even causing local extinction; this may lead to reduced community diversity (e.g., Wethey 1985).

3. Disturbance may increase environmental heterogeneity, potentially providing a basis for increased specialization and resource partitioning (Grubb 1977; Platt and Weis 1977; Denslow 1980a,b; Tilman 1982).

These effects cumulatively may produce a general pattern: Habitats in which there is very little disturbance tend to have low diversity, because they tend to be dominated by one or a small number of abundant species; habitats in which there is a great deal of disturbance also tend to have low diversity,

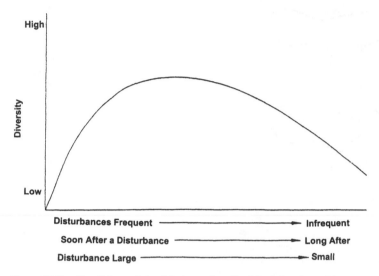

Figure 15.2. The "intermediate disturbance hypothesis" of diversity maintenance. From Connell (1978).

because few species can maintain viable populations in the face of continued high mortality. All other factors being equal, therefore, the highest number of species can coexist at intermediate levels of disturbance, an idea codified as the "intermediate disturbance hypothesis" (Connell 1975, 1978, 1979; see also Lubchenco 1978; Fox 1979; Grime 1979; Huston 1979, 1994; Miller 1982; Tilman 1982; del Moral 1983; Denslow 1985; Petraitis et al. 1989) (figure 15.2).

Disturbance also acts as an important selective agent (e.g., Suchanek 1981; Sousa 1985). It is in fact continuous with "normal" environmental fluctuations (White and Pickett 1985)—"perturbations" as used here—and with all environmental changes and variation that occur throughout ecological, and ultimately geological, time scales. In the present context, the question is, What are the long-term consequences of such environmental perturbations for species diversity? To answer this, we need an understanding of how species arise.

The Process of Allopatric Speciation

Until relatively recently, the prevailing view of allopatric speciation—especially of its geography and ecology—among most biologists was that articulated by Mayr (1942, 1963): Allopatric speciation (which Mayr believed to be the dominant process of speciation, at least in animals) occurred by the

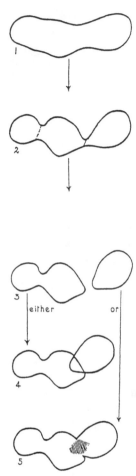

Figure 15.3. The classical allopatric theory of speciation, indicating the necessary stages of isolate formation (steps 2 to 3), isolate persistence (during step 3) or not (step 4), and isolate differentiation (during step 3, resulting in sympatry without interbreeding in step 5). From Mayr (1942).

spatial separation of a daughter population from a parent population (figure 15.3). The daughter stayed isolated long enough to diverge genetically from the parent, under natural selection or drift or both, so that when spatial co-occurrence was reestablished, interbreeding no longer regularly took place.

Over the last two decades or so, several elements of this view have been strongly challenged. Leaving aside whether at least some degree of allopatry is actually required for most speciation to occur (see, e.g., Tauber and Tauber 1989; Bush 1993), critics have complained, inter alia, that defining species by their relationship to each other does not explain what they are unto themselves, or how they arose (e.g., Paterson 1985; Masters 1993), nor does it focus adequate attention on cohesion within species (Templeton 1989; Cracraft 1989). A focus on reproductive isolation may be misguided, furthermore, because gene flow among closely related taxa, long accepted as "good" species, may be higher than previously believed (Endler 1977; Howard 1994).

Despite all this criticism, however, it still appears that most zoologists continue to agree that something resembling Mayr's concept of allopatric speciation applies widely to diversification in most animal groups (e.g., Coyne et al. 1988; Otte and Endler 1989; Eldredge 1993; Knowlton 1993; Coyne 1994; Hebert and Wilson 1994; Palumbi 1994). It is clear that different taxa show somewhat different patterns (e.g., more hybridization, faster divergence, less character displacement or reinforcement), but it still seems reasonable to conclude that a substantial proportion of animal species arise by a process of geographic separation, persistence, and genetic divergence of populations.

If we thus accept, at least for the sake of the present discussion, some version of speciation resembling Mayrian allopatry, we are still left with a significant epistemological problem (which, indeed, would also apply to any mode of speciation we discussed). When we ask the question, What causes speciation? what are we really asking? At what point in this process are we asking about cause? Are the causes the same at each point? Much of the speciation literature consists of taking this process apart. But if one attempts to synthesize a general theory for causes of speciation from this literature, it quickly can appear as though each worker has looked at a different subset of the actual problem.

For this reason, we prefer to see allopatric speciation as divided (conceptually and in nature) into at least three more-or-less discrete stages (Allmon 1992) (figure 15.4). Every instance of allopatric speciation must involve (1) separation of one population from another, (2) persistence of these populations, and (3) genetic differentiation of at least one of these populations. This three-stage framework for examining speciation is useful because it forces us to focus our attention, and questions, on one process at a time, and to ask how a given factor affects each particular aspect of the overall process.

For example, examining a particular episode of diversification using this three-stage framework compels one to examine what factors intrinsic and extrinsic to the taxa involved affected them at each stage (Allmon 1994; McKinney and Allmon 1995). Do new species arise because of increased isolate survival [because of "key" innovations (Heard and Hauser 1995), or because they diversify in an empty or otherwise "permissive" environment (e.g., Grant and Grant 1995)], or increased isolate formation [because of changing dispersal ability (Hansen 1978; Palumbi 1993) or increased vicariance (Cracraft 1985, 1992)]? The macroevolutionary implications can be quite different for different answers to these questions (Stanley 1986b; Allmon 1992, 1994; Johnson et al. 1995).

In this context, what is the effect of disturbance on speciation?

Disturbance and the Origins of Diversity

Reasoning from the three-stage framework (figure 15.4), some expectations about the long-term effects of disturbance are possible. Because it creates habitat patchiness, disturbance is an important potential source of vicariance—important in the origin of species because it creates isolated populations. Disturbance is also potentially important in affecting the survival of the patches it creates; more disturbance might lead to increased extinction of isolated populations. Some level of disturbance therefore might balance creation of isolates and their destruction. We suggest that the expectation should be that, in a manner analogous to the case of diversity maintenance addressed by Connell (e.g., 1978), speciation should be expected to be maximal at intermediate levels of disturbance over ecological and into evolutionary time (figure 15.5). For allopatric speciation, then, an intermediate disturbance hypothesis of speciation can be proposed: *For a given clade and a given type of disturbance, origin of species should be highest at an intermediate level of that disturbance.*

Is there any support for this idea that origination (speciation) is maximal

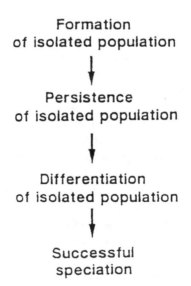

Figure 15.4. Formalization of the stages illustrated in figure 15.3 into a three-stage framework for examining speciation. From Allmon (1992).

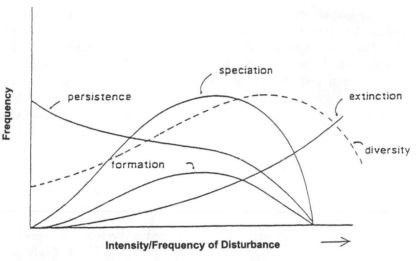

Figure 15.5. Qualitative relationships among processes leading to speciation (isolate formation and persistence), successful speciation itself, extinction of established species, and species diversity. From Allmon (1992).

at intermediate levels of disturbance? A number of authors have discussed similar ideas before, but much less explicitly and not in the context of a general theory of speciation.

Simpson (1944, p. 89) argued that occurrence of abundant speciation in a group might result from "a decline in the adaptive status of the ancestral populations, and consequent centrifugal selection and fragmentation of groups imperfectly adapted by tending more or less toward a variety of different types." Lewis (1962) suggested that speciation might be more likely during times of adverse environmental conditions because harsh conditions might eliminate entire populations by "catastrophic selection"; if followed by divergent selection, such vicariance might lead to new species.

Stanley (1979) suggested that environmental "adversity would manifest itself most readily in suppressing speciation, rather than in accelerating the extinction of established species [since] . . . small populations involved in the speciation process should be more susceptible to extirpation or to restraint from expansion than a full-fledged species should be vulnerable to total extinction" (1979, pp. 197 to 198). He elaborated and expanded this idea in 1986: "High rates of speciation are actually promoted by less severe environmental deterioration—deterioration severe enough to elevate extinction rates to a moderate degree but not so severe as to cause wholesale extinction" (1986, p. 104), i.e., "intermediate." Stanley designated as the fission effect "the general phenomenon in which hostile environmental conditions on a broad geographic scale accelerate speciation, while also causing extinction" (Stanley 1986b, p. 105; see also Stanley 1990c).

Although not explicitly addressing the effects of varying intensities of disturbance, Vrba's "turnover pulse hypothesis" (Vrba 1985, 1993) focuses on environmental perturbations as the ultimate source of most speciation events. By implication, these perturbations must be of a magnitude sufficient to cause vicariance, but not so severe as to lead to extinction. (See Eldredge 1996, for further discussion.)

A Simulation Approach

The Model

We set up a simple computer model to examine the effects of disturbance on species origin. The model operates in a two-dimensional grid of cells (the "world"), representing a geographic range occupied by organisms. Each cell can be occupied by only one population of one species at a time, simulating the interactions of closely related species that compete for similar resources and respond in similar ways to disturbances on particular temporal and spatial scales. A given species may have populations occupying anywhere between a single cell and all cells.

As defined in this paper (and its model), disturbance is total within a unit area; therefore "intensity" of disturbance is defined by its geographic extent or size—i.e., how many unit areas (cells) are affected. Duration of a disturbance is shorter than the lifetime of the biological entities under consideration (e.g., species). Disturbances are by definition rare, because of the empirical inverse relationship between disturbance frequency and intensity (see figure 15.1, and below).

This definition for disturbance differs somewhat from others in the literature (e.g., White and Pickett 1985, p. 7; Huston 1994, p. 215), chiefly in its attempt to be applicable to a wider variety of temporal scales (i.e., in evolutionary time). In defining disturbance as total for a given taxon in a given area, it also differs from the usage in McKinney and Allmon (1995), where *intensity* (degree of disturbance) was distinguished from *extent* (geographic area).

The definition used here assumes an inverse relationship between frequency and intensity of disturbance (see figure 15.1). This relationship may lack indisputable experimental proof within biological systems (Petraitis et al. 1989, p. 413), but it is clearly demonstrable in the physical world (Raup 1991a). In the real world, physical disturbances of varying sorts are distributed according to specific functions; it so happens that disturbances of large size have lower frequency (Raup 1991a, 1992; see also Dott 1983; Hsü

1989). Empirically, the intensity of a disturbance thus essentially determines its frequency.

The model has the following parameters: grid size, disturbance probability, speciation probability, competition probability, immigration probability, speciation model, and initial number of species (one or many). As the model runs, a given proportion of occupied cells are disturbed in each time step; i.e., the species occupying those cells is eliminated from those cells. Other parameters can be varied in the model, including probability of differentiation, probability that one species can outcompete an adjacent species and move into an occupied cell, and probability that a species can immigrate into an adjacent empty cell. *Speciation model* refers to whether speciation (i.e., isolate differentiation) is population size independent, size dependent, or size and distance dependent.

The algorithm behind the model operates on a set of two-dimensional matrices, representing (1) cell occupied or not, (2) a unique number representing the species occupying the cell, and (3) a number representing the population occupying the cell, where a population is defined as an interconnected set of cells occupied by the same species. In a single time step, the algorithm (1) counts the number of species present, (2) steps through each cell in the first matrix (occupied or not) and randomly decides if that cell is going to be disturbed in that time step by applying the disturbance probability, (3) steps through each empty cell and determines if that cell will be occupied by immigration from an adjacent occupied cell, (4) steps through each occupied cell and determines if each cell adjacent to a cell occupied by a different species will be competitively overrun by its neighbor, (5) walks through the third matrix using a binary tree walk to identify and number each discreet (geographically separate) population (identified with the second and third matrices), and (6) uses the speciation probability to determine if each discreet population will speciate or not. This probability can optionally be influenced by population size and distance of the population from other populations of the same species. The algorithm then calculates the results of each step, including the number of species, number of originations, number of extinctions, number of populations, and running averages, and it then outputs results to screen and a file. The program is written in Microsoft QuickBasic 4.0, and run on a Pentium-based microcomputer. The code is available from the second author.

Results

Determining Model Parameters

We are interested in exploring behavior in a steady-state system and thus in the equilibrium values of diversity (i.e., origination and extinction) that can be obtained from a typical situation over some number of time steps of the model. Several issues need to be considered with regard to initial conditions, beginning with original species number. The model can be started from either a world occupied by a single species or a world containing many species. In a world filled with a single species (when speciation rate is greater than zero and immigration rate is greater than or equal to disturbance rate) the number of species will increase over time, ultimately fluctuating around some equilibrium value. In a world filled with many species, the number of species will rise or decline over time and eventually fluctuate around some equilibrium value. By running the model, we found that filling the world with an initial set of species each occupying a rectangle of four adjacent cells would always equilibrate (or clearly head towards extinction) within less than 100 time steps. Conversely, starting with a world filled by a single species, certain starting conditions could take considerably longer to reach equilibrium conditions. Based on observations of many model runs under different parameter values, we decided to use an initial world filled with many species (one quarter of the number of cells) and use averages of values (e.g., number of species, number of originations per time step) over time steps 100 to 200 of a model run as approximations of steady-state values (figure 15.6A).

The second issue to consider is grid size. Because speciation is affected by the number of isolated populations, the total diversity in the model can be influenced by edge effects; it is easier to isolate a population adjacent to an outer boundary or in a corner than a population in the center of the map. Thus, small grid sizes could show results skewed by these effects. In addition, because small grid sizes have small numbers of cells, small random fluctuations can produce considerable variance in the results of different runs. Small grids also provide statistically small sample sizes. Because the algorithm is computationally intensive (taking about 2.5 minutes to run 200 time steps on a 30 by 30 grid), we wish to identify a minimum grid size that gives replicable results. To do this, we ran the model for 200 time steps (taking as output the average diversity over time steps 100 to 200) ten times and calcu-

Figure 15.6. (Opposite) Determining model parameters. **(A)** Typical time series in the model with immigration > disturbance, showing that diversity levels out between 100 and 200 time steps. 30 × 30 grid; 5 percent hits; speciation probability, 0.1; immigration probability, 0.2; competition probability, 0.05. **(B)** Effect of grid size on replicability in the model, showing that the coefficient of variation (CV) of average diversity from ten runs (of 200 steps each) drops below 3 to 5 percent at a grid size of 30 by 30 cells. **(C)** Relationship between grid size and diversity, showing that at larger grid sizes, average diversity shows a roughly linearly increasing relationship (density averaged over the second hundred time steps).

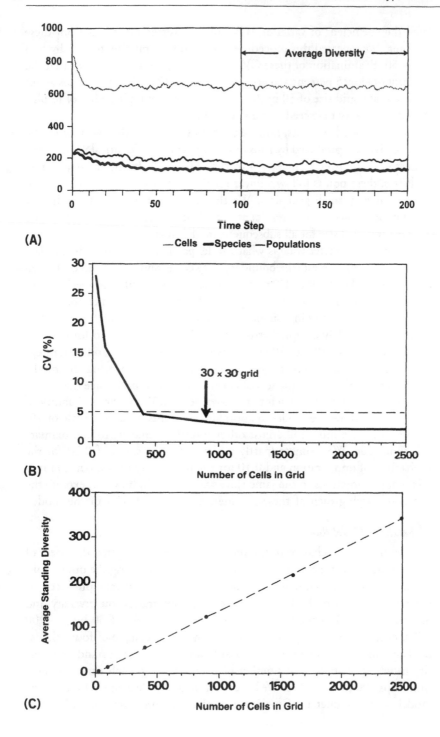

(A)

Time Step

— Cells   — Species   — Populations

(B)

Number of Cells in Grid

(C)

Number of Cells in Grid

lated the coefficient of variation (CV) of the average diversities from these ten runs. We repeated this procedure with grid sizes ranging from 5 by 5 to 50 by 50. Examination of these CV values (see figure 15.6B) shows that the CV drops below 5 percent at a grid size of about 25 by 25 and reaches about 3 percent at a grid size of 30 by 30. Thus, in terms of replicability of results, it is not necessary to exceed a grid size of 30 by 30.

The issue of edge effects, however, remains. To assess this, we plotted the average diversity produced by runs of the model (using identical parameters) against grid size over a range of 5 by 5 to 50 by 50. In the range of larger grid sizes, these points follow a linear relationship (see figure 15.6C), which deviates at very small grid sizes. We thus suggest that edge effects have a minimal influence on our results at a grid size of 30 by 30. We therefore adopted this grid size for all subsequent model runs.

The final parameter issue is what model parameters might best simulate the time scale of interest (the boundary between evolutionary and ecological time scales at about 10^4 to 10^5 years). Consider a probability of speciation of 0.50; with this value, there is a 0.5 probability that each distinct population will form a new species in a single time step. This is a very high probability. If we use the widely accepted estimate for the time it takes to form a new species of between 5000 to 50,000 years (e.g., Stanley 1979), our time steps would represent durations longer than 50,000 years. To bring the model closer to the scale of infrequent ecological events, we must reduce the probability of speciation. We therefore use a value of 0.01 for the probability of speciation to allow the model to approximate events on a time scale of 10^4 to 10^5 years. We must still adjust other model parameters to approximate this time scale. We estimate (partly arbitrarily) that values of 0.50 for the probability of immigration and 0.10 for the probability of competition represent approximations to the same time scale. We will, however, explore the effects of varying each of these parameters on the behavior of the model.

Results of Model Runs

To investigate the behavior of the model, we have investigated the effect of altering each model parameter independently while holding the others constant. We have also examined the more complicated landscape produced by the effects of varying both disturbance and immigration on diversity and origination. In all of the following runs, we used a grid of 30 by 30 (900 cells), occupied by 225 species (each occupying a rectangle of four cells) at the beginning of the first time step, and taking an average of standing diversity (number of species) or number of originations in each time step over time steps 100 through 200. We have examined how diversity behaves in the model as the parameters of speciation probability, immigration probability,

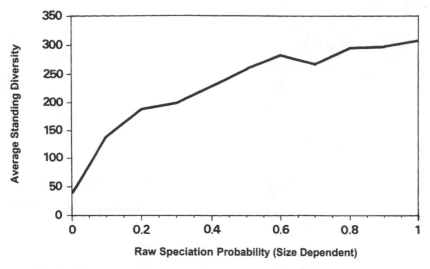

Figure 15.7.   Relationship between probability of species origination and standing diversity, using the size-dependent model of differentiation. 30 × 30 grid; 5 percent hits; immigration probability, 0.2; competition probability, 0.05.

competition, and speciation model are varied. The results discussed here do not fully examine the consequences of setting given parameters to zero. For example, setting the immigration rate to zero will always result in a steady-state world in which all populations have gone extinct, except when the disturbance rate is also zero. As noted above, we are more interested in the behavior of the model with parameter values that approximate real-world conditions at time scales of interest.

When all other parameters are kept constant, increases in the probability of differentiation result in the entirely predictable result of an increase in steady-state diversity (figure 15.7). This relationship is not linear: it shows a more rapid increase in diversity at low values of this probability than at high values. This reduction of increase in diversity at high speciation probabilities may be attributable to the greater time available for disturbance and competition to fragment species at lower rates of speciation per time step.

Altering the probability of immigration while holding other parameters constant reveals a clear threshold behavior. When immigration probabilities are much less than disturbance probability, the model will progress to an empty steady-state world. This threshold does not lie exactly at the point where disturbance is equal to immigration, as the model applies the disturbance probability to empty cells as well as occupied cells. If immigration is set to zero, the model can be used to explore cases in which destruction of habitat by disturbance is permanent at the time scale of interest (e.g., the

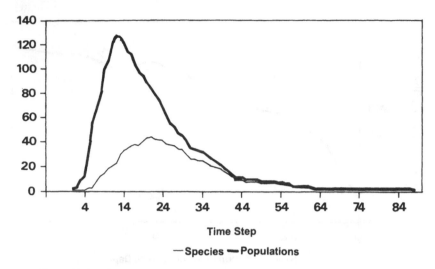

Figure 15.8. Time series tracking of species diversity in a single clade, showing the relationship between number of new populations formed and number of new species formed. 30 × 30 grid; 10 percent hits; speciation probability, 0.10; immigration probability, 0; competition probability, 0.05.

effect of regression on the biota of an epicontinental sea). For example, a run of the model starting with a single species occupying all cells of the map, which is disturbed over time but never allowed to immigrate into disturbed cells, shows a rise in diversity as disturbance fragments populations, but then a decline in diversity as the extinction rate rises above the origination rate for new species (figure 15.8). Returning to an examination of the effects of varying immigration probability, it can be seen that as immigration is increased, the average standing diversity decreases. This is attributable to the ability of greater immigration to reunite isolated populations before they have an opportunity to speciate. As the probability of competitive exclusion is increased, average diversity rapidly decreases (figure 15.9). This can readily be attributed to the competitive success of a single or small number of species of high competitive success.

Varying the probability of disturbance in the model can lead to both highest levels of diversity and highest originations at intermediate levels of disturbance. Maximal diversity at some intermediate value of disturbance is seen with all three of the speciation models that we examined (figure 15.10A). As disturbance is increased from zero, the average standing diversity in a model run increases to a maximal value and then declines (the position of this maximal value depends on other model parameters). Ultimately, as disturbance becomes much greater than immigration, the steady state diversity drops to zero. All three speciation models show this pattern of increasing, then declining diversity with increased disturbance probability. The speciation model that incorporates both size and distance, however, has a much flatter (platykurtic) curve than the others (figure 15.10A). The different behavior of this model is even more obvious in a plot of average number of originations per time step against disturbance probability (figure 15.10B). In this graph, originations in the size-independent and size-dependent speciation models are similar to those seen with diversity (figure 15.10A), with maximal values at low, but not minimal, values of disturbance. The size-and-distance-dependent speciation model, however, shows a maximal number of originations at a high level of disturbance, with a very rapid drop-off to zero above that value (figure 15.10B). This is attributable to increased disturbance producing greater gaps between isolated populations and thus increasing the chance that any given population will speciate. This relationship holds up to a threshold at which disturbance is sufficiently larger than immigration to drive all species to extinction.

The occurrence of maximal diversity and origination at intermediate levels of disturbance is not universal in the model. High immigration probabilities produce patterns in which increased disturbance produces decreased diversity. In figure 15.11A, the average standing diversity has been plotted

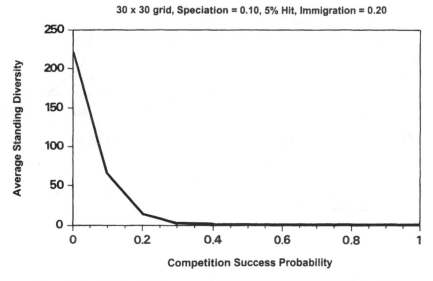

30 x 30 grid, Speciation = 0.10, 5% Hit, Immigration = 0.20

Figure 15.9.   Relationship between competition and diversity. Probability of competition success is the chance of one species occupying a cell adjacent to that of another species, assessed for each boundary cell of one species at a time. 30 × 30 grid; 5 percent hits; speciation probability, 0.10; immigration probability, 0.20.

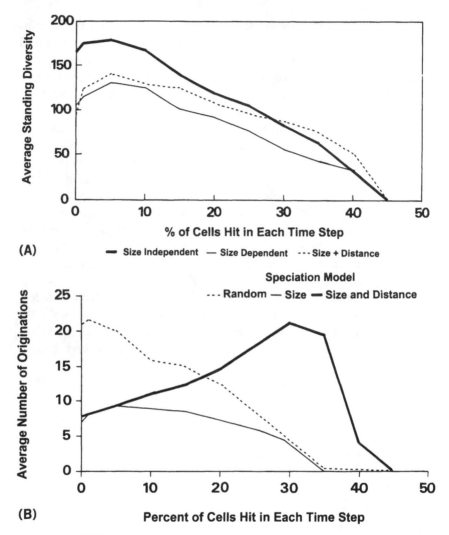

Figure 15.10. Relationship between disturbance and species origination and total diversity using each of the three models of differentiation. **(A)** Disturbance versus diversity. 30 × 30 grid; speciation probability, 0.01; immigration probability, 0.20; competition probability, 0.05. **(B)** Disturbance versus species origination.

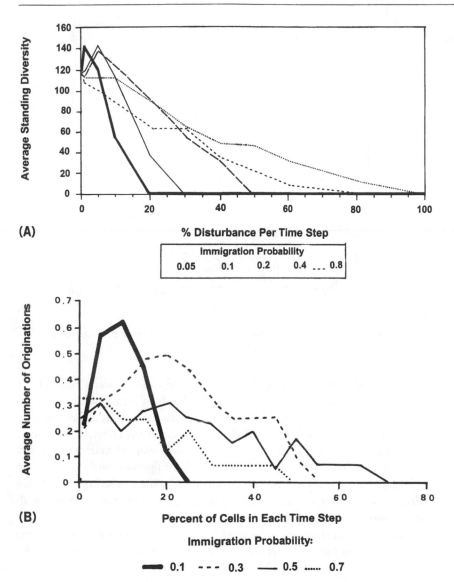

Figure 15.11. Relationship between disturbance, immigration, and species origination and total diversity. **(A)** Disturbance and diversity at varying levels of immigration. 30 × 30 grid; speciation probability, 0.10; competition probability, 0.05. **(B)** Disturbance and species origination at varying levels of immigration.

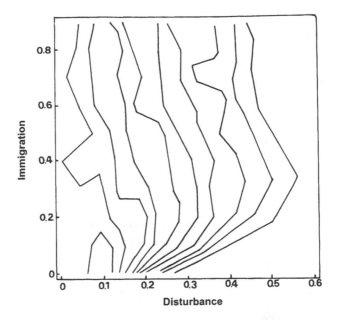

Figure 15.12.   Three-dimensional contour plot of diversity versus immigration and disturbance. Lighter tones indicate higher species diversity.

against the disturbance probability for runs of the model over a range of five different immigration probabilities. At low levels of immigration, maximal diversity is found at intermediate levels of disturbance, but as immigration probability is increased, maximal diversity is observed at zero disturbance. This pattern can also be plotted as a contoured map of maximal diversity values on axes of disturbance probability and immigration probability (figure 15.12). Here, a maximal diversity peak at intermediate values of disturbance disappears as immigration probability increases. A similar result is seen with origination (figure 15.11B).

Setting the parameters of the model to values that approximate the time scale of interest (10^4 to 10^5 years) and examining the effect of different scales of disturbance leads to a result in which both diversity and origination reach maxima at some intermediate value of disturbance (figure 15.13).

Discussion

Realism

Is our model an exact description of the natural world? Of course not, in part because it is a model, in part because it is a simple model. Our principal objective is to assess the influence of disturbance on speciation. We have done so by constructing a model that allows us to vary those parameters we consider to be most relevant to the temporal and spatial scales in which we are interested. At the same time, however, we have oversimplified several significant aspects of the real world.

First, our model does not include any consideration of long-distance dispersal. Despite implications to the contrary (e.g., Nelson and Platnick 1981; Lynch 1989; Vrba 1993), long-distance dispersal does occur in nature and is an important source of founding of new incipient species (e.g., Williams 1972; Teeter 1973; Kay 1984; Palumbi 1994). If incorporated into our model, long-distance dispersal might (1) drop a species into a previously empty cell, (2) drop a species into a cell previously occupied by another species, resulting in the exclusion of one of the two, or (3) drop a species into a cell previously occupied by the same species, resulting in gene flow. The first two processes might be expected to increase speciation probability; the third would be expected to decrease it. Modeling this process would significantly increase the complexity of the model and the difficulty of interpreting the results.

Second, we have set probability of competitive exclusion (P_C) and immigration (P_I) equal in all species in a single run. We also determine competitive outcomes in the model cell by cell, so that species A may displace species

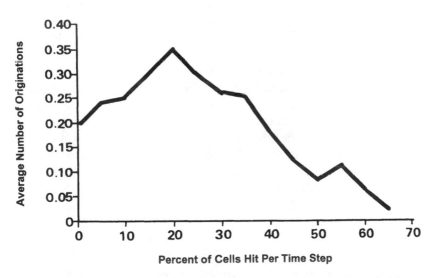

Figure 15.13. Relationship between species origination and disturbance postulated in our model, with disturbance measured as percent of cells hit, or disturbed, per unit time. This plot is the result of an actual model run with a 30 by 30 grid, the size-dependent model of differentiation. Speciation probability, 0.01; immigration probability, 0.50; competition probability, 0.10.

B in one cell, but the reverse can occur in the adjacent cell. It remains to be seen whether these conditions apply frequently in nature (see, e.g., Connell and Keough 1985; Grant 1986b; Schoener 1986).

Third, the model employs a perfectly homogeneous environment, without preexisting habitat patches. The only patches in the model are those created by disturbance. This condition is clearly unrealistic in many environments and communities (e.g., White and Pickett 1985). In a single run of our model, however, we are examining only a single disturbance-induced scale of patch; we are looking only at disturbance and patchiness that are relevant to speciation. We consider other forms and scales of patchiness (e.g., habitat heterogeneity) to exist at a scale finer than that of cells in the model.

Fourth, in a given run of the model, we are considering only species with very similar "niches"; i.e., the model states that two species cannot occupy the same cell at the same time. In runs beginning with one species, all subsequently appearing species by definition belong to one higher taxon. In runs beginning with more than one species, all species may or may not be closely related, but all are presumed to share similar ecological preferences. This is useful in simplifying the definitions of disturbance intensity, cell sizes, and time steps, but it is not an accurate simulation of a diverse natural community. To include species from ecologically very dissimilar clades, however, would have required making the model substantially more complicated. To represent multiple species occurring in the same space at the same time, a greater number of computational matrices would be required, resulting in a dramatic increase in computational time.

Finally, in each run of the model, every time step has the same intensity of disturbance. This is not realistic, but it makes it much easier to interpret results. If we had altered the disturbance intensity within single runs, we would have needed to examine substantially larger time-series to understand the long-term/steady-state behavior of the model. By restricting disturbances to a single scale, it is possible to observe the effect of that disturbance over a relatively short number of time steps (i.e., 100). Our model is a simplification of the world in that we are considering only a single scale of disturbance at a time; the length of our time steps is in fact defined by the intensity of the disturbance (see figure 15.1). As discussed earlier, in the real world, disturbances of varying intensity may happen during the same time interval.

This is not so much a departure of the model from reality as it is an attempt to isolate a single temporal/spatial scale of disturbance. For example, if one thought that the 1000-year flood was of a scale relevant to speciation events, the model allows us to consider a consecutive series of 1000-year floods without having to model the fine-scale noise of 100-year, 10-year, and annual floods. Likewise, a single run of the model will not accidentally be affected by the random insertion of a 10,000-year flood (see figure 15.1).

These qualifications notwithstanding, the model does allow us to examine the relationship between disturbance and the origin of diversity. Most details of this relationship, although certainly not new, have perhaps been underappreciated.

Disturbance and Speciation

Consider our basic hypothesis: Maximal levels of origination and diversity should occur at intermediate levels of disturbance. Two results from our modeling are notable. First, under certain circumstances, the intermediate disturbance hypothesis of speciation is confirmed. Second, when immigration levels are high, diversity and origination are maximal at low levels of disturbance.

Figure 15.10A shows that, if immigration is low, regardless of the speciation model used (i.e., dependence on isolate size or distance of separation), maximal diversity is attained at some intermediate level of disturbance. Five percent may seem like a small rather than an intermediate level; what is important, however, is that above and below this level, diversity declines. Figure 15.10B shows that the same is true of origination. Note that for the size-and-distance-dependent speciation model, the maximal level of origination occurs at a level of disturbance more truly "intermediate." In other words, when probability of immigration is low, the chief determinants of speciation are the rate of formation of isolated populations by vicariance and their preservation or extinction by environmental conditions (cf. figure 15.4). Maximal levels of both species diversity and species origination occur at moderate to low levels of disturbance, which are not so low that no isolates are created, but not so high that many isolates cannot survive to differentiate.

The three-stage framework (see figure 15.4) allows us to understand the difference between the effects of disturbance on species origination and standing diversity (see figure 15.10). Figure 15.10B shows that the size-and-distance-dependent model of speciation yields the peak level of origination at a higher level of disturbance than the other two models. This can be interpreted as resulting from the creation of more distant isolates by higher levels of disturbance. For all three speciation models, more isolates form with increased disturbance; the size-and-distance-dependent model increases the probability that more of these isolates will differentiate to become new species. Comparison of figures 15.10A and 15.10B indicates, furthermore, that, for the size-and-distance-dependent model of differentiation, many of the species originating at higher levels of disturbance do not survive for long; i.e., extinction of established species is high at these levels of disturbance.

When immigration is increased, even slightly, high species origination and species diversity occur at extremely low levels of disturbance. That is, at high levels of immigration, the intermediate disturbance hypothesis of speciation does not hold (see figures 15.11, 15.12). This makes biological sense in the context of the allopatric model of speciation: in a homogeneous environment, gene flow (immigration) should retard speciation.

Consideration of the effects of a heterogenous environment and/or long-distance dispersal on speciation in light of these results may be instructive. Dispersal in a patchy environment would favor speciation by establishing isolated populations some distance from the parental range. In such a setting, maximal speciation could occur with higher levels of immigration at moderate levels of disturbance. This result suggests a potentially very useful clarification of what may be called *Stanley's rule* for the relationship between dispersal potential and speciation probability: "For species that are characterized by stable, relatively continuous geographical distributions, effective dispersal will retard rate of speciation by opposing the formation of isolates. . . . On the other hand, for species characterized by patchy or unstable populations, effective dispersal will promote speciation by generating isolates" (Stanley 1990c, p. 116). Disturbance (vicariance) and immigration (dispersal) interact to create and preserve isolated populations, some of which may go on to become new species. In a homogeneous habitat, in a clade with low dispersal probability, speciation will be controlled largely by disturbance. In a patchy habitat, disturbance can still play a major role in controlling speciation frequency, even in clades with higher dispersal probabilities. The model also suggests that if probability of competitive displacement is high, then speciation is low, in large part because isolated populations are at high risk of being excluded (i.e., driven to extinction) by arriving competitors. If one accepts the somewhat controversial assumption that similar species are more likely to compete with each other than species that are more different (e.g., MacArthur and Levins 1967), this result leads to the intriguing suggestion that within a clade, speciation events that produce sister species that are relatively very different from each other will have a higher probability of succeeding than will events that produce sisters that are relatively similar (Eldredge 1995, 1996). [If true, this is an unexpected confirmation of Darwin's "principle of divergence" (1859, p. 112).]

Implications

An intermediate disturbance hypothesis of speciation offers the promise of insight into the ecological context of macroevolutionary change (Allmon 1994). If, for example, different periods of time or different environments are or have been characterized by different spectra of intensity of environmental disturbance, we would expect that, all other things being equal, they will have different rates of species extinction and origination and, therefore, different standing diversities. Over the Phanerozoic, we would expect particular times and/or particular environments to have had different characteristic rates of disturbance and therefore different rates of origination and different standing diversities. Given a spectrum of disturbance intensity, we would expect that times or environments characterized by disturbance at intermediate intensities would witness the highest rates of species origination. A particularly clear example of this has recently been provided by deMenocal (1995, p. 58), who suggests that speciation in hominids was maximal during an interval of intermediate climatic variability.

This expected relationship may be complicated, however, by the fact that species origin and species extinction may show different reactions to environmental disturbance (see figure 15.5): Species extinction increases continuously with disturbance; species origination increases and then decreases with increasing disturbance. This possibility has been noted by Stanley (1979, 1986b, 1990c). This result produces a modification of the original (largely qualitative) version of the hypothesis shown in figure 15.5. On geological time scales, the levels of disturbance that allow for maximal origination of species may be too high to allow for maintenance of those species. Maximal diversity should thus occur at lower levels of disturbance than maximal origination. Connell's (1978) "intermediate" levels of disturbance that allow for maintenance of high diversity may therefore be lower than the intermediate levels that allow for origin of high diversity. Yet both can be arrayed along the same disturbance axis.

This is perhaps clearer with an example. Consider a long period of time, over which the characteristic rates of disturbance vary, for example a period of 20 million years. During part of this time, environmental fluctuation occurs with roughly 100,000-year periodicity, and during the rest of the time it occurs with roughly 50,000-year periodicity. Given our model of species origin alone, we would expect to see different levels of both standing diversity and species origination during these two intervals. Incorporating the effects of disturbance on the maintenance of species diversity (i.e., Connell's intermediate disturbance hypothesis), however, we may find that disturbances of both periodicities constrain the maximal standing diversity (by raising extinction) to some lower level. Maintenance thus "caps" the maximal sustainable diversity. We therefore might see constant diversity over these two intervals, even though rates of species origination were different. Allmon et al. (1993) have discussed an example of just such a pattern from the fossil record.

The issue of diversity maintenance versus origination is further complicated by the crucial question of scale. As ecology is now widely acknowledging, the pattern detected depends to a large degree on the spatial and temporal scale at which it is viewed (e.g., Aronson and Plotnick, chapter 18). Our findings here must therefore be interpreted in terms of "intermediate to what?" Spatially, for example, endemic species will be subjected to intermediate disturbances more often because small disturbances are more common. We therefore would predict, on average, higher rates of species origination and extinction for endemic species. This agrees with a large body of ecological (Gaston 1994) and paleontological (Eldredge 1992; McKinney and Allmon 1995) data. We might expect a lower limit to the scale of endemism where this would apply. This is supported by Rosenzweig's (1995) findings (using fragmentation simulations) that species with very small ranges are less likely to undergo range fragmentation than species with more moderate ranges. His reasoning is that very small ranges are less likely to experience significant fragmentation because they will be missed by most disturbances.

We can take this reasoning further because endemism (and rarity in general) is correlated with many species-level traits that also influence susceptibility to disturbance. Brown (1995), for example, has long argued that species tend to be rare because of narrow niche occupancy. Such specialists would, in theory, be more susceptible to environmental changes in general, regardless of the spatial scale (extent) involved, and thus they would experience higher rates of speciation (Eldredge 1992). The existence of such species-level traits must be considered because they are the intrinsic factors that determine whether a disturbance will have an effect. A generalist species that can tolerate a greater variety of dietary or other changes may be unaffected by the same environmental change that devastates a specialist species. Furthermore, there is growing evidence that such intrinsic influences of disturbance susceptibility are phylogenetically inherited. For example, narrow niche-breadth (Brown 1995) and rarity (Gaston 1994) tend to be nonrandomly distributed among higher taxa at a variety of scales in the phylogenetic hierarchy. This agrees with Boucot's (1983) observations from fossil data that stenotopic, rare species tend to co-occur in the same genera and exhibit faster rates of taxonomic and morphological evolution than abundant species.

We should not, however, limit the role of scale to maximal speciation rates. It is the dynamic between species origination and extinction that is crucial. As many papers in this book have noted, diversity growth (or decline) is equal to origination minus extinction (and see especially Rosenzweig 1995). In terms of our model (e.g., figure 15.5), the question thus becomes whether the higher speciation rates of endemic species are compensated, or even overcompensated, by very high rates of species extinction. More rigorously, if global diversification (D) equals origination (O) minus extinction

(E), then is D, on average, higher in endemic species or widespread species? Or, is D higher in stenotopic compared to eurytopic species? The role of phylogenetic (intrinsic) traits, such as abundance or niche-breadth, will help determine the answer to this.

Perhaps the most thorough attempt thus far to analyze the role of such phylogenetically inherited traits on net diversification rate is by Marzluff and Dial (1991). In their review of life-history correlates of taxonomic diversity, they show that some species-level traits such as small body size, short generation time, wide dispersal, and high abundance seem to promote higher rates of net diversification (D). These are classic "r-selected" or disturbance-adapted traits that seem to increase the ratio of speciation relative to extinction rates (Marzluff and Dial 1991). Thus, although wide dispersal and high abundance (for example) are often viewed as promoting lower absolute speciation rates than poor dispersal and low abundance (e.g., Stanley 1990c), D is actually higher in species with wide dispersal and high abundance. This is because such species also have much lower extinction rates (e.g., Stanley 1990c). This conforms to our model because the hollow curve of disturbance predicts that very large disturbances must be extremely rare (McKinney and Allmon 1995). The lower speciation rates of the "disturbance-adapted" taxa are more than offset by their lower extinction rates. It also conforms to our model (and results) in that disturbance-adapted species would be predicted to benefit most from disturbance, including intermediate levels of disturbance.

Conclusions

The number of species in a given community is a result of the interaction of (1) processes that gave rise to those species—in that community or elsewhere—in the first place, (2) events and conditions that have allowed access to that community for species that arose elsewhere, and (3) processes that permit them, and have permitted them, to coexist in that community in the past and in the present. Within ecology, much more attention has been given to the second and third (processes of geographical ecology and mechanisms of the *maintenance* of species diversity) than to the first (mechanisms of the *origin* of that diversity). If we are ever to understand the macroevolutionary significance of ecological interactions, we must come to a better understanding of the role of those interactions in the origin of species. The abundant research on the effects of ecological disturbance in communities provides a window into this problem.

Under particular conditions, especially at relatively low levels of immigration, speciation is most likely to occur under "intermediate" levels of environmental disturbance, in a manner analogous to the intermediate levels of disturbance that under certain circumstances permit maximal maintenance of diversity within some communities.

This study represents a first attempt at an explicit model that connects the effects of vicariance, dispersal, and environmental patchiness. It offers insights that can be applied to field studies of speciation in both fossils and living organisms. We tend to find no more than what we look for, and if we do not look for—and distinguish carefully among—the effects of vicariance, dispersal, and habitat grain, we will continue to see speciation as a fuzzy and unknowable collection of unrelated processes.

16

Turnover Dynamics Across Ecological and Geological Scales

Gareth J. Russell

In this chapter, I will argue for turnover as an underappreciated and under-utilized property of all biological systems. Currently, study of turnover is piecemeal. Turnover is defined, measured, and modeled in many different ways, depending on the discipline, the scale of measurement, and whether this measurement is over time or space. Where models are similar, they seem to have been arrived at independently. I will demonstrate the following:

1. Turnover may be found in all biological systems.
2. Turnover can be measured in a similar way over time and space.
3. Turnover can be measured in a similar way at different scales.
4. In some circumstances, turnover can be modeled in a similar way at different scales.

None of these propositions is new. Together, however, they make a compelling case for turnover as one of the most useful phenomena for comparative biological study.

I will illustrate these propositions using the most commonly recognized type of biological turnover—the difference in species composition of two communities. My arguments are divided into four main sections. In the first section, I discuss the nature of turnover—what is it, and where might we expect to find it? In the second section, I discuss the measurement of turnover and show that the methods currently in use have much in common. In the third section, I review some recent attempts to model temporal turnover in a particular type of community, and I show that these models share a key mathematical technique. In the fourth section, I apply one of these ecological models to a paleontological data set and show that the model is useful at this scale also.

There are other types of turnover. Since I am an ecologist by training, this chapter demonstrates a bias towards ecological examples—for this I apologize. Fortunately, this predisposition is balanced by Gilinsky's chapter (chapter 9), in which he addresses turnover at different scales from his perspective as a paleontologist. I start with ecological scales of space and time and expand my view to the longer time spans and often wider spaces of the fossil record. Gilinsky begins with higher taxonomic levels—families and orders—at a global scale, and then focuses down on the species, decades, and hectares that constitute much of ecology. But turnover is more general still. In the concluding section of this chapter, I will argue that turnover is a property of all biological systems, and that methods for measuring and modeling turnover should be the same wherever it is found.

Identifying Turnover

I propose that in ecology, turnover be defined simply as the difference in composition between two community censuses. This utilitarian approach does not distinguish between space and time, and I will try to show that when measuring community differences, there is no reason to make such a distinction.

Temporal Turnover

The first ecologists to study biological communities held polarized opinions. Some maintained that communities develop over time through a predictable "succession" of stages, culminating in a persistent set of species called the climax (Cowles 1899; Clements 1916, 1936). A succession is driven by interactions between species; for example, an early-colonizing species might modify the soil in such a way as to allow the later colonization of a less tolerant species. Clearly, communities in different environments will have different successional sequences and reach different climax compositions. As a corollary, early versions of the theory implied that all communities in the same environment will exhibit the same successional sequence and reach the same climax.

Others proposed that communities are the chance result of the independent dynamics of species (e.g., Gleason 1926). In this view, the ranges occupied by species are determined largely by environmental factors, with little biotic interaction. The community at a particular place consists of those species whose ranges intersect there, and there is no predictable succession or climax.

We now know that neither of these views is wholly correct. Partly predictable successional sequences do occur. At the same time, species do move, often independently, in response to environmental changes. The composition of a community is determined not only by climate and soil conditions, but also by a unique sequence of species replacements that is, at least in part, a function of species interactions. This sequence is commonly called an assembly history, distinguishing it from a more deterministic succession (for example, see Drake 1991; Luh and Pimm 1993). But all these viewpoints imply that communities change their composition over time. Even the staunchest supporters of succession acknowledge that communities are constantly being "knocked back" from their climax state by environmental forces, after which they return to the climax through a process of "secondary" succession. All types of community change over time can be classified under the heading of temporal turnover.

MacArthur and Wilson (1963, 1967) proposed a theory that combines these opposing viewpoints and makes testable predictions about temporal turnover. Using islands as their example system, they suggested that communities develop toward an equilibrium of species richness, but not of species composition. According to their theory, community development, unlike a successional sequence, is not wholly deterministic, but a function of probabilistic processes of colonization and local extinction. Equilibrium is reached not when these processes cease, but when they balance each other. Turnover may still be considerable. This theory has been supported by empirical observation of colonization and extinction (e.g., Simberloff 1969, 1974; Simberloff and Wilson 1969), and under certain assumptions, models based on these processes have yielded good predictions of both species richness and turnover (e.g., Diamond 1969, 1971, 1984a,b; Diamond and May 1977; Hunt and Hunt 1974; Jones and Diamond 1976; Lynch and Johnson 1974; Reed 1980; Schoener and Spiller 1987; Terborgh and Faaborg 1973; Williamson 1983).

At longer time scales, increased study of fossils during the eighteenth and nineteenth centuries (e.g., Cuvier 1812; see Mayr 1982 for a review) revealed that the communities of the past were made up of very different species from those we know today. Temporal turnover is evident in almost all fossil communities (although see Brett and Baird 1995). More recent analyses (e.g.,

Sepkoski 1981) also indicate that the total number of species has varied dramatically. The origination and extinction of species are responsible for these changes.

Spatial Turnover

If we census the community found at two different places and find different species compositions, then according to the definition presented earlier, we are seeing turnover. But the existence of spatial turnover is perhaps not obvious. Turnover is often defined, either explicitly or implicitly, as changes in a "community." Some might argue that in the temporal example we are indeed censusing the "same" community (albeit at different times), but that in the spatial example we are censusing different communities. It is easy to visualize the sequence that leads from a set of species at time t to a different set of species at time $t + n$. Over time, species come and go in the area we are observing. There is an obvious continuity. It is less easy to visualize the sequence that leads from a set of species in one place to a set of species in another. Why might we make this distinction? To see temporal turnover, we need only sit and wait, but to see spatial turnover, we must move ourselves. Because of this, we tend to view space and time differently. If we do move from one community to another, however, species "appear" and "disappear" just as they do over time. I suggest that if we are concerned with differences in species composition, then calling two communities separated by time "the same," and two communities separated by space "different," is not a useful distinction.

Both kinds of community comparison also have this in common: on average, the greater the separation in time or space, the greater the difference in species composition. (This pattern becomes noisy but still recognizable if there are cyclic appearances and disappearances of particular species in time, or alternating community patches in space). So again, it makes sense to treat temporal and spatial turnover a similar manner. In the remainder of this chapter, I will use the term *temporal turnover* for changes in community composition over time, the term *spatial turnover* for changes in community composition over space.

Measuring Turnover

Standardizing Measurements

Ecological turnover is a result of the appearances and disappearances of species. The most widely used index of temporal turnover for two censuses performed at different times has been

$$\text{Turnover rate} = \frac{\begin{array}{c}\text{Number of appearances}\\ + \text{ number of disappearances}\end{array}}{\begin{array}{c}\text{Total number of species present}\\ \times \text{ census interval}\end{array}} \qquad [1]$$

(e.g., Diamond and May 1977; Schoener 1983). This index consists of two parts: a measure of change (the numerator) and a method of standardization (the denominator). The measure of change is straightforward—simply add the number of species that either appear or disappear when comparing the two censuses. The standardization has two parts. First, divide by the sum of the number of species in both censuses. This converts the change into a proportion of the total set of species (allowing us to compare the turnover in communities of different size). Then, divide by the census interval. The index now describes the proportion of the community that changed per time unit (which is why it is called turnover *rate*).

Recently, Russell et al. (1995) proposed the form

$$\text{Turnover} = \frac{\begin{array}{c}\text{Number of extinctions}\\ + \text{ number of immigrations}\end{array}}{\text{Total number of species present}}. \qquad [2]$$

This index standardizes turnover by community size but not by interval (and so is not a rate). Their reason for not dividing by census interval is that the relationship of turnover to interval is asymptotically nonlinear. Diamond and May showed that for breeding birds on islands, turnover *rate* declines hyperbolically to a slope of -1. Gingerich (1983) noticed that metrics that calculate paleontological extinction rate (a component of turnover rate) also show a negative relationship with interval.

Figure 16.1 shows a plot of observed turnover (as measured by equation 16.2) against census interval. The data are yearly censuses of breeding bird populations from the island of Skokholm. [We will see more of these data later in this chapter. They are the same data used by Diamond and May (1977), who used equation 1 to calculate turnover.] The figure shows the turnover for every possible pairwise comparison of years. The data are clearly asymptotic. In this particular data set, we can observe year-to-year turnover directly, since there are many pairs of censuses separated by only 1 year. But this may not always be the case, and in a later section I will show that calculation of turnover *rate* is misleading when the only census data available are separated by long intervals.

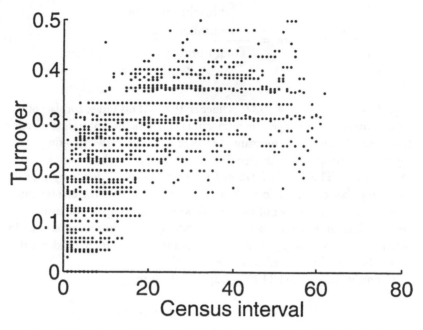

Figure 16.1. Observed turnover on the island of Skokholm, calculated using the index of equation 2. Points represent the turnover for every possible pairwise combination of census years.

In paleontology, there are a variety of turnover indices. Oliver and Pedder (1994) identify three alternatives (shown in equations 3, 4, and 5), attributed to Roth (1987), McGhee (1988), and Oliver (1990), respectively. I have changed the words for easier comparison with the ecological indices above, and to prevent prejudgment about the processes involved.

$$\text{Turnover rate} - \frac{\text{Number of appearances} + \text{Number of disappearances}}{\text{Length of stratigraphic stage}} \qquad [3]$$

$$\text{Turnover rate} = \frac{\text{Number of appearances} - \text{Number of disappearances}}{\text{Length of stratigraphic stage}} \qquad [4]$$

$$\text{Turnover rate} = \frac{\text{Number of disappearances}}{\text{Number of genera present (in stratigraphic stage)}} \times 100 \qquad [5]$$

These, like the ecological indices, are measures of community turnover. They should not be confused with taxonomic turnover (see Simpson 1944; Stanley 1979; Sepkoski 1981, 1984, 1987a; Gilinsky, chapter 9). Their similarity to the ecological indices is immediately apparent. The first model is most similar; the numerator is the same, but it is standardized by time and not by community size. The second model is like the first, but the number of disappearances is *subtracted* from the number of appearances. Both models use as their interval the stage: a somewhat arbitrary period of time that may be anything from 1 million to 20 million years in length. Standardization thus gives the turnover rate per 1 million years. The third model is described by Oliver and Pedder (1994) as being equivalent to what they call "proportional extinctions." In this case, the number of disappearances is standardized by community size, except that size is defined in terms of the number of genera rather than species. Indeed, extinction indices show a similar distribution of forms to turnover indices, from the simple number of extinctions (E) through extinction rate (E/t, where t is time) and proportional extinctions (E/D where D is diversity) to a proportional extinction rate [$E/(D \times t)$] (Foote 1994).

There is, however, an important difference between all these paleontological indices and the ecological indices of equations 1 and 2. In ecological turnover, there is one census at the beginning of the interval and one at the end. In the paleontological indices, *all* appearances and disappearances within the interval are counted. We might call the indices *pairwise* and *cumulative* turnover, respectively. Species that appear and disappear within the interval are included in paleontological turnover but not in ecological turnover.

The difference between pairwise and cumulative turnover is profound. In cumulative turnover, the relationship between appearances-plus-disappearances and census interval will be positive and approximately linear—as we look at a greater time span, we will see a proportional increase in the total number of appearance and disappearance events within that time. Dividing by census interval becomes sensible in these circumstances. On the other hand, dividing by the total number of species becomes difficult to justify. In pairwise turnover, standardizing by the number of species (or genera) means that the maximal possible turnover—when every taxon in an initial census is not present in a second—is unity. In cumulative turnover, the total number of appearances plus disappearances can be greater than the total number of species. Dividing by the number of species could produce a turnover that is greater than unity (and hence not a proportion).

There is, of course, nothing to stop us from calculating cumulative turnover in ecology, or pairwise turnover in the fossil record. But from here onward, I will restrict my discussion to pairwise turnover.

So, we can measure turnover, using a similar index, over days, years, decades, millennia, or less consistent units such as stratigraphic stages. What if we want to measure turnover across kilometers, patches, or biomes? Existing approaches to community change over space reveal a unique set of indices. Cody (1975) defined *alpha diversity* as the number of species at a given time in "a small patch of habitat within an extensive and uniform stand of the same type" (see also Whittaker 1960, 1970). If the composition of the community changes over space, then the rate at which we pick up new species along a transect provides a measure called *beta diversity* (the differences between local communities, or patches). We can also measure *gamma diversity* (the differences between larger areas, such as biogeographic provinces). Confusingly, alpha diversity is a measure from one area, whereas beta and gamma diversity are spatially comparative measures. But rather than calculating beta or gamma diversity, there is no reason why we cannot use the index of turnover shown in equation 2. This index works for any kind of community difference.

Intracensus Turnover—Time Averaging and Space Averaging

Whether turnover is spatial or temporal, there are a number of practical difficulties associated with measuring it in the field. Temporal turnover is the difference in species composition of two communities at different points in time. Spatial turnover is the difference in species composition of two communities at different points in space. But what is a point in time, or a point in space?

Censusing a community takes a certain amount of time, and the longer the time, the more likely that there will be some changes in species composition while the census is being conducted—some *intracensus* turnover. Suppose we want to measure the turnover (i.e., the *intercensus* turnover) of a community over an interval of one year. Ideally, we would conduct our census on the same day in each of two consecutive years. But what if, for some reason, each census takes a month to complete? Clearly, there may be some turnover as the census is being taken. If so, each census will tend to pick up more species than if the species composition of the community remained constant during the census period. The species lists from both the first and second censuses will be longer than if we had just looked at one day. The intercensus turnover will therefore tend to underestimate the true yearly turnover.

In paleontological studies, the sampling has been done for us by the physical processes of sedimentation, and the problem of intracensus turnover is acute enough to have the special name of *time averaging*. Time averaging refers to the fact that even relatively thin slices of rock typically contain sediments that differ in age by thousands of years (see Schindel 1982 for a discussion of the effects of different sampling schemes). The greater the time-averaging, the more intracensus turnover and the less (apparent) intercensus turnover there will be. It would perhaps be wise if all pairwise turnover studies include, as a summary statistic, the ratio of the intracensus interval to the intercensus interval. The smaller this ratio, the better.

So, in an ideal world, we would measure temporal turnover by censusing a defined area over the smallest possible interval of time. By analogy, we should measure spatial turnover by censusing the smallest possible area (over a defined time interval). Ecologists often talk about "point diversity," but typically they mean the number of species in a certain "homogeneous" area (i.e., alpha diversity). In principle, we could actually measure point diversity. The procedure might be something like this: We observe a very small area, in the form of a column of small diameter rising from the ground, and for a fixed length of time we record the species of every organism that intersects the column, even if only for an instant. This procedure is not as bizarre as it sounds—plant ecologists already employ a similar approach in the form of a point transect. If we want to include more mobile organisms in our census, though, the procedure is rather inefficient—we might have to wait a very long time before getting a decent sample of species. So instead, the census method used is exactly the same as for measuring temporal turnover—a count of all the species in a certain area.

Once again, we have the problem of intracensus turnover. Suppose we census two 100-ha plots separated in space by 200 m. If the plots are squares, 1 km on each side, do we mean that the edge of one plot and the edge of the other are 200 m apart? Or that the *centers* of each plot are 200 m apart? in which case there would be an overlap of 80 percent. This problem might be called space averaging, and as with time averaging our goal should be to minimize the ratio of intracensus to intercensus turnover. In this case, we would try to minimize the ratio of the area sampled to the distance between the samples.

Counting Everything

Empirical studies of community turnover are beset by incomplete species lists. Sepkoski (1994) estimated the total number of species present during the Cenozoic as about 600,000. The number of fossil species from this period is about 40,000, which is less than 7 percent. Schopf (1978) has pointed out that species with the highest probability of forming identifiable fossils, such as intertidal, sediment-dwelling macrofauna, can reach about 40 percent representation, but for many terrestrial species the probability of fossilization is close to zero. The problem of inadequate sampling can be reduced by instead counting higher taxa such as genera or families. The dynamics of such taxa correlate fairly well with those of their constituent species (Sepkoski 1992b; Bambach and Sepkoski 1992b; Jablonski 1995), but they have a much more consistent preservational record. Unfortunately, the species or population remains the dominant unit of study in ecology, and emphasis on different taxonomic levels remains a stumbling block for the integration of paleontology with ecology.

The limited sampling provided by the fossil record has profound consequences for the interpretation of paleoecological community dynamics. If fossils represent a random sample of *individuals,* the species content of this sample will be biased towards those species that are most abundant (McKinney 1996a). The smaller the sample, the stronger the bias; Sepkoski (1994) estimates that 90 percent of fossil individuals come from the most abundant 0.5 percent of fossilizable species. If the sampling is over a large area, there will also be a bias towards those species that are widespread. These may be the same species that were abundant: In both fossil and contemporary data there is a correlation between abundance and geographic range (Russell and Lindberg 1988; Gaston 1994; Brown 1995).

So in most circumstances, paleoecologists are studying only the most abundant and/or widespread species (even if the fossils include 40 percent of the extant species at that time). Much ecological evidence suggests that such species are less prone to extinction (Gaston 1994 and references therein). The same is true of populations at a more local scale (Goodman 1987; Pimm et al. 1993). The often overlooked consequence of this is that the fossil record describes the dynamics of a peculiarly persistent subset of species. Turnover in this subset will be much lower than in the community as a whole (McKinney et al. 1996), and this may explain some examples of apparent community stasis in the fossil record (e.g., Brett and Baird 1995; Alroy, chapter 12).

Ecologists often accuse paleontologists of poor quality data (personal observation; see also Jackson 1988). These critics overlook the equally obvious fact that no ecological field studies ever consider an entire, natural community. Through constraints of time, budget, and taxonomic knowledge, we study birds on islands, bacteria and protists in laboratory flasks, or the complete but artificial communities of a computer program. [The closest approach to a field study of a complete community is the recording by Simberloff and Wilson (1969) of the colonization of arthropod species onto several defaunated mangrove islands.] Assemblages such as "birds" may represent far less than 7 percent of the total species richness, given the large number of unknown invertebrate and bacterial species we believe to inhabit the soil (World Conservation Monitoring Centre 1992). Indeed, if the total number of species on the planet is conservatively estimated at 30 million (Pimm et al. 1995), the 9770 recognized extant bird species (Sibley and Monroe 1990, 1993) represent a meager 0.03 percent.

Should we ignore the turnover of assemblages and wait until we know the details of every species in a community? The current rate at which these very species are being lost (Pimm et al. 1995) does not allow us that luxury. Conservation biologists, for practical reasons, often turn to single species as indicators of the "health" of entire communities (e.g., Thomas et al. 1973; Soulé and Kleppel 1988). An assemblage can provide information, if not on the whole community, then at least on community-*level* dynamics, providing a more powerful diagnostic tool than the status of a single species (Fry et al. 1986; Block et al. 1986; Kremen 1992).

Separating Space and Time

It may be difficult to distinguish between spatial and temporal turnover. If there is a limitation on the number of personnel available to collect census data for a study of spatial turnover, it may not be possible to census all the different areas at the same time. The same census takers may have to move from one site to another, sampling them sequentially. While a careful experimental design can minimize this problem, there will always be an element of temporal turnover in the data.

In paleontological data, there is always some error associated with age estimates (whether by radiometric methods or by stratigraphic correlation). When comparing fossils from different areas, it is sometimes hard to be sure that we are looking at exactly the same time period. If we are not, we will again introduce temporal noise into any spatial patterns.

One type of fossil sampling method is to take all the fossils from a vertical column of rock (usually called a core). The age of the rock, and hence of the fossils trapped within, is an increasing function of depth. So, by sampling at different depths in the core, we should detect any temporal turnover. But this may not be all we are detecting. Sediments deposited on a sea floor can be brought by an ocean current from several hundred kilometers away. A shift in ocean currents can change the source of sediment to a new area, far enough away from the first to have a completely different species composition. If one of these changes has occurred in the area sampled by a core, different parts of the core will effectively be sampling different areas (as well as different times). It is often possible to detect substantial changes in sediment origin, especially if sampling is not limited to a core but covers a larger area (so that a picture of spatial changes may be built up). But at a fine scale, undetected vagaries of sedimentation will introduce spatial noise into a temporal pattern.

As mentioned, spatial and temporal dynamics are often correlated (again, see Brown 1995 for a review of the ecological evidence). This is a mixed blessing. It may be virtually impossible to filter out the unwanted noise while leaving the pattern of interest, but at the same time the noise may be similar enough to the pattern that the combined signal may still represent the pattern.

Interpreting Turnover

There are two aspects to interpreting turnover. One aspect is understanding what processes are involved. What are the "appearances" and "disappearances" that we observe? The second aspect is linking these processes—showing how the observed patterns of turnover arise from them.

Changing Scales Means Changing Processes

Appearances and disappearances come in two forms. An appearance can represent the *colonization* of a preexisting entity into the area of observation, or the *origination* of something new. Similarly, a disappearance can represent *local extinction* from just the area of observation, or *global extinction*. At the ecological scale of bird assemblages on islands, the entity is usually a population of breeding pairs of a given species. In this case, appearances represent the formation of one or more breeding pairs where previously there were none. The individuals may have come from another population (on another island or on the mainland), or they may have been already present on the island as part of a nonbreeding population. Disappearances represent the dissolution of all breeding pairs of that species, resulting from any combination of (1) death, (2) movement of some individuals away, or (3) the failure of, or decision by, a pair not to breed in a given year.

At the paleoecological scale of an assemblage of foraminiferan species in a stratigraphic core, appearances can describe either colonization (to the area sampled by the core) or the origination of a new species. [Or perhaps both: Gould and Eldredge (1977; also see Gould 1980) have argued that most apparently rapid transitions from one form to another in the fossil record are actually records of invasions from an isolated population that has undergone

TABLE 16.1   Processes contributing significantly to turnover
at combinations of temporal and spatial scale

	Ecological time	Geological time
Local community	Immigration Local extinction	Immigration Local extinction
Wider community	Immigration	Immigration Local extinction
	Local extinction (Global extinction)	Origination Global extinction
Global community		Origination
	(Global extinction)	Global extinction

Note: Parentheses indicate the special case of mass extinctions.

evolutionary change in an undetected location before reappearing to replace the maternal population—a colonization following an unseen origination.) Disappearances from the core can represent local extinction or global extinction.

The important difference between these processes is that colonizations and local extinctions can happen repeatedly, whereas originations and global extinctions cannot. In the next few paragraphs, I will show that if we observe temporal turnover at certain scales, we may simultaneously observe the effects of colonizations, originations, local extinctions, and global extinctions!

We saw that for bird breeding assemblages on small islands, where the time scale is measured in years or decades and the spatial scale in kilometers, appearance usually represents colonization and disappearance usually represents local extinction. Originations of new species are rare enough that on an ecological time scale of decades, their influence is negligible. Also, very few species are endemic to areas of only a few hectares. Distant islands or otherwise isolated areas have disproportionately more endemics, but in general, an extinction from a very small area will be a local extinction. Community dynamics at these scales are the result of only two processes (table 16.1, top of the left column).

If we extend the time scale to thousands of years or more, it is tempting to think that the first assumption becomes increasingly inappropriate—that appearances may also be due to the evolutionary origination of new species within the area of observation. But if the area is small enough, such evolutionary events are still rare. Soulé (1980) has documented the smallest islands on which autochthonous speciation is known to have occurred for various taxa. For birds and mammals, that island is Madagascar, with an area of about 600,000 km². (This is also why, as mentioned earlier, most apparent originations observed in the fossil record are probably colonizations; the originations occurred elsewhere.) So we still need consider only two processes (see table 16.1, top of the right column).

What if we return to a short, ecological time scale but consider a large area (say, a continent)? At this short time scale, there will still be negligible originations, even in a large area, but does global extinction become more likely with the increased area? It does, but with the following caveat. Global diversity has increased since the beginning of life, so the total number of originations must have been more than the total number of global extinctions (Raup 1994). So, if originations are rare over short time periods, global extinctions must *on average* be rarer still (Gilinsky and Good 1991). But there is a complication—there have been episodes in the history of life when the rate of extinction has, for a short while, vastly exceeded the rate of origination. During these so-called "mass extinction" events, global extinctions may be relatively common even over short time periods. Thus there are either two or three processes contributing to turnover (see table 16.1, middle of left column).

Consider both a large spatial scale and an extended time scale (millions of years). Originations and global extinctions now play an important part. But, if the spatial scale is anything less than global, colonization and local extinction must still play a role. There are four processes contributing to turnover (see table 16.1, middle of right column).

Finally, if we extend the spatial scale to include the entire planet, all appearances must be the result of origination, and all extinctions must be global. There can be only two processes. Still, over short time periods their impact will be low except for occasional peaks in rate of global extinction (see table 16.1, bottom of left column). It is only over long intervals that both will have a significant impact (see table 16.1, bottom of right column).

I present table 16.1 as food for thought. It suggests that at intermediate scales of space and time, temporal turnover requires a complex, four-parameter explanation, but that at restricted scales of space and/or time and for the global community at geological time scales, just one process of appearance and one process of disappearance are important. There is the intriguing hint that models of turnover at these extreme scales might look rather similar. Just such a conclusion has already been reached for models of diversity rather than turnover: Sepkoski (1991b,c) has attempted to model the overall pattern of diversification of life on earth using a logistic model that is more usually applied to population dynamics.

Linking Space and Time (Metapopulation Models Predict Both Temporal and Spatial Turnover)

The term *turnover* need not be applied solely to communities. I am going to preempt the concluding section of this chapter a little by briefly considering a different form of turnover. The spatial distribution (range) of most species changes over time. If we represent the range as a formally defined metapopulation (Levins 1970), or even if we divide a continuous population into an array of arbitrary locations (rasterization), then we will be able to measure a presence/absence pattern of local populations over time (Maurer 1994; Maurer and Nott, chapter 3). This might be called *range turnover* (Curnutt, personal communication, 1994).

Metapopulations, in either the formal or loose sense described earlier (Harrison 1994 and chapter 2), provide simultaneous explanations of spatial and temporal community turnover. The metapopulation of one species may exhibit turnover of its local populations (range turnover). A given geographical area will contain portions of the metapopulations of *many* species. Assuming that their ranges do not overlap completely, different subsets of the area (patches) contain different subsets of species (local communities). As we move from patch to patch, we encounter different communities. But if we stay in one patch for a while, the set of species will also change. The idea of species as metapopulations predicts both spatial and temporal turnover. Figure 16.2 represents this graphically: At each point in time, the distribution of each of species 1 through 5 is patchy, and for species 1, 2, 3, and 5, this patchy distribution changes between times y and $y + n$.

Spatial and temporal turnover are united in the idea of the "shifting mosaic," identified empirically by Watt (1947) and now a cornerstone of landscape ecology. Perhaps the utility of metapopulation models lies as much in their ability to unite these different forms of community change as in their ability to explain the dynamics of any one species.

Modeling Turnover

Might mathematical models of turnover, as a function of appearance and disappearance processes, be similar at different scales? In this section, I will present some models that were developed for ecological data. Significantly, all the models use a similar and characteristic mathematical technique for dealing with repeatable appearances and disappearances.

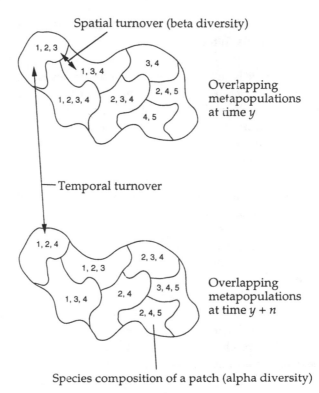

Figure 16.2. Metapopulation theory explains both temporal and spatial turnover.

Markov Expansions for Repeatable Events

For many kinds of appearance and disappearance (sometimes called birth and death) processes, we can express the probability that each event will occur over a given interval of time in terms of the probability that it will occur over a unit interval of time. For example, over the last 400 years, mammal species have had, on average, a yearly extinction probability of about 3.5×10^{-5}. We can easily calculate the probability that a species will become extinct in n years by simply multiplying this probability by n. If a mammal species is extant in 1996, the probability that it will have disappeared by 2050 is $54 \times 3.5 \times 10^{-5} \approx 0.002$.

As I noted above, such global extinctions can happen only once. A more complicated example is provided by breeding populations of birds on islands. In this case, "disappearance" is simply the failure by a particular species to breed on an island when it had bred before. The species is still breeding somewhere, and it may return to the island to breed again, so the disappearance is a local extinction. If a species breeds in a particular year, we can describe the probability that it will not breed at a point n years into the future, in terms of the probabilities of local extinction and of colonization over *one* year. The expression, which is called a Markov expansion, is

$$\delta_n = \frac{\delta_1}{\lambda_1 + \delta_1}\left[1 - (1 - \lambda_1 - \delta_1)^n\right] \qquad [6]$$

where δ_1 is the probability of observing a local extinction over 1 year, λ_1 is the probability of observing a colonization over 1 year, and δ_n is the probability of observing a local extinction when comparing two years separated by n years.

Note that this is *not* the probability that a local extinction will occur at any time during this interval. It is the probability that a species will be breeding in the first year and not in the year that is n years later. It may have come and gone many times in between. We saw the same distinction when comparing methods of measuring turnover in a previous section—paleontologists tend to look at a span of time, ecologists tend to compare two moments in time.

Since colonization may also be temporary (it may be followed by a local extinction), then a similar expression exists for λ_n, the probability of observing a colonization over an interval n:

$$\lambda_n = \frac{\lambda_1}{\lambda_1 + \delta_1}\left[1 - (1 - \lambda_1 - \delta_1)^n\right] \qquad [7]$$

Both expressions use the same two variables, λ_1 and δ_1. Note that λ_1 and δ_1 are not expected to be the same, even if the number of species in the assemblage remains approximately constant. This is because the extinction probability applies to the species in the assemblage, and the colonization probability to the species *not* in the assemblage. Also, the probability of a species being present in year y and in year $y + n$ is simply $(1 - \delta_n)$, and the probability of a species being present in year y and in year $y + n$ is $(1 - \lambda_n)$.

There is one more wrinkle to this story. All the breeding pairs of a species might cease breeding in one year, but be replaced *before the next year* by more pairs of the same species. This is especially likely if the number of breeding pairs is small. In this case, a local extinction has clearly occurred, but unless we can recognize individual birds, we will not observe it. Let us express the probability of a "real" local extinction over 1 year as μ_1, to distinguish it from the probability of an "observed" local extinction (δ_1). Although we do not see some of the "real" extinctions, we can still estimate the value of μ_1. To observe a local extinction, not only must the breeding population fail, but *no new population must colonize.* Thus δ_1, the probability of observing a local extinction, is given by $\mu_1(1 - \lambda_1)$, so that μ_1 is given by $\delta_1/(1 - \lambda_1)$. In the next section, I will show that if we observe a species over time we can estimate λ_1, δ_1, and hence also μ_1.

The distinction between "real" and observed local extinctions applies to only certain kinds of data, and it is useful only if we want to know the probability that a given set of individuals will fail. Often we simply want the probability that the population will fail; whether or not there is turnover of individuals is not an issue. In such cases, we calculate only δ_1.

Why is this mathematical technique so useful? Unit probabilities such as λ_1 and δ_1 are standardized by time, and they allow us to compare different species or communities. If we are lucky, our census data will be from adjacent time periods and we will be able to estimate λ_1 and δ_1 directly (e.g., Rosenzweig and Clark 1994). It is much more likely that our data will be irregularly spaced, and it may never include adjacent time units. Markov expansions can be used *backwards* to estimate unit probabilities from observations over longer intervals (Clark and Rosenzweig 1994). There are three types of models.

Maximum Likelihoods Applied to Species

The first type of model addresses the species in a community separately. By looking at the patterns of appearances and disappearances over time, we can estimate λ_1 and δ_1 for each species. We will then have distributions of these probabilities, and in principle we could then describe the behavior of the community in terms of these distributions. We can estimate λ_1 and δ_1 by fitting the Markov expressions shown earlier to the sequences of presence and absence for each species, using a maximum likelihood approach. (Likelihood is the probability of obtaining the observed data, given a particular model.)

Suppose we have the following sequence:

Time unit	1	5	6	9	11	12	16	18
Presence or absence?	A	P	P	A	P	A	A	P

This sequence has as its likelihood function

$$L = \lambda_4 \cdot (1 - \delta_1) \cdot \delta_3 \cdot \lambda_2 \cdot \delta_1 \cdot (1 - \lambda_4) \cdot \lambda_2 \qquad [8]$$

which is to say, the product of the probabilities of each transition. I have shown that we can represent all of these transition probabilities in terms of just δ_1 and λ_1. The maximum likelihood estimators (MLEs) are obtained by finding those values of λ_1 and δ_1 that produce the largest value of L.

Such MLEs are obtained by using a nonlinear maximization algorithm that searches the space of all possible values if λ_1 and δ_1, "homing in" on those that provide the highest value of L. There are either two or three constraints on the possible values. Since λ_1 and δ_1 are probabilities, they are constrained to be within the ranges $0 \leq \lambda_1 \leq 1$ and $0 \leq \delta_1 \leq 1$. For mathematical reasons, there is also a constraint that $(\lambda_1 + \delta_1) \leq 1$. Furthermore, if we are then going to calculate μ_1 from $\delta_1 / (1 - \lambda_1)$, then δ_1 must lie in the range $0 \leq \delta_1 \leq (1 - \lambda_1)$.

Estimates obtained by this method vary widely in the confidence we can place in them, depending on the quality of the data. Figure 16.3 (A through C) shows the likelihood surfaces for combinations of colonization and extinction probability for three the bird species—buzzard (*Buteo buteo*), robin (*Erithacus rubecula*), and skylark (*Alauda arvensis*)—on the island of Skokholm (see Russell et al. 1995 for a more complete description of these data). The first points of interest are the extremely small values of the likelihoods for the buzzard and robin. This is because Skokholm has census data for many years, so that these likelihoods are the product of many individual probabilities. If our data are extensive, likelihood values can be fantastically low, but it is not their absolute values that matter. The point of this, and indeed all, statistical techniques is to find the *best* model that describes the data. What is important is the value of the maximum likelihood *relative to the likelihoods for other combinations of parameters*. I could normalize the likelihoods, or use logarithms, but no particular purpose would be served in this case.

The next point is that the confidence we place in these MLEs can vary widely. For example, the surface for the buzzard shows a clear maximum, but for the robin the surface is more of a ridge—there is considerable uncertainty as to the value of δ_1, the extinction probability. The reason is that there are only two years when robins were present, and so only two years when we could observe an extinction. The skylark shows an even more extreme surface topology—in this case the MLE of δ_1 is zero, whereupon all the possible values of λ_1 have the same likelihood. This surface is produced because skylarks are *always* present on Skokholm. It is clear that the probability of extinction when present must be very low, but we can discern nothing about the probability of colonization because the species has to be absent before it can colonize! It is possible to calculate formal confidence intervals for maximum likelihood estimates of this kind, although the procedure is rather complicated.

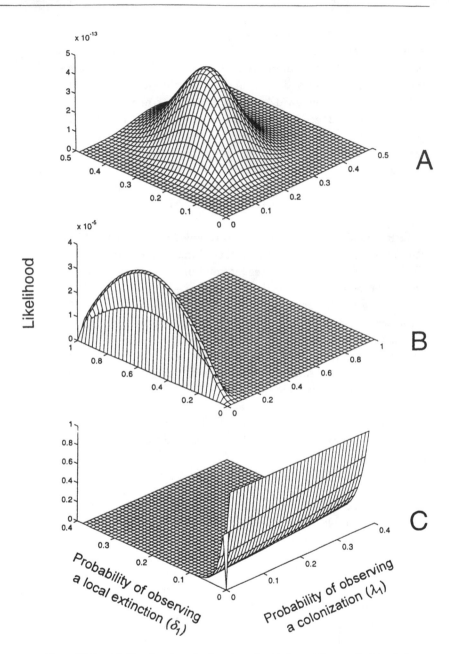

Figure 16.3. Likelihood surfaces for three breeding bird populations on the island of Skokholm. A, buzzard; B, robin; C, skylark. See text for details.

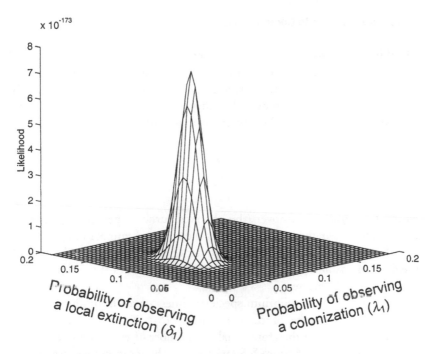

Figure 16.4. Likelihood surface for the entire bird community on Skokholm.

Maximum Likelihoods Applied to Communities

The second type of modeling approach addresses the community as a whole. We effectively assume that a community can be characterized by average species values. There will be an average appearance probability and an average disappearance probability for the community. Why do this, instead of the species-by-species approach? We might not have enough data to analyze species separately. Clark and Rosenzweig (1994) applied the maximum likelihood model (as described earlier) to entire plant assemblages on small Australian islands, rather than to the species. The number of years for which census data are available is so few that reliable estimates for each species are unobtainable. (Even in the Skokholm data set, we had to eliminate some estimates for some species for this reason.) When we consider an entire assemblage, we search for those estimates of λ_1 and δ_1 that, if applied to *every species* in the assemblage, give the highest probability of obtaining the assemblage data. The likelihood surface usually shows an unambiguous maximum: Figure 16.4 shows the likelihood surface for the bird community on Skokholm.

Turnover Applied to Communities

The third type of approach makes the same assumption about average probabilities as the second approach (with the same advantages and disadvantages). But instead of estimating these average probabilities directly, we can use a model of the expected turnover that they give rise to. The description of turnover in equation 2 can be written symbolically as

$$T_n = \frac{E_n + I_n}{S_y + S_{y+n}} \qquad [9]$$

where T_n is *apparent* turnover (the observed turnover per species between two censuses separated by n years), E_n is the observed number of extinctions over this interval, I_n is the observed number of colonizations, y is the time of the first census and S_y and S_{y+n} are the numbers of species at the times of the first and second censuses.

If we assume an equilibrium species richness, three things are approximately true: The number of species is constant ($S_y = S_{y+n} = S_{eq}$), the number of colonizations must therefore equal the number of extinctions ($I_n = E_n$), and we may estimate E_n from $\delta_n S_{eq}$ where δ_n is the mean per-species extinction probability: the probability that a species from a given community breeds in year y but fails to breed in year $y + n$. Hence,

$$T_{n,\,eq} = \frac{E_n + I_n}{S_y + S_{y+n}} = \frac{2E_n}{2S_{eq}} = \frac{2\delta_n S_{eq}}{2S_{eq}} = \delta_n \qquad [10]$$

where we can represent δ_n in terms of λ_1 and δ_1 as given by equation 2.

Russell et al. also present a "nonequilibrium" model that takes into account change in species number over time (a phenomenon that will inflate turnovers, especially at longer intervals). If ß is the slope of a plot of ln(species number) against time (an exponential, rather than linear, regression is used to prevent the possibility of a negative number of species), then

$$T_{n,\,ne} = \frac{\frac{\beta}{|\beta|}(\exp^{\beta n} - 1) + 2\delta_n \exp^{\frac{\beta}{|\beta|}\beta n}}{1 + \exp^{\beta n}} \qquad [11]$$

where, again, δ_n is given by equation 2.

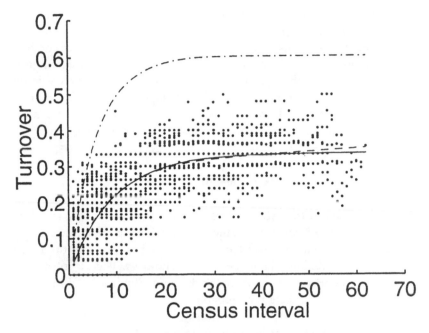

Figure 16.5.   The predictions of three models about turnover on the island of Skokholm. *Solid line,* equilibrium turnover model; *dashed line,* nonequilibrium turnover model; *dash-dot line,* maximum likelihood model.

Over the period for which data are available, Skokholm shows a 4.8 per-cent decrease in species number per decade ($\beta = 0.0048$). I calculate turnover for the Skokholm data of figure 16.1 using all possible pairwise combinations of censuses. I fit both the models using a nonlinear fitting algorithm, except that in this case the model is a regression, so I minimize the sum of squares of the residuals. Figure 16.5 shows the observed turnover on Skokholm, the least-squares fit of the equilibrium model and of the nonequilibrium model. In addition, it shows the turnover predictions based on the estimates of λ_1 and δ_1 obtained from the maximum likelihood approach of the previous section.

The Relative Merits of the Models

The turnover models, especially the nonequilibrium model, provide a good fit to the data for Skokholm except at the shortest intervals, where they underestimate (see Russell et al. 1995 for other examples of the fit of this model). The estimates of λ_1 and δ_1 obtained by the community-level maximum likelihood approach make good predictions *only* at the shortest intervals. The reason probably lies in the fact the turnover models are fitted to the data as shown—to every pairwise combination of years—whereas the maximum likelihood models are fitted only to sequential pairwise combinations (see equation 8). Which model is better depends on the questions we want to ask. But in either case there is a caveat. The results of the species-by-species approach suggest that distributions of appearance and disappearance probabilities are highly skewed. In this case, "average" values for a community may not be useful statistics. They may, however, allow us to compare communities.

Markov Expansions for Nonrepeatable Events

What if our data are from scales where it appears that originations and disappearances are global extinctions? Since, as table 16.1 suggests, there are two processes, any model should have two probabilities, just like the models above. The probabilities will not be the same, however. There is still a probability that a species (or other taxon) will disappear. But instead of a probability of recolonization, there is a probability that a taxon will split into two. Gilinsky and Good (1991; see also Gilinsky, chapter 9) show that these probabilities can be used to predict changes in the total number of taxa (rather than turnover). Since an increase is assumed to be the result of branching evolution of one of the taxa already present, both appearances and disappearances are a function of the number of taxa present. If P_0 is the probability of a taxon going extinct over a given interval, and P_2 is the probability of a taxon splitting into two, then the probability of going from S_0 to S_1 entities over that same interval is the coefficient of the x^{S_1} term in the expanded form of the following expression:

$$(P_0 + P_1 x + P_2 x^2)^{S_0} \qquad [12]$$

where $P_1 = 1 - P_0 - P_2$. As in the other Markov expansions, this expression can be used backwards. By fitting it to an observed diversity path, whose overall likelihood is equal to the product of the probabilities of each transition, we can obtain maximum likelihood estimates of P_0 and P_2.

Identifying "Lazarus" Species: A Case Study in Cross-Scale Modeling

The preceding sections suggest that some models, developed for particular entities and scales, might be applied in different contexts. In this section, I will present one example where a model developed in ecology may be usefully applied to paleontological data. These data consist of records of 102 species of foraminifera from a stratigraphic core (McKinney and Frederick 1992; McKinney and Allmon 1995). The core samples are from 1032 to 915 feet in depth, spanning the late Eocene and early Oligocene (Priabonian and Rupelian stages, a period of a few million years). The exact ages of the samples are unavailable, so there are two ways of treating the time component of these data. We can assume equal spacing, in which case we assign the time-step value of one to the first (deepest) sample, two to the second, and so on. Or, we can assume that the depth of the samples is proportional to their age, in which case we assign to each sample a measure of depth in whatever units we like. In this case I assume equal spacing, as if the samples were in consecutive "years."

One of the many problems encountered by paleontologists is that of "Lazarus" taxa: taxa which disappear from the fossil record for an extended period of time, only to reappear unexpectedly and show that they were not globally extinct at all (Jablonski 1986a). The reappearance of a taxon after any period of time indicates that the absence was either a local extinction (in the area sampled by the core) or a sampling artifact. If we are interested simply in whether the taxon was extant *somewhere,* we can ignore this distinction and interpolate between two appearances in the fossil record. Absences are still a problem, however, at the beginning and end of a stratigraphic sequence. If the taxon disappears before the end and does not reappear, is it truly extinct or not? Would it be found in this end period if our sample were larger? Would it reappear if our record were longer? If the taxon is absent from the earliest part of the sequence, did it originate before or during the time covered by the sample? The shorter the record available, the worse the problem becomes.

Marshall (1994) has addressed this problem for continuous samples, using the distribution of gaps within the record to generate confidence intervals for gaps at the beginning or end. Unfortunately, his method is not appropriate for discrete samples such as the foraminifera counts in the example data set used here. The maximum likelihood model of Clark and Rosenzweig does, however, provide a method for evaluating apparent extinctions. By estimating per time-unit probabilities of disappearance and appearance based on discrete samples, we can quantify the likelihood of an apparent extinction. Consider the foraminiferan species *Reussella moodyensis,* which has the following fossil sequence:

P A A P P A A P A P A A A A A P A A A A A A A A A,

where A stands for absence and P for presence. We would like to know if the species has truly become extinct at the end of this record, or if this might result from inadequate sampling. The simplest procedure is to obtain the maximum likelihood estimators for that part of the sequence that is bounded by presences:

P A A P P A A P A P A A A A A P,

because in this region, we know that absences are not true extinctions. Using the stage-based approach, $\lambda_1 = 0.333$, and $\delta_1 = 0.667$. We can then estimate the likelihood of the final sequence of ten absences as $(1 - 0.333)^9 = 0.026$.

The smaller this likelihood, the more likely it is that the species has really become extinct. We might choose to define truly extinct species as those for which this likelihood is below a certain threshold (say, 0.05). We would then suspect the others—those for which the likelihood is above the threshold—of being Lazarus species. Of the 102 foraminiferan species in our example data set, 60 are not present in the final (most recent) sample. Six of these (including *R. moodyensis*) have a likelihood for their final sequence of absences of less than 0.05, and we may choose to regard them as truly extinct. The remaining 54 species would probably be found at some point after their last recorded disappearances if more samples were analyzed or if the data extended to more recent times (i.e., if the absence is caused by local extinction from the area sampled by the core).

Similarly, 58 foraminiferan species are not present in the first (oldest) sample. Of these, ten show initial sequences of absence with probabilities of less than 0.05. We might conclude that only these species originated during this period, and that the rest were just locally extinct.

This approach could also be used to identify "Elvis taxa" (Erwin and Droser 1993). These are fossils that appear to represent the reemergence of a taxon that was presumed to be extinct, i.e., they appear to be Lazarus taxa. They are, however, different taxa that bear a strong similarity to the extinct form, perhaps because of convergent evolution as they evolved to fill its vacant niche. Using the same procedure as for Lazarus taxa, we can quantify the likelihood that a reemergence is indeed the original taxon (as opposed to an impersonator). Again, we estimate the likelihood of the period of absence before the supposed "return" of the taxon in question. If the likelihood is below a certain threshold, we can consider it probably extinct and investigate the new form more closely as a possible Elvis taxon.

If long data sets are available, we might be able to calibrate the thresholds by comparing predictions made from a short subset of the data with what the longer time series reveals about the existence of Lazarus and Elvis taxa.

The Future of Turnover

At the beginning of this chapter, I suggested that turnover is an underutilized phenomenon, that it has the potential to allow comparisons of biological dynamics across time and space and at different scales. This can happen only if there are standards in measurement and terminology. We need not fix on just one kind of index or one kind of model. Rather, we should be explicit about *pairwise* versus *cumulative* turnover, and about *intracensus* versus *intercensus* turnover.

Later, I suggested that for communities at the combined extremes of the spatial and temporal scales, just two processes are important. These processes are different, but in each case there is one "appearance" and one "disappearance" process. This suggests that any mechanistic models developed to model change at these extremes should look rather similar. Differences will be related to those between colonization and origination, and between local and global extinction. Indeed, sometimes little distinction between these is made (Harrison, chapter 2). Schopf and Ivany (chapter 10) find evidence for such similarity of extremes in their review of empirical patterns in paleobiology. I can envisage a set of related models, tailored for different circumstances but allowing estimation of the same parameters (probably λ_1 and δ_1). We could then make even wider comparisons.

Another cross-scale link that needs to be made is that between the species-by-species approach and the community-level approach. We need to know just how a community-level phenomenon such as turnover is a function of species-level probabilities of appearance and disappearance. I suspect that the next generation of cross-scale models will invoke these distributions, rather than average values.

Finally, the link between two-parameter models of (1) colonization and local extinction and (2) origination and global extinction needs investigating. Could the paleontological model presented by Gilinsky and Good predict turnover? Could the maximum likelihood approach of Clark and Rosenzweig predict changes in the number of species? Could there be an expression that combines the probabilities of repeatable and nonrepeatable events?

Ecology as a "Null Model" for Paleontology

Our understanding of ecological turnover can inform our understanding of paleontological turnover. For example, the nonlinear model of ecological turnover presented in equation 12 predicts that the turnover of breeding species on small islands will approach its maximum, unity, after 1000 to 10,000 years (Russell et al. 1995). If our island data were in the form of fossils, spread over intervals of millions of years, then we would expect to see a turnover of unity even for the shortest available interval, and certainly for the longest. We should be amazed that we find any of the same species in rock samples separated by "geological" intervals of time. Ecological theory thus provides a "null model" for the fossil record. That we do find the same species is partly explained by the sampling bias of the fossil record discussed earlier—we tend to see only the most abundant, widespread, and hence persistent species. Again, ecological theory provides an expectation of paleontological patterns. It makes sense to try to first interpret paleontological data using ecological principles. Only when we find a pattern that ecological theory does not predict need we search for alternative explanations.

Turnover Everywhere

I have also suggested that we can study turnover at any level in the hierarchy of life. The same basic processes—appearances and disappearances—are the foundation of any dynamic set of objects. In principle, we could measure the spatial or temporal turnover of populations across a species' range: of cells in the body of an individual, of proteins in a cell membrane, of carbon dioxide molecules in the sea, or of ozone molecules in the upper atmosphere. Who knows what similarities might emerge? Patterns and theories of turnover are fundamental to our understanding of the natural world.

17

Catastrophic Fluctuations in Nutrient Levels as an Agent of Mass Extinction: Upward Scaling of Ecological Processes?

Ronald E. Martin

Mass extinctions are most parsimoniously explained by purely physical mechanisms (e.g., global temperature change, sea level fall, anoxia). Perhaps this is because from our own "organismic" perspective, it is "difficult for us to view processes at higher [ecosystem, biosphere] levels . . . as emerging out of results of processes at *lower* biotic . . . levels . . ." (Salthe 1985, p. 219; my italics; see also Valentine 1973a). These "ultimate" physical mechanisms have probably varied in relative importance depending on the particular tectonic, paleoceanographic, and paleoclimatic setting during each extinction episode (Cracraft 1992). But significant ("catastrophic") changes in nutrient (and presumably food) concentrations in the photic zone ("proximate cause") would have occurred no matter what the relative importance of each ultimate physical agent. Sluggish rates of ocean circulation would have caused nutrients to be sequestered in deep, anoxic ocean waters, thereby starving marine ecosystems in the photic zone (e.g., much of the Cambro-Devonian and possibly Late Permian; figure 17.1; table 17.1). Global cooling (accompanied by sea-level fall), such as during the Late Ordovician or Late Devonian (e.g., Newell 1967; Stanley 1988), on the other hand, would have increased nutrient input to the photic zone from the deep sea and from land, and it could have generated major instabilities in pelagic and benthic food webs (Margalef 1968; Hallock 1987). A major transgression (rise) in sea level could have caused ecologic instability by bringing anoxic, nutrient-rich waters onto shallow shelves (e.g., Hallam 1989; Wignall and Hallam 1992).

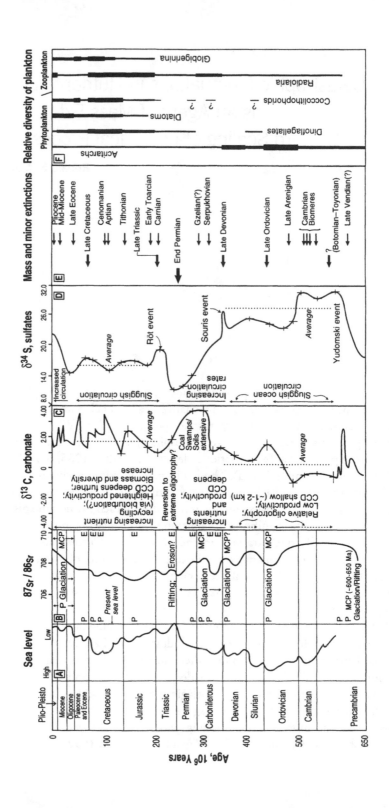

Figure 17.1. (Opposite)   Sea level, stable isotope, and lithologic indices of nutrient fluxes and productivity, extinction episodes, and major changes in plankton assemblages through the Phanerozoic. **(A)** Sea level (Hallam 1992). **(B)** $^{87}Sr/^{86}Sr$ ratios (Holser et al. 1988; Vendian portion based on Kaufman et al. 1993). The greater the $^{87}Sr/^{86}Sr$ ratio, presumably the greater the rate of continental weathering (Raymo 1991) and nutrient input to seas from the continents. The $^{87}Sr/^{86}Sr$ curve is in general agreement with runoff rates (a function of latitudinal position of the continents and relative area of continents and oceans) calculated by Tardy et al. (1989). Deviation between present $^{87}Sr/^{86}Sr$ ratios and calculated runoff rates may reflect tectonics and glaciation (Raymo 1991; Richter et al. 1992) and deforestation and agriculture (Mackenzie and Morse 1992). P, phosphorite peak (Cook and Shergold 1984); MCP episodes (eutrophication episodes) after Worsley et al. (1986); E, additional eutrophication, discussed in text (not formally recognized as MCP episodes by Worsley et al. 1986). **(C)** $\delta^{13}C$ (primarily from Holser et al. 1988; late Precambrian–Early Cambrian from Magaritz et al. 1986; cf. Kaufman et al. 1993; Cretaceous–Recent after Renard 1986). Each datum point (center of cross-hair) represents single analyses aggregated at 25×10^6-year intervals; cross-hairs indicate 1 standard error of the mean for each interval (Holser et al. 1988). Positive values indicate increased marine or terrestrial photosynthesis, whereas negative values indicate decreased productivity or erosion and oxidation of marine or terrestrial C_{org} reservoirs (release of ^{12}C-rich CO_2). General shift toward $\delta^{13}C$ positive values through the Phanerozoic suggests increased marine primary productivity and increased marine C:P burial ratios. Averages (*dotted lines*) fitted visually. **(D)** $\delta^{34}S$ (mainly after Holser et al. 1988; "events" after Claypool et al. 1980). Each datum point (center of cross-hair) represents single analyses aggregated at 25×10^6-year intervals; cross-hairs indicate 1 standard error of the mean for each interval. High average values suggest extensive SO_4^- reduction in widespread anoxic basins (Cambro-Devonian, Mesozoic). Gradual shift toward lower values in late Paleozoic suggests increasing deep-water overturn rates and oxygenation. Pronounced excursions to high values (events) suggest mixing into the photic zone of nutrient-rich anoxic waters previously stored in isolated basins (Claypool et al. 1980). See text for discussion of Neogene $\delta^{34}S$ excursion. Averages fitted visually. **(E)** Mass and minor extinction episodes of the Phanerozoic. Thickness of *arrows* indicates intensity (modified from Sepkoski (1992b; cf. Sepkoski 1986; late Early Cambrian extinction based on Signor 1992, 1994). **(F)** Geologic range and relative diversity of selected plankton groups (from Martin 1995). Shift from phytoplankton characteristic of presumed "superoligotrophic" conditions during the Cambro-Devonian (acritarchs) to taxa characteristic of mesotrophic conditions in the Mesozoic (dinoflagellates) to diatoms (which prefer nutrient-rich waters) in the Neogene suggests secular rise in nutrient availability and productivity through the Phanerozoic (geologic ranges and diversity based on references in Martin 1995; see text for further discussion).

TABLE 17.1. Nutrient levels and biotic turnover through the Phanerozoic

Interval	General nutrient level	Mass extinction[a]	Ultimate cause of extinction	Evidence for nutrient-related perturbations during extinction
Late Precambrian–Devonian	"Superoligotrophic" (but increasing through time as a result of nutrient-related perturbations) *Evidence:* Dominance of acritarchs (cysts); inferred habitats of plankton; cryptic habitats and low metabolic rates of modern representatives of fossil benthos; shallow bioturbation; increasing tiering of epifaunal suspension feeders	Late Precambrian–Early Cambrian	Glaciation/Global cooling?	Positive δ^{13}C shifts; MCP episode(s)
		Late Ordovician	Glaciation/Global cooling; sea-level fall?	Positive δ^{13}C shifts; MCP episode; shift to heavy Sr isotope ratios; decline and recovery of acritarchs
		Late Devonian	Glaciation/Global cooling? Anoxia?/Sea-level fall?	Positive δ^{13}C shifts; MCP episode; shift to heavy Sr isotope ratios; final decline of acritarchs after Devonian

Permo-Carboniferous	"Submesotrophic" *Evidence:* Near absence of acritarchs; alga-dominated reefs; increasing epifaunal tiering and depth of bioturbation; increased durophagous predation; further phosphogenic episodes	Late Permian	"Murder on the Orient Express" (sea-level fall and shelf destruction through assembly of Pangea; rapid global warming; oxidation of shelfal organic reservoirs and extrusion of Siberian Traps; anoxia?)	Gradual negative shift in $\delta^{13}C$ followed by strong negative $\delta^{13}C$ shift near end of Permian; possible decline in epifaunal tiering; increased deposit feeding; rise of gymnosperms (reversion to "superoligotrophy"?)
Mesozoic–Cenozoic	"Mesotrophic" (Mesozoic) increasing to "Eutrophic" (Cenozoic) *Evidence:* Rise of dinoflagellates and other plankton; increased predation and depth of bioturbation; rise of angiosperms in Cretaceous; eutrophication and MCP episodes	Late Triassic Late Cretaceous	Impact?/Anoxia? Cooling?/Sea-level fall?/Impact	Positive $\delta^{13}C$ shift; shift to heavy Sr ratios Ditto (followed by strong negative $\delta^{13}C$ shift presumably related to impact)

*See figure 17.1 for minor extinctions.
MCP, marine carbon-to-phosphorus ratio.

Micropaleontologists were among the first to suggest fluctuations in trophic resources (nutrients, food) as an agent of extinction. Based on the work of Bramlette (1965), Tappan (1968, 1971, 1982, 1986; see also Tappan and Loeblich 1971, 1973) emphasized sequestration of nutrients in terrestrial organic carbon (C_{org}) reservoirs (e.g., coal swamps) and in the deep sea as a result of decreased deep-ocean turnover rates [the nutrient reduction hypothesis of Erwin (1993a)]. As a result, marine primary productivity and atmospheric pO_2 declined, and marine food webs collapsed (e.g., loss of suspension feeders), during the end-Devonian, end-Permian, and end-Cretaceous. Unfortunately, many of Tappan's assertions were based on less-detailed stable isotope curves than are now available (see Fig. 17.1), and her interpretations conflicted with the fossil record (Erwin 1993a, pp. 228–230; Martin 1995, 1996) because she based her conclusions on a strict analogy to ecologic succession (Odum 1969). Consequently, her views have been treated in recent accounts of mass extinction as being largely of historical interest only or they have been ignored altogether. Nevertheless, a number of workers have recently interpreted shifts in $\delta^{13}C$ toward negative (lighter) values that often *follow* the main phase of extinction (whatever the ultimate cause) to indicate decreased rates of marine primary production *during* the extinction (e.g., Hsü 1986; Magaritz 1989; Zachos et al. 1989; Paul and Mitchell 1994; Wang et al. 1994; see also Vermeij 1987a, 1989, p. 346, 1994; Hansen et al. 1993).

But significant turnover in microfossil and macrofossil assemblages and increased amounts of phosphorus-rich sediments frequently occur in association with *enhanced* marine photosynthesis (positive $\delta^{13}C$ shifts) *prior* to final ecosystem collapse, all of which suggest elevated nutrient levels and food availability *during* extinction (Martin 1995, 1996). Extinction may, then, have been exacerbated from either too much *or* too little food, and perhaps also changes in the quality of food available. Indeed, Hallock (1987) defined a *trophic resource continuum* (TRC), the expansion and contraction of which provides a mechanism for biotic diversification and extinction, respectively. The TRC reflects the relative proportions of oligotrophic (nutrient-poor) and eutrophic (nutrient-rich) conditions through the Phanerozoic. According to the TRC hypothesis, reduced rates of ocean turnover produce regionally intensified oligotrophy and eutrophy (i.e., expansion of the TRC); lower nutrient (food) supplies, in turn, result in longer food chains (the "polytaxy" of Fischer and Arthur 1977; Valentine 1973a; Lipps 1979), slower-growing organisms, and relatively stable population sizes (Margalef 1968, 1971; Hallock 1987). Conversely, increased ocean turnover rates cause overall nutrient availability to increase and the TRC to shrink. Food chains of such high productivity regimes are short (Margalef 1968, 1971) and diversity low (the "oligotaxy" of Fischer and Arthur 1977); increased nutrient levels presumably allow fast-growing opportunistic species to outcompete slower-reproducing ones (Hallock 1987). Much of the resultant primary production is presumably used inefficiently compared to nutrient-poor, diverse (and presumably efficient) ecosystems, which destabilizes nutrient-rich ecosystems and causes large swings in population size (Margalef 1968).

The TRC is based on the effects of nutrient-related perturbations detected on ecological spatiotemporal scales by long-term studies of ecosystems. Moreover, the relationship between biodiversity, ecosystem dynamics, and nutrient cycling is receiving increasing scrutiny on ecological spatiotemporal scales (e.g., Rosenzweig and Abramsky 1993; Abrams and Roth 1994a,b; Hastings and Higgins 1994; Naeem et al. 1994). Based on carbon isotope, lithologic (phosphorite), and fossil evidence, I suggest that large-scale perturbations in nutrient-related phenomena scale upward from ecological phenomena to much longer, and much larger, geological (evolutionary) scales. I also explore how nutrient perturbations may scale upward through their effects on reproductive strategies.

Nutrients and Extinction Through the Phanerozoic

The Cambro-Devonian

Stable Isotopes

Surface nutrient levels may have been lower during much of the Cambro-Devonian than those of oligotrophic waters today (e.g., oceanic gyres), and therefore unable to sustain large phytoplankton populations (what I have termed superoligotrophic conditions; Martin 1995, 1996). During the latest Precambrian and early-to-mid Paleozoic, ice caps occurred only in the Late Ordovician and perhaps late Precambrian to Early Cambrian, and Late Devonian (Frakes et al. 1992). Thus, ocean circulation during much of this time was presumably relatively sluggish (Fischer and Arthur 1977; Berry and Wilde 1978), as suggested by high values of the $\delta^{34}S$ curve (the ratio of heavy ^{34}S to light ^{32}S isotopes) and extensive carbon-rich (black) shale deposition (see figure 17.1; table 17.1). Most nutrients would have been sequestered below the photic zone (when glaciers were absent) or trapped near shore by high sea level (Holser et al. 1988; see Martin 1996, for an alternative interpretation). Low rates of marine photosynthesis and C_{org} burial during much of the Cambro-Devonian are also suggested by relatively light (negative) ratios of ^{13}C to ^{12}C in marine carbonates ($\delta^{13}C$; see figure 17.1). Berner and Raiswell (1983) calculated low rates of C_{org} burial (relative to modern values) for much of the Cambro-Devonian. Moreover, continental weathering rates were probably too low until the Devonian (Knoll and James 1987) to supply significant levels of nutrients to the photic zone.

The *overall* low productivity conditions during the Cambro-Devonian were punctuated, however, by relatively short intervals of heightened nutrient availability. Positive $\delta^{13}C$ shifts in the late Precambrian (Vendian) through Early Cambrian, apparently in conjunction with glaciation, suggest increased nutrient availability during this time (Brasier 1989a,b; 1992a,b). A pronounced excursion of $\delta^{34}S$ to high values in the late Precambrian (Yudomski "event"; see figure 17.1) also suggests mixing into the photic zone of nutrient-rich anoxic waters that may have been stored in isolated basins or deep in the oceans. Significant biotic turnover also occurred during the late Precambrian (Brasier 1989a; McMenamin and McMenamin 1990, pp. 128–135, 148–160). The broad Late Cambrian $\delta^{13}C$ shift (see figure 17.1) occurred in conjunction with a major transgression, which may have stimulated marine photosynthesis by releasing nutrients trapped in shelfal sediments (Broecker 1982; Compton et al. 1993). Positive shifts of the $\delta^{13}C$ curve in the Late Ordovician and Late Devonian are associated with positive excursions in the $\delta^{18}O$ curve (Orth et al. 1986; Popp et al. 1986), glaciation (positive or heavy oxygen isotope ratios suggest the presence of extensive ice), and sea-level fall (Frakes et al. 1992). Another positive $\delta^{34}S$ shift (Souris "event") occurred in the Late Devonian (see figure 17.1). The relatively small Late Devonian $\delta^{13}C$ shift (as opposed to the strong Late Ordovician rise) shown in figure 17.1 may have resulted from terrestrial C_{org} input (soils were well developed by this time) during sea-level fall (Algeo et al. 1995) or less well developed glaciers (Frakes et al. 1992) and slower deep-ocean turnover rates than in the Late Ordovician. Joachimski and Buggisch (1993) report much larger carbon isotope shifts from sections in central Europe.

Strontium isotope ratios ($^{87}Sr/^{86}Sr$) of marine carbonates hint that episodic influxes of nutrients from land also contributed to significant changes in $\delta^{13}C$ through the Cambro-Devonian. Increased $^{87}Sr/^{86}Sr$ ratios in the late Precambrian, Late Ordovician, and Late Devonian (see figure 17.1) suggest increased runoff (nutrient input) from the continents as a result of orogeny, glaciation, and sea-level fall (e.g., Raymo 1991). Increased weathering rates no doubt enhanced nutrient flux from the continents by the Late Devonian (Knoll and James 1987; Algeo et al. 1995).

Lithologic Evidence

Lithologic evidence supports the hypothesis of episodic eutrophication during the Cambro-Devonian. Worsley et al. (1986) recognized a series of step-like increases in marine carbon-to-phosphorus (MCP) burial ratios that presumably reflect increased nutrient availability and permanently enhanced marine productivity (C_{org} burial rates) during the Phanerozoic (see figure 17.1, table 17.1). According to this scenario, extensive phosphorite deposition during MCP episodes resulted from intensified recycling of phosphorus (and other nutrients such as nitrogen) back to the photic zone as a result of glaciation, better oxygenation of shallow waters, and enhanced rates of bioturbation (phosphorus scavenging; Worsley et al. 1986; see also Thayer 1983).

According to these authors, the first MCP episode of the Phanerozoic began in the late Precambrian (about 600 to 650 Ma; see figure 17.1). This episode is associated with possible glaciation, mixing of nutrient-rich (presumably anoxic) waters into the photic zone (Yudomski $\delta^{34}S$ "event"), and the pronounced positive excursions in $\delta^{13}C$ described previously (see figure 17.1). A second MCP episode (again in association with a positive $\delta^{13}C$ excursion) occurred in the Late Ordovician. Worsley et al. (1986) do not indicate an MCP episode in the Late Devonian, but there is a small rise in the frequency of phosphorites that approximately corresponds to the carbon and sulfur isotope excursions noted previously.

Fossil Evidence

Microfossil and other biotic indices support these interpretations. Low *overall* phytoplankton densities during the Cambro-Devonian are suggested (counterintuitively) by the record of acritarch diversity (Martin 1995, 1996). Acritarchs were the dominant phytoplankton of Cambro-Devonian seas (Tappan 1980), and although of uncertain affinities, they are normally considered to be cysts of marine eukaryotic unicellular algae resistant to inimical conditions (Tappan 1980). Many workers consider high acritarch diversity in ancient rocks to indicate nutrient-rich conditions and high productivity (e.g., Tappan 1968, 1971, 1980, 1982, 1986), but modern plankton diversity is lowest in nutrient-rich regimes and highest in oligotrophic waters. Thus, *high* cyst diversity prior to the Carboniferous (see figure 17.1) may reflect primarily superoligotrophic conditions rather than nutrient-rich conditions.

Fossil zooplankton also suggest adaptation to overall low phytoplankton densities during much of the Cambro-Devonian (Martin 1995, 1996). Graptolites, for example, appear to have lived in low-oxygen, nutrient-rich waters just below the photic zone, and they may have migrated upward to feed on occasional phytoplankton blooms caused by intrusions of deeper nutrient-rich waters into the photic zone (Berry et al. 1987). Conodonts may have lived similarly, as they also exhibited depth-stratification (Clark 1987). Radiolarians apparently survived by living either in highly productive shallow waters, in oligotrophic oceanic gyres with symbiotic algae, or in deeper subphotic layers of the open ocean as detritivores and bacterivores (Casey 1993).

Intervals of relatively low acritarch diversity during the Cambrian and Silurian presumably reflect elevated nutrient levels following the late Precambrian and Late Ordovician eutrophication episodes, respectively (i.e., abundant nutrients precluded cyst formation; see figure 17.1). Through the Cambrian, following the late Precambrian eutrophication episode, levels of dissolved nutrients presumably declined as they were progressively incorporated into plankton (and other) biomass along food chains. As nutrients were progressively sequestered into biomass, cyst diversity rose to a peak in the Ordovician (with a return to low levels of dissolved nutrients). A similar decrease and increase in acritarch diversity followed the Late Ordovician MCP episode in the Silurian and Devonian, respectively (see figure 17.1).

Studies of fossil macrobenthos also suggest that food availability was relatively low, but increasing, during much of the Cambro-Devonian. Suspension-feeders were poorly represented before the Late Cambrian and may have fed predominantly on bacteria-size or smaller food particles; after the Late Cambrian and Early Ordovician, suspension-feeding started to become more prevalent, suggesting increasing food supplies in the water column (Signor and Vermeij 1994). Based on experiments with modern brachiopods, which today characterize oligotrophic habitats, Paleozoic brachiopods had very low food (energy) requirements and were able to survive intervals of starvation of 2 years or more (Rhodes and Thayer 1991). Other relict Paleozoic taxa appear to have low metabolic rates and survive today in oligotrophic refugia (Vermeij 1987a, 1994; Rhodes and Thayer 1991); food supplies are presumably too low in such refugia to sustain the high metabolic levels of competitors and predators that arose much later (Thayer 1992; Bambach 1993; Signor and Vermeij 1994).

The Permo-Carboniferous

The trend of increasing nutrient levels and marine productivity continued in the Permo-Carboniferous with the spread of terrestrial forests, as suggested by the broad rise in $^{87}Sr/^{86}Sr$ ratios and another MCP episode (see figure 17.1; the "submesotrophic" conditions of Martin 1995, 1996). Increased rates of terrestrial photosynthesis and continental weathering may have drawn down atmospheric CO_2, thereby contributing to southern hemisphere glaciation, sea-level fall, and enhanced ocean turnover rates, as suggested by the trend toward lower $\delta^{34}S$ values during the Permo-Carboniferous (increased oxygenation of deep waters; see figure 17.1).

The marine fossil record also indicates heightened productivity. After the

Devonian, acritarchs are only a relatively minor component of the microfossil record (see figure 17.1), which suggests permanently elevated nutrient levels (cf. Pitrat 1970, and Tappan 1970). Other biotic evidence suggesting higher nutrient levels and plankton densities after the Devonian includes the decline of the Fusulinacea (Tappan and Loeblich 1988) and the prevalence of calcareous algae (Hallock 1987). Increased food supplies high in the water column and deep in sediment (detritus) are also suggested by extensive tiering in epifaunal suspension-feeding communities and increasing depth of bioturbation (Bottjer and Ausich 1986; Sepkoski et al. 1991; Bambach 1993; Bottjer and Droser 1994). Moreover, increased durophagous (shell-crushing) predation beginning in the Late Devonian hints at lengthening food chains and increasing metabolic rates (Signor and Brett 1984; Vermeij 1987a; Bambach 1993; Signor and Vermeij 1994; Martin 1995, 1996).

The secular trend of rising nutrient and food availability through the Paleozoic appears to have been ended by a reversion of the oceans toward a superoligotrophic state. The strong negative $\delta^{13}C$ shift in the latest Permian (see figure 17.1) was interpreted by Holser et al. (1991) as indicating erosion and oxidation of shelfal C_{org} reservoirs that culminated, at the Permo-Triassic boundary, in well-developed anoxic waters. Anoxia, by itself, does not, however, appear to have caused the end-Permian extinctions, because extinctions occurred before the spread of anoxic waters (Erwin 1993a, pp. 242–248). Instead, rapid global warming, caused perhaps by the extrusion of huge volumes of basaltic lava in Siberia ("Siberian Traps"; Erwin 1993a; Renne et al. 1995), and perhaps also salinity stratification of the oceans (Holser et al. 1991), may have lowered deep-water overturn rates, so that dissolved nutrients were sequestered below the photic zone as anoxia was developing (see also Małkowski et al. 1989).[1] Indeed, only four out of 11 source rocks correspond to upwelling in the Late Permian, whereas three out of four organic-rich rocks are related to upwelling in the Early Permian (Parrish 1987).

As the negative $\delta^{13}C$ shift declined during this time, so did the $^{87}Sr/^{86}Sr$ ratios. But sea-level fall (see figure 17.1) and erosion of continents should, seemingly, have added nutrients to the oceans, increased $^{87}Sr/^{86}Sr$ ratios, and stimulated marine photosynthesis. There are several possible explanations for this dilemma. First, the Permian was quite arid, and interior drainage may have been extensive (e.g, Parrish 1993). Second, any strontium delivered to the oceans may have been concentrated in the light isotope because of erosion of the basaltic Siberian Traps (Holser and Magaritz 1992; Renne et al. 1995). Third, during the Permian, gymnosperm forests were replacing the

---

1. Martin (1997) discuss the recent hypothesis of Knoll et al. (1996) regarding the cause of end-Permian extinctions; see also discussion of Knoll et al.'s paper in *Science* 274:1549–1552.

lycopsids of the Carboniferous; evergreen leaf litter releases nutrients slowly (Tappan 1986; Knoll and James 1987) and may have retarded weathering of continental rocks.

Changes in the marine benthos are consistent with lowered productivity. If suspended food supplies decreased during the Permian, deposit-feeding would have become increasingly advantageous (cf. Sheehan and Hansen 1986; Jablonski and Raup 1995); indeed, extinction rates for brachiopods exceeded those of deposit-feeding bivalves. During the Late Permian, there appears to have been a decline in the height of epifaunal suspension-feeding communities (Bottjer and Ausich 1986; Thayer 1983), which may have been augmented by biological "bulldozing" of suspension feeders (Thayer 1983). Reversion to superoligotrophy might also account for the possible selection against taxa with planktotrophic larvae during end-Permian extinctions (Valentine and Jablonski 1986).

The Meso-Cenozoic

Carbon isotope values started to rise again in the Triassic, suggesting enhanced marine primary production as a result of the diversification of plankton (see figure 17.1). The strontium isotope curve exhibits a sharp rise across the Permo-Triassic boundary and closely approaches the average Phanerozoic value (see figure 17.1). Despite the overall aridity of the Permo-Triassic, Tardy et al. (1989) indicate a slight rise in continental runoff across the Permo-Triassic boundary, and Holser and Magaritz (1992) suggested increased erosion at this time. Besides erosion, the rise in the strontium curve across the Permo-Triassic boundary is also consistent with upwelling in incipient seaways (associated with rifting of Pangea) or overturn or release of anoxic marine waters in which terrestrially derived nutrients may have accumulated (note sharp $\delta^{34}S$ spike or Röt "event"). A slight rise in the strontium and carbon isotope curves also occurs in the Late Triassic (see figure 17.1).

During the rest of the Mesozoic, the $\delta^{13}C$ displays a series of sharp positive excursions that correspond to intervals of heightened C_{org} burial [note $\delta^{13}C$ spikes or oceanic anoxic events (OAEs); see figure 17.1]. Despite relatively sluggish ocean circulation during the Mesozoic (note $\delta^{34}S$ average; Fischer and Arthur 1977; Frakes et al. 1992), enhanced circulation or sea-level transgression may have stimulated dramatic rises in marine productivity at times (e.g., Föllmi et al. 1993; see Martin 1995 for further references). Indeed, extensive phosphorites sometimes occur in the vicinity of OAEs (see figure 17.1). Nutrient cycling on shelves may also have accelerated in response to rising bioturbation rates through the Mesozoic (Bambach 1993; Signor and Vermeij 1994; Martin 1995, 1996).

Discussions of end-Cretaceous extinctions have emphasized the sudden negative shift in $\delta^{13}C$ and a rapid reduction in marine photosynthesis ("Strangelove ocean") as evidence for an impact (e.g., Hsü 1986; Zachos et al. 1989). Nevertheless, significant climate change was occurring prior to the end of the Cretaceous. Based on a positive shift in $\delta^{13}C$ during the early–late Maastrichtian transition, Barrera (1994) concluded that marine productivity was increasing in response to global cooling and enhanced oceanic overturn. Sea level was also falling and $^{87}Sr/^{86}Sr$ ratios (and presumably nutrient input) increasing during this time (see figure 17.1; cf. Bramlette 1965). Moreover, angiosperms were expanding in the Cretaceous, which may have increased nutrient fluxes to shallow seaways since angiosperm leaf litter decays more rapidly than gymnosperm litter (Tappan 1986; Knoll and James 1987; Vermeij 1987a).

Strong positive excursions in $\delta^{13}C$ occurred again in the Cenozoic, beginning near the Eo-Oligocene boundary. These shifts reflect increased marine productivity as a result of formation of polar ice caps, enhanced deep-water overturn, and enhanced continental erosion (through sea-level fall) and nutrient input from land [note rise in $^{87}Sr/^{86}Sr$ curve (figure 17.1); see also Hallock et al. 1991; Brasier 1995]. Also beginning about this time was another series of MCP boosts (see figure 17.1). The $\delta^{34}S$ shift to higher values in the Neogene (see figure 17.1), despite *increased* deep-water overturn and presumed oxygenation, suggests that increased C_{org} rain rates combined with falling sea-level and sediment influx (rapid burial) to make C_{org} increasingly available for sulfate (SO_4^-) reduction (cf. interpretation of Cambro-Devonian $\delta^{34}S$ values).

The diversification of marine plankton beginning in the Mesozoic has been attributed to sea-level rise and the resultant increase in water column stratification and habitat availability (Lipps 1970). Among the predominant groups of Mesozoic plankton are the dinoflagellates, which are often preserved as cysts, and the modern representatives of which tend to prefer mesotrophic (intermediate) nutrient levels (Kilham and Kilham 1980). Diversification of dinoflagellate cysts, unlike that of acritarchs, during the Mesozoic may signal *heightened* nutrient levels [although not as high as today's (Bralower and Thierstein 1984)] that may have fueled the rise in plankton diversity as pelagic habitats reappeared. Diatoms, which today prefer nutrient-rich ("eutrophic") conditions (Kilham and Kilham 1980), diversified explosively in the Miocene (Tappan 1980) in response to further eutrophication. Rates of bioturbation, predation, and (apparently) metabolism continued to rise through the Cenozoic, presumably in response to increasing surface productivity (e.g., Vermeij 1987a; Bambach 1993; Signor and Vermeij 1994).

The rise in productivity in the Meso-Cenozoic was accompanied by a dramatic rise in abundance and diversity of the bivalve-rich Modern fauna (Bambach 1993). Modern bivalves require much higher food levels than do modern brachiopods (Rhodes and Thayer 1991; Rhodes and Thompson 1993; Bambach 1993). The diversification of the Modern fauna began in Cambro-Ordovician near-shore environments, where nutrients were presumably concentrated, after which it radiated into progressively deeper habitats, presumably in response to increasing nutrient (food) availability away from shore (Bambach 1993). The locus of siliceous ooze deposition also shifted across the shelf during the Phanerozoic, with radiolarians eventually giving way to diatoms (see figure 17.1) (Maliva et al. 1989; see Martin 1995, 1996, for further discussion). Calcareous plankton appear to have followed a similar path, which may account for enigmatic occurrences of calcareous nannofossil-like objects from the Paleozoic (see figure 17.1) (Martin 1995, 1996). Rising food levels in the Mesozoic may have also permitted reefbuilding taxa to sustain growth rates sufficient to build the multiserial skeletons necessary for extensive photosymbiosis (Wood 1993), and they may have fueled the diversification of antifouling predators and grazers that presumably contributed to the success of reef ecosystems (Vermeij 1987a; Wood 1993; Martin 1995, 1996).

Implications for Mass and Minor Extinctions

Mass Extinctions

Marked changes in carbon isotope, lithologic, and microfossil indices during the Phanerozoic support the hypothesis of catastrophic fluctuations in nutrient and food levels as a proximate agent of mass extinction. Since Tappan's (1968, 1971, 1982, 1986) work, mass extinction scenarios have typically emphasized a rapid *decrease* in rates of marine photosynthesis during extinction. The negative $\delta^{13}C$ shift is normally preceded by a significant positive shift (see figure 17.1), however, which suggests heightened productivity (and ecosystem instability) *during* collapse of marine food chains; negative $\delta^{13}C$ shifts may represent only the final phase of collapse of marine food webs. Moreover, unlike extinction caused by physical change alone, relatively rapid "rediversification following a change to a fluctuating resource regime is simply not possible, for the extinctions bring diversity into approximate equilibrium with the regime, and restabilization is required to permit another diversity increase" (Valentine 1973a, p. 399).

Interestingly, soft- and hard-bottom (reef) communities differed in their response to massive eutrophication (Valentine 1973a, p. 399). Frame-building reef ecosystems are characterized by highly stenotopic taxa (Margalef 1971; Hallam 1992) and react to environmental stresses relatively quickly (Jablonski 1991; Kauffman and Fagerstrom 1993; cf. Johnson et al. 1995). Reefs are also presumably characterized by a high degree of "connectance" (Boucot 1983, 1990b; see also Pimm 1984a,b), and perturbations (including those produced by marked increases in nutrient fluxes) no doubt cascade through them rapidly (Plotnick and McKinney 1993), permitting invasions of opportunistic species (Hallock and Schlager 1986; Hallock 1987, 1988a).

On the other hand, level- (or soft-) bottom communities were not as severely affected by mass extinction (Sheehan 1985). Level-bottom communities are presumably subject to more frequent disturbance than reef communities and prone to lower diversity than reefs (Margalef 1968, see also May 1972). According to this view, soft-bottom communities would maintain a lower level of integration (connectance) than reef communities and therefore be less susceptible to extinction. Nevertheless, if eutrophic conditions were to become sufficiently pronounced, as they apparently did in the geologic past, then nutrient-related perturbations could contribute to the widespread simultaneous extinction of both hard- *and* soft-bottom communities (cf. Sheehan 1985).

Regardless of the ultimate extinction mechanism, following mass extinction, a period of recovery ensued in both hard- and soft-bottom communities, estimated by Sheehan (1985) to have lasted at least several million years (the "macroevolutionary lags" of Jablonski and Bottjer 1990b). It is during the postextinction recovery that a disturbed thermodynamically open system, such as a disturbed ecosystem, would begin to evolve toward a "mature" state. In so doing, energy flow tends to decrease (i.e., it is used more efficiently; cf. previous discussion of TRC), and the connectance ("complicatedness") of the system increases [decreased entropy and increased entification and emergence of new "levels" such as species (Salthe 1985)]; as the system evolves, it channels more energy into maintaining its own internal stability. As the system becomes more highly connected, however, it also becomes increasingly susceptible to disturbance (Prigogine 1980; Salthe 1985; Kauffman and Johnsen 1991; see also McKinney 1989; Stanley 1990b). Indeed, during each lag, marine communities were highly unstable and underwent rapid turnover (Sheehan 1975, 1982, 1985), which may have delayed the reappearance of reef communities (relative to soft-bottom ones) because incipient reef-builders would have been quickly eliminated by fluctuations in the resource regime and by accompanying outbursts of opportunistic species (e.g., Hallock 1987, 1988a). Similar fluctuations may explain apparent blooms of Cretaceous plankton during the Danian (Hollander et al. 1993) that have been attributed to upward reworking from below the Cretaceous–

Tertiary boundary (cf. Smit 1982; Canudo et al. 1991; see also MacLeod and Keller 1994).

Minor Extinctions

I have suggested that nutrient input occurred repeatedly but in a catastrophic manner. Filippelli and Delaney (1992), on the other hand, concluded that phosphorus accumulation rates in ancient phosphorite deposits and modern environments are comparable; ancient fluxes were said to fall within the range of fluxes off the modern Peru margin. If modern and ancient phosphorus fluxes are indeed comparable, then nutrient fluxes to the photic zone, and corresponding increases in productivity, biomass, and diversity of the marine biosphere, must have occurred much more frequently through the Phanerozoic than is indicated by MCP episodes alone. In other words, MCP episodes are detectable in the geologic record because of suitable tectonic and paleoceanographic circumstances (e.g., Compton et al. 1993; Ruttenberg and Berner 1993), and they are the most extreme versions of much more numerous phosphogenic intervals. If this is the case, nutrient input to the oceans, and the consequent disturbance, may have been much more frequent than I have described here.

Thus nutrient-related disturbance may be scale independent (cf. Reice 1994); i.e., a continuum appears to exist between background and mass extinctions (Raup 1986, 1991a, 1994; Jablonski 1989, 1991; Boyajian 1986, 1988, 1991; McKinney 1995; see also Allen and Starr 1982). Nutrient perturbations might account, then, not only for mass extinctions, but also for minor biotic turnovers such as (1) the Late Cambrian–Early Ordovician extinctions, via transgression and release of nutrients trapped in sediment (Broecker 1982; Compton et al. 1993; Loch et al. 1993); turnover of trilobites during this interval appears to represent "a difference in degree rather than in kind" (Fortey 1989, p. 334); (2) turnover of planktic graptolite communities in the early-to-mid Silurian (Wenlock; Wilde and Berry 1984; Melchin 1994) in association with increased rates of wind-driven upwelling and marine productivity (Moore et al. 1993); and (3) the frequent extinction–radiation cycles of ammonites, which appear to be associated with regressive–transgressive couplets and turnover of anoxic water (House 1989) (see also figure 17.1). Although many extinction scenarios emphasize global cooling, overturn of nutrient-rich waters caused by *warming* and destabilization of the water column (Wilde and Berry 1984) may have contributed to minor or background extinctions when glaciers were either poorly developed or absent (e.g., most of the Cambro-Ordovician, mid-Silurian, most of the Mesozoic; Frakes et al. 1992).

Upward Scaling of Eutrophication and Extinction

Observations on Approximately 1- to 10-Year Scales

If background, minor, and mass extinction do indeed form a continuum, then nutrient perturbations, and their effect on trophic resource and ecosystem stability, ought to scale upward from ecologic to geologic spatiotemporal scales. Upward scaling of ecological processes is, of course, one of the principal limitations to interpreting ancient fossil assemblages (e.g., Pimm 1991; Harrison, chapter 2; Maurer and Nott, chapter 3; cf. Aronson 1994). Processes of ecologic succession, for example, are not necessarily equivalent to those of community evolution (Miller 1986; cf. Copper 1988), and such an approach must be used with caution, as it is potentially misleading (e.g., Karlson and Hurd 1993). Indeed, based on a strict analogy of ecologic succession proposed by others, Tappan (1971) described a sequence of Phanerozoic communities that was at odds with the fossil record (see Martin 1995, 1996, for further discussion).

Nevertheless, various authors have argued strongly that ecological processes scale upward to those that act on evolutionary scales (Vermeij 1987a, 1994; Jackson 1988; Sepkoski 1992b; Aronson 1994; Lawton et al. 1994; McKinney and Allmon 1995; Aronson and Plotnick, chapter 18). Abrams and Roth (1994a,b) concluded, for example, that in model food chains consisting of three species (trophic levels), nutrient enrichment may adversely affect high-level consumers (see also DeAngelis 1992). They found that if density dependence is strong enough, disturbance causes sudden changes in dynamics [e.g., from cycles to apparent chaos and back again (Hastings and Higgins 1994)], and the time required to reach final population dynamics approaches thousands of generations [tens of thousands of years (i.e., geologic scales)].

How, then, could disturbance scale upward? In an attempt to construct a plausible scenario for upward scaling of eutrophication-related phenomena, and their role in extinction on geological scales, I draw upon the coral reef literature, bearing in mind the caveats just discussed, and the less-than-satisfactory state of knowledge regarding the taxonomy and life histories of many modern coral species (e.g., Wallace and Willis 1994; Johnson et al. 1995).

In fringing reefs located on the west side of Barbados (West Indies), Tomascik and Sander (1985; 1987a,b) found that growth rates of the common frame-building coral *Montastrea annularis* (Ellis and Solander) increased with rising concentrations of nutrients up to a maximum, and then decreased in response to reduced levels of light and photosymbiosis (see also Rosenzweig and Abramsky 1993). They (1987a) concluded that anthropogenic eutrophication (sewage) stresses Barbados reefs via its effect on feeding strategies (see also Reice 1994, pp. 429–431), and that *Porites astreoides* Lamarck, *P. porites* (Pallas), *Siderastrea radians* (Pallas), and *Agaricia agaricites* (Linnaeus) were the most abundant species on polluted reefs because they are all less susceptible to fouling. Elevated nutrient levels cause increased density of plankton populations and turbidity, and they allow the invasion of bioeroders and opportunistic species that overwhelm reefs (e.g., Highsmith 1980; Birkeland 1987; Hallock 1987, 1988a). Hermatypic reef corals adapted to low light levels (e.g., *Agaricia*) may also move upward into shallower waters under these conditions (Acevedo and Morelock 1988).

Similar processes appear to occur naturally on regional scales. Margalef (1968, 1971) noted the segregation of well-developed hermatypic coral reefs and highly productive pelagic ecosystems in the Caribbean and Pacific (cf. Liddell and Ohlhorst 1988). Wilkinson (1987) suggested that nutrient levels in the Caribbean are higher than in the Indo-West Pacific, based on the much greater sponge biomass (suspension feeders) of West Indian reefs. Wilkinson and Cheshire (1988) noted similar phenomena across the shelf of the Great Barrier Reef, with "heterotrophic communities" dominating near shore and "phototrophic communities" dominating offshore. Hallock (1988b) and Brasier and Green (1993) came to similar conclusions based on foraminiferal assemblages, which are dominated by more eurytopic species in the Caribbean (Martin 1986). Higher nutrient levels in the Caribbean may result from localized upwelling (Margalef 1971) and river input to a region that is much more landlocked than the Indo-West Pacific (Highsmith 1980; Birkeland 1982; maps in Berger 1989).

Elevated nutrient levels and pelagic productivity also appear to affect life history traits. *Porites astreoides, P. porites, Siderastrea radians,* and *Agaricia agaricites* all reproduce by brooding (planulating) larvae to an advanced stage before release (Szmant 1986; Richmond and Hunter 1990). Indeed, sexual recruits in the Caribbean are dominated by *Agaricia* and *Porites* (Sammarco 1985, 1987; see also Bak and Engel 1979; Neigel and Avise 1983; Rylaarsdam 1983; Jackson and Hughes 1985; Birkeland 1977, 1987). Such a reproductive strategy appears to correlate with small, opportunistic species capable of survival in stressed environments (i.e., eurytopy) (Szmant 1986; Edinger and Risk 1995; cf. Johnson et al. 1995). Although larvae of brooding coral species can be carried long distances (Richmond 1989; Richmond and Hunter 1990), recruitment success appears to be more important than dispersal (Edinger and Risk 1995). This pattern seems to hold for other marine invertebrates except molluscs, which mostly produce planktotrophic larvae ("broadcasters"), a trait that appears to confer extinction resistance by increasing dispersal range (Jablonski 1986a,b; 1989; Edinger and Risk 1995).

In contrast to the Caribbean, in the Indo-West Pacific, *Porites* broadcasts gametes into the sea, as do most other hermatypic coral genera there; like *Porites*, however, other genera are often brooders in the Caribbean (Richmond and Hunter 1990). Similarly, in the large, branching, frame-building genus *Acropora*, newly broadcasted spat often represent 50 to 80 percent of juveniles in the Indo-West Pacific (Sammarco 1987), but *Acropora* in the Caribbean is uncommon among juveniles and typically reproduces by asexual fragmentation and recementation of branches. Asexual reproduction is prominent in coral populations living near their physiological limits (Richmond and Hunter 1990; see also Scott et al. 1988), and some coral populations have been reported to be sterile at their geographic limits (Richmond and Hunter 1990).

Geological Spatiotemporal Scales

Alternative community states that are indicative of disturbance have been described for modern reef habitats (e.g., Knowlton et al. 1990; Knowlton 1992; see also May 1977; Hughes 1989; Done 1992). These alternative states may persist for a decade or more, but based on the fossil record of Caribbean Pleistocene reefs, Jackson (1992) concluded that *Acropora*-dominated reef communities are typical of the Caribbean in the Pleistocene, whereas other community types are less evident; i.e., the "domains of attraction" are different for alternative community states (Jackson 1992, p. 725; see also Aronson and Precht 1997). Nevertheless, if regional or global environmental conditions changed sufficiently, such disturbed alternate community types might persist for sufficiently long times or be sufficiently widespread to be the norm and to be preserved in the fossil record.

"Environmental limits" and "biotic potential" can be envisioned as placing upper and lower boundaries, respectively, on a "constraint envelope" (analogous to a niche hypervolume) in which complex ecological systems (ecosystems, communities) operate over broad spatiotemporal scales (O'Neill et al. 1989). In the present case, physical factors such as temperature and nutrients are assumed to form the upper environmental limits on the constraint envelope, while feeding and reproductive strategies and environmental tolerance determine the lower limits. The ecosystem that exists within the constraint envelope is metastable (O'Neill et al. 1989); i.e., it moves along a pathway ("manifold") to new community states (stable, local attractors) as some limiting factor—such as nutrient levels—changes. If the upper environmental limits should change because of, say, global cooling, then increased rates of deep-ocean overturn and increased nutrient input to the photic zone would result. Because they occur at a higher hierarchical level, such physical limiting factors change on slower (geological) time scales than the biological ones (Allen and Starr 1982; Salthe 1985; O'Neill et al.

1989). The ecosystem moves along the manifold, which consists of a series of local attractors to which the faster dynamics of the biological system (measured state variables such as productivity and biodiversity) are asymptotically stable (O'Neill et al. 1989). If the trajectory should move toward an unstable region (bifurcation), the rate of return of the faster (biological) dynamics to the manifold should decrease as the point of instability is approached (increased return times or decreased resilience), and there should be increased correlation of "demographic stochasticity" (birth, death, and reproductive rates; population densities; etc.) and environmental conditions (nutrients, temperature) among widely distributed local and regional populations as the TRC shrinks (Harrison and Quinn 1989; Johnson et al. 1995). In the case of the Caribbean and perhaps also the eastern equatorial Pacific, which is a high productivity region (Berger 1989), the manifold may have shifted toward instability as a result of the closure of the Isthmus of Panama about 3.5 to 3.0 million years ago, and the establishment of the present oceanographic regime and rates of nutrient input to each region (Glynn and Colgan 1992; see also Cracraft 1992). Since that time, reefs in the eastern Pacific, especially, have been much more temporally and spatially discontinuous, as compared to the western Atlantic and Indo-West Pacific, and less likely to be preserved in the fossil record (Glynn and Colgan 1992).

If the environmental limits should change sufficiently, however, the entire constraint envelope itself (and its contained community states) shift through time, which could cause radical changes in the constraint structure (extinction). These changes could also have been accompanied by global temperature and sea-level change and biogeographic restriction, leaving relatively few refugia to serve as sources of larvae or other propagules (Harrison and Quinn 1989; Johnson et al. 1995; Harrison, chapter 2). Reef diversity declined, for example, during the Early Oligocene after a cooling episode in the Late Eocene that established the deep-ocean "psychrosphere" (Kennett 1982; Kauffman and Fagerstrom 1993). Based on the record of reef biota and phosphorites, Edinger and Risk (1994) suggested that further biogeographic restriction and extinction of coral taxa in the Caribbean occurred across the Oligo-Miocene boundary via nutrient input and global cooling (a result of Antarctic glaciation; Kennett 1982). Edinger and Risk (1995) concluded that brooding coral genera preferentially survived the extinction, not because of greater recruitment success, but because of the ecological correlates of eurytopy (species sorting). Declining reef diversity occurred again following Antarctic ice sheet formation in the Middle Miocene and continued into the Pliocene (Kennett 1982; Kauffman and Fagerstrom 1993). Such changes could presumably have occurred on even larger scales during mass extinctions.

Nutrient History, Reproductive Strategies, and Biodiversity

There is a great difference in modern coral diversity between the Indo-West Pacific and Caribbean regions. Approximately 37 percent of young (Late Miocene–Recent) coral genera are restricted to the Indo-West Pacific and they are much more species-rich than Caribbean genera (Rosen 1984, p. 221; Fagerstrom 1987, p. 167). These differences are most easily ascribed to the very high geographic complexity of the Indo-West Pacific region [vicariance (Rosen 1984; Sammarco 1987, pp. 146–147; Cracraft 1992)]. The availability of numerous local (and genetically variable) sources of larvae and other propagules has no doubt also promoted high diversity in this region on geological time scales (e.g., Ricklefs and Schluter 1993a; see also Cornell and Karlson 1996).

Assuming that the presumed effects of nutrient-related perturbations on reproductive patterns scale upward, the history of nutrient-related disturbance may have also played a role in generating modern regional coral diversity patterns (Sammarco 1987, pp. 144–146; Hughes 1989). With the closure of the western Tethys (Mediterranean), and the rise of the Isthmus of Panama, the western Atlantic and eastern Pacific were not only cut off from significant sources of larvae or rafted individuals (Frost 1977; Sammarco 1987, p. 146; Jackson 1986), they were also presumably exposed to chronically higher nutrient levels, which could have affected life history traits and rates of speciation. In the Caribbean and eastern Pacific, an increased incidence of both brooding and asexual reproduction in response to elevated nutrient levels beginning in the Pliocene may have decreased the chances of coral speciation in these regions because brooded larvae settle rapidly, thereby promoting philopatry and inbreeding (Jackson 1985, 1986); brooded larvae also persist for long periods in the plankton, thereby promoting gene flow (Richmond 1989; Harriott 1992; see also Szmant 1986; Harrison and Wallace 1990). Thus while brooding (and asexual fragmentation) in corals may decrease the chances of extinction of coral genera (cf. Johnson et al. 1995; see also Maurer and Nott, chapter 3; McKinney 1995), they may also decrease rates of speciation (cf. Richmond 1989).

Phanerozoic Diversity Trends

Upward scaling of nutrient-related phenomena may have contributed to a long-term secular rise in marine diversity of the Phanerozoic. Despite mass extinctions, there was an overall decline in background extinction rates

(Raup and Sepkoski 1982), and a corresponding rise in biomass and diversity, through the Phanerozoic (Benton 1979; Bambach 1993; Martin 1995, 1996). This decline has been attributed to the accumulation of geographically widespread (environmentally tolerant), species-rich clades (e.g., Flessa and Jablonski 1985; Boyajian 1986; Gilinsky, chapter 9; see also Vermeij 1987a, pp. 400–401) and more pronounced latitudinal environmental gradients and provinciality through time (e.g., Valentine and Jablonski 1983).

Bambach (1993) reviewed a variety of evidence, which, although mostly anecdotal, provided robust patterns indicative of rising marine productivity, biomass, and metabolism: (1) biotic diversity and ecosystem complexity increased through the Phanerozoic; even without increasing productivity, more resources must now be consumed than in the past because resources from previously unexploited ecospace came to be utilized; (2) total biomass of marine consumers has increased; (3) more energetic modes of life have replaced more passive ones, and taxa with inferred higher rates of metabolism originated in settings with high rates of food supply and spread into waters with previously lower rates of food availability (see also Vermeij 1987a, 1994; Thayer 1992); and (4) increasing tiering above and below the sediment–water interface suggests greater food availability, as does the diversification of plankton through time (see also Martin 1995, 1996).

Stable isotope, lithologic, and microfossil indices also suggest that marine biomass and diversity were being ratcheted upward through the Phanerozoic by a secular—seemingly steplike—rise in nutrient and food availability. The scenario presented herein suggests that the manifold branched repeatedly each time the deep-ocean turnover rates moved past a critical threshold (O'Neill et al. 1989). This resembles the "fluctuations" in thermodynamically open physical systems that give rise to new configurations through bifurcations of the system (Prigogine 1980; see also May and Oster 1976; Föllmi et al. 1993; see also Ulanowicz 1980; Johnson 1988; Wicken 1988). The catastrophic eutrophication that occurred during most mass (and perhaps minor) extinction episodes increased the carrying capacity of the marine biosphere during the postextinction interval (Bambach 1993; Martin 1995, 1996). Following each extinction, nutrients recycled to the photic zone were incorporated into larger amounts of biomass (increased C:P ratios) and larger populations of organisms (Bambach 1993). Possibly, this property was found in certain species prior to massive eutrophication (Salthe 1985); opportunistic species tend to reproduce rapidly, and it is these taxa that tend to survive extinction and rediversify in the postrecovery period. Bambach (1993) concluded that since feeding efficiency does not increase with time within established modes of life, increased effectiveness of resource utilization occurs only by the development of new major taxa. Because of their larger populations, new groups were better able to spread and diversify following extinction. It may not be entirely fortuitous, for example, that the diversification of rapidly growing, branching corals that today dominate modern undisturbed reef crests (e.g., *Acropora, Porites, Pocillopora*) are rare or unknown from the Paleozoic (Copper 1974; Frost 1977; Fagerstrom 1987, p. 167; Wood 1993).

Others have denied that such trends in productivity and biomass have occurred (e.g., Van Valen 1976). Schidlowski (1991), for example, sees no trend in the carbon isotope curve or organic carbon burial rates that is indicative of rising productivity. This may be the result of looking at data over long periods of time; the variation exhibited by a population (in this case, carbon isotope values) increases the longer the population is sampled (Pimm 1991). Moreover, it is possible that as productivity increased through the Phanerozoic, increased rates of bioturbation, for example, may have destroyed much of the carbon produced (Garrels et al. 1976), so that the fraction of organic carbon buried through time remained more or less constant. Garrels et al. (1976) described a similar scenario involving numerical simulation of increased rates of *anthropogenic* erosion and nutrient input. Ironically, anthropogenic eutrophication of the oceans may be accelerating a trend started by nature hundreds of millions of years ago.

An *Ideal* Agenda for Future Research

Upward scaling of nutrient-related phenomena can be tested using the geologic record. The durations of different ecosystem states in the fossil record (relative sizes of the domains of attraction) should be positively correlated to the overall stability of the ecosystem. If alternative community states exist in response to changes in the temperature-, nutrient-controlled constraint envelope, then the stability (as measured by duration in the fossil record) and recovery time of "normal" communities should decrease as extinction proceeds and extinction rates intensify (as the manifold moves toward unstable regions). The reverse should occur as the manifold moves away from an unstable region following extinction. Soft-bottom communities ought to persist longer and recover earlier than hard-bottom communities. Opportunistic suspension-feeding or detritus-feeding species should become more prominent as the extinction progresses toward final ecosystem collapse.

The hypothesis of large-scale nutrient/productivity perturbations as agents of mass extinction should be tested by continuous, detailed paleontological and stable isotope analyses of multiple sections. The dynamics of fossil taxon abundance (McKinney and Frederick 1992; McKinney and Allmon 1995), prior to, during, and following extinction in each section, should be compared with detailed records of geochemical indices, bearing in mind the pitfalls of taxonomy (e.g., Fortey 1989), taphonomy (e.g., Frost 1977; Scoffin 1992; Martin 1993), sampling (e.g., Buzas et al. 1982; Signor and Lipps 1982; Koch 1987), and time series (McKinney and Oyen 1989). As the extinction progresses, there should be increased correlation between taxon abundance (demographic stochasticity) and changes in productivity, as evidenced by stable isotopes or other indicators. Less common taxa may be more indicative of environmental change than abundant ones (McKinney et al. 1996).

Whatever the scale of disturbance, sections spanning the disturbance should be compared with intervals of little or no biotic turnover [evolutionary "stasis" (e.g., Boucot 1983, 1990b; DiMichele 1994; Brett 1995; Lieberman et al. 1995)] from the same sorts of environments, when possible, as the control. Perhaps intervals of apparent stasis are characterized not so much by stasis as by disturbances insufficient to cause biotic turnover of common taxa (McKinney et al. 1996); conversely, chronic low levels of disturbance may keep a community in stasis (e.g., Potts 1984).

Ideally, sections would be of relatively high sedimentation rate so as to provide the best temporal resolution of paleontologic and geochemical events. Complete sections that span *mass* extinction intervals are, of course, atypical, so that large, taxonomically consistent databases will probably have to serve in their stead in many cases (e.g., Johnson et al. 1995). If nutrient-related disturbances scale upward, though, then they ought to be detectable across at least some minor extinctions, which are much more numerous. Even smaller-scale perturbations may be recorded in the skeletons (e.g., corals) themselves (Dunbar et al. 1994), although "vital effects" must be assessed (Druffel et al. 1990).

By comparing multiple sections for each time slice, the continuity of section [determined via graphic correlation (e.g., MacLeod 1995; Martin and Fletcher 1995)] and the influence of local, regional, and global factors can also be assessed. The particular paleoceanographic regime in existence during each time slice at a particular locale, for example, must be evaluated for its overprint on larger-scale phenomena.

Conclusion

Upward scaling of nutrient- and productivity-related phenomena forms a plausible scenario for extinction on local, regional, and global spatiotemporal scales. Based on stable isotope, phosphorite, and fossil records, mass, and perhaps many minor, marine extinctions were accompanied by "catastrophic" changes in nutrient availability and productivity. Although this paper has emphasized the effects of eutrophication on reef-building taxa, extinction may be caused by *decreased* nutrient levels and productivity, such as in the Late Permian. Following each disruption, the carrying capacity of the marine biosphere was ratcheted up, and overall productivity, biomass, and diversity of the marine biosphere increased as background extinction rates declined. Despite extinction, the history of the biosphere is one of an increasingly diverse global ecosystem driven by the increasing availability of food.

Acknowledgments

This paper is an outgrowth of taphonomic studies supported by NSF Grants EAR-8815997 and EAR-9017864. My thanks to Mike McKinney for the invitation to contribute to this volume, and for his support and encouragement along the way (including preprints of other authors contributing to this volume). The manuscript also benefitted from the patience and open-mindedness of two anonymous reviewers.

18

Scale-Independent Interpretations of Macroevolutionary Dynamics

Richard B. Aronson and Roy E. Plotnick

Most paleobiologists acknowledge that small-scale interactions between organisms can sum to produce large-scale patterns that directly reflect those individual interactions. Vermeij (1977, 1978, 1987a) popularized this idea by showing that increasing predation caused the morphology of gastropod shells to vary in predictable, similar ways on multiple scales of time and space. A growing body of evidence supports scale-independent, or fractal, models of biological interaction, diversification, and extinction (Burlando 1993; Aronson 1994; McKinney 1995; Perry 1995; Rosenzweig, chapter 14). In apparent opposition to this evidence is the concern that our perceptions of pattern and process depend on our scale of observation (Miller 1990b; Schopf and Ivany, chapter 10; and many others). The "effects of scale" remain a subject of continuing controversy. Newly proposed physical and mathematical models of punctuated dynamics, as well as current paleontological interest in the structure and history of biological communities, bring these issues of scale into sharp focus.

We contend that the effects of scale, as deduced from the theory of punctuated equilibrium and its ecological companion theory, coordinated stasis, have often been misconstrued. We further argue that the debate over stability versus flux in marine communities is muddied by confusion over the appropriate scales and levels of analysis. These problems have arisen because pattern and process have been considered from the wrong perspective. This is largely a result of the conflation of two concepts of scale: physical or spatial extent (the usage we prefer) and level in an organizational hierarchy. Although spatial extent and position in a hierarchy are often correlated, they need not be (Allen and Hoekstra 1992).

Three critical questions must be addressed when evaluating the effects of scale.

1. Do similar patterns and processes occur over a wide range of spatial and temporal scales? Complete scale independence implies that patterns and processes observed at one spatial or temporal scale within a system are similar to patterns and processes occurring at all larger and smaller scales (this similarity can be statistical rather than exact). In contrast, scale dependence implies that as observational scales change, the observed patterns and processes change.

2. Do similar patterns and processes occur across a range of hierarchical levels? For example, do the same general processes that control species diversity also control generic diversity?

3. To what extent do processes at one scale or level control processes acting at another higher or lower scale or level? According to Gould (1985; Gould and Eldredge 1993), the most significant contribution of the theory of punctuated equilibrium lies not in the observation of stasis and change, but rather in the recognition that patterns of speciation are strongly influenced by processes occurring at scales larger than that of the individual or population. Gould (1985) suggested that sporadically occurring processes and events at the largest scales, such as mass extinctions, are powerful and perhaps overriding forcing functions in macroevolution.

Paleontological studies rarely consider more than one spatiotemporal scale or one level of organization (see Schopf and Ivany, chapter 10). Yet observations at a single scale or level give an incomplete picture of the dynamics of the entire system. Interestingly, when all scales are observed simultaneously, both scale-dependent and scale-independent aspects can be seen in the dynamics of biological systems.

What Is Scale?

Valentine and Jablonski (1993) argued, "Change, rather than stability, is the normal lot of communities over ecological and evolutionary time." At the same time, Brett et al. (1990) stated, "The fossil record suggests that long periods of stasis may characterize . . . the taxonomic composition of benthic assemblages as well as the morphology of their component species." What is the cause of this apparent discrepancy? One possibility is that it stems from observing the behavior of systems at different scales—looking at changes in community membership on local scales versus considering the dynamics of entire regional biotas (Jackson 1994a).

To address this problem, it is first necessary to utilize a consistent approach to the description of scale. It is especially important to separate spatiotemporal scale from hierarchical level. We advocate the approach to ecological system description discussed in detail by Allen and Hoekstra (1992; see Miller 1990b for a similar discussion).

The most commonly encountered way of portraying scale in paleontology is by use of a bivariate diagram, in which temporal duration is plotted against spatial extent, usually on a log-log scale. For example, Schopf (1972) plotted the "dimensions of paleontologically important items" on a diagram of this sort, with such entities as "individuals" and "communities" occupying non-overlapping regions. Similarly, Delcourt and Delcourt (1991) divided their diagram into a series of domains, bounded by specific temporal and spatial limits. Their "meso-scale" domain, for example, was defined by spatial scales of 10^6 to 10^{10} m^2 and time ranges of 500 to 10,000 years. The spatiotemporal extent of environmental disturbances, biotic responses, and biotic units (e.g., individual trees and stands of trees) were then plotted on this diagram.

The implication of these diagrams is that there are particular spatiotemporal ranges corresponding to units in the ecological or evolutionary hierarchy. Following Allen and Hoekstra (1992), we argue instead that ecological and evolutionary units, and the physical environmental processes that affect them, may not have characteristic spatiotemporal scales. For example, individual organisms can have size scales ranging from that of a bacterium to that of an elephant, and longevities ranging from minutes to millennia. In addition, entities higher on the traditional ecological hierarchy (organism-population-community-ecosystem-biome-biosphere) may not necessarily be larger than those lower down. An individual elephant is larger than the bacterial populations in a mud puddle. In fact, as Allen and Hoekstra stated (1992, p. 51), "Apart from organism and biosphere levels, there is plenty of room for entities from almost any type of ecological system to be contained within an entity belonging to any other class of systems."

This is not to deny that a particular entity, such as a trilobite population, has a characteristic scale. Rather, Allen and Hoekstra (1992) pointed out that ecological entities are not recognized on the basis of their extent but on the criteria used in their definition. These ecological criteria include (among others) organisms, populations, communities, ecosystems, landscapes, and biomes. A paleontologist might want to add such entities as clades to the list of criteria. The criteria are observer dependent and scale independent; for example, populations of bacteria and elephants are recognized as populations for similar reasons, regardless of the vast difference in scale. Allen and Hoekstra's approach requires that we explicitly state the criteria that define the type of entity we wish to study (e.g., population, community, monophyletic group) and then use those criteria to delineate the relevant spatiotemporal scales of that entity for the organisms under investigation. The relevant scales of observation are not defined a priori; rather, they are a consequence of the criteria used and the organisms examined.

These concepts must be kept in mind when considering system stability (or instability) and its relationship to external environmental change. What can be viewed as a system disruption at one scale or with reference to one criterion can be either unimportant or an aspect of stability at another (Wu and Loucks 1995). The replacement of one brachiopod species by another changes community composition, but it may have little or no effect on the pattern of energy flow in the corresponding ecosystem. Allen and Hoekstra (1992, p. 116) stated, "Instability is not a property of the system itself but an aspect of the mode of system description. . . . A tree crashing to the forest floor can either be seen as the local tree exhibiting instability or as a healthy, normal process of replacement on a forested landscape." As another example, the "intermediate disturbance hypothesis" [diversity is highest when disturbance balances competitive exclusion (Connell 1978)] suggests nonequilibrium at smaller scales and dynamic equilibrium at larger scales.

Abiotic disturbances themselves may lack a characteristic scale; they show many aspects of fractal structure (see next section). Examples include climatic change (Fluegemann and Snow 1989) and sea-level fluctuations (Hsui et al. 1993). Disturbances thus should be defined relative to the type of system disturbed and to the scales of that system. As pointed out by White and Pickett (1985), "Disturbance to bryophyte communities on stream-side boulders can occur on a spatial . . . and temporal . . . scale that is irrelevant to the disturbance regime of the forest community growing on the same site."

Finally, it must be remembered that the resolution of the fossil and stratigraphic records does not allow us to observe directly the following categories:

1. all types of ecological units,
2. representatives of all scales of particular criteria, such as small and transient communities and ecosystems,
3. all scales of particular kinds of disturbance, especially the higher-frequency (lower-intensity) ones, and
4. all scales of biotic response to disturbance, again especially the higher frequencies.

We agree with Miller (1990b) that a primary goal of taphonomy should be to clarify the relationship between ecological entities and their expression in the fossil record.

Self-organized Criticality as an Evolutionary Model

If ecological and evolutionary phenomena display scale-independent aspects, then they may be described using fractal models. Many natural patterns show evidence of self-similar structure across multiple scales of observation: they "appear" the same when viewed at different scales. For example, a portion of an ammonite suture resembles the entire suture when both are drawn to the same scale (Lutz and Boyajian 1995). As these phenomena are viewed in greater detail, finer and finer irregularities are recognized. Therefore, their total measured size increases as the precision of measurement increases. The total length of an ammonite suture measured with a crude ruler is much less than that measured with a precision digitizer. Spatial objects that show such scaling behavior are known as fractals, and the parameter that measures the change in measured size caused by changes in precision is called the fractal dimension (Mandelbrot 1983; Korvin 1992). A number of studies have documented the fractal nature of geological and ecological patterns (Plotnick and Prestegaard 1995; Solé and Manrubia 1995).

Many time series also exhibit self-similar behavior. A time series can often be viewed as an additive series of sinusoidal waves, with the amplitude of the waves being an inverse function of the frequency. Such a series is termed $1/f$ noise, since its power spectrum shows a decline of power proportional to frequency. $1/f$ noise has been found in a variety of natural systems, including geophysical well-logs and physiological rhythms (Goldberger and West 1987; Schroeder 1991; Plotnick and Prestegaard 1995). The abundance spectra of many natural populations display $1/f$ noise, although the corresponding mathematical population models generally do not (Cohen 1995). McKinney and Allmon (1995) gave examples of paleontologically relevant parameters that suggest $1/f$ noise and imply scale-independent dynamics.

Bak et al. (1987, 1988, 1990; Bak and Chen 1991) proposed a class of simple dynamical models that generate both fractal spatial patterns and $1/f$ temporal patterns. The behavior of these models is known as self-organized criticality (SOC). SOC and related models have received significant attention from physicists and earth scientists (Hwa and Kardar 1989; Kertész and Kiss 1990; Liu et al. 1990; Drossel and Schwabl 1992; Grieger 1992; Hüssner and Roessler 1995; Stølum 1996). Bak and coworkers have applied SOC to a variety of natural systems, including earthquakes, forest fires, and what they term punctuated equilibrium (Bak and Chen 1989; Bak and Sneppen 1993; Bak et al. 1994). A variation on the SOC idea was used by Plotnick and McKinney (1993) to model ecosystem evolution and collapse.

Bak and colleagues used a simple model, the dynamics of a sandpile, to illustrate self-organized criticality. The system begins with an empty, finite grid. Grains of sand are sprinkled, one by one, at randomly chosen sites (cells) on the grid. If the difference in the "height" of sand between neighboring cells exceeds a specified value, sand is transported from the highest pile to adjacent lower piles. Any sand that reaches the edge of the grid is considered to have left the system. Initially, the movement of sand is local and short in duration, so that the pile as a whole continues to build in height and overall steepness. Eventually, however, the steepness of the pile reaches a point where the addition of a single sand grain produces a large movement— an avalanche. At this stage, the total height of the pile no longer continues to increase. Instead, repeated additions produce new critical states and subsequent avalanches of different sizes. Sand leaves the system in short, irregular bursts. The spatial arrangement of the avalanches is fractal and the time series of sand movements shows a 1/f power spectrum. The analogy of the system's behavior to the familiar concept of angle of repose should be obvious.

The sandpile illustrates the general features of SOC models:

1. They are cellular automata (Schroeder 1991). The space occupied by the system is divided into a grid of discrete cells. Each cell can be in one of a relatively limited set of states. At each discrete time step of the simulation, the state of each cell is updated, based on the states of itself and its local neighbors. Probably the most familiar cellular automata model is The Game of Life.

2. They have extended spatial degrees of freedom (Bak et al. 1987). At each iteration, the state of each cell can be modified by adding a sand grain, independent of the state of its neighbors. The cell selected and the direction of change are random in most models. As a result, these models differ significantly from deterministic nonlinear models, such as the logistic growth system (Peitgen and Saupe 1988).

3. They are open and dissipative. Material can enter and leave the system and "energy" is used up. In the case of the sandpile, potential energy is converted to kinetic energy and then dissipated in the avalanches.

4. In the early stages of system evolution, changes caused by modifications of individual cells have local and short-term effects; only close neighbors are affected. At a particular discrete stage, however, long-range interactions develop; at this point, the modification of a single cell produces effects that cascade, or avalanche, through large parts of the system. These systems

thus show a "phase transition" when their behavior changes. The stage in system evolution at which this change occurs is the "critical point." In this regard, SOC models are similar to percolation models (Plotnick and McKinney 1993; Plotnick and Gardner 1993), where a transition from short- to long-range interactions occurs at a discrete critical level.

5. Regions in the system affected by avalanches suddenly collapse back to a state in which only local interactions occur. These regions can then slowly build to a state in which long-range interactions are once again possible.

6. The system as a whole reaches and remains at a state where some regions are at or near the critical state and others are well below it.

7. The evolution of the system is driven only by its internal dynamics; there is no external controlling parameter. The system thus "self-organizes" to reach its critical state.

Maddox (1994) reviewed several papers that attempted to model punctuated evolutionary dynamics in terms of self-organized criticality. Bak et al. (1989; Bak and Sneppen 1993; Bak 1996) and Kauffman (1991, 1992; see also Waldrop 1990; Depew and Weber 1995) explicitly argued that evolution is a punctuated, scale-independent, self-critical process.

The Bak-Sneppen (1993) model for the evolution of ecological systems is very simple. An ecosystem is represented as consisting of a finite number (e.g., 64) of entities or "species." The species are arranged along a line. The ends of the line are connected so that each entity, including those at the ends of the line, is connected to two neighbors. Each species is then randomly assigned a "barrier height" with values between 0 and 1. The barrier heights essentially represent the overall fitness of each species. Species with high barriers are those closest to optimal fitness on the adaptive landscape and therefore the species most resistant to change.

The model begins by selecting the entity with the lowest barrier height—the species with the lowest fitness—and randomly changing it to another value between 0 and 1. The assumption is that the least fit species are those most likely to change. Taken alone, this rule would lead to a more or less monotonic increase in average barrier height until it reached a value of nearly 1.

However, the model further specifies that each species is linked to its immediate neighbors to the left and right. When the species with the lowest barrier height is changed, then the barrier heights of its two neighbors are also changed by random amounts, regardless of their previous values. Here the assumption is that a change in the fitness of one species will lead to changes in the fitness of the species interacting with it. Since only the immediate neighbors are affected in each iteration, the changes do not propagate further.

The behavior of the model is straightforward. Initially, the position of the species with the lowest barrier can be anywhere along the line. Since the initial average barrier height is roughly 0.5, the probabilities that the new barrier height of a neighboring site (species) will represent an increase or decrease from its previous value are roughly equal. As a result, the probability that the next site chosen will be one of the two neighbors is relatively low. The successive positions of the sites with the lowest barrier heights are essentially independent.

As the system evolves, the average barrier height increases to greater than 0. 5. It thus becomes far more likely that a change in the value of one of the neighboring sites will be negative and that this site will be the next site chosen. The locations of the chosen sites thus become correlated. Eventually, a change in a single site leads to an avalanche of changes in nearby sites. It is this phenomenon that Bak and Sneppen (1993) analogized to punctuated equilibrium.

Compared to most models familiar to paleobiologists, the Bak-Sneppen model is highly abstract in that the entities and their behavior are not derived directly from measured properties of real ecological or evolutionary units. Grassberger (1995), while doubting that the Bak-Sneppen model truly represents punctuated equilibrium, stated, "Experience in critical phenomena suggests that any model which is simple enough to be paradigmatic will sooner or later find an application." Similarly, Maddox (1994) indicated that although the model produces punctuated patterns that might be akin to punctuated equilibrium, "whether it represents the real world is less obvious." He went on to say, however, that the model might still be robust enough to have relevance for understanding a general class of dynamical behaviors to which evolutionary processes might belong.

We generally agree with this view. Although the term *punctuated equilibrium* is a misnomer for the results of this model, the model nevertheless suggests a possible mechanism for stasis-punctuation patterns in ecological and evolutionary systems. What is especially intriguing is that the behavior of the system depends on its internal organization rather than on external perturbations. This is consistent with results from the percolation-based model of Plotnick and McKinney (1993). Nevertheless, excluding external perturbations oversimplifies the real world. Any realistic model for long-term ecosystem behavior should accommodate both internal organization and multiscaled external disturbances that act to disrupt that organization.

Whether or not it falls under the rubric of self-organized criticality, punctuated equilibrium can be viewed as scale dependent or scale independent, according to one's perspective. From the perspective of the individual organism, scale is important because the punctuated dynamics of speciation, whether externally or internally driven (Kirkpatrick 1982; Gould and Eldredge 1993), occur at scales beyond the scale of adaptation to current conditions. Moving up the phylogenetic hierarchy, cladogenesis occurs sporadically at scales beyond that of individual speciation events. At the largest scales, diversification is stimulated in the aftermath of mass extinctions (e.g., Sepkoski 1984; Erwin 1990). Debate continues over the reasons for rapid diversification at high taxonomic levels at the beginning of the Phanerozoic. Several hypotheses, involving externally or internally driven global phase shifts as causes of the "Cambrian explosion" (Stanley 1973a; McMenamin 1990; Valentine et al. 1991a; Knoll 1994; Schopf 1994; Valentine 1995), imply self-critical behavior.

Just as individuals are affected by things happening at scales larger than themselves, so too are species and higher-order taxa affected by processes at still larger scales. The scale-dependent, punctuated dynamics of diversification are similar at multiple scales. In other words, when viewed from a system-wide perspective, they are scale independent.

Extinction, Geographic Range, and Species Selection

A similar scale-independence argument can be made for the dynamics of extinction. Even at small, ecological scales, individual organisms experience infrequent, larger-scale changes to which they cannot adapt, and which kill them regardless of how well they are otherwise adapted to the small-scale vagaries of their environment. Good examples of ecological-scale catastrophes include lightning strikes and hurricanes. Sometimes, individuals simply fail internally and catastrophically, dying (of cancer, heart attacks, etc.) for no external reason that is immediately apparent. Scale is therefore important to the individual even in a purely ecological context.

Now consider a higher level of organization. A hypothetical species of marine invertebrate splits into daughter species with different life history characteristics in a punctuated fashion (Wray 1995). A daughter species that acquires mechanisms to ensure wider larval dispersal may avoid extinction caused by small-scale events such as lightning strikes and hurricanes. Nevertheless, larger, regional events may cause the extinction of the species, regardless of how well it is able to survive local events.

Finally, there is the largest scale to consider. At times, species and the lineages to which they belong experience global-scale events, such as asteroid impacts and worldwide environmental degradation. Entire lineages may disappear in mass extinctions, regardless of how extinction resistant they are during normal, "background" times (Jablonski 1986a).

Although there is some evidence for qualitative differences in extinction selectivity between background times and times of accelerated extinction (Jablonski 1986a; Johnson et al. 1995), there is also evidence that mass extinctions are merely times of increased background extinction. In some cases, the same biological features confer extinction resistance on higher taxa regardless of extinction intensity (Erwin 1989; Stanley and Yang 1994; reviewed in McKinney 1995). High species richness confers extinction resistance at all extinction intensities in these clades. These situations clearly suggest that scale-independent processes are operating.

In other cases, traits such as high species richness, which confer extinction resistance on higher taxa during background times, no longer do so during mass extinctions (Jablonski 1986a). Superficially, these cases suggest rule changes and scale dependence. However, McKinney (1995) raised the possibility that similar rule changes in extinction selectivity also occur at lower taxonomic levels, blurring the distinction between mass and background extinctions. Specifically, monotypic echinoid genera become less prone to extinction relative to polytypic genera as the overall extinction rate increases. In other words, there is a greater bias toward the extinction of monotypic genera at lower overall extinction rates, just as there is a greater bias toward the extinction of species-poor higher taxa during background times.

Another example of an apparent rule change during mass extinction involves the geographic ranges of species; broad geographic range at the species level confers extinction resistance on species and genera during background times, but not during times of mass extinction (Jablonski 1986a; but see Erwin 1989). Yet broad geographic range at the genus level confers extinction resistance on genera during mass extinction times (Jablonski 1986a, 1995). The rule has not changed, only the hierarchical level on which it operates. The rule is, resistance to extirpation by disturbance at a particular scale is enhanced by broad spatial distribution at the level of biological organization corresponding to that spatial scale.

Correlations of larval dispersal mechanisms with the geographic ranges of species are intimately tied to the debate over species selection. Jablonski (1987) argued that individual organisms cannot be characterized as having the geographic range of the species. He therefore claimed that geographic range is an emergent property at the species level. Since broad geographic range confers extinction resistance during times of background extinction, species selection must be operative. As an alternative, Aronson (1994) offered an interpretation of geographic range based on the dispersal and selection of genotypes, which, unlike individuals, do have geographic ranges.

The more widely the individual genotypes of a species are dispersed, the greater the extinction resistance of the genotypes and the species.

The converse of the argument for geographic range and extinction resistance involves the evolutionary loss of mechanisms for wide dispersal (see Havenhand 1995 for a recent review). A change from planktotrophic to non-planktotrophic larval development in marine invertebrates is thought to increase diversification and extinction rates through the isolation of subpopulations (but see Hedgecock 1986). Again, the action would appear to occur at an organizational level higher than that of the individual. In at least one case, however, the higher diversity of poorly dispersing species within a molluscan clade does not require species selection as an explanation (Lieberman et al. 1993; Lieberman and Vrba 1995).

As Williams (1992) discussed, there is nothing sacred (or profane) about selection at the individual level or the species level. Jablonski's (1987) claim of species selection based on geographic range may well be correct. The important point is that similar mechanisms of selection at higher and lower hierarchical levels are equally plausible. As with punctuation, the processes involved in the tempo and selectivity of extinction may have scale-dependent aspects to them; nevertheless, they appear to operate in analogous ways at multiple scales, suggesting self-organized criticality.

Coordinated Stasis

Self-critical models have also been applied to the "evolution" of communities and ecosystems (Allen 1985; Kauffman and Johnsen 1991; Drake et al. 1992; Plotnick and McKinney 1993; Rand and Wilson 1993; Newman and Roberts 1995; Solé and Manrubia 1995; Patterson and Fowler 1996). Punctuated ecological dynamics are viewed as the consequence of self-organized criticality with, or possibly without (Sneppen et al. 1995), external perturbations. Ecosystems reach threshold states, and then state changes occur because effects on one species or functional group propagate through the community via trophic and other ecological connections. Vrba (1985) presaged these ideas with her turnover-pulse hypothesis, a threshold model of stasis and synchronous change across lineages.

In parallel with recent theoretical developments, empirical work has shifted the focus of punctuated equilibrium from the individual and the lineage to the community and the paleocommunity. DiMichele (1994) summarized research on the subject:

> Perhaps the clearest documentation of hierarchy in the marine realm comes from the work of Brett and colleagues on the Silurian and Devonian

of New York State. . . . These authors recognize several levels of organization. Species can be grouped into guilds, which are functional groups; any one guild tends to be filled for millions of years by the same or closely related species (Brett et al. 1990). Biofacies or recurrent assemblages are habitat-specific species groups (roughly communities) that persist for 5–6 my with limited turnover or introduction. Groups of biofacies form regional faunas . . . that also persist for long time spans (Brett and Baird [1995]). Biofacies . . . display abrupt changes during ecologically brief restructuring events. At a global level, regional faunas are parts of still longer-ranging, and geographically widespread ecological-evolutionary units.

Brett and Baird (1995) and Brett et al. (1996) reviewed paleontological evidence of coordinated stasis and rapid change at multiple scales in marine systems. Similar dynamics occur in terrestrial floral assemblages of the Paleozoic (DiMichele and Hook 1992; DiMichele and Phillips 1994). As in the stasis and punctuation that characterize the evolution of individual taxa, habitat tracking prolongs the survivorship of particular paleocommunity types when the environment changes (Eldredge 1989; Gould and Eldredge 1993; Brett and Baird 1995). Eldredge (1989) drew the analogy between the rapid turnovers that punctuate these usually static systems and the geologically sudden biosphere-scale turnovers associated with mass extinctions. To the paleontological data may be added evidence of punctuated dynamics on an ecological scale in living communities (Smith 1994a). The arguments for system-wide scale independence developed in the previous sections apply equally well to assemblages of taxa as to the taxa themselves.

Community Integration

Evidence for the punctuated dynamics of ecosystems is intimately related to the revival of the old ecological debate over community structure. In the Clementsian view, the community is modeled as a tightly integrated superorganism composed of codistributed species; the Gleasonian perspective holds that the community is a collection of independently distributed species (Clements 1916, 1936; Gleason 1926; reviewed in Simberloff 1980; Springer and Miller 1990; Price 1984; McIntosh 1995; and many others). Coordinated stasis has been taken as strong evidence for the Clementsian model. The argument is that communities exist for long periods as tightly integrated entities. The component species (or at least closely related, morphological equivalents) are so strongly interdependent that they resist minor disturbances. Beyond a threshold intensity of disturbance, however, interspecific linkages are broken and the community collapses. (DiMichele and Phillips 1994; Brett and Baird 1995; Morris et al. 1995). Once again, the implication is that the important causes are those acting at scales and levels above that of the individual and the population.

Aberhan (1993, 1994) showed that, over evolutionary time, more frequent environmental changes resulted in the more frequent replacement of species of marine mollusks. Here the replacement species were morphological equivalents, and presumably ecological equivalents, of the species they replaced, but they were not necessarily closely related. Other paleontological evidence, primarily from Cenozoic marine and terrestrial systems, strongly contradicts the notion of tight community integration, pointing instead to the independent behaviors of individual taxa (Taylor 1978; Davis 1986; Graham 1986; Webb 1987; Huntley and Webb 1989; Paulay 1990; Valentine and Jablonski 1993; Buzas and Culver 1994; Johnson et al. 1995; Roy et al. 1995; FAUNMAP Working Group 1996).

Evidence for individualistic behavior also exists on an ecological scale of decades to centuries. In a multidecadal study of the response of an intertidal community in California to climate change, most species showed individualistic responses, while one species closely tracked the species on which it is an epibiont (Barry et al. 1995b). Individualistic, temperature-related range shifts have been observed in response to El Niño events in the eastern Pacific and the multidecadal Russell Cycle in the western English Channel (Southward 1980; Holme 1984; Glynn 1988; Southward et al. 1988). Thus, habitat tracking by entire communities in the Paleozoic may boil down to the separate movements of component species with similar autecologies, producing a signal that is intensified by time averaging (Bennington and Bombach 1996; W. Miller 1996; A. Miller 1997). At the largest scales, however, disturbances cause major community reorganizations, with entirely new suites of taxa invading particular habitats (Vrba 1985; Copper 1988; Sepkoski 1991b; DiMichele and Phillips 1994; Brett and Baird 1995). Brett et al. (1996) therefore suggested that coordinated stasis may occur at scales and levels above those of the community and paleocommunity.

It is interesting to note in the context of the community integration debate that the initial formulation of the "onshore-offshore hypothesis" was cast in terms of the origin and spread of novel communities as units (Jablonski et al. 1983). Later, more refined versions of the hypothesis recognized that individual taxa originate onshore and spread offshore (Bottjer and Jablonski 1988; Jablonski and Bottjer 1991; McNamara 1994). These separate dynamics sum to produce the movements of community types.

Allen and Hoekstra (1992) argued that the Clementsian and Gleasonian models are not mutually exclusive. Just as an individualistic analysis shows novel taxa originating onshore and spreading offshore, so a paleocommunity-scale analysis shows novel paleocommunities originating onshore and spreading offshore. At the largest spatiotemporal scales, a biosphere-scale analysis shows novel "evolutionary faunas" (global assemblages of taxa) originating onshore and spreading offshore (Sepkoski 1984, 1991b; Sepkoski and Miller 1985). At small scales, onshore-offshore gradients in predation over several centuries by a single species, *Homo sapiens,* and the resulting gradients of change in benthic communities, constitute an ecological analogue of the evolutionary patterns (Aronson 1990).

In the terrestrial realm, DiMichele and Aronson (1992) provided evidence that late Paleozoic floras show environmental patterns of origination and spread that are similar to the marine pattern. Evolutionary novelties appear first in marginal, seasonally stressed, upland habitats. The novel floras later spread to basinal lowlands, displacing the incumbent floras. While DiMichele (1994) took these patterns to imply tight community integration within the floras, this view must be balanced by an equally plausible individualistic interpretation. Again, Allen and Hoekstra's (1992) scale-independent reconciliation of the two views is germane, as it is for interpreting the disturbance-driven geographic shifts that Brett and Baird (1995) observed in their taxonomically static, Paleozoic marine communities.

Coral Reefs

"Community unity" will no doubt be hotly debated for years to come. At present, the community integration issue is being argued primarily from case histories rather than from a general, theoretical perspective. Even particular cases, however, are subject to multiple interpretations. As an example, Jackson (1992, 1994a, 1994b) proposed that living reef corals of the Caribbean form integrated communities, displaying coordinated stasis and punctuated change. Many paleobiologists now view coral reefs as an exception to the individualistic dynamics generally observed in Neogene paleocommunities (Pandolfi 1996; Jablonski and Sepkoski 1996). Some fossil and Recent evidence is consistent with this view, but other evidence directly contradicts such an interpretation.

Over the past two decades, many coral reefs around the world have shown changes that are widely believed to represent degradation. Reef-building corals generally have declined in cover and diversity, and fleshy noncoralline macroalgae (henceforth "macroalgae") have increased (Done 1992; Wilkinson 1993; Ginsburg 1994). This change has been particularly pronounced in the Caribbean region (Rogers 1985; Porter and Meier 1992; Ginsburg

1994; Hughes 1994). It is unknown whether the replacement of corals by macroalgae is the result of new types of natural or human impacts, or whether these changes are part of a natural, long-term trend or cycle. Hurricanes, coral bleaching, diseases of corals and sea urchins, over-fishing, eutrophication, pollution, sedimentation, and other factors may all be involved (Gladfelter 1982; Lessios 1988; Bythell et al. 1989; Rogers 1990; Glynn 1993; Liddell and Ohlhorst 1993; Hughes 1994).

The western Atlantic coral fauna passed through an extinction filter at the end of the Pliocene as a result of climatic cooling, and the surviving fauna persisted virtually intact through the Pleistocene glacial cycles (Budd et al. 1994, 1996; Jackson 1994b). Coral zonation patterns typical of living Caribbean reefs, with *Acropora palmata* (elkhorn coral) dominant in the shallowest reef zones and *A. cervicornis* (staghorn coral) dominant at intermediate depths (5 to 25 m on living reefs), have been recognized in Pleistocene and Holocene reef deposits throughout the Caribbean (Mesolella 1967; Macintyre et al. 1977; Geister 1980, 1983; Boss and Liddell 1987; Macintyre 1988). Jackson (1992) concluded from these observations that characteristically zoned, coral-dominated communities "are the norm, rather than a chance event," and that macroalgal dominance has occurred rarely or not at all in the past. Knowlton (1992) suggested that coral-dominated and macroalgae-dominated reefs represent alternative community states, each of which is resistant to conversion to the other.

Coral mortality, caused by hurricanes, coral diseases, and other disturbances, opens space for colonization. If they are present in sufficient numbers, herbivores limit macroalgal growth and promote the recruitment of corals to newly available substratum (Carpenter 1986; Lewis 1986; Steneck 1988; Knowlton 1992). These herbivores include parrotfish (Labridae, formerly Scaridae) and surgeonfish (Acanthuridae), which feed selectively, as well as sea urchins (Echinoidea), which are nonselective herbivores.

The blackspined urchin, *Diadema antillarum,* was the most important nonselective herbivore on most Caribbean reefs until 1983 (Hay 1984; Carpenter 1986). In 1983–84 a disease-induced mass mortality removed *D. antillarum* from reefs throughout the Caribbean region (Lessios 1988). In the absence of nonselective herbivory by *D. antillarum,* macroalgae flourished on overfished reefs such as the one at Discovery Bay, Jamaica, from which parrotfish and surgeonfish had been largely removed (Hughes 1994). Even on many lightly fished and protected reefs, however, certain macroalgal species—those that are unpalatable to selectively feeding, herbivorous fishes—increased in cover (Levitan 1988; Knowlton 1992). The macroalgae prevented coral recruitment by overgrowing and shading newly settled corals (e.g., Hughes 1994; see also Sammarco 1980, 1982). Although the generality of this scenario remains to be established, it appears that the combined effects of coral mortality and reduced nonselective herbivory were primarily responsible for the decreased coral cover and increased macroalgal cover seen throughout the Caribbean in recent years. Nutrient loading on some reefs has further promoted macroalgal growth (Tomascik and Sander 1987; Tomascik 1991). Hatcher (1984) observed a similar coral-resistant changeover to macroalgal dominance as the direct result of a ship grounding on the Great Barrier Reef.

In the community integration interpretation, coral species have persisted in predictable associations, forming long-lived communities (see also Pandolfi 1996). The conjunction of recent disturbances has disrupted these associations, switching Caribbean reef communities to the rare (and perhaps unique) state of macroalgal dominance that we see today (Jackson 1992, 1994a, 1994b; Stemann and Johnson 1995). This line of reasoning does not a priori exclude the possibility of previous periods of macroalgal dominance. After all, in its original conception, coordinated stasis was applied to time-averaged assemblages on a large scale, and it explicitly included sporadic community disruption on smaller scales (Brett et al. 1990). Times of macroalgal dominance, if they occurred, were brief periods of noise compared to the signal periods during which corals were dominant.

From a paleontological perspective, the fundamental problem with this scenario is taphonomic control. Hard corals and macroalgae have very different preservation potentials; corals are often preserved virtually intact, but macroalgae are not preserved at all (Kauffman and Fagerstrom 1993). Therefore, the observation of coral zonation in the fossil record does not by itself prove that corals, or their characteristic zonation scheme, were dominant on most Caribbean reefs most of the time. Macroalgae may in fact have been "the norm" for long intervals (cf. Woodley 1992). The conclusion that the past few decades are unusual would be legitimate if both endpoints of a recently observed community transition had a high preservation potential but only one endpoint—the norm—was actually observed in the fossil record. Aronson and Precht (1997) avoided this problem of preservation potential in a study of reef community transition in Belize.

Along the flanks of shoaling reefs in the Belizean Barrier Reef lagoon, *Acropora cervicornis* was the dominant space-occupier at a 3 to 15 m depth from at least as far back as the 1970s. At one of these reefs, Channel Cay, the dominance of *A. cervicornis* lasted until 1986. From 1986 to 1993, virtually all the *A. cervicornis* at Channel Cay was wiped out by white band disease (WBD). WBD is a presumptive bacterial infection that has destroyed *A. cervicornis* populations throughout the Caribbean; it is one of the factors that has led to macroalgal dominance on many Caribbean reefs. The cover of *A.*

cervicornis dropped from approximately 70 percent in 1986 to near 0 percent in 1993 and has remained approximately 0 percent ever since.

The *A. cervicornis* at Channel Cay was not replaced by macroalgae; macroalgae covered only 1 to 10 percent of the substratum from 1986 to 1995. Rather, *Agaricia* spp. (lettuce coral) recruited to the rubble of dead *A. cervicornis*. *Agaricia* spp., which had previously been minor constituents of the sessile biota (10 percent cover in 1986), replaced *A. cervicornis* as the most common occupant of space on the reef (56 percent cover in 1995). The combined cover of the 12 other coral species present at Channel Cay, which never exceeded 9 percent, showed no significant change during the period 1986 to 1995. Over that 10-year period, a layer of *Agaricia* rubble with a mean thickness of 22 cm accumulated beneath the living community at Channel Cay.

A similar replacement sequence was observed on all other reefs within the surrounding 250-km² area. The transition to another coral rather than to macroalgae was mediated by two factors:

1. herbivory by an echinoid, *Echinometra viridis,* and
2. the reproductive biology of the family Agariciidae.

Echinometra viridis, which was unusually abundant at Channel Cay, is known to promote the recruitment of *Agaricia* spp. by consuming algae (Sammarco 1982). The recruitment of *Agaricia* spp. was also favored by their brooding mode of reproduction (Bak and Engel 1979; Hughes 1989, 1994). It is interesting to note parenthetically that brooding may confer extinction resistance on coral species on regional, evolutionary scales as well as on local, ecological scales (Edinger and Risk 1995).

Because *Agaricia* preserves well and the reef sediments at Channel Cay are uncemented and uncompacted, any earlier *Acropora*-to-*Agaricia* transitions would be visible in the fossil record as layers of *Agaricia* rubble. However, dated Holocene cores from Channel Cay were dominated by *Acropora cervicornis* and showed no evidence of a previous replacement sequence of this type in at least the past 3400 to 3800 years (Westphall 1986; Aronson and Precht 1997). If this coral-to-coral replacement was unique, then it may be that recent coral-to-macroalgal transitions in the Caribbean are unique. Yet there is no guarantee that *A. cervicornis* did not at some time in the past give way to algae rather than to *Agaricia* at Channel Cay. Regardless of the uniqueness of the transition, the data from Channel Cay certainly do not prove that Caribbean coral reef communities are assemblages of strongly interactive species that rise and fall together. When *Acropora cervicornis* disappeared from Channel Cay, the new dominants were former subordinate competitors, and the majority of coral species were unaffected by the transition. The Channel Cay reef, like most communities, fell between the Clementsian and Gleasonian extremes. This result suggests that coral species declined together on most Caribbean reefs as a common response to external (regional) environmental changes, and not specifically in response to each other. The rapid, opportunistic replacement of *Acropora* by other corals or macroalgae on a regional (Caribbean-wide) scale is strongly reminiscent of the rapid radiations of surviving taxa into vacant ecospace following mass extinctions (reviewed in Jablonski and Sepkoski 1996).

Discussion and Conclusions

What is the relationship between community integration and the mechanisms by which communities remain static for long periods and then suddenly change? If communities are indeed integrated wholes, then it is reasonable a priori to model them as self-critical systems, which reach thresholds and undergo punctuated state changes, such as those provoked by extinctions. However, such tightly integrated communities, in which the component species are highly connected, will not persist for long periods, because of destabilizing positive feedback (May 1973; Allen and Hoekstra 1992, p. 117). Although there is some minimal level of connectance necessary to produce punctuated patterns (Plotnick and McKinney 1993), coordinated stasis may in fact be evidence against tight community integration.

Even if, as we have argued, communities are to some extent happenstance assemblages of species, that does not preclude interspecific interactions (McIntosh 1995). Those interactions may be strong and they may be important. Yet strong, important interactions do not imply that the local extinction of even the most important, connected species will cause the local extinction of numerous other species, as paleobiological models tend to assume (e.g., Plotnick and McKinney 1993; Patterson and Fowler 1996). Rather, the removal of a "keystone" species—a top predator, for example—generally alters the relative abundances of other species in the interaction web (for recent reviews, see Estes and Duggins 1995; Menge 1995).

Extinctions of multiple species on local or regional scales may be "coordinated" when they do occur, but temporal conjunction does not necessarily mean that the extinctions are caused by other extinctions. Even if species within communities are largely unconnected, a drastic decline in nutrient levels, for example, will have dire, punctuated consequences, as many consumer species disappear for the same reason: declining primary productivity. Likewise, the fact that predator–prey interactions within ecological communities sum to produce larger-scale patterns of prey morphology and distribution does not contradict the notion of individualistic dynamics; escalation apparently has taken place diffusely rather than as strict coevolution between particular pairs of species (Vermeij 1977, 1978, 1987a, 1994; Aronson 1992, 1994; Jablonski and Sepkoski 1996).

Punctuated equilibrium has been a focus of strenuous attack. At this point, few would dispute the claim that life moves in fits and starts. More interesting is the observation of episodic behavior on a number of scales (Raup 1986; Crowley and North 1988). The punctuated patterns, and the processes that cause those patterns, are analogous at scales from the gene through the higher-level taxon (Williams 1992; McKinney and Allmon 1995; Harrison, chapter 2). Punctuated dynamics also scale up from populations and paleopopulations to the communities and paleocommunities they constitute.

Debate continues over the extent to which higher-level patterns result from emergent properties at those levels (e.g., Jablonski 1986a, 1987; Levinton 1988; Aronson 1994; Lieberman and Vrba 1995; McKinney 1995; Rosenzweig, chapter 14). However, this debate may not be crucial to solving the problems of scale (Williams 1992). Rather, if the theories and observations reviewed here connote scale dependence on particular scales, they also connote scale independence from a global, system-wide perspective. Punctuated equilibrium, coordinated stasis, and other macroevolutionary constructs may have more to offer as theories centered on scale independence than as theories of scale dependence. Scale-independent or not, macroevolution must be understood within an ecological context (Allmon 1994). A new "macroecology" (Brown 1995) would meet this need by examining ecological patterns and processes at the scales at which macroevolutionary phenomena occur.

Summary

Interpretations of natural phenomena often depend on the spatial and temporal scales at which those phenomena are observed. As a prime example from paleobiology, some observations indicate that communities are in perpetual flux, whereas other observations suggest long periods of community stasis followed by rapid change. To reconcile these conflicting views, it is first necessary to distinguish scale from hierarchical level, since the two are not always correlated. The spatial or temporal scale of a phenomenon does not necessarily correspond to a particular level within the ecological or evolutionary hierarchy. Both the hierarchical level of interest and the particular organisms studied determine the appropriate scale of observation. Stability and the impact of disturbance depend on both scale and hierarchical level; disturbances that have a major impact on bacterial populations at a small spatial scale go unnoticed by the individual elephant, which is an entity at a lower hierarchical level but a larger scale.

Exposition of the effects of scale would be a relatively simple matter were it not for the fact that many patterns and processes have scale-independent as well as scale-dependent aspects. Physical disturbances, which probably exert a strong influence on the evolution of taxa and the dynamics of communities, may not have characteristic scales; a particular type of disturbance may operate in a similar fashion at many or all scales. The effects of external disturbance or internal dynamics on individual taxa can be imitated by models of self-organized criticality, in which complex systems repeatedly reach threshold states and then undergo sudden changes at all scales. These fractal descriptions may explain the observation of punctuated evolution at multiple scales and taxonomic levels. Higher-order influences (scale dependence) are recognized, but they are recognized at multiple scales and taxonomic levels (scale independence).

Like diversification, extinction at any particular scale of interest is mediated by processes at larger scales. Scale independence emerges when the entire system is considered and all scales are viewed simultaneously. Extinction resistance and selection do not differ qualitatively among scales; there is nothing special about them at the species level.

Similar considerations apply to coordinated stasis, the community-level analogue of punctuated equilibrium. Communities are characterized by stasis and sudden change at multiple scales, suggesting self-organized criticality. The perception of continuous community flux at smaller scales—apparent scale dependence—is in reality the coalescence of small-scale cycles of stasis and punctuation—actual scale independence.

Some authors have interpreted coordinated stasis to imply tight community integration (i.e., strong mutual dependence among the species in a community). However, individualistic models provide equally cogent or superior explanations of observed threshold patterns of community change at multiple scales and levels, including onshore-offshore macroevolutionary dynamics. The onshore origin and offshore movements of individual novel taxa simply sum to produce the same movements at the community level. In a similar fashion, the sudden, and perhaps unique, change from coral to macroalgal dominance on Caribbean reefs over the past two decades has been regarded by some as evidence of interdependence among species of reef corals. Yet a field study on a reef in Belize provided evidence that stasis and change do not necessarily imply tight community integration. This result suggests that coral species on Caribbean reefs declined together as a common response to external conditions, not specifically in response to each other. Again, the behaviors of individual taxa summed to produce a community-level pattern. Recognition of the simultaneously scale-dependent and scale-independent character of ecological and evolutionary entities provides valuable insight into the mechanisms behind the patterns.

Acknowledgments

We thank Marty Buzas, Bill DiMichele, Jim Drake, Ken Heck, Mike McKinney, Rich Mooi, Bill Precht, and John Valentine for valuable discussion and comments on earlier drafts of this chapter. Partial support was provided by grants from the National Science Foundation to R.B.A. (OPP-9413295) and to (EAR-9506639) This is Contribution No. 276 of the Dauphin Island Sea Lab.

References

Aberhan, M. 1993. Faunal replacement in the Early Jurassic of northern Chile: Implications for the evolution in Mesozoic benthic shelf ecosystems. *Palaeogeography, Palaeoclimatology, Palaeoecology* 103:155–177.

——. 1994. Guild-structure and evolution of Mesozoic benthic shelf communities. *Palaios* 9:516–545.

Abrams, P. A. and J. D. Roth. 1994a. The responses of unstable food chains to enrichment. *Evolutionary Ecology* 8:150–171.

——. 1994b. The effects of enrichment of three-species food chains with nonlinear functional responses. *Ecology* 75:1118–1130.

Acevedo, R. and J. Morelock. 1988. Effects of terrigenous sediment influx on coral reef zonation in southwestern Puerto Rico. *Proceedings 6th International Coral Reef Conference* 2:189–194.

Agassiz, A. 1888. *Three Cruises of the U.S. Coast and Geodetic Survey Steamer "Blake," V. II.* Boston: Houghton Mifflin.

——., L. F. de Pourtales, and T. Lyman. 1878. Reports on the dredging operations of the U.S. Coast Survey Steamer Blake, II [echini, corals, crinoids, and ophiurans]. *Museum of Comparative Zoology Bulletin* 5:181–238.

Alberch, P. 1980. Ontogenesis and morphological diversification. *American Zoologist* 20:653–667.

——. 1985. Problems with the interpretation of developmental sequences. *Systematic Zoology* 34:46–58.

——. 1989. The logic of monsters: Evidence for internal constraint in development and evolution. *Geobios, mémoire spécial* 12:21–57.

Algeo, T., R. A. Berner, J. B. Maynard, and S. E. Scheckler. 1995. Late Devonian oceanic anoxic events and biotic crises: "Rooted" in the evolution of vascular land plants? *GSA Today* 5:45.

Allen, P. M. 1985. Ecology, thermodynamics, and self-organization: Towards a new understanding of complexity. In R. E. Ulanowicz and T. Platt, eds., *Ecosystem Theory for Biological Oceanography*, pp. 3–26. Ottawa: Canadian Bulletin of Fisheries and Aquatic Sciences 213, Department of Fisheries and Oceans.

Allen, T. F. H. and T. W. Hoekstra. 1991. Roles of heterogeneity in scaling of ecological systems under analysis. In J. Kolasa and S. T. A. Pickett, eds., *Ecological Heterogeneity*, pp. 47–68. New York: Springer-Verlag.

——. 1992. *Toward a Unified Ecology.* New York: Columbia University Press.

Allen, T. F. H. and T. B. Starr. 1982. *Hierarchy: Perspectives for Ecological Complexity.* Chicago: University of Chicago Press.

Allison, P. A. and D. E. G. Briggs. 1993. Paleolatitudinal sampling bias, Phanerozoic species diversity, and the end-Permian extinction. *Geology* 21:65–68.

Allmon, W. D. 1992. A causal analysis of stages in allopatric speciation. *Oxford Surveys in Evolutionary Biology* 8:219–257.

——. 1994. Taxic evolutionary paleoecology and the ecological context of macroevolutionary change. *Evolutionary Ecology* 8:95–112.

Allmon, W. D., G. Rosenberg, R. W. Portell, and K. S. Schindler. 1993. Diversity of Atlantic coastal plain mollusks since the Pliocene. *Science* 260:1626–1629.

Alroy, J. 1989. *Biochronological Parsimony Analysis.* B. A. thesis, Reed College.

——. 1992. Conjunction among taxonomic distributions and the Miocene mammalian biochronology of the Great Plains. *Paleobiology* 18:326–343.

——. 1994a. Appearance event ordination: A new biochronologic method. *Paleobiology* 20:191–207.

——. 1994b. *Quantitative Mammalian Biochronology, Biogeography, and Diversity History of North America.* Ph.D. thesis, University of Chicago.

——. 1996. Constant extinction, constrained diversification, and uncoordinated stasis in North American mammals. *Palaeogeography, Palaeoclimatology, Palaeoecology* 127:285–312.

Andrén, H. 1994. Can one use nested subset pattern to reject the random sample hypothesis? Examples from boreal bird communities. *Oikos* 70:489–491.

Andrewartha, H. G. and L. C. Birch. 1954. *The Distribution and Abundance of Animals.* Chicago: University of Chicago Press.

Antonovics, J. and P. H. van Tienderen. 1991. Ontoecogenophyloconstraints? The chaos of constraint terminology. *Trends in Ecology and Evolution* 6:166–168.

Archibald, J. D. 1993. The importance of phylogenetic analysis for the assessment of species turnover: A case history of Paleocene mammals in North America. *Paleobiology* 19:1–27.

———. 1996. *Dinosaur Extinction and the End of an Era.* New York: Columbia University Press.

Arnold, E. N. 1994. Investigating the origins of performance advantage: Adaptation, exaptation, and lineage effects. In P. Eggleton and R. I. Vane-Wright, eds., *Phylogenetics and Ecology,* pp. 124–168. Linnean Society Symposium Series No. 17. London: Academic Press.

———. 1995. Identifying the effects of history on adaptation: Origins of different sand-diving techniques in lizards. *Journal of the Zoological Society of London* 235:351–388.

Arnold, S. J. 1992. Constraints on phenotypic evolution. *American Naturalist* 140: S85–S107.

Aronson, R. B. 1990. Onshore-offshore patterns of human fishing activity. *Palaios* 5:88–93.

———. 1992. Biology of a scale-independent predator-prey interaction. *Marine Ecology Progress Series* 89:1–13.

———. 1994. Scale-independent biological interactions in the marine environment. *Oceanography and Marine Biology: An Annual Review* 32:435–460.

Aronson, R. B. and W. F. Precht. 1997. Stasis, biological disturbance, and community structure of a Holocene coral reef. *Paleobiology* 23:326–346.

Atmar, W. and B. D. Patterson. 1993. The measure of order and disorder in the distribution of species in fragmented habitat. *Oecologia* 96:373–382.

Ausich, W. I., T. W. Kammer, and T. K. Baumiller. 1994. Demise of the middle Paleozoic crinoid fauna: A single extinction event or rapid faunal turnover? *Paleobiology* 20:345–361.

Babin, C. 1993. Rôle des plates-formes gondwaniennes dans les diversifications des mollusques bivalves durant l'Ordovicien. *Société Géologique de France, Bulletin* 164:141–153.

Bader, R. S. and J. S. Hall. 1960. Osteometric variation and function in bats. *Evolution* 14:8–17.

Bak, P. 1996. *How Nature Works.* New York: Springer-Verlag.

Bak, P. and K. Chen. 1989. The physics of fractals. *Physica D* 38:5–12.

———. 1991. Self-organized criticality. *Scientific American* 264(1):46–53.

Bak, P., K. Chen, and M. Creutz. 1989. Self-organized criticality and the 'Game of Life.' *Nature (London)* 342:780–782.

Bak, P., H. Flyvbjerg, and K. Sneppen. 1994. Can we model Darwin? *New Scientist,* 12 March 1994, 36–39.

Bak, P. and K. Sneppen. 1993. Punctuated equilibrium and criticality in a simple model of evolution. *Physical Review Letters* 71:4083–4086.

Bak, P., C. Tang, and K. Wiesenfeld. 1987. Self-organized criticality: An explanation of 1/f noise. *Physical Review Letters* 59:381–384.

———. 1988. Self-organized criticality. *Physical Review A* 38:364–374.

Bak, R. P. M. and M. S. Engel. 1979. Distribution, abundance and survival of juvenile hermatypic corals (Scleractinia) and the importance of life history strategies in the parent coral community. *Marine Biology* 54:341–352.

Baldwin, B. G. 1993. Molecular phylogenetics of calycadenia (Compositae) based on its sequences of nuclear ribosomal DNA: Chromosomal and morphological evolution reexamined. *American Journal of Botany* 80:222–238.

Bambach, R. 1985. Classes and adaptive variety: The ecology of diversification in marine faunas through the Phanerozoic. In J. W. Valentine, ed., *Phanerozoic Diversity Patterns: Profiles in Macroevolution*, pp. 191–253. Princeton: Princeton University Press.

Bambach, R. K. 1977. Species richness in marine benthic habitats through the Phanerozoic. *Paleobiology* 3:152–167.

——. Phanerozoic marine communities. In D. M. Raup and D. Jablonski. eds., *Patterns and Processes in the History of Life*, pp. 407–428. Berlin: Springer-Verlag.

——. 1993. Seafood through time: Changes in biomass, energetics, and productivity in the marine ecosystem. *Paleobiology* 19:372–397.

Bambach, R. K. and J. B. Bennington. 1996. Do communities evolve? A major question in evolutionary paleoecology. In D. H. Erwin, D. Jablonski, and J. Lipps, eds., *Evolutionary Paleobiology: Essays in Honor of James W. Valentine*, pp. 123–160. Chicago: University of Chicago Press.

Bambach, R. K. and J. J. Sepkoski, Jr. 1992. Historical evolutionary information in the traditional Linnean hierarchy. *Proceedings of the Fifth North American Paleontological Convention*, Paleontological Society Special Publication No. 6, p. 16.

Barnosky, A. D. 1989. The late Pleistocene event as a paradigm for widespread mammal extinction. In S. K. Donovan, ed., *Mass Extinctions: Processes and Evidence*, pp. 235–254. London: Bellhaven Press.

Barrera, E. 1994. Global environmental changes preceding the Cretaceous-Tertiary boundary: Early-late Maastrichtian transition. *Geology* 22:877–880.

Barry, J. C., L. J. Flynn, and D. R. Pilbeam. 1990. Faunal diversity and turnover in a Miocene terrestrial sequence. In R. M. Ross and W. D. Allmon, eds., *Causes of Evolution: A Paleontological Perspective*, pp. 381–421. Chicago: University of Chicago Press.

Barry, J. C., M. E. Morgan, L. J. Flynn, D. Pilbeam, L. L. Jacobs, E. H. Lindsay, S. M. Raza, and N. Solounias. 1995a. Patterns of faunal turnover and diversity in the Neogene Siwaliks of Northern Pakistan. *Palaeogeography, Palaeoclimatology, Palaeoecology* 115 (4):209–226.

Barry, J. P., C. H. Baxter, R. D. Sagarin, and S. E. Gilman. 1995b. Climate-related, long-term faunal changes in a California rocky intertidal community. *Science* 267:672–675.

Barry, J. P. and P. K. Dayton. 1991. Physical heterogeneity and the organization of marine communities. In J. Kolasa and S. T. A. Pickett eds., *Ecological Heterogeneity*, pp. 270–320. New York: Springer-Verlag.

Bartholomew, G. A. 1981. A matter of size: An examination of endothermy in insects and terrestrial vertebrates. In B. Heinrich, ed., *Insect Thermoregulation*, pp. 45–78. New York: Wiley.

Batschelet, E. 1979. *Introduction to Mathematics for Life Scientists*. Third edition. Berlin, Heidelberg: Springer-Verlag.

Baumiller, T. K. 1993. Survivorship analysis of Paleozoic Crinoidea: Effect of filter morphology on evolutionary rates. *Paleobiology* 19:304–321.

Beeby, A. 1993. *Applying Ecology*. London: Chapman & Hall.

Beerbower, J. R. 1952. The chronofauna and quantum evolution. *Evolution* 7:178.

Bengtsson, J. 1989. Interspecific competition increases local extinction rate in a metapopulation system. *Nature* 340:713–715.

———. 1991. Interspecific competition in metapopulations. *Biological Journal of the Linnaean Society* 42:219–37.

———. 1994. Temporal predictability in forest soil communities. *Journal of Animal Ecology* 63:653–665.

Bennett, P. M. and I. P. Owens. 1977. Variation in extinction risk among birds: Chance or evolutionary predisposition? *Proceedings of the Royal Society of London* B264:401–408.

Bennington, J. B. and R. K. Bambach. 1996. Statistical testing for paleocommunity recurrence: Are similar fossil assemblages ever the same? *Palaeogeography, Palaeoclimatology, Palaeoecology* 127:107–133.

Benson, R. H. 1979. In search of lost oceans: A paradox in discovery. In J. Gray and A. Boucot, eds., *Historical Biogeography*, pp. 379–389. Corvallis: Oregon State University Press.

Benton, M. J. 1979. Increase in total global biomass over time. *Evolutionary Theory* 4:123–128.

———. 1995. Diversification and extinction in the history of life. *Science* 268:52–58.

———. 1996. On the non-prevalence of competitive replacement in the evolution of tetrapods. In D. Jablonski, D. H. Erwin, and J. Lipps, eds., *Evolutionary Paleobiology*, pp. 185–210. Chicago: University of Chicago Press.

Benton, M. J., ed. 1993. *The Fossil Record 2*. London: Chapman & Hall.

Berger, W. H. 1989. Global maps of ocean productivity. In W. H. Berger, V. S. Smetacek, and G. Wefer, eds., *Productivity of the Ocean: Past and Present*, pp. 429–455. New York: John Wiley.

Berner, R. A. and R. Raiswell. 1983. Burial of organic carbon and pyrite sulfur in sediments over Phanerozoic time: A new theory. *Geochimica et Cosmochimica Acta* 47:855–862.

Berry, W. B. N. and P. Wilde. 1978. Progressive ventilation of the oceans—An explanation for the distribution of the lower Paleozoic black shales. *American Journal of Science* 278:257–275.

Berry, W. B. N., P. Wilde, and M. S. Quinby-Hunt. 1987. The oceanic non-sulfidic oxygen minimum zone: A habitat for graptolites? *Bulletin of the Geological Society of Denmark* 35:103–114.

Birkeland, C. 1977. The importance of rate of biomass accumulation in early successional stages of benthic communities to the survival of coral recruits. *Proceedings 3rd International Coral Reef Symposium* 1:15–26.

——. 1982. Terrestrial runoff as a cause of outbreaks of *Acanthaster planci* (Echinodermata: Asteroidea). *Marine Biology* 69:175–185.

——. 1987. Nutrient availability as a major determinant of differences among
coastal hard-substratum communities in different regions of the tropics. In
C. Birkeland, ed., *Comparison Between Atlantic and Pacific Tropical Marine
Coastal Ecosystems: Community Structure, Ecological Processes, and Productivity*,
pp. 45–97. Paris: UNESCO.

Blackburn, T. M. and K. J. Gaston. 1995. Reply. *Trends in Ecology and Evolution.*
10:205.

Blalock, H. M., Jr. 1964. *Causal Inferences in Nonexperimental Research.* New York:
Norton.

Block, W. M., L. A. Brennan, and R. J. Gutiérrez. 1986. The use of guilds and guild-
indicator species for assessing habitat suitability. In J. Verner, M. L. Morrison,
and C. J. Ralph, eds., *Wildlife 2000: Modeling Habitat Relationships of Terrestrial Vertebrates*, pp. 109–113. Madison: University of Wisconsin Press.

Blum, H. F. 1951. *Time's Arrow and Evolution.* Princeton: Princeton University
Press.

Bodmer, R. E., J. F. Eisenberg, and K. H. Redford. 1997. Hunting and the likelihood of extinction of Amazonian mammals. *Conservation Biology* 11:
460–467.

Bonner, J. T. 1988. *The Evolution of Complexity by Means of Natural Selection.*
Princeton: Princeton University Press.

Bormann, F. H. and G. E. Likens, eds. 1979. *Pattern and Process in a Forested Ecosystem.* New York: Springer-Verlag.

Boss, S. K. and W. D. Liddell. 1987. Back-reef and fore-reef analogs in the Pleistocene of north Jamaica: Implications for facies recognition and sediment flux
in fossil reefs. *Palaios* 2:219–228.

Boswell, M. T. and G. P. Patil. 1971. Chance mechanisms generating the logarithmic
series distribution used in the analysis of number of species and individuals.
In G. P. Patil, E. C. Pielou, and W. E. Waters, eds., *Statistical Ecology*, Vol. 3.,
pp. 99–130. University Park, Pa.: Pennsylvania State University Press.

Botkin, D. B. 1990. *Discordant Harmonies.* Oxford: Oxford University Press.

Bottjer, D. J. and W. I. Ausich. 1986. Phanerozoic development of tiering in soft
substrata suspension-feeding communities. *Paleobiology* 12:400–420.

Bottjer, D. J. and M. L. Droser. 1994. The history of Phanerozoic bioturbation. In
S. K. Donovan, ed., *The Palaeobiology of Trace Fossils*, pp. 155–176. Chichester: John Wiley and Sons.

Bottjer, D. J. and D. Jablonski. 1988. Paleoenvironmental patterns in the evolution
of post-Paleozoic benthic marine invertebrates. *Palaios* 3:540–560.

Boucot, A. J. 1975. *Evolution and Extinction Rate Controls.* Amsterdam: Elsevier.

——. 1978. Community evolution and rates of cladogenesis. In M. K. Hecht, W. C. Steere, and B. Wallace, eds., *Evolutionary Biology, Vol. 11,* pp. 545–654. Mt. Kisco, N.Y.: Plenum Press.

——. 1981. *Principles of Benthic Marine Paleoecology.* New York: Academic Press.

——. 1983. Does evolution take place in an ecological vacuum? *Journal of Paleontology* 57:1–30.

——. 1990a. *Evolutionary Paleobiology of Behavior and Coevolution.* Amsterdam: Elsevier.

——. 1990b. Community evolution: Its evolutionary and biostratigraphic significance. In W. Miller, ed., *Paleocommunity Temporal Dynamics: The Long-term Development of Multispecies Assemblies,* pp. 48–70. Paleontological Society Special Publication No. 5.

——. 1996. Epilogue. *Palaeogeography, Palaeoclimatology, Palaeoecology* 127: 339–349.

——. 1997. Comment on "Long term faunal stasis without evolutionary coordination: Jurassic benthic marine paleocommunities, Western Interior, United States." *Geology* 25(5):472.

Boyajian, G. E. 1986. Phanerozoic trends in background extinction: Consequence of an aging fauna. *Geology* 14:955–958.

——. 1988. Mass vs. background extinction: No difference on the basis of taxon age distributions. *Geological Society of America Abstracts with Programs* 20:A105.

——. 1991. Taxon age and selectivity of extinction. *Paleobiology* 17:49–57.

Boyajian, G. and T. Lutz. 1992. Evolution of biological complexity and its relation to taxonomic longevity in the Ammonoidea. *Geology* 20:983–986.

Boyce, M. S. 1992. Population viability analysis. *Annual Review of Ecology and Systematics* 23:481–506.

Bralower, T. J. and H. R. Thierstein. 1984. Low productivity and slow deep-water circulation in mid-Cretaceous oceans. *Geology* 12:614–618.

Bramlette, M. N. 1965. Massive extinctions in biota at the end of Mesozoic time. *Science* 148:1696–1699.

Brandon, R. N. 1985. Adaptive explanations: Are adaptations for the good of replicators or interactors? In D. J. Depew and B. H. Weber, eds., *Evolution at a Crossroads: The New Biology and the New Philosophy of Science,* pp. 81–96. Cambridge: MIT Press.

——. 1992. Environment. In E. F. Keller and E. A. Lloyd, eds., *Keywords in Evolutionary Biology,* pp. 81–86. Cambridge: Harvard University Press.

Brasier, M. D. 1989a. On mass extinctions and faunal turnover near the end of the Precambrian. In S. K. Donovan, ed., *Mass Extinctions: Processes and Evidence,* pp. 73–88. New York: Columbia University Press.

——. 1992a. Global ocean-atmosphere change across the Precambrian-Cambrian transition. *Geological Magazine* 129:161–168.

——. 1992b. Paleoceanography and changes in the biological cycling of phosphorus across the Precambrian-Cambrian boundary. In J. H. Lipps and P. W. Bengston, eds., *Origin and Early Evolution of the Metazoa,* pp. 483–523. Mt. Kisco, N.Y.: Plenum Press.

——. 1995. Fossil indicators of nutrient levels. 2: Evolution and extinction in relation to oligotrophy. In D. W. J. Bosence and P. A. Allison, eds., *Marine Palaeoenvironmental Analysis from Fossils,* pp. 133–150. London: Geological Society of London Special Publication No. 83.

Brasier, M. D. and O. R. Green. 1993. Winners and losers: Stable isotopes and microhabitats of living Archaiadae and Eocene *Nummulites* (larger foraminifera). *Marine Micropaleontology* 20:267–276.

Brasier, M. D., J. D. Marshall, G. A. F. Carden, D. B. R. Robertson, D. G. F. Long, T. Meidla, L. Hints, and T. F. Anderson. 1994. Bathymetric and isotopic evidence for a short-lived Late Ordovician glaciation in a greenhouse period. *Geology* 22:295–298.

Bretsky, P. W. and S. W. Bretsky. 1976. The maintenance of evolutionary equilibrium in Late Ordovician benthic marine invertebrate faunas. *Lethaia* 9: 223–233.

Bretsky, P. W., Jr. 1969. Central Appalachian Late Ordovician communities. *Geological Society of America Bulletin* 80:193–212.

——. 1970. Upper Ordovician ecology of the central Appalachians. *Peabody Museum Natural History Bulletin* 34:150–178 (plus plates).

Brett, C. B. 1995. Stasis: Life in the balance. *Geotimes* 40:18–20.

Brett, C. E. and G. C. Baird. 1992. Coordinated stasis and evolutionary ecology of Silurian-Devonian marine biotas in the Appalachian Basin. *Abstracts with Programs, GSA Annual Meeting, Cincinnati,* p. A139.

——. 1995. Coordinated stasis and evolutionary ecology of Silurian to middle Devonian faunas in the Appalachian Basin. In D. H. Erwin and R. L. Anstey, eds., *New Approaches to Speciation in the Fossil Record,* pp. 285–315. New York: Columbia University Press.

Brett, C. E., L. C. Ivany, and K. M. Schopf. 1996. Coordinated stasis: An overview. *Palaeogeography, Palaeoclimatology, Palaeoecology* 127:1–20.

Brett, C. E., K. B. Miller, and G. C. Baird. 1990. A temporal hierarchy of paleoecological processes within a Middle Devonian epeiric sea. In W. Miller III, ed., *Paleocommunity Temporal Dynamics: The Long-Term Development of Multispecies Assemblages,* pp. 178–209. Lawrence, Kansas: Paleontological Society Special Publication No. 5.

Briggs, D. E. G., R. A. Fortey, and M. A. Wills. 1992. Morphological disparity in the Cambrian. *Science* 256:1670–1673.

Briggs, J. C. 1995. *Global Biogeography.* Amsterdam: Elsevier.

Broecker, W. S. 1982. Ocean chemistry during glacial time. *Geochimica et Cosmochimica Acta* 46:1689–1705.

Brooks, D. R. and D. H. McLennan. 1991. *Phylogeny, Ecology, and Behavior: A Research Program in Comparative Biology.* Chicago: University of Chicago Press.

———. 1992. Historical ecology as a research program. In R. L. Mayden, ed., *Systematics, Historical Ecology and North American Freshwater Fishes*, pp. 76–113. Palo Alto: Stanford University Press.

———. 1993. Historical ecology: Examining phylogenetic components of community evolution. In R. E. Ricklefs and D. Schluter, eds., *Species Diversity in Ecological Communities*, pp. 267–280. Chicago: University of Chicago Press.

Brown, J. H. 1971. Mountaintop mammals: Nonequilibrium insular biogeography. *American Naturalist* 105:467–478.

———. 1978. The theory of island biogeography and the distribution of boreal birds and mammals. *Great Basin Naturalist Memoirs* 2:209–227.

———. 1984. On the relationship between abundance and distribution of species. *American Naturalist* 124:255–279.

———. 1986. Two decades of interaction between the MacArthur-Wilson model and the complexities of mammalian distributions. *Biological Journal of the Linnean Society* 28:231–251.

———. 1988. Species diversity. In A. A. Myers and P. S. Giller, eds. *Analytical Biogeography*, pp. 57–90. London: Chapman & Hall.

———. 1994. Organisms and species as complex adaptive systems: Linking the biology of populations with the physics of ecosystems. In C. G. Jones and J. H. Lawton, eds., *Linking Species and Ecosystems*, pp. 16–24. New York: Chapman & Hall.

———. 1995. *Macroecology.* Chicago: University of Chicago Press.

Brown, J. H. and M. A. Bowers. 1984. Patterns and processes in three guilds of terrestrial vertebrates. In D. R. Strong, Jr., D. Simberloff, L. G. Abele, and A. B. Thistle eds., *Ecological Communities*, pp. 282–296. Princeton: Princeton University Press.

Brown, J. H. and A. C. Gibson. 1983. *Biogeography.* St. Louis: Mosby.

Brown, J. H. and R. C. Lasiewski. 1972. Metabolism of weasels: The cost of being long and thin. *Ecology* 53:939–943.

Brown, J. H., P. A. Marquet, and M. L. Taper. 1993. Evolution of body size: Consequences of an energetic definition of fitness. *American Naturalist* 142:573–584.

Brown, J. H. and B. A. Maurer. 1987. Evolution of species assemblages: Effects of energetic constraints and species dynamics on the diversification of the North American terrestrial avifauna. *American Naturalist* 130:1–17.

———. 1989. Macroecology: The division of food and space among species on continents. *Science* 243:1145–1150.

Brown, J. H., D. W. Mehlman, and G. C. Stevens. 1995. Spatial variation in abundance. *Ecology* 76:2028–2043.

Brown, J. H. and P. F. Nicoletto. 1991. Spatial scaling of species composition: Body masses of North American land mammals. *American Naturalist* 138:1478–1512.

Brown, W. L., Jr., and E. O. Wilson. 1956. Character displacement. *Systematic Zoology* 7:49–64.

Budd, A. F., K. G. Johnson, and T. A. Stemann. 1996. Plio-Pleistocene turnover and extinctions in the Caribbean reef coral fauna. In J. B. C. Jackson, A. F. Budd, and A. G. Coates, eds., *Evolution and Environment in Tropical America,* pp. 168–204. Chicago: University of Chicago Press.

Budd, A. F., T. A. Stemann, and K. G. Johnson. 1994. Stratigraphic distributions of genera and species of Neogene to Recent Caribbean reef corals. *Journal of Paleontology* 68:951–977.

Burian, R. 1988. Challenges to the evolutionary synthesis. *Evolutionary Biology* 23: 247–269.

Burlando, B. 1993. The fractal geometry of evolution. *Journal of Theoretical Biology* 163:161–172.

Bush, G. L. 1993. A reaffirmation of Santa Rosalia, or why are there so many kinds of small animals? In D. R. Lees and D. Edwards, eds., *Evolutionary Patterns and Processes,* pp. 229–150. London: Linnean Society of London and Academic Press.

Buss, L. W. 1988. Diversification and germ-line determination. *Paleobiology* 14: 313–321.

Buzas, M. A. and S. J. Culver. 1994. Species pool and dynamics of marine paleocommunities. *Science* 264:1439–1441.

Buzas, M. A., C. F. Koch, S. J. Culver, and N. Sohl. 1982. On the distribution of species occurrence. *Paleobiology* 8:142–150.

Bythell, J. C., E. H. Gladfelter, W. B. Gladfelter, K. E. French, and Z. Hillis. 1989. Buck Island Reef National Monument—changes in modern reef community structure since 1976. In D. K. Hubbard, ed., *Terrestrial and Marine Geology of St. Croix, U.S. Virgin Islands,* pp. 145–153. St. Croix, U.S. Virgin Islands: Special Publication Number 8, West Indies Laboratory.

Caccone, A., G. D. Amato, and J. R. Powell. 1988. Rates and Patterns of scnDNA and mrDNA divergence within the *Drosophila melanogaster* subgroup. *Genetics* 118:671–683.

Caccone, A. and J. R. Powell. 1987. Molecular evolutionary divergence among North American cave crickets. II DNA-DNA hybridization. *Evolution* 41: 1215–1238.

Cadle, J. E. and H. W. Greene. 1993. Phylogenetic patterns, biogeography, and the ecological structure of Neotropical snake assemblages. In R. E. Ricklefs and D. Schluter, eds., *Species Diversity in Ecological Communities,* pp. 281–293. Chicago: University of Chicago Press.

Calder, W. A., III. 1984. *Size, Function, and Life History.* Cambridge, Mass.: Harvard University Press.

Canudo, J. I., G. Keller, and E. Molina. 1991. Cretaceous-Tertiary boundary extinction pattern and faunal turnover at Agost and Caravaca, S. E. Spain. *Marine Micropaleontology* 17:319–341.

Carlson, S. J. 1992. Evolutionary trends in the articulate brachiopod hinge mechanism. *Paleobiology* 18:344–366.

Carpenter, R. C. 1986. Partitioning herbivory and its effects on coral reef algal communities. *Ecological Monographs* 56:345–363.

Carr, T. R. and J. A. Kitchell. 1980. Dynamics of taxonomic diversity. *Paleobiology* 6:427–443.

Casey, R. E. 1993. Radiolaria. In J. H. Lipps, ed., *Fossil Prokaryotes and Protists*, pp. 249–284. Boston: Blackwell Scientific Publications.

Caswell, H. and J. E. Cohen. 1993. Local and regional regulation of species-area relations: A patch occupancy model. In R. Ricklefs and D. Schluter, eds., *Species Diversity in Ecological Communities: Historical and Geographical Perspectives*, pp. 99–107. Chicago: University of Chicago Press.

Chang, H., D. Wang, and F. J. Ayala. 1989. Mitochondrial DNA variation in the *Drosophila nasuta* subgroup of species. *Journal of Molecular Evolution* 28: 337–348.

Charlesworth, B. 1995. Making evolution seem complicated. *BioEssays* 17:363–365.

Cifelli, R. L. 1981. Patterns of evolution among the Artiodactyla and Perissodactyla (Mammalia). *Evolution* 35:433–400.

Cisne, J. L. 1974. Evolution of the world fauna of free-living arthropods. *Evolution* 28:337–366.

Clark, C. W. and M. L. Rosenzweig. 1994. Extinction and colonization processes: Parameter estimates from sporadic surveys. *American Naturalist* 143:583–596.

Clark, D. L. 1987. Phylum Conodonta. In R. S. Boardman, A. H. Cheetham, and A. J. Rowell, eds., *Fossil Invertebrates*, pp. 636–662. Palo Alto, Calif.: Blackwell Scientific Publications.

Clarke, G. M. 1995. Relationships between developmental stability and fitness: Application for conservation biology. *Conservation Biology* 9:18–24.

Claypool, G. E., W. T. Holser, I. R. Kaplan, H. Sakai, and I. Zak, I. 1980. The age curves of sulfur and oxygen isotopes in marine sulfate and their mutual interpretation. *Chemical Geology* 28:199–260.

Clements, F. E. 1916. Plant succession, an analysis of the development of vegetation. *Carnegie Institute of Washington Publication* 242:1–512.

———. 1936. Nature and structure of the climax. *Journal of Ecology* 24:252–284.

Cliff, A. D. and J. K. Ord. 1973. *Spatial Autocorrelation*. London: Pion.

———. 1981. *Spatial Processes: Models and Applications*. London: Pion

Cody, M. L. 1975. Towards a theory of continental species diversity. In M. L. Cody and J. M. Diamond, eds., *Ecology and Evolution of Communities*, pp. 214–257. Cambridge: Harvard University Press, Belknap Press.

Cody, M. 1983. The land birds. In T. J. Case and M. L. Cody, eds., *Island Biogeography in the Sea of Cortez*, pp. 21–245. Berkeley: University of California.

Cohen, J. E. 1995. Unexpected dominance of high frequencies in chaotic nonlinear population models. *Nature* 378:610–612.

Cole, F. R., D. M. Reeder, and D. E. Wilson. 1994. A synopsis of distribution patterns and the conservation of mammal species. *Journal of Mammalogy* 75: 266–276.

Colwell, R. K. and J. A. Coddington. 1994. Estimating terrestrial biodiversity through extrapolation. *Philosophical Transactions of the Royal Society of London* B345:101–118.

Compton, J. S., D. A. Hodell, J. R. Garrido, and D. J. Mallinson. 1993. Origin and age of phosphorite from the south-central Florida Platform: Relation of phosphogenesis to sea-level fluctuations and $\delta^{13}C$ excursions. *Geochimica et Cosmochimica Acta* 57:131–146.

Connell, J. H. 1971. On the role of natural enemies in preventing competitive exclusion in some marine animals and in rain forest trees. In P. J. van den Boer and G. Gradwell, eds., *Dynamics of Populations*, pp. 298–312. Wageningen, Netherlands: Centre for Agricultural Publishing and Documentation.

———. 1975. Some mechanisms producing structure in natural communities. In M. L. Cody and J. M. Diamond, eds., *Ecology and Evolution of Communities*, pp. 460–490. Cambridge: Harvard University Press, Belknap Press.

———. 1978. Diversity in tropical rain forests and coral reefs. *Science* 199:1302–1310.

———. 1979. Tropical rain forests and coral reefs as open non-equilibrium systems. In R. M. Anderson et al., eds., *Population Dynamics*, pp. 141–163. Oxford: Blackwells.

Connell, J. H. and M. J. Keough. 1985. Disturbance and patch dynamics of subtidal marine animals on hard substrata. In S. T. A. Pickett and P. S. White, eds., *The Ecology of Natural Disturbance and Patch Dynamics*, pp. 125–152. Orlando: Academic Press.

Connell, J. H. and E. Orias. 1964. The ecological regulation of species diversity. *American Naturalist* 98:399–414.

Connell, J. H. and W. P. Sousa. 1983. On the evidence needed to judge ecological stability or persistence. *American Naturalist* 121:789–824.

Connor, E. F. and E. D. McCoy. 1979. The statistics and biology of the species-area relationship. *American Naturalist* 113:791–833.

Cook, P. J. and J. H. Shergold. 1984. Phosphorus, phosphorites and skeletal evolution at the Precambrian-Cambrian boundary. *Nature* 308:231–236.

Copper, P. 1974. Structure and development of Early Paleozoic reefs. *Proceedings Second International Coral Reef Symposium* 1:365–386.

———. 1988. Ecological succession in Phanerozoic reef ecosystems: Is it real? *Palaios* 3:136–151.

Cornell, H. V. and R. H. Karlson. 1996. Species richness of reef-building corals determined by local and regional processes. *Journal of Animal Ecology* 65:233–241.

Courtillot, V. and Y. Gaudemer. 1996. Effects of mass extinctions on biodiversity. *Nature* 381:146–148.

Cowen, R. and W. L. Stockton. 1978. Testing for evolutionary equilibria. *Paleobiology* 4:195–200.

Cowles, H. C. 1899. The ecological relations of the vegetation on the sand dunes of Lake Michigan. *Botanical Gazette* 27:95–117, 167–202, 281–308, 361–391.

Coyne, J. A. 1994. Ernst Mayr and the origin of species. *Evolution* 48:19–30.

Coyne, J. A., H. A. Orr, and D. J. Futuyma. 1988. Do we need a new species concept? *Systematic Zoology* 37:190–200.

Cracraft, J. 1985. Biological diversification and its causes. *Annals of the Missouri Botanical Garden* 72:794–822.

———. 1989. Speciation and its ontology: The empirical consequences of alternative species concepts for understanding patterns and processes of differentiation. In D. Otte and J. A. Endler, eds., *Speciation and Its Consequences*, pp. 28–59. Sunderland, Mass.: Sinauer.

———. 1990. The origin of evolutionary novelties: Pattern and process at different hierarchical levels. In M. Nitecki, ed., *Evolutionary Innovations*, pp. 21–44. Chicago: University of Chicago Press.

———. 1992. Explaining patterns of biological diversity: Integrating causation at different spatial and temporal scales. In N. Eldredge, ed., *Systematics, Ecology and the Biodiversity Crisis*, pp. 59–76. New York: Columbia University Press.

Crane, P. R. and S. Lidgard. 1989. Angiosperm diversification and paleolatitudinal gradients in cretaceous floristic diversity. *Science* 246:675–678.

Crowley, T. J. and G. R. North. 1988. Abrupt climate change and extinction events in Earth history. *Science* 240:996–1002.

Curnutt, J. C., S. L. Pimm, and B. A. Maurer. 1996. Population variability of sparrows in space and time. *Oikos* 76:131–144.

Currie, D. J. 1993. What shape is the relationship between body size and population density? *Oikos* 66:353–358.

Cutler, A. H. 1991. Nested faunas and extinction in fragmented habitats. *Conservation Biology* 5:496–505.

———. 1994. Nested biotas and biological conservation: Metrics, mechanisms, and meaning of nestedness. *Landscape and Urban Planning* 28:73–82.

Cuvier, G. 1812. *Recherches sur les ossemens fossiles des quadrupèdes, etc.* Paris: Déterville.

Dafni, J. 1986. A biomechanical model for the morphogenesis of regular echinoid tests. *Paleobiology* 12:143–160.

———. 1988. A biomechanical approach to the ontogeny and phylogeny of echinoids. In C. R. C. Paul and A. B. Smith, eds., *Echinoderm Phylogeny and Evolutionary Biology*, pp. 175–188. Oxford: Clarendon Press.

Damuth, J. 1971. Population density and body size in mammals. *Nature* 290:699–700.

Darlington, P. J., Jr. 1957. *Zoogeography: The Geographical Distribution of Animals*. New York: Wiley.

Darwin, C. 1859. *On the Origin of Species by Means of Natural Selection*. London: John Murray.

David, B. 1989. Jeu en mosaïque des hétérochronies: Variation et diversité chez les Pourtalesiidae (échinides abissaux). *Geobios, mémoire spécial* 12:115–131.

Davidson, E. H., K. J. Peterson, and R. A. Cameron. 1995. Origin of bilaterian body plans: Evolution of developmental regulatory mechanisms. *Science* 270:1319–1325.

Davis, M. B. 1986. Climatic instability, time lags, and community disequilibrium. In J. Diamond and T. J. Case, eds., *Community Ecology,* pp. 269–284. New York: Harper and Row.

———. 1989. Retrospective studies. In G. E. Likens, ed., *Long-Term Studies in Ecology,* pp. 71–89. New York: Springer-Verlag.

———. 1994. Ecology and paleoecology begin to merge. *Trends in Ecology and Evolution* 9:357–358.

Dayton, P. K. 1971. Competition, disturbance, and community organization: The provision and subsequent utilization of space in a rocky intertidal community. *Ecological Monographs* 41:351–389.

———. 1984. Processes structuring some marine communities: Are they general? In D. R. Strong, Jr., D. Simberloff, L. G. Abele, and A. B. Thistle, eds., *Ecological Communities,* pp. 181–197. Princeton: Princeton University Press.

Dayton, P. K. and R. R. Hessler. 1972. The role of biological disturbance in maintaining diversity in the deep sea. *Deep-Sea Research* 19:351–389.

Dayton, P. K., M. J. Tegner, P. E. Parnell, and P. B. Edwards. 1992. Temporal and spatial patterns of disturbance and recovery in a kelp forest community. *Ecological Monographs* 62:421–445.

DeAngelis, D. L. 1992. *Dynamics of Nutrient Cycling and Food Webs.* New York: Chapman & Hall.

DeAngelis, D. L. and J. C. Waterhouse. 1987. Equilibrium and non-equilibrium concepts in ecological models. *Ecological Monographs* 57:1–21.

del Moral, R. 1983. Competition as a control mechanism in subalpine meadows. *American Journal of Botany* 70:232–245.

Delcourt, H. R. and P. A. Delcourt. 1991. *Quaternary Ecology.* London: Chapman & Hall.

deMenocal, P. B. 1995. Plio-Pleistocene African climate. *Science* 270:53–59.

Denslow, J. S. 1980a. Gap partitioning among tropical rainforest trees. *Biotropica* 12(suppl.):47–55.

———. 1980b. Patterns of plant species diversity during succession under different disturbance regimes. *Oecologia* 46:18–21.

———. 1985. Disturbance-mediated coexistence of species. In S. T. A. Pickett and P. S. White, eds., *The Ecology of Natural Disturbance and Patch Dynamics,* pp. 307–324. Orlando, Fla.: Academic Press.

———. 1987. Tropical rainforest gaps and tree species diversity. *Annual Review of Ecology and Systematics* 18:431–451.

Depew, D. J. and B. H. Weber. 1995. *Darwinism Evolving: Systems Dynamics and the Genealogy of Natural Selection.* Cambridge, Mass.: MIT Press.

DeSalle, R. and E. Carew. 1992. Phyletic phenocopy and the role of developmental genes in morphological evolution in the Drosophilidae. *Journal of Evolutionary Biology* 5:363–374.

Desmond, A. 1982. *Archetypes and Ancestors: Palaeontology in Victorian London 1850–1875.* Chicago: University of Chicago Press.

Dewey, J. 1909. The influence of Darwinism on philosophy. *Popular Science Monthly* 75(1):90–98.

Dial, K. P. and J. M. Marzluff. 1988. Are the smallest organisms the most diverse? *Ecology* 69:1620–1624.

——. 1989. Nonrandom diversification within taxonomic assemblages. *Systematic Zoology* 38:26–37.

Diamond, J. M. 1969. Avifaunal equilibria and species turnover rates in the channel islands of California. *Proceedings of the National Academy of Sciences of the U.S.A.* 64:57–63.

——. 1971. Comparison of faunal equilibrium turnover rates on a tropical and a temperate island. *Proceedings of the National Academy of Sciences of the U.S.A.* 68:2742–2745.

——. 1972. Biogeographic kinetics: Estimation of relaxation times for avifaunas of southwest Pacific islands. *Proceedings National Academy of Science U.S.A.* 69:3199–3203.

——. 1975. Assembly of species communities. In M. L. Cody and J. M. Diamond, eds., *Ecology and Evolution of Communities,* pp. 342–444. Cambridge: Harvard University Press.

——. 1984a. Historic extinctions: A rosetta stone for understanding prehistoric extinctions. In M. H. Niteki, ed., *Extinctions,* pp. 191–246. Chicago: University of Chicago Press.

——. 1984b. "Normal" extinctions of isolated populations. In P. S. Martin and R. G. Klein, eds., *Quaternary Extinctions: A Prehistoric Revolution,* pp. 824–862. Tucson: University of Arizona Press.

Diamond, J. M. and R. M. May. 1977. Species turnover rates on islands: Dependence on census interval. *Science* 197:226–270.

DiMichele, W. A. 1993. Persistent ecological association and Clementsian ecosystem dynamics in Late Paleozoic terrestrial environments. *Abstracts with Programs, GSA Annual Meeting, Boston,* p. A389.

——. 1994. Ecological patterns in time and space. *Paleobiology* 20:89–92.

DiMichele, W. A. and R. B. Aronson. 1992. The Pennsylvanian-Permian vegetational transition: A terrestrial analogue to the onshore-offshore hypothesis. *Evolution* 46:807–824.

DiMichele, W. A. and R. W. Hook. 1992. Paleozoic terrestrial ecosystems. In A. K. Behrensmeyer, J. D. Damuth, W. A. DiMichele, R. Potts, H.-D. Sues, and S. L. Wing, eds., *Terrestrial Ecosystems Through Time: Evolutionary Paleoecology of Terrestrial Plants and Animals,* pp. 205–325. Chicago: University of Chicago Press.

DiMichele, W. A. and T. L. Phillips. 1992. Consequences of extinction in tropical peat-forming vegetation of the Middle to Late Pennsylvanian (Westphalian-Stephanian). *Abstracts with Programs, GSA Annual Meeting, Cincinnati,* p. A120.

———. 1994. The response of hierarchically structured ecosystems to long-term climatic change: A case study using tropical peat swamps of Pennsylvanian age. In *Effects of Past Global Change on Life,* pp. 134–155. National Research Council, Studies in Geophysics, Washington, D. C.: National Academy Press.

Dobzhansky, T. 1973. Nothing in biology makes sense except in the light of evolution. *American Biology Teacher* 73 (March):125–129.

Done, T. J. 1992. Phase shifts in coral reef communities and their ecological significance. *Hydrobiologia* 247:121–132.

Dony, J. G. 1963. The expectation of plant records from prescribed areas. *Watsonia* 5:377–385.

d'Orbigny, A. 1849–1852. *Cours Elementaire de Paleontologie et de Geologie Stratigraphique, Masson, Paris:* 299, 382, 383–841.

———. 1850–1852. *Prodrome de Paleontologie, Masson, Paris,* 394, 427, 197, 190.

Dott, R. H. 1983. Episodic sedimentation—how normal is normal? *Journal of Sedimentary Petrology* 53:5–23.

Downes, J. A. 1980. The post-glacial colonization of the North Atlantic islands. *Memoirs of the Entomological Society of Canada* 144:55–92.

Doyle, J. A. and M. J. Donoghue. 1993. Phylogenies and angiosperm diversification. *Paleobiology* 19:141–167.

Drake, J. A. 1990. Communities as assembled structures: Do rules govern pattern? *Trends in Ecology and Evolution* 5:159–164.

———. 1991. Community-assembly mechanics and the structure of an experimental species ensemble. *American Naturalist* 137:1–26.

Drake, J. A., G. J. Whitteman, and G. R. Huxel. 1992. Development of biological structure: Critical states, and approaches to alternative levels of organization. In J. Eisenfeld, D. S. Levine, and M. Witten, eds., *Biomedical Modeling and Simulation,* pp. 457–463. Amsterdam: Elsevier Science Publishers.

Drossel, B. and F. Schwabl. 1992. Self-organized critical forest fire model. *Physical Review Letters* 69:1629–1632.

Druffel, E. R. M., R. B. Dunbar, G. M. Wellington, and S. A. Minnis. 1990. Reef-building corals and identification of ENSO warming episodes. In P. W. Glynn, ed., *Global Ecological Consequences of the 1982–83 El Niño-Southern Oscillation,* pp. 233–253. New York: Elsevier.

Dubinsky, K., ed. 1990. *Coral Reefs.* Amsterdam: Elsevier.

Dunbar, R. B., G. M. Wellington, M. W. Colgan, and P. W. Glynn. 1994. Eastern Pacific sea surface temperature since 1600 A. D.: The $\delta^{18}O$ record of climate variability in Galápagos corals. *Paleoceanography* 9:291–315.

Eble, G. J. 1995a. A new look at temporal patterns of evolutionary innovation: Cumulative origination functions. *Geological Society of America Abstracts with Program* 27:A-428.

——. 1995b. Testing the role of development in evolutionary radiations. *Fifth Congress of the European Society for Evolutionary Biology Abstracts* 99:86.

——. 1996. Development and non-development morphospaces in evolutionary paleobiology. *Paleontological Society Special Publication* 8:111.

——. 1997. Disparity in ontogeny and paleontology. *Journal of Vertebrate Paleontology* 17:44A.

Edinger, E. N. and M. J. Risk. 1994. Oligo-Miocene extinction and geographic restriction of Caribbean corals: Roles of turbidity, temperature, and nutrients. *Palaios* 9:576–598.

——. 1995. Preferential survivorship of brooding corals in a regional extinction. *Paleobiology* 21:200–219.

Eggleton, P. and R. I. Vane-Wright. 1994. Some principles of phylogenetics and their implications for comparative biology. In P. Eggleton and R. I. Vane-Wright, eds., *Phylogenetics and Ecology*, pp. 345–366. Linnean Society Symposium Series No. 17. London: Academic Press.

Eldredge, N. 1985. *Unfinished Synthesis: Biological Hierarchies and Modern Evolutionary Thought.* New York: Oxford University Press.

——. 1989. *Macroevolutionary Dynamics: Species, Niches, and Adaptive Peaks.* New York: McGraw-Hill.

——. 1992. Where the twain meet: Causal intersections between the genealogical and ecological realms. In N. Eldredge, ed., *Systematics, Ecology and the Biodiversity Crisis*, pp. 1–14. New York: Columbia University Press.

——. 1993. What, if anything, is a species? In W. H. Kimbel and L. B. Martin, eds., *Species, Species Concepts, and Primate Evolution*, pp. 3–20. Mt. Kisco, N. Y.: Plenum Press.

——. 1995. Species, speciation, and the context of adaptive change in evolution. In D. H. Erwin and R. L. Anstey, eds., *New Approaches to Speciation in the Fossil Record*, pp. 39–63. New York: Columbia University Press.

——. 1996. Hierarchies in evolution. In D. Jablonski, D. H. Erwin, and J. Lipps, eds., *Evolutionary Paleobiology*, pp. 42–61. Chicago: University of Chicago Press.

Eldredge, N. and S. J. Gould. 1972. Punctuated equilibria: An alternative to phyletic gradualism. In T. J. M. Schopf, ed., *Models in Paleobiology*, pp. 82–115. San Francisco: Freeman Cooper.

Ellers, O. 1993. A mechanical model of growth in regular sea urchins: predictions of shape and a developmental morphospace. *Proceedings of the Royal Society of London* B254:123–129.

Emlet, R. B. 1989. Apical skeletons of sea urchins (Echinodermata: Echinoidea): Two methods for inferring mode of larval development. *Paleobiology* 15: 223–254.

———. 1995. Developmental mode and species geographic range in regular sea urchins (Echinodermata: Echinoidea). *Evolution* 49:476–489.

Endler, J. A. 1977. *Geographic Variation, Speciation and Clines.* Princeton: Princeton University Press.

Erwin, D. H. 1989. Regional paleoecology of Permian gastropod genera, southwestern United States and the end-Permian mass extinction. *Palaios* 4:424–438.

———. 1990. The end-Permian mass extinction. *Annual Review of Ecology and Systematics* 21:69–91.

———. 1993a. *The Great Paleozoic Crisis: Life and Death in the Permian.* New York: Columbia University Press.

———. 1993b. The origin of metazoan development: A paleobiological perspective. *Biological Journal of the Linnean Society* 50:255–274.

———. 1994. Early introduction of major morphological innovations. *Acta Paleontologica Polonica* 38:281–294.

Erwin, D. H. and M. L. Droser. 1993. Elvis taxa. *Palaios* 8:623–624.

Erwin, D. H. and J. W. Valentine. 1984. "Hopeful monsters," transposons, and Metazoan radiation. *Proceedings of the National Academy of Sciences U.S.A.* 81:5482–5483.

Erwin, D. H., J. W. Valentine, and J. J. Sepkoski, Jr. 1987. A comparative study of diversification events: The early Paleozoic versus the Mesozoic. *Evolution* 41:1177–1186.

Estes, J. A. and D. O. Duggins. 1995. Sea otters and kelp forests in Alaska: Generality and variation in a community ecological paradigm. *Ecological Monographs* 65:75–100.

Fagerstrom, J. A. 1987. *The Evolution of Reef Communities.* New York: John Wiley.

Farrell, B. D. and C. Mitter. 1993. Phylogenetic determinants of insect/plant community diversity. In R. E. Ricklefs and D. Schluter, eds., *Species Diversity in Ecological Communities,* pp. 253–266. Chicago: University of Chicago Press.

FAUNMAP Working Group. 1996. Spatial response of mammals to late Quaternary environmental fluctuations. *Science* 272:1601–1606.

Feller, W. 1971. *An Introduction to Probability Theory and Its Applications,* Vol II, 2nd edition. New York: John Wiley and Sons.

Felsenstein, J. 1978. The number of evolutionary trees. *Systematic Zoology* 27:27–22.

———. 1985. Phylogenies and the comparative method. *American Naturalist* 125:1–15.

Filippelli, G. M. and M. L. Delaney. 1992. Similar phosphorus fluxes in ancient phosphorite deposits and a modern phosphogenic environment. *Geology* 20:709–712.

Fischer, A. G. and M. A. Arthur. 1977. Secular variations in the pelagic realm. In H. E. Cook and P. Enos, eds., *Deep-water Carbonate Environments,* pp. 19–50. Tulsa: Society of Economic Paleontologists and Mineralogists Special Publication No. 25.

Fisher, D.C. 1986. Progress in organismal design. In D. M. Raup and D. Jablonski, eds., *Patterns and Processes in the History of Life,* pp. 99–117. Berlin: Springer-Verlag.

Fisher, R. A., A. S. Corbet, and C. B. Williams. 1943. The relation between the number of species and the number of individuals in a random sample of an animal population. *Journal of Animal Ecology* 12:42–58.

Flessa, K. W. 1975. Area, continental drift and mammalian diversity. *Paleobiology* 1:189–194.

Flessa, K. W. and J. Imbrie. 1973. Evolutionary pulsations: Evidence from Phanerozoic diversity patterns. In D. H. Tarling and S. K. Runcorn, eds., *Implications of Continental Drift to the Earth Sciences,* vol. 1, pp. 247–285. London: Academic Press.

Flessa, K. W. and D. Jablonski. 1985. Declining Phanerozoic background extinction rates: Effect of taxonomic structure? *Nature* 313:216–218.

Flessa, K. W. and J. S. Levinton. 1975. Phanerozoic diversity patterns: Tests for randomness. *Journal of Geology* 83:239–248.

Flessa, K. W. and J. J. Sepkoski, Jr. 1978. On the relationship between Phanerozoic diversity and changes in habitable area. *Paleobiology* 4:359–366.

Flower, B. P. and J. P. Kennett. 1994. The middle Miocene climatic transition: East Antarctic ice sheet development, deep ocean circulation and global carbon cycling. *Palaeogeography, Palaeoclimatology, Palaeoecology* 108:537–555.

Fluegemann, R. H., Jr. and R. S. Snow. 1989. Fractal analysis of long-range paleoclimatic data: Oxygen isotope record of Pacific core V28–239. *Pure and Applied Geophysics* 131:309–313.

Flynn, J. J. 1986. Faunal provinces and the Simpson Coefficient. *Contributions to Geology, University of Wyoming, Special Paper* 3:317–338.

Föllmi, K. B., H. Weissert, and A. Lini A. 1993. Nonlinearities in phosphogenesis and phosphorus-carbon coupling and their implications for global change. In R. Wollast, F. T. Mackenzie, and L. Chou, eds., *Interactions of C, N, P and S Biogeochemical Cycles and Global Change,* pp. 447–474. Berlin: Springer-Verlag.

Foote, M. 1988. Survivorship analysis of Cambrian and Ordovician trilobites. *Paleobiology* 14:258–271.

——. 1991. Analysis of morphological data. In N. L. Gilinsky and P. W. Signor, eds., *Analytical Paleobiology,* pp. 59–86. Knoxville, Tenn.: The Paleontological Society.

——. 1992. Rarefaction analysis of morphological and taxonomic diversity. *Paleobiology* 18:1–16.

——. 1993a. Contributions of individual taxa to overall morphological disparity. *Paleobiology* 19:403–419.

——. 1993b. Discordance and concordance between morphological and taxonomic diversity. *Paleobiology* 19:185–204.

———. 1994. Temporal variation in extinction risk and temporal scaling of extinction metrics. *Paleobiology* 20:424–444.

———. 1995. Morphological diversification of Paleozoic crinoids. *Paleobiology* 21: 273–299.

———. 1996a. Models of morphological diversification. In D. Jablonski, D. H. Erwin, and J. H. Lipps, eds., *Evolutionary Paleobiology,* pp. 62–86. Chicago: University of Chicago Press.

———. 1996b. Perspective: Evolutionary patterns in the fossil record. *Evolution* 50: 1–11.

———. 1997. In press. The evolution of morphological diversity. *Annual Review of Ecology and Systematics* 28.

Foote, M. and S. J. Gould. 1992. Cambrian and Recent morphological disparity. *Science* 258:1816.

Foote, M. and D. M. Raup. 1996. Fossil preservation and the stratigraphic ranges of taxa. *Paleobiology* 22:121–140.

Fortey, R. A. 1989. There are extinctions and extinctions: Examples from the Lower Paleozoic. *Philosophical Transactions of the Royal Society of London* B325: 327–355.

Foster, J. B. 1964. The evolution of mammals on islands. *Nature* 202:234–235.

Fox, B. J. 1983. Mammal species diversity in Australian heathlands: The importance of pyric succession and habitat diversity. In F. J. Kruger, D. T. Mitchell, and J. U. M. Jarvis, eds., *Mediterranean-Type Ecosystems: The Role of Nutrients,* pp. 473–489. Berlin: Springer.

Fox, J. F. 1979. Intermediate disturbance hypothesis. *Science* 204:1344–1345.

Fox, R. C. 1990. The succession of Paleocene mammals in western Canada. *Geological Society of America Special Paper* 243:51–70.

Frakes, L. A., J. E. Francis, and J. I. Syktus. 1992. *Climate Modes of the Phanerozoic.* Cambridge: Cambridge University Press.

Frost, S. H. 1977. Cenozoic reef systems of Caribbean—Prospects for paleoecologic synthesis. In S. H. Frost, M. P. Weiss, and J. B Saunders, eds., *Reefs and Related Carbonates—Ecology and Sedimentology,* pp. 93–110. Tulsa: American Association of Petroleum Geologists Studies in Geology No. 4.

Fry, M. E., R. J. Risser, H. A. Stubbs, and J. P. Leighton. 1986. Species selection for habitat evaluation procedures. In J. Verner, M. L. Morrison, and C. J. Ralph, eds., *Wildlife 2000: Modeling Habitat Relationships of Terrestrial Vertebrates,* pp. 105–108. Madison: University of Wisconsin Press.

Futuyma, D. 1986. *Evolutionary Biology.* Sunderland, Mass.: Sinauer.

Futuyma, D. J. 1992. History and evolutionary processes. In M. H. Nitecki and D. V. Nitecki, eds., *History and Evolution,* pp. 103–130. Albany: State University of New York Press.

Futuyma, D. J., M. C. Keese, and S. J. Scheffer. 1993. Genetic constraints and the phylogeny of insect-plant associations: Responses of *Ophraella communa* (Coleoptera: Chrysomelidae) to host plants of its congeners. *Evolution* 47: 888–905.

Futuyma, D. J. and G. Moreno. 1988. The evolution of ecological specialization. *Annual Review of Ecology and Systematics* 19:207–233.

Garrels, R. M., A. Lerman, and F. T. Mackenzie. 1976. Controls of atmospheric O_2 and CO_2: Past, present, and future. *American Scientist* 64:306–315.

Gaston, K. J. 1994. *Rarity.* London: Chapman & Hall.

Gaston, K. J. and T. M. Blackburn. 1997. Evolutionary age and risk of extinction in the global avifauna. *Evolutionary Ecology* 11:557–565.

Gauch, H. G., Jr. 1982. *Multivariate Analysis in Community Ecology.* Cambridge, England: Cambridge University Press.

Gazin, C. L. 1962. A further study of the lower Eocene mammalian faunas of southwestern Wyoming. *Smithsonian Miscellaneous Collections* 144:1–98.

———. 1976. Mammalian faunas zones of the Bridger middle Eocene. *Smithsonian Contributions to Paleobiology* 2:1–16.

Geister, J. 1980. Calm-water reefs and rough-water reefs of the Caribbean Pleistocene. *Acta Palaeontologica Polonica* 25:541–556.

———. 1983. Holozäne westindische Korallenriffe: Geomorphologie, Ökologie und Fazies. *Facies* 9:173–284.

Gentry, A. H. 1989. Speciation in tropical forests. In L. B. Holm-Nielsen, I. C. Nielsen, and H. Balslev, eds., *Tropical Forests. Botanical Dynamics, Speciation, and Extinction,* pp. 113–134. London: Academic Press.

George, M., Jr. and O. A. Ryder. 1986. Mitochondrial DNA variation in the genus *Equus. Molecular Biological Evolution* 3:535–546.

Ghiold, J. 1988. Species distributions of irregular echinoids. *Biological Oceanography* 6:79–162.

Gilinsky, N. L. 1986. Species selection as a causal process. *Evolutionary Biology* 20: 249–273.

———. 1991a. Estimating probabilities of origination and extinction. In E. C. Dudley, ed., *The Unity of Evolutionary Biology,* pp. 237–255. Portland, Oregon: Dioscorides Press.

———. 1991b. The pace of taxonomic evolution. In N. L. Gilinsky and P. W. Signor, eds., *Analytical Paleobiology,* pp. 157–174. Knoxville, Tenn.: The Paleontological Society.

———. 1994. Volatility and the Phanerozoic decline of background extinction intensity. *Paleobiology* 20:445–458.

Gilinsky, N. L. and R. K. Bambach. 1987. Asymmetrical patterns of origination and extinction in higher taxa. *Paleobiology* 13:427–445.

Gilinsky, N. L. and I. J. Good. 1991. Probabilities of origination, persistence, and extinction of families of marine invertebrate life. *Paleobiology* 17:145–166.

Gingerich, P. D. 1980. Early Cenozoic paleontology and stratigraphy of the Bighorn Basin, Wyoming. *University of Michigan Papers on Paleontology* 24.

———. 1983. Rates of evolution: Effects of time and temporal scaling. *Science* 222: 159–161.

———. 1989. New earliest Wasatchian mammalian fauna from the Eocene of northwestern Wyoming. *University of Michigan Papers on Paleontology* 28.

Ginsburg, R. N., compiler. 1994. *Proceedings of the Colloquium on Global Aspects of Coral Reefs: Health, Hazards and History, 1993.* Miami: Rosenstiel School of Marine and Atmospheric Science, University of Miami.

Gittleman, J. L. 1982. The phylogeny of parental care in fishes. *Animal Behavior* 29:936–941.

———. 1994. Are the pandas successful specialists or evolutionary failures? *BioScience* 44:456–464.

Gittleman, J. L., C. G. Anderson, M. Kot, and H.-K. Luh. 1995. Comparative tests of evolutionary lability and rates using molecular phylogenies. In P. H. Harvey, A. J. Leigh Brown, J. Maynard Smith, S. Nee, eds., *New Uses for New Phylogenies.* Oxford: Oxford University Press.

Gittleman, J. L., C. G. Anderson, M. Kot, and H.-K. Luh. 1996. Comparing behavioral and morphological evolution: Using molecular phylogenies to measure phylogeny, lability and rates. In E. P. Martins, ed., *Phylogenies and the Comparative Method in Animal Behavior,* pp. 166–205. Oxford: Oxford University Press.

Gittleman, J. L. and M. Kot. 1990. Adaptation: Statistics and a null model for estimating phylogenetic effects. *Systematic Zoology* 39:227–241.

Gittleman, J. L. and H.-K. Luh, 1992. On comparing comparative methods. *Annual Review of Ecology and Systematics* 23:383–404.

Gittleman, J. L. and A. Purvis. 1997. In press. Body size and species richness in carnivores and primates. *Proceedings of the Royal Society of London.*

Gladfelter, W. B. 1982. White band disease in *Acropora palmata:* Implications for the structure and growth of shallow reefs. *Bulletin of Marine Science* 32:639–643.

Gleason, H. A. 1926. The individualistic concept of plant association. *Bulletin of the Torrey Botanical Club* 53:7–26.

Glen, S. M. 1990. Regional analysis of mammal distributions among Canadian Parks: Implications for park planning. *Canadian Journal of Zoology* 68:2457–2464.

Glynn, P. W. 1988. El Niño-Southern Oscillation 1982–1983: Population, community, and ecosystem responses. *Annual Review of Ecology and Systematics* 19: 309–345.

———. 1993. Coral reef bleaching: Ecological perspectives. *Coral Reefs* 12:1–17.

Glynn, P. W. and M. W. Colgan. 1992. Sporadic disturbances in fluctuating coral reef environments: El Niño and coral reef development in the eastern Pacific. *American Zoologist* 32:707–718.

Goldberger, A. L. and B. J. West. 1987. Fractals in physiology and medicine. *Yale Journal of Biology and Medicine* 60:421–435.

Goodman, D. 1987. The demography of chance extinction. In M. E. Soulé, ed.,

Viable Populations for Conservation, pp. 11–34. Cambridge: Cambridge University Press.

Goodwin, B. C. 1990. Structuralism in biology. *Science Progress (Oxford)* 74: 227–244.

——. 1994. *How the Leopard Changed Its Spots.* New York: Charles Scribner's Sons.

Gorman, O. T. 1992. Evolutionary ecology vs. historical ecology: Assembly, structure, and organization of stream fish communities. In R. L. Mayden, ed., *Systematics, Historical Ecology and North American Freshwater Fishes,* pp. 659–688. Palo Alto: Stanford University Press.

Gould, S. J. 1977. Eternal metaphors of paleontology. In A. Hallam, ed., *Patterns of Evolution,* pp. 1–26. Amsterdam: Elsevier.

——. 1980. Is a new and general theory of evolution emerging? *Paleobiology* 6: 119–130.

——. 1981. Palaeontology plus ecology as palaeobiology. In R. M. May, ed., *Theoretical Ecology: Principles and Applications,* pp. 218–236. Sunderland, Mass.: Sinauer.

——. 1984. Covariance sets and ordered geographic variation in *Cerion* from Aruba, Bonaire and Curaçao: A way of studying nonadaptation. *Systematic Zoology* 33:217–237.

——. 1985. The paradox of the first tier: An agenda for paleobiology. *Paleobiology* 11:2–12.

——. 1988. Trends as changes in variance: A new slant on progress and directionality in evolution. *Journal of Paleontology* 62:319–329.

——. 1989a. *Wonderful Life: The Burgess Shale and the Nature of History.* New York: Norton.

——. 1989b. A developmental constraint in *Cerion*, with comments on the definition and interpretation of constraint in evolution. *Evolution* 43:516–539.

——. 1990. Speciation and sorting as the source of evolutionary trends, or 'things are seldom what they seem.' In K. J. MacNamara, ed., *Evolutionary Trends,* pp. 3–27. Tucson: University of Arizona Press.

——. 1991. The disparity of the Burgess Shale arthropod fauna and the limits of cladistic analysis: Why we must strive to quantify morphospace. *Paleobiology* 17:411–423.

——. 1993. How to analyze Burgess Shale disparity—a reply to Ridley. *Paleobiology* 19:522–523.

——. 1994. A developmental constraint in *Cerion,* with comments on the definition and interpretation of constraint in evolution. *Evolution* 43:516–539.

——. 1996. *Full House.* New York: Harmony.

Gould, S. J. and C. B. Calloway. 1980. Clams and brachiopods—ships that pass in the night. *Paleobiology* 3:23–40.

Gould, S. J. and N. Eldredge. 1977. Punctuated equilibria: The tempo and mode of evolution reconsidered. *Paleobiology* 3:115–151.

——. 1993. Punctuated equilibrium comes of age. *Nature (London)* 366:223–227.

Gould, S. J., N. L. Gilinsky, and R. Z. German. 1987. Asymmetry of lineages and the direction of evolutionary time. *Science* 236:1437–1441.

Gould, S. J., D. M. Raup, J. J. Sepkoski, Jr., T. J. M. Schopf, and D. S. Simberloff. 1977. The shape of evolution: A comparison of real and random clades. *Paleobiology* 3:23–40.

Gould, S. J. and E. S. Vrba. 1982. Exaptation—a missing term in the science of form. *Paleobiology* 8:4–15.

Graham, R. W. 1986. Response of mammalian communities to environmental changes during the Late Quaternary. In J. Diamond and T. J. Case, eds., *Community Ecology*, pp. 300–313. New York: Harper and Row.

Grandcolas, P. 1993 The origin of biological diversity in a tropical cockroach lineage. *Acta Oecologia* 14:259–270.

Grant, B. R. and P. R. Grant. 1993. Evolution of Darwin's finches caused by a rare climatic event. *Proceedings of the Royal Society of London* B:111–117.

Grant, P. R. 1986a. *Ecology and Evolution of Darwin's Finches*. Princeton: Princeton University Press.

——. 1986b. Interspecific competition in fluctuating environments. In J. Diamond and T. J. Case, eds., *Community Ecology*, pp.173–191. New York: Harper and Row.

Grant, P. R. and B. R. Grant. 1995. The founding of a new population of Darwin's finches. *Evolution* 49:229–240.

Grassberger, P. 1995. The Bak-Sneppen model for punctuated evolution. *Physics Letters A* 200:277–282.

Grayson, D. K. and S. D. Livingston. 1993. Missing mammals on Great Basin Mountains: Holocene extinctions and inadequate knowledge. *Conservation Biology* 7:527–532.

Grene, M. 1990. Is evolution at a crossroads? *Evolutionary Biology* 24:51–81.

Grieger, B. 1992. Quaternary climatic fluctuations as a consequence of self-organized criticality. *Physica A* 191:51–56.

Grime, J. P. 1979. *Plant Strategies and Vegetation Processes*. New York: Wiley.

Grubb, P. J. 1977. The maintenance of species-richness in plant communities: The importance of the regeneration niche. *Biological Reviews* 52:107–145.

Guyer, C. and J. B. Slowinski. 1993. Adaptive radiation and the topology of large phylogenies. *Evolution* 47:253–263.

Hall, B. K. 1992. *Evolutionary Developmental Biology*. London: Chapman & Hall.

Hallam, A. 1989. The case for sea-level change as a dominant causal factor in mass extinction of marine invertebrates. *Philosophical Transactions of the Royal Society of London* B325:437–455.

——. 1992. *Phanerozoic Sea-Level Changes*. New York: Columbia University Press.

——. 1994. *An Outline of Phanerozoic Biogeography*. Oxford: Oxford University Press.

Hallock, P. 1987. Fluctuations in the trophic resource continuum: A factor in global diversity cycles? *Paleoceanography* 2:457–471.

———. 1988a. The role of nutrient availability in bioerosion: Consequences to carbonate buildups. *Palaeogeography, Palaeoclimatology, Palaeoecology* 63: 275–291.

———. 1988b. Interoceanic differences in foraminifera with symbiotic algae: A result of nutrient supplies? *Proceedings of the 6th International Coral Reef Symposium* 3:251–255.

Hallock, P., I. Premoli Silva, and A. Boersma. 1991. Similarities between planktonic and larger foraminiferal evolutionary trends through Paleogene paleoceanographic changes. *Palaeogeography, Palaeoclimatology, Palaeoecology* 83: 49–64.

Hallock, P. and W. Schlager. 1986. Nutrient excess and the demise of coral reefs and carbonate platforms. *Palaios* 1:389–398.

Hansen, T. A. 1978. Larval dispersal and species longevity in Lower Tertiary gastropods. *Science* 199:885–888.

Hansen, T. A. 1988. Early Tertiary radiation of marine molluscs and the long-term effects of the Cretaceous-Tertiary extinction. *Paleobiology* 14:37–51.

Hansen, T. A., B. R. Farrell, and B. Upshaw. 1993. The first 2 million years after the Cretaceous-Tertiary boundary in east Texas: Rates and paleoecology of the molluscan recovery. *Paleobiology* 19:251–265.

Hanski, I. 1989. Metapopulation dynamics: Does it help to have more of the same? *TREE* 4:113–114.

———. 1991. Single-species metapopulation dynamics: Concepts, models and observations. *Biological Journal of the Linnean Society* 42:17–38.

Hanski, I. and M. Gilpin. 1991. Metapopulation dynamics: Brief history and conceptual domain. *Biological Journal of the Linnean Society* 42:3–16.

Hanski, I. and M. E. Gilpin, eds. 1997. *Metapopulation Biology: Ecology, Genetics and Evolution.* New York: Academic Press.

Hanski, I., M. Kuussaari, and M. Nieminen. 1994. Metapopulation structure and migration in the butterfly Melitaea cinxia. *Ecology* 75:747–762.

Harding, E. F. 1971. The probabilities of rooted tree-shapes generated by random bifurcation. *Advances in Applied Probability* 3:44–77.

Harland, W., R. L. Armstrong, A. V. Cox, L. E. Craig, A. G. Smith, and D. G. Smith. 1990. *A Geologic Time Scale, 1989.* Cambridge: Cambridge University Press.

Harries, P. J., E. G. Kauffman, and T. A. Hansen. 1996. Models for biotic survival following mass extinction. In M. B. Hart, ed., *Biotic Recovery from Mass Extinction Events,* pp. 41–60. London: Geological Society.

Harriott, V. J. 1992. Recruitment patterns of scleractinian corals in an isolated subtropical reef system. *Coral Reefs* 11:215–219.

Harrison, P. L. and C. C. Wallace. 1990. Reproduction, dispersal, and recruitment of scleractinian corals. In Z. Dubinsky, ed., *Coral Reefs,* pp. 133–208. New York: Elsevier.

Harrison, S. 1991. Local extinction in a metapopulation context: An empirical evaluation. In M. E. Gilpin and I. Hanski, eds., *Metapopulation Dynamics: Empirical and Theoretical Investigations,* pp. 73–88. London: Academic Press.

———. 1992. Species diversity, spatial scale and global change. In P. Kareiva, R. Huey, and J. Kingsolver, eds., *Biotic Interactions and Global Change,* pp. 388–401. Sunderland, Mass.: Sinauer.

———. 1994. Metapopulations and conservation. In P. J. Edwards, N. R. Webb, and R. M. May, eds., *Large-Scale Ecology and Conservation Biology,* pp. 111–128. Oxford: Blackwell Scientific.

Harrison, S. H. and A. Hastings. 1996. Genetic and evolutionary consequences of metapopulation structure. *Trends in Ecology and Evolution* 11:180–184.

Harrison, S., D. D. Murphy, and P. R. Ehrlich. 1988. Distribution of the bay checkerspot butterfly, *Euphydryas editha bayensis:* evidence for a metapopulation model. *American Naturalist* 132:360–82.

Harrison, S. and J. F. Quinn. 1989. Correlated environments and the persistence of metapopulations. *Oikos* 56:293–298.

Harrison, S., S. J. Ross, and J. H. Lawton. 1992. Beta diversity on geographic gradients in Britain. *Journal of Animal Ecology* 61:141–148.

Harrison, S. and A. D. Taylor. 1997. Empirical evidence for metapopulation dynamics. In I. Hanski and M. E. Gilpin, eds., *Metapopulation Biology: Ecology, Genetics and Evolution,* pp. 27–42. New York: Academic Press.

Harrison, S., C. D. Thomas, and T. M. Lewinsohn. 1995. Testing a metapopulation model of coexistence in the insect community on ragwort (*Senecio jacobaea*). *American Naturalist* 145:545–561.

Harvey, P. H. 1996. Phylogenies for ecologists. *Journal of Animal Ecology* 65:255–263.

Harvey, P. H., A. J. Leigh Brown, J. Maynard Smith, and S. Nee, eds. 1996. *New Uses for New Phylogenies,* pp. 1–17. Oxford: Oxford University Press.

Harvey, P. H., R. M. May, and S. Nee. 1994. Phylogenies without fossils. *Evolution* 48:523–529.

Harvey, P. H. and M. D. Pagel. 1991. *The Comparative Method in Evolutionary Biology.* Oxford: Oxford University Press.

Hastings, A. 1991. Structured models of metapopulation dynamics. *Biological Journal of the Linnean Society* 42:57–71.

Hastings, A. and K. Higgins. 1994. Persistence of transients in spatially structured ecological models. *Science* 263:1133–1136.

Hastings, A. M. and S. Harrison. 1994. Metapopulation dynamics and genetics. *Annual Review of Ecology and Systematics* 25:167–188.

Hatcher, B. G. 1984. A maritime accident provides evidence for alternate stable states in benthic communities on coral reefs. *Coral Reefs* 3:199–204.

Havenhand, J. N. 1995. Evolutionary ecology of larval types. In L. McEdward, ed., *Ecology of Marine Invertebrate Larvae,* pp. 79–122. Boca Raton: CRC Press.

Hay, M. E. 1984. Patterns of fish and urchin grazing on Caribbean coral reefs: Are previous results typical? *Ecology* 65:446–454.

Hayward, T. L. and J. A. McGowan. 1979. Pattern and structure in an oceanic zooplankton community. *American Zoologist* 19:74–83.

Heard, S. B. and D. L. Hauser. 1995. Key evolutionary innovations and their ecological mechanisms. *Historical Biology* 10:151–173.

Hebert, P. D. N. and C. C. Wilson. 1994. Provincialism in plankton: Endemism and allopatric speciation in Australian Daphnia. *Evolution* 48:1333–1349.

Hecht, A. D. and B. Agan. 1972. Diversity and age relationships in recent and Miocene bivalves. *Systematic Zoology* 21:308–312.

Hedgecock, D. 1986. Is gene flow from pelagic larval dispersal important in the adaptation and evolution of marine invertebrates? *Bulletin of Marine Science* 39:550–564.

Hedgpeth, J. W., ed. 1957. Treatise on Marine Ecology and Paleoecology, Volume I: Ecology. *Geological Society of America Memoir 67.* Washington D. C.: Waverly Press.

Hellberg, M. 1994. Relationships between inferred levels of gene flow and geographic distance in a philopatric coral, *Balanophyllia elegans. Evolution* 48:221–229.

Hengeveld, R. 1990. *Dynamic Biogeography.* Cambridge: Cambridge University Press.

Hengeveld, R. and J. Haeck. 1981. The distribution of abundance. II. Models and implications. *Proceedings of the Koninklijke Nederlandse Akademie van Wetenshappen,* C84:257–284.

———. 1982. The distribution of abundance. I. Measurements. *Journal of Biogeography* 9:303–316.

Hennig, W. 1966. *Phylogenetic Systematics.* Urbana, Ill.: University of Illinois Press.

Herbold, B. 1984. Structure of an Indiana stream fish association: choosing an appropriate model. *American Naturalist* 124:561–572

Hey, J. 1991. The structure of genealogies and the distribution of fixed differences between DNA sequence samples from natural populations. *Genetics* 128:831–840.

———. 1992. Using phylogenetic trees to study speciation and extinction. *Evolution* 46:627–640.

Hickman, C. S. 1993. Theoretical design space: A new program for the analysis of structural diversity. *Neues Jahrbuch für Geologie und PaläontologieAbhandlungen* 190:183–190.

Highsmith, R. C. 1980. Geographic patterns of coral bioerosion: A productivity hypothesis. *Journal of Experimental Marine Biology and Ecology* 46:177–196.

Ho, M.-W. 1990. An exercise in rational taxonomy. *Journal of Theoretical Biology* 147:43–57.

———. 1992. Development, rational taxonomy and systematics. *Rivista di Biologia— Biology Forum* 85:193–211.

Hoffman, A. 1989. *Arguments on Evolution.* Oxford: Oxford University Press.

Holland, S. M. 1995. The stratigraphic distribution of fossils. *Paleobiology* 21:92–109.

Hollander, D. J., J. A. McKenzie, and K. J. Hsü. 1993. Carbon isotope evidence for unusual plankton blooms and fluctuations of surface CO_2 in "Strangelove ocean" after terminal Cretaceous event. *Palaeogeography, Palaeoclimatology, Palaeoecology* 104:229–238.

Holling, C. S. 1993. Cross-scale morphology, geometry, and dynamics os ecosystems. *Ecological Monographs* 62:447–502.

Holme, N. A. 1984. Fluctuations of *Ophiothrix fragilis* in the western English Channel. *Journal of the Marine Biological Association of the United Kingdom* 64: 351–378.

Holser, W. T. and M. Magaritz. 1992. Cretaceous/Tertiary and Permian/Triassic boundary events compared. *Geochimica et Cosmochimica Acta* 56:3297–3309.

Holser, W. T., M. Schidlowski, F. T. Mackenzie, and J. B. Maynard. 1988. Biogeochemical cycles of carbon and sulfur. In C. B. Gregor, R. M. Garrels, F. T. Mackenzie, and J. B. Maynard, eds., *Chemical Cycles in the Evolution of the Earth*, pp. 105–173. New York: John Wiley.

Holser, W. T., H.-P. Schönlaub, K. Boeckelmann, and M. Magaritz, M. 1991. The Permian-Triassic of the Gartnerkofel-1 Core (Carnian Alps, Austria): Synthesis and conclusions. *Abhandlungen der Geologischen Bundesanstalt* 45: 213–232.

Holt, R. D. 1977. Predation, apparent competition, and the structure of prey communities. *Theoretical Population Biology* 12:197–229.

———. 1993. Ecology at the mesoscale: The influence of regional processes on local communities. In R. E. Ricklefs and D. Schluter, eds., *Species Diversity in Ecological Communities: Historical and Geographical Perspectives*, pp. 77–88. Chicago: University of Chicago Press.

Holt, R. D., J. H. Lawton, K. J. Gaston, and T. M. Blackburn. 1997. On the relationship between range size and local abundance: Back to basics. *Oikos* 78: 183–190.

Holterhoff, P. F. 1994. Biofacies recurrence and asymmetry in cyclic sequences: Crinoids from the Upper Pennsylvanian (Missourian) Lansing Group, Midcontinent, N. A. *Abstracts with Programs, GSA Annual Meeting, Seattle*, p. A519.

———. 1996. Crinoid biofacies in Upper Carboniferous cyclothems, Midcontinent, North America: Faunal tracking and the role of regional processes in biofacies recurrence. *Palaeogeography, Palaeoclimatology, Palaeoecology* 127:47–82.

House, M. R. 1989. Ammonoid extinction events. *Philosophical Transactions of the Royal Society of London* B325:307–326.

Howard, D. J. 1994. Reinforcement: The origin, dynamics, and fate of an evolutionary hypothesis. In R. J. Harrison, ed., *Hybrid Zones and the Evolutionary Process*, pp. 78–104. Oxford: Oxford University Press.

Hsü, K., L. Jiliang, C. Haihong, W. Qingchen, S. Shu, and A. M. C. Sengör. 1990. Tectonics of South China: Key to understanding West Pacific geology. *Tectonophysics* 183:9–39.

Hsü. K. J. 1986. Environmental changes in times of biotic crisis. In D. M. Raup and

D. Jablonski, eds., *Patterns and Processes in the History of Life*, pp. 297–312. Berlin: Springer-Verlag.

——. 1989. Catastrophic extinctions and the inevitability of the improbable. *Journal of the Geological Society of London* 146:749–754.

Hsui, A., K. A. Rust, and G. D. Klein. 1993. A fractal analysis of Quaternary, Cenozoic-Mesozoic, and Late Pennsylvanian sea-level changes. *Journal of Geophysical Research* 98:21963–21967.

Hubbell, S. P. and R. B. Foster. 1986. Biology, chance, and history and the structure of tropical rain forest tree communities. In J. M. Diamond and T. J. Case, eds., *Community Ecology*, pp. 314–329. New York: Harper and Row.

Hughes, N. C. 1990. Morphological plasticity and genetic flexibility in a Cambrian trilobite. *Geology* 19:913–916.

Hughes, R. G. 1986. Theories and models of species abundance. *American Naturalist* 128:879–899.

Hughes, T. P. 1989. Community structure and the diversity of coral reefs: The role of history. *Ecology* 70:275–279.

——. 1994. Catastrophes, phase shifts and large-scale degradation of a Caribbean coral reef. *Science* 265:1547–1551.

Hulbert, R. C., Jr. 1993. Taxonomic evolution in North American Neogene horses (subfamily Equinae): The rise and fall of an adaptive radiation. *Paleobiology* 19:216–234.

Humphries, C. J., P. H. Williams, and R. I. Vane-Wright. 1995. Measuring biodiversity value for conservation. *Annual Review of Ecology and Systematics* 26: 93–111.

Hunt, A. 1993. Effects of contrasting patterns of larval dispersal on the genetic connectedness of local populations of two intertidal starfish, *Patiriella calcar* and *Patiriella exigua*. *Journal of Experimental Marine Biology and Ecology* 92: 179–86.

Hunt, G. L. and M. W. Hunt. 1974. Trophic levels and turnover rates: The avifauna of Santa Barbara Island, California. *Condor* 76:363–369.

Huntley, B. and T. Webb III. 1988. *Vegetation History*. Boston: Kluwer Academic Publishers.

——. 1989. Migration: Species' response to climatic variations caused by changes in the earth's orbit. *Journal of Biogeography* 16:5–19.

Hurlbert, S. H. 1971. The nonconcept of species diversity: A critique and alternative parameters. *Ecology* 52:577–586.

Hüssner, H. and J. Roessler. 1995. Self-organization in geological and (paleo-) biological systems. *Neues Jahrbuch für Geologie und Paläontologie, Abhandlungen* 195:159–180.

Huston, M. A. 1979. A general hypothesis of species diversity. *American Naturalist* 113:81–101.

——. 1994. *Biological Diversity. The Coexistence of Species on Changing Landscapes*. Cambridge: Cambridge University Press.

Hutchinson, G. E. 1959. Homage to Santa Rosalia: Or, why are there so many kinds of animals? *American Naturalist* 93:145–159.

———. 1965. *The Ecological Theater and the Evolutionary Play.* New Haven: Yale University Press.

Hutchinson, G. E. and R. H. MacArthur. 1959. A theoretical ecological model of size distributions among species. *American Naturalist* 93:117–123.

Hwa, T. and M. Kardar. 1989. Fractals and self-organized criticality in dissipitative dynamics. *Physica D* 38:198–202

Ivany, L. C. 1997. *Faunal stability and environmental change in the middle Eocene Gulf Coastal Plain.* Dissertation, Harvard University, Cambridge, MA.

Ivany, L. C., C. R. Newton, and H. T. Mullins. 1994a. Benthic invertebrates of a modern carbonate ramp: A preliminary survey. *Journal of Paleontology* 68: 417–433.

Ivany, L. C. and K. M. Schopf, eds. 1996. New perspectives on faunal stability in the fossil record. Theme issue, *Palaeogeography, Palaeoclimatology, Palaeoecology* 127(1–4):1–361.

Ivany, L. C., K. M. Schopf, and C. E. Brett. 1994b. Biofacies and coordinated stasis. *Abstracts with Programs, GSA Annual Meeting, Seattle,* p. A453.

Jablonski, D. 1983. Onshore-offshore patterns in the evolution of Phanerozoic shelf communities. *Science* 222:1123–1125.

———. 1986a. Background and mass extinctions: The alternation of macroevolutionary regimes. *Science* 231:129–133.

———. 1986b. Causes and consequences of mass extinctions: A comparative approach. In D. K. Elliott, ed., *Dynamics of Extinction,* pp. 183–229. New York: John Wiley.

———. 1986c. Larval ecology and macroevolution in marine invertebrates. *Bulletin of Marine Science* 39:565–587.

———. 1987. Heritability at the species level: Analysis of geographic ranges of Cretaceous mollusks. *Science* 238:360–363.

———. 1989. The biology of mass extinction: A palaeontological view. *Philosophical Transactions of the Royal Society of London* B325:357–368.

———. 1991. Extinctions: A paleontological perspective. *Science* 253:754–757.

———. 1995. Extinctions in the fossil record. In J. H. Lawton and R. M. May, eds., *Extinction Rates,* pp. 25–44. Oxford: Oxford University Press.

Jablonski, D. and D. J. Bottjer. 1990a. The ecology of evolutionary innovation: The fossil record. In M. Nitecki, ed., *Evolutionary Innovations,* pp. 253–288. Chicago: University of Chicago Press.

———. 1990b. The origin and diversification of major groups: Environmental patterns and macroevolutionary lags. In P. D. Taylor and G. P. Larwood, eds., *Major Evolutionary Radiations. Systematics Association Special Volume No. 42,* pp. 17–57. Oxford: Clarendon Press.

———. 1990c. Onshore-offshore trends in marine invertebrate evolution. In R. M.

Ross and W. D. Allmon, eds., *Causes of Evolution: A Paleontological Perspective,* pp. 21–75. Chicago: University of Chicago Press.

——. 1991. Environmental patterns in the origin of higher taxa: The post-Paleozoic fossil record. *Science* 252:1831–1833.

——. 1995. Extinctions in the fossil record. In J. H. Lawton and R. M. May, eds., *Extinction Rates,* pp. 25–44. Oxford: Oxford University Press.

Jablonski, D., K. W. Flessa, and J. W. Valentine. 1985. Biogeography and paleobiology. *Paleobiology* 11:75–90.

Jablonski, D. and R. A. Lutz. 1983. Larval ecology of marine benthic invertebrates. *Biological Reviews* 58:21–89.

Jablonski, D. and D. M. Raup. 1995. Selectivity of end-Cretaceous Marine bivalve extinctions. *Science* 268:389–391.

Jablonski, D. and J. J. Sepkoski, Jr. 1996. Paleobiology, community ecology, and scales of ecological pattern. *Ecology* 77(5):1367–1378.

Jablonski, D., J. J. Sepkoski, Jr., D. J. Bottjer, and P. M. Sheehan. 1983. Onshore-offshore patterns in the evolution of Phanerozoic shelf communities. *Science* 222:1123–1125.

Jackson, J. B. C. 1974. Biogeographic consequences of eurytopy and stenotopy among marine bivalves and their evolutionary significance. *American Naturalist* 108:541–560.

——. 1985. Distribution and ecology of clonal and aclonal benthic invertebrates. In J. B. C. Jackson, L. W. Buss, and R. E. Cook, eds., *Population Biology and Evolution of Clonal Organisms,* pp. 297–355. New Haven: Yale University Press.

——. 1986. Modes of dispersal of clonal benthic invertebrates: Consequences for species' distributions and genetic structure of local populations. *Bulletin of Marine Science* 39:588–606.

——. 1988. Does ecology matter? (book review). *Paleobiology* 14:307–312.

——. 1992. Pleistocene perspectives on coral reef community structure. *American Zoologist* 32:719–731.

——. 1994a. Community unity? *Science* 264:1412–1413.

——. 1994b. Constancy and change of life in the sea. *Philosophical Transactions of the Royal Society of London* B344:55–60.

Jackson, J. B. C. and T. P. Hughes. 1985. Adaptive strategies of coral-reef invertebrates. *American Scientist* 73:265–274.

Jacobs, D. K. 1990. Selector genes and the Cambrian radiation of Bilateria. *Proceedings of the National Academy of Sciences U.S.A.* 87:4406–4410.

Jaeger, J.-J. 1994. The evolution of biodiversity among the Southwest European rodent (Mammalia, Rodentia) communities: Pattern and process of diversification and extinction. *Palaeogeography, Palaeoclimatology, Palaeoecology* 111: 305–336.

Janzen, D. H. 1970. Herbivores and the number of tree species in tropical forests. *American Naturalist* 104:501–528.

———. 1977. Why are there so many species of insects? *Proceedings XV Congress Entomologists, Entomology Society of America,* pp. 84–94.

Järvinen, O., C. Babin, R. K. Bambach, E. Flügel, F. T. Fürsich, D. J. Futuyma, K. J. Niklas, A. L. Panchen, D. Simberloff, A. J. Underwood, and K. F. Weidlich. 1986. The neontological-paleontological interface of community evolution: How do the pieces in the kaleidoscopic biosphere move? In D. M. Raup and D. Jablonski, eds., *Patterns and Processes in the History of Life,* pp. 331–350. Berlin: Springer-Verlag.

Joachimski, M. M. and W. Buggisch. 1993. Anoxic events in the late Frasnian— Causes of the Frasnian-Famennian faunal crisis? *Geology* 21:675–678.

Johnson, K. G., A. F. Budd, and T. A. Stemann. 1995. Extinction selectivity and ecology of Neogene Caribbean reef-corals. *Paleobiology* 21:52–73.

Johnson, L. 1988. The thermodynamic origin of ecosystems: A tale of broken symmetry. In B. H. Weber, D. J. Depew, and J. D. Smith, eds., *Entropy, Information, and Evolution: New Perspectives on Physical and Biological Evolution,* pp. 75–105. Cambridge: MIT Press.

Jones, H. L. and J. M. Diamond. 1976. Short-time-base studies of turnover in breeding bird populations on the Californian channel islands. *Condor* 78: 526–529.

Karlson, R. H. and L. E. Hurd. 1993. Disturbance, coral reef communities, and changing ecological paradigms. *Coral Reefs* 12:117–125.

Karr, J. R. and K. E. Freemark. 1985. Disturbance and vertebrates: An integrative perspective. In S. T. A. Pickett and P. S. White, eds., *The Ecology of Natural Disturbance and Patch Dynamics,* pp. 153–168. Orlando: Academic Press.

Kauffman, E. G. and J. A. Fagerstrom. 1993. The Phanerozoic evolution of reef diversity. In R. E. Ricklefs and D. Schluter, eds., *Species Diversity in Ecological Communities: Historical and Geographical Perspectives,* pp. 315–329. Chicago: University of Chicago Press.

Kauffman, S. A. 1985. Self-organization, selective adaptation and its limits. In D. J. Depew and B. H. Weber, eds., *Evolution at a Crossroads,* pp. 169–207. Cambridge: MIT Press.

———. 1989. Cambrian explosion and Permian quiescence: Implications of rugged fitness landscapes. *Evolutionary Ecology* 3:274–281.

———. 1991. Antichaos and adaptation. *Scientific American* 265(2):78–84.

———. 1993. *The Origins of Order: Self-Organization and Selection in Evolution.* New York: Oxford University Press.

———. 1995. *At Home in the Universe.* Oxford: Oxford University Press.

Kauffman, S. A. and S. Johnsen. 1991. Co-evolution to the edge of chaos: Coupled fitness landscapes, poised states, and co-evolutionary avalanches. In C. G. Langton, C. Taylor, J. D. Farmer, and S. Rasmussen, eds., *Artificial Life II: Santa Fe Institute Studies in the Sciences of Complexity,* pp. 325–369. Reading, Mass.: Addison-Wesley.

———. 1991. Coevolution to the edge of chaos: Coupled fitness landscapes, poised states, and coevolutionary avalanches. *Journal of Theoretical Biology* 149: 467–505.

Kauffman, S. A. and S. Levin. 1987. Towards a general theory of adaptive walks on rugged landscapes. *Journal of Theoretical Biology* 128:11–45.

Kaufman, A. J., S. B. Jacobsen, and A. H. Knoll. 1993. The Vendian record of Sr and C isotopic variations in seawater: Implications for tectonics and paleoclimate. *Earth and Planetary Science Letters* 120:409–430.

Kay, E. A. 1984. Patterns of speciation in the Indo-West Pacific. In F. Radovsky, P. Raven, and S. H. Sohmer, eds., *Biogeography of the Tropical Pacific*, pp. 15–31. Honolulu: Bishop Museum.

Kelt, D. A. 1997. In press. Assembly of local communities: Consequences of an optimal body size for the organization of competitively structured communities. *Biological Journal of the Linnean Society* 61.

Kennett, J. P. 1982. *Marine Geology.* Englewood Cliffs, N.J.: Prentice-Hall.

Kerr, R. A. 1994. Who profits from ecological disaster? *Science* 266:28–30.

Kertész, J. and L. B. Kiss. 1990. The noise spectrum in the model of self-organized criticality. *Journal of Physics A* 23:L433–L440

Kidwell, S. M. and D. W. J. Bosence. 1991. Taphonomy and time-averaging of marine shelly faunas. In P. A. Allison and D. E. G. Briggs, eds., *Taphonomy: Releasing the Data Locked in the Fossil Record*, pp. 115–209. Mt. Kisco, N.Y.: Plenum Press.

Kihm, A. J. 1984. Early Eocene mammalian faunas of the Piceance Creek Basin, northwestern Colorado. Ph.D. thesis, University of Colorado.

Kilham, P. and S. S. Kilham. 1980. The evolutionary ecology of phytoplankton. In I. Morris, ed., *The Physiological Ecology of Phytoplankton*, pp. 571–597. Berkeley: University of California Press.

Kim, K.-J, B. L. Turner, and R. K. Jansen. 1992. Phylogenetic and evolutionary implications of interspecific chloroplast DNA variation in Krigia (Asteraceae-Lactuceae). *Systematic Botany* 17:449–469.

Kingman, J. F. C. 1982. The coalescent. *Stochastic Processes and Applications* 13: 235–248.

Kirkpatrick, M. 1982. Quantum evolution and punctuated equilibria in continuous genetic characters. *American Naturalist* 119:833–848.

Kirkpatrick, M. and M. Slatkin. 1993. Searching for evolutionary patterns in the shape of a phylogenetic tree. *Evolution* 47:1171–1181.

Kitchener, D. J., A. Chapman, J. Dell, B. G. Muir, and M. Palmer. 1980a. Lizard assemblage and reserve size and structure in the Western Australian wheatbelt—some implications for conservation. *Biological Conservation* 17: 25–62.

Kitchener, D. J., A. Chapman, B. G. Muir, and M. Palmer. 1980b. The conservation value for mammals of reserves in the Western Australian wheatbelt. *Biological Conservation* 18:179–207.

Kitchener, D. J., J. Dell, B. G. Muir, and M. Palmer. 1982. Birds in Western Australian wheatbelt reserves—implications for conservation. *Biological Conservation* 22:127–163.

Knoll, A. H. 1984. Patterns of extinction in the fossil record of vascular plants. In M. H. Nitecki, ed., *Extinctions*, pp. 21–68. Chicago: University of Chicago Press.

———. 1986. Patterns of change in plant communities through geologic time. In J. Diamond and T. J. Case, eds., *Community Ecology*, pp. 126–141. New York: Harper and Row.

———. 1994. Proterozoic and Early Cambrian protists: Evidence for accelerating evolutionary tempo. *Proceedings of the National Academy of Sciences of the U.S.A.* 91:6743–6750.

Knoll, A. H., R. K. Bambach, D. E. Canfield, and J. P. Grotzinger. 1996. Comparative Earth history and Late Permian mass extinction. *Science* 273:452–457.

Knoll, M. A. and W. C. James. 1987. Effect of the advent and diversification of vascular land plants on mineral weathering through geologic time. *Geology* 15: 1099–1102.

Knowlton, N. 1992. Thresholds and multiple stable states in coral reef community dynamics. *American Zoologist* 32:674–682.

———. 1993. Sibling species in the sea. *Annual Review of Ecology and Systematics* 24: 189–216.

Knowlton, N., J. C. Lang, and B. D. Keller. 1990. Case study of natural population collapse: Post-hurricane predation on Jamaican staghorn corals. *Smithsonian Contributions to the Marine Sciences* 31:1–25.

Koch, C. F. 1987. Prediction of sample size effects on the measured temporal and geographic distribution patterns of species. *Paleobiology* 13:100–107.

Koch, C. F. and J. P. Morgan. 1988. On the expected distribution of species' ranges. *Paleobiology* 14:126–138.

Koch, P. L., J. C. Zachos, and P. D. Gingerich. 1992. Correlation between isotope records in marine and continental carbon reservoirs near the Paleocene/Eocene boundary. *Nature* 358:319–322.

Kodrick-Brown, A. and J. H. Brown. 1993. Incomplete data sets in community ecology and biogeography: A cautionary tale. *Ecological Applications* 3:736–742.

Korvin, G. 1992. *Fractal Models in the Earth Sciences.* Amsterdam: Elsevier.

Krajewski, C. 1990. Relative rates of single-copy DNA evolution in cranes. *Molecular Biological Evolution* 7:65–73.

Kremen, C. 1992 Assessing the indicator properties of species assemblages for natural areas monitoring. *Ecological Applications* 2:203–217.

Kubo, T. and Y. Iwasa. 1995. Inferring the rates of branching and extinction from molecular phylogenies. *Evolution* 49:694–704.

KurtYn, B. 1959. On the longevity of mammalian species in the Tertiary. *Commentationes Biologicae Societas Scientiarum Fennica* 21:1–14.

KurtYn, B. and E. Anderson. 1980. *Pleistocene Mammals of North America.* New York: Columbia University Press.

Laan, R. and B. Verboom. 1990. Effect of pool size and isolation on amphibian communities. *Biological Conservation* 54:251–62.

Labandeira, C. C. and J. J. Sepkoski, Jr. 1993. Insect diversity in the fossil record. *Science* 261:310–315.

Lamarck, J. B. P. A. M. 1809 [1984]. *Zoological Philosophy.* Chicago: University of Chicago Press.

Lamb, T., C. Lydeard, R. B. Walker, and J. W. Gibbons. 1994. Molecular systematics of map turtles (*Graptemys*): A comparison of mitochondrial restriction site versus sequence data. *Systematic Biology* 43:543–559.

Lande, R. 1986. The dynamics of peak shifts and the pattern of morphological evolution. *Paleobiology* 12:343–354.

———. 1993. Risks of population extinction from demographic and environmental stochasticity, and random catastrophes. *American Naturalist* 142:911–927.

Larwood, G. P, ed. 1988. *Extinction and Survival in the Fossil Record.* Oxford: Oxford University Press.

Lauder, G. V. and K. F. Liem. 1989. The role of historical factors in the evolution of complex organismal functions. In D. B. Wake and G. Roth, eds., *Complex Organismal Functions: Integration and Evolution in Vertebrates,* pp. 63–78. New York: John Wiley and Sons.

Lawton, J. H. 1990. Species richness and population dynamics of animal assemblages. Patterns in body size: Abundance and space. *Philosophical Transactions of the Royal Society of London* B330:283–291.

———. 1995. Population dynamic principles. In J. H. Lawton and R. M. May, eds., *Extinction Rates,* pp. 147–163. Oxford: Oxford University Press.

Lawton, J. H., T. M. Lewisohn, and S. G. Compton. 1993. Patterns of diversity for the insect herbivores on bracken. In R. Ricklefs and D. Schluter, eds., *Species Diversity in Ecological Communities: Historical and Geographical Perspectives,* pp. 178–184. Chicago: University of Chicago Press.

Lawton, J. H., S. Nee, A. J. Letcher, and P. H. Harvey. 1994. Animal distributions: Patterns and processes. In P. J. Edwards, R. M. May, and N. R. Webb, eds., *Large-Scale Ecology and Conservation Biology,* pp. 41–58. Oxford: Blackwell.

Lawton, J. H. and G. L. Woodroffe. 1991. Habitat and the distribution of water voles: Why are there gaps in a species' range? *Journal of Animal Ecology* 60: 79–91.

Legendre P. 1993. Spatial autocorrelation: Trouble or new paradigm? *Ecology* 74: 1659–1673.

Legendre, S. and J.-L. Hartenberger. 1992. Evolution of mammalian faunas in Europe during the Eocene and Oligocene. In D. R. Prothero and W. A. Berggren, eds. *Eocene-Oligocene Climatic and Biotic Evolution,* pp. 516–528. Princeton: Princeton University Press.

Leigh, E. G., Jr. 1981. The average lifetime of a population in a varying environment. *Journal of Theoretical Biology* 90:213–239.

Leitner, W. A. and M. L. Rosenzweig. 1997. Nested species-area curves and stochastic sampling: A new theory. *Oikos* 79:503–512.

Lessios, H. A. 1988. Mass mortality of *Diadema antillarum* in the Caribbean: What have we learned? *Annual Review of Ecology and Systematics* 19:371–393.

Levin, S. A. 1992. The problem of pattern and scale in ecology. *Ecology* 73:1943–1967.

Levins, R. 1969. Some demographic and genetic consequences of environmental heterogeneity for biological control. *Bulletin of the Entomological Society of America* 15:237–240.

——. 1970. Extinction. *Lectures on Mathematics in the Life Sciences* 2:75–107.

Levinton, J. 1988. *Genetics, Paleontology, and Macroevolution.* Cambridge: Cambridge University Press.

——. 1995. Life in the tangled lane. *Evolution* 49:575–577.

Levitan, D. R. 1988. Algal-urchin biomass responses following mass mortality of *Diadema antillarum* Philippi at Saint John, U.S. Virgin Islands. *Journal of Experimental Marine Biology and Ecology* 119:167–178.

Lewis, H. 1962. Catastrophic selection as a factor in speciation. *Evolution* 16:257–271.

Lewis, S. M. 1986. The role of herbivorous fishes in the organization of a Caribbean reef community. *Ecological Monographs* 56:183–200.

Lewontin, R. 1983. The organism as the subject and object of evolution. *Scientia* 118:63–82.

Liddell, W. D. and S. L. Ohlhorst. 1988. Comparison of western Atlantic coral reef communities. *Proceedings 6th International Coral Reef Symposium* 3:281–286.

——. 1993. Ten years of disturbance and change on a Jamaican fringing reef. *Proceedings of the 7th International Coral Reef Symposium, Guam* 1:144–150.

Lidgard, S. 1986. Ontogeny in animal colonies: A persistent trend in the bryozoan fossil record. *Science* 232:230–232.

Lieberman, B. S., W. D. Allmon, and N. Eldredge. 1993. Levels of selection and macroevolutionary patterns in the turritellid gastropods. *Paleobiology* 19:205–215.

Lieberman, B. S., C. E. Brett, and N. Eldredge. 1995. A study of stasis and change in two species lineages from the Middle Devonian of New York state. *Paleobiology* 21:15–27.

Lieberman, B. S. and E. S. Vrba. 1995. Hierarchy theory, selection, and sorting: A phylogenetic perspective. *BioScience* 45:394–399.

Liem, K. F. 1990. Key evolutionary innovations, differential diversity, and symecomorphosis. In M. Nitecki, ed., *Evolutionary Innovations*, pp. 147–170. Chicago: University of Chicago Press.

Lillegraven, J. A. 1972. Ordinal and familial diversity of Cenozoic mammals. *Taxon* 21:261–274.

Lipps, J. H. 1970. Plankton evolution. *Evolution* 24:1–22.

——. 1979. Ecology and Paleoecology of planktic foraminifera. In J. H. Lipps,

W. H. Berger, M. A. Buzas, R. G. Douglas, and C. A. Ross, eds., *Foraminiferal Ecology and Paleoecology,* pp. 62–104. Tulsa: Society of Economic Paleontologists and Mineralogists Short Course No. 6.

Liu, S. H., T. Kaplan, and L. J. Gray. 1990. Geometry and dynamics of deterministic sand piles. *Physical Review A* 42:3207–3212.

Loch, J. D., J. H. Stitt, and J. R. Derby. 1993. Cambrian-Ordovician boundary interval extinctions: Implications of revised trilobite and brachiopod data from Mount Wilson, Alberta. *Journal of Paleontology* 67:497–517.

Lomolino, M. V. 1985. Body size of mammals on islands: The island rule reexamined. *American Naturalist* 125:310–316.

——. 1989. Interpretations and comparisons of constants in the species-area relationship: An additional caution. *American Naturalist* 133:277–280.

Lord, J., M. Westoby, and M. Leishman. 1995. Seed size and phylogeny in six temperate floras: Constraints, niche conservatism, and adaptation. *American Naturalist* 146:349–364.

Losos, J. B. 1995. Community evolution in Greater Antillean Anolis lizards: Phylogenetic patterns and experimental tests. *Philosophical Transactions of the Royal Society of London* B349:69–75.

Løvtrup, S. 1987. On species and other taxa. *Cladistics* 3:157–177.

——. 1988. Epigenetics. In C. J. Humphries, ed., *Ontogeny and Systematics,* pp. 189–227. New York: Columbia University Press.

Lubchenco, J. 1978. Plant species diversity in a marine intertidal community: Importance of herbivore food preference and algal competitive abilities. *American Naturalist* 112:23–39.

Luh, H.-K. and S. L. Pimm. 1993. The assembly of ecological communities: A minimalist approach. *Journal of Animal Ecology* 62:749–765.

Lutz, T. M. and G. E. Boyajian. 1995. Fractal geometry of ammonoid sutures. *Paleobiology* 21:329–342.

Lynch, J. D. 1989. The gauge of speciation: On the frequencies of modes of speciation. In D. Otte and J. A. Endler, eds., *Speciation and Its Consequences,* pp. 527–553. Sunderland, Mass.: Sinauer.

Lynch, J. F. and N. K. Johnson. 1974. Turnover and equilibria in insular avifaunas, with special reference to the California channel Islands. *Condor* 76:370–384.

MacArthur, R. H. 1964. Environmental factors affecting bird species diversity. *American Naturalist* 98:387–397.

——. 1969. Patterns of communities in the tropics. *Biological Journal of the Linnean Society* 1:19–30.

MacArthur, R. H. and R. Levins. 1967. The limiting similarity, convergence and divergence of coexisting species. *American Naturalist* 101:377–385.

MacArthur, R. H. and E. O. Wilson. 1963. An equilibrium theory of insular zoogeography. *Evolution* 17:373–387.

——. 1967. The theory of islands biogeography. *Monographs in Population Biology* 1:1–203, Princeton: Princeton University Press.

MacFadden, B. J. 1986. Fossil horses from "Eohippus" (*Hyracotherium*) to Equus: Scaling, Cope's Law, and the evolution of body size. *Paleobiology* 12:355–369.

MacFadden, B. J. and R. C. Hulbert, Jr. 1988. Explosive speciation at the base of the adaptive radiation of Miocene grazing horses. *Nature* 336:466–468.

Macintyre, I. G. 1988. Modern coral reefs of western Atlantic: New geological perspective. *American Association of Petroleum Geologists Bulletin* 72:1360–1369.

Macintyre, I. G., R. B. Burke, and R. Stuckenrath. 1977. Thickest recorded Holocene reef section, Isla Pérez core hole, Alacran Reef, Mexico. *Geology* 5: 749–754.

Mackenzie, F. T. and J. W. Morse. 1992. Sedimentary carbonates through Phanerozoic time. *Geochimica et Cosmochimica Acta* 56:3281–3295.

MacLeod, N. 1995. Graphic correlation of new Cretaceous/Tertiary (K/T) boundary sections/cores from Denmark, Alabama, Mexico, and the southern Indian Ocean: Implications for a global sediment accumulation model. In K. Mann, H. R. Lane, and J. A. Stein, eds., *Graphic Correlation/Composite Standard— The Method and Its Application.* Tulsa: Society of Economic Paleontologists and Mineralogists Special Publication No. 53.

MacLeod, N. and G. Keller. 1994. Comparative biogeographic analysis of planktic foraminiferal survivorship across the Cretaceous/Tertiary (K/T) boundary. *Paleobiology* 20:143–177.

Maddox, J. 1994. Punctuated equilibrium by computer. *Nature (London)* 371:197.

Magaritz, M. 1989. ^{13}C minima follow extinction events: A clue to faunal radiation. *Geology* 17:337–340.

Magaritz, M., W. T. Holser, and J. L. Kirschvink. 1986. Carbon-isotope events across the Precambrian-Cambrian boundary on the Siberian platform. *Nature* 320:258–259.

Maliva, R. G., A. H. Knoll, and R. Siever. 1989. Secular change in chert distribution: A reflection of evolving biological participation in the silica cycle. *Palaios* 4: 519–532.

Małkowski, K., M. Gruszczynski, A. Hoffman, and S. Halas. 1989. Oceanic stable isotope composition and a scenario for the Permo-Triassic crisis. *Historical Biology* 2:289–309.

Mandelbrot, B. 1983. *The Fractal Geometry of Nature.* New York: Freeman.

Margalef, R. 1968. *Perspectives in Ecological Theory.* Chicago: University of Chicago Press.

———. 1971. The pelagic ecosystem of the Caribbean Sea. *Symposium on Investigations and Resources of the Caribbean Sea and Adjacent Regions,* pp. 483–498. Paris: UNESCO.

Mark, G. A. and K. W. Flessa. 1977. A test for evolutionary equilibria: Phanerozoic brachiopods and Cenozoic mammals. *Paleobiology* 3:17–22.

Marquet, P. A. and M. L. Taper. 1998. In press. On size and area: Patterns of mammalian body size extremes across landmasses. *Evolutionary Ecology* 12.

Marshall, C. R. 1994. Confidence intervals on stratigraphic ranges: Partial relaxation of the assumption of randomly distributed fossil horizons. *Paleobiology* 20: 459–469.

Marshall, C. R., E. C. Raff, and R. A. Raff. 1994. Dollo's law and the death and resurrection of genes. *Proceedings of the National Academy of Sciences U.S.A.* 91:12283–12287.

Marshall, L. G., S. D. Webb, J. J. Sepkoski, Jr., and D. M. Raup. 1982. Mammalian evolution and the Great American Interchange. *Science* 215:1351–1357.

Martin, R. E. 1986. Habitat and distribution of the foraminifer *Archaias angulatus* (Fichtel and Moll) (Miliolina, Soritidae). *Journal of Foraminiferal Research* 16:201–206.

——. 1993. Time and taphonomy: Actualistic evidence for time-averaging of benthic foraminiferal assemblages. In S. M. Kidwell and A. K. Behrensmeyer, eds., *Taphonomic Approaches to Time Resolution in Fossil Assemblages*, pp. 33–56. Paleontological Society Short Course No. 6.

——. 1995. Cyclic and secular variation in microfossil biomineralization: Clues to the biogeochemical evolution of Phanerozoic oceans. *Global and Planetary Change* 11:1–23.

——. 1996. Secular increase in nutrient levels through the Phanerozoic: Implications for productivity, biomass, and diversity of the marine biosphere. *Palaios* 11:209–220.

——. 1997. In press. *One Long Experiment: Scale and Process in Earth History.* New York: Columbia University Press.

Martin, R. E. and R. R. Fletcher. 1995. Graphic correlation of Plio-Pleistocene sequence boundaries, Gulf of Mexico: Oxygen isotopes, ice volume, and sea level. In K. Mann, H. R. Lane, and J. A. Stein, eds., *Graphic Correlation/ Composite Standard—The Method and Its Application*, pp. 235–248. Tulsa: Society of Economic Paleontologists and Mineralogists Special Publication No. 53.

Marzluff, J. M. and K. P. Dial. 1991. Life history correlates of taxonomic diversity. *Ecology* 72:428–439.

Masters, J. C. 1993. Primates and paradigms: Problems with the identification of genetic species. In W. H. Kimbel and L. B. Martin, eds., *Species, Species Concepts, and Primate Evolution*, pp. 43–64. Mt. Kisco, N. Y.: Plenum Press.

Maurer, B. A. 1985. On the ecological and evolutionary roles of competition. *Oikos* 45:300–302.

——. 1989. Diversity-dependent species dynamics: Incorporating the effects of population-level processes on species dynamics. *Paleobiology* 15:133–146.

——. 1990. The relationship between distribution and abundance in a patchy environment. *Oikos* 58:181–189.

——. 1994. *Geographical Population Analysis: Tools for the Analysis of Biodiversity.* Oxford: Blackwell Scientific.

——. 1997. In press. *Untangling Ecological Complexity.* Chicago: University of Chicago Press.

Maurer, B. A., J. H. Brown, and R. D. Rusler. 1992. The micro and macro in body size evolution. *Evolution* 46:939–953.

Maurer, B. A. and M.-A. Villard. 1994. Geographic variation in abundance of North American birds. *Research & Exploration* 10:307–317.

May, R. M. 1972. Will a large complex system be stable? *Nature* 238:413–414.

——. 1973. *Stability and Complexity in Model Ecosystems.* Princeton: Princeton University Press.

——. 1975. Patterns of species abundance and diversity. In M. L. Cody and J. M. Diamond, eds., *Ecology and Evolution of Communities,* pp. 81–120. Cambridge: Harvard University Press, Belknap Press.

——. 1977. Thresholds and breakpoints in ecosystems with a multiplicity of stable states. *Nature* 269:471–477.

——. 1978. The dynamics and diversity of insect faunas. In L. A. Mound and N. Waloff, eds., *Diversity of Insect Faunas,* pp. 188–204. Oxford, England: Blackwell.

——. 1986. The search for patterns in the balance of nature: Advances and retreats. *Ecology* 67:1115–1126.

——. 1988a. How many species are there on Earth? *Science* 241:1441–1449.

——. 1988b. Levels of organization in ecology. In J. M. Cherrett, ed., *Ecological Concepts: The Contribution of Ecology to an Understanding of the Natural World,* pp. 339–363. Boston: Blackwell Scientific Publications.

——. 1994. The effects of spatial scale on ecological questions and answers. In P. J. Edwards, R. M. May, and N. R. Webb, eds., *Large-scale Ecology and Conservation Biology,* pp. 1–17. Oxford: Blackwell Scientific Publications.

May, R. M. and G. F. Oster. 1976. Bifurcations and dynamic complexity in simple ecological models. *American Naturalist* 110:573–599.

Maynard Smith, J. 1989. The causes of extinction. *Philosophical Transactions of the Royal Society of London* B325:241–252.

Maynard Smith, J., R. Burian, S. Kauffman, P. Alberch, J. Cambell, B. Goodwin, R. Lande, D. Raup, and L. Wolpert. 1985. Developmental constraints and evolution: A perspective from the Mountain Lake Conference on development and evolution. *Quarterly Review of Biology* 60:265–287.

Mayr, E. 1942. *Systematics and the Origin of Species.* New York: Columbia University Press.

——. 1954. Change of genetic environment and evolution. In J. Huxley, ed., *Evolution as a Process,* pp. 157–180. London: Allen and Unwin.

——. 1959. Isolation as an evolutionary factor. *Proceedings of the American Philosophical Society* 103:221–230.

——. 1963. *Animal Species and Evolution.* Cambridge: Harvard University Press.

——. 1969. Bird speciation in the tropics. *Biological Journal of the Linnaean Society* 1:1–17.

——. 1970. *Populations, Species, and Evolution.* Cambridge: Harvard University Press, Belknap Press.

——. 1982. *The Growth of Biological Thought*. Cambridge: Harvard University Press, Belknap Press.

McCauley, D. E. 1991. Genetic consequences of local population extinction and recolonization. *Trends in Ecology and Evolution* 6:5–8.

McDonald, H. A. and J. H. Brown. 1992. Using montane mammals to model extinctions due to global change. *Conservation Biology* 6:409–415.

McGhee, G. R. 1988. The Late Devonian extinction event: Evidence for abrupt ecosystem collapse. *Paleobiology* 14:250–257.

McGhee, G. R., Jr. 1981. Evolutionary replacement of ecological equivalents in Late Devonian benthic marine communities. *Palaeogeography, Palaeoclimatology, Palaeoecology* 34:267–283.

——. 1991. Theoretical morphology: The concept and its applications. In N. L. Gilinsky and P. W. Signor, eds., *Analytical Paleobiology*, pp. 87–102. Knoxville: The Paleontological Society.

——. 1996. *The Late Devonian Mass Extinction*. New York: Columbia University Press.

McGowran, J. A. and P. W. Walker. 1979. Structure in the copepod community of the North Pacific central gyre. *Ecological Monographs* 49:195–226.

McIntosh, R. P. 1995. H. A. Gleason's 'individualistic concept' and theory of animal communities: A continuing controversy. *Biological Reviews of the Cambridge Philosophical Society* 70:317–357.

McKerrow, W. S. 1978. *The Ecology of Fossils*. Cambridge: MIT Press.

McKinney, M. L. 1988a. Heterochrony in evolution: An overview. In M. L. McKinney, ed., *Heterochrony in Evolution: A Multidisciplinary Approach*, pp. 327–340. Mt. Kisco, N. Y.: Plenum Press.

——. 1989. Periodic mass extinctions: Product of biosphere growth dynamics? *Historical Biology* 2:273–287.

——. 1990a. Classifying and analysing evolutionary trends. In K. J. MacNamara, ed., *Evolutionary Trends*, pp. 28–58. Tucson: University of Arizona Press.

——. 1990b. Trends in body size evolution. In K. J. McNamara, ed., *Evolutionary Trends*, pp. 75–118. Tucson: University of Arizona Press.

——. 1995. Extinction selectivity among lower taxa: Gradational patterns and rarefaction error in extinction estimates. *Paleobiology* 21:300–313.

——. 1996a. The biology of fossil abundance. *Revista Española de Paleontología* 11(2):125–133.

——. 1996b. How do rare species avoid extinction? A paleontological view. In W. Kunin and K. J. Gaston, eds. *The Biology of Rarity*, pp. 110–129. London: Chapman & Hall.

——. 1997. Extinction vulnerability and selectivity: Combining ecological and paleontological views. *Annual Review of Ecology and Systematics* 28:495–516.

——. 1998. In press. a. Branching models predict loss of many bird and mammal orders within centuries. *Animal Conservation*.

———. 1998. In press. b. On predicting biotic homogenization: A test with marine biota. *Global Ecology and Biogeography Letters.*

McKinney, M. L., ed. 1988b. *Heterochrony in Evolution: A Multidisciplinary Approach.* Mt. Kisco, N. Y.: Plenum Press.

McKinney, M. L. and W. D. Allmon. 1995. Metapopulations and disturbance: From patch dynamics to biodiversity dynamics. In D. H. Erwin and R. L. Anstey, eds., *New Approaches to Speciation in the Fossil Record,* pp. 123–183. New York: Columbia University Press.

McKinney, M. L. and D. Frederick. 1992. Extinction and population dynamics: New methods and evidence from Paleogene foraminifera. *Geology* 20: 343–346.

McKinney, M. L. and J. L. Gittleman. 1995. Ontogeny and phylogeny: Tinkering with covariation in life history, morphology, and behavior. In K. J. McNamara, ed., *Evolutionary Change and Heterochrony,* pp. 21–48. New York: Wiley.

McKinney, M. L., J. L. Lockwood, and D. Frederick. 1996. Does ecological and evolutionary stasis include rare species? *Palaeogeography, Palaeoclimatology, Palaeoecology* 127:191–207.

McKinney, M. L. and K. J. McNamara. 1991. *Heterochrony: The Evolution of Ontogeny.* Mt. Kisco, N. Y.: Plenum Press.

McKinney, M. L. and C. W. Oyen. 1989. Causation and nonrandomness in biological and geological time series: Temperature as a proximal control of extinction and diversity. *Palaios* 4:3–15.

McKitrick, M. C. 1993. Phylogenetic constraint in evolutionary theory: Has it any explanatory power? *Annual Review of Ecology and Systematics* 24:307–330.

McLennan, D. A. 1991. Integrating phylogeny and experimental ethology: From pattern to process. *Evolution* 45:1773–1789.

McMenamin, M. A. S. 1990. The origins and radiation of the early Metazoa. In K. Allen and D. Briggs, eds., *Evolution and the Fossil Record,* pp. 73–98. Washington, D. C.: Smithsonian Institution Press.

McMenamin, M. A. S. and D. L. S. McMenamin. 1990. *The Emergence of Animals: The Cambrian Breakthrough.* New York: Columbia University Press.

McNamara, K. J. 1986. The role of heterochrony in the evolution of Cambrian trilobites. *Biological Reviews* 61:121–156.

———. 1988. The abundance of heterochrony in the fossil record. In M. L. McKinney, ed., *Heterochrony in Evolution: A Multidisciplinary Approach,* pp. 287–325. Mt. Kisco, N. Y.: Plenum Press.

———. 1994. The significance of gastropod predation on patterns of evolution and extinction in Australian Tertiary echinoids. In B. David, A. Guille, J.-P. Féral, and M. Roux, eds., *Echinoderms Through Time: Proceedings of the 8th International Echinoderm Conference, Dijon,* pp. 785–793. Rotterdam: A. A. Balkema.

McShea, D. W. 1992. A metric for the study of evolutionary trends in the complexity of serial structures. *Biological Journal of the Linnean Society* 45:39–55.

———. 1993. Arguments, tests, and the Burgess Shale—A commentary on the debate. *Paleobiology* 19:399–402.

———. 1994a. Investigating Mechanisms of Large-Scale Evolutionary Trends. *Evolution* 48:1747–1763.

———. 1994b. Mechanisms of large-scale evolutionary trends. *Evolution* 48:1747–1763.

———. 1996. Metazoan complexity and evolution: Is there a trend? *Evolution* 50:477–492.

Meffe, G. K. and C. R. Carroll. 1994. *Principles of Conservation Biology.* Sunderland, Mass: Sinauer.

Melchin, M. J. 1994. Graptolite extinction at the Llandovery-Wenlock boundary. *Lethaia* 27:285–290.

Menge, B. A. 1995. Indirect effects in marine rocky intertidal interaction webs: Patterns and importance. *Ecological Monographs* 65:21–74.

Mesolella, K. J. 1967. Zonation of uplifted Pleistocene coral reefs on Barbados, West Indies. *Science* 156:638–640.

Meyer, A., P. A. Ritchie, and K.-E. Witte. 1995. Predicting developmental processes from evolutionary patterns: A molecular phylogeny of the zebrafish (*Danio rerio*) and its relatives. *Philosophical Transactions of the Royal Society of London* B349:103–111.

Mikkelson, G. M. 1993. How do food webs fall apart? A study of changes in trophic structure during relaxation on habitat fragments. *Oikos* 67:539–547.

Miller, A. I. 1988a. Spatial resolution in subfossil molluscan remains: Implications for paleobiological analyses. *Paleobiology* 14:91–103.

———. 1988b. Spatio-temporal transitions in Paleozoic Bivalvia: An analysis of North American fossil assemblages. *Historical Biology* 1:251–273.

———. 1989. Spatio-temporal transitions in Paleozoic Bivalvia: A field comparison of Upper Ordovician and upper Paleozoic bivalve-dominated fossil assemblages. *Historical Biology* 2:227–260.

———. 1990a. Bivalves. In K. J. McNamara, ed., *Evolutionary Trends*, pp. 143–161. London: Belhaven Press.

———. 1996a. Counting fossils in a Cincinnatian storm bed: Spatial resolution in the fossil record. In C. E. Brett, ed., *Paleontologic Events: Stratigraphic, Ecologic, and Evolutionary Implications.* New York: Columbia University Press.

———. 1997. Coordinated stasis or coincident relative stability? *Paleobiology* 23(2):155–164.

Miller, A. I. and H. Cummins. 1993. Using numerical models to evaluate the consequences of time-averaging in marine fossil assemblages. In S. M. Kidwell and A. K. Behrensmeyer, eds., *Taphonomic Approaches to Time Resolution in Fossil Assemblages*, pp.150–168. Paleontological Society Short Courses in Paleontology No. 6.

Miller, A. I. and M. Foote. 1996. Calibrating the Ordovician radiation of marine life: Implications for Phanerozoic diversity trends. *Paleobiology* 22:304–309.

Miller, A. I., G. Llewellyn, K. M. Parsons, H. Cummins, M. R. Boardman, B. J. Greenstein, and D. K. Jacobs. 1992. Effects of Hurricane Hugo on molluscan skeletal distributions, Salt River Bay, St. Croix, U.S. Virgin Islands. *Geology* 20:23–26.

Miller, A. I. and S. Mao. 1995. Association of orogenic activity with the Ordovician radiation of marine life. *Geology* 23:305–308.

Miller, A. I. and J. J. Sepkoski, Jr. 1988. Modeling bivalve diversification: The effect of interaction on a macroevolutionary system. *Paleobiology* 14:364–369.

Miller, R. R. 1948. The cyprinodont fishes of the Death Valley system of eastern California and southwestern Nevada. *University of Michigan Museum of Zoology Miscellaneous Publication* 42:1–80.

Miller, T. E. 1982. Community diversity and interactions between the size and frequency of disturbance. *American Naturalist* 120(4):533–536.

Miller, W. 1986. Paleoecology of benthic community replacement. *Lethaia* 19: 225–231.

Miller, W., III. 1990b. Hierarchy, individuality, and paleoecosystems. In W. Miller III, ed., *Paleocommunity Temporal Dynamics: The Long-Term Development of Multispecies Assemblages,* pp. 31–47. Knoxville: Paleontological Society Special Publication No. 5. University of Tennessee Press.

——. 1993. Models of recurrent fossil assemblages. *Lethaia* 26:182–183.

——. 1996b. Ecology of coordinated stasis. *Palaeogeography, Palaeoclimatology, Palaeoecology* 127:177–190.

Miller, W., III, ed. 1990c. Paleocommunity temporal dynamics: The long term development of multispecies assemblies. *Paleontological Society Special Publication 5.* Knoxville: The University of Tennessee Press.

Mitchell, B. R. 1983. *International Historical Statistics: The Americas and Australia.* Detroit: Gale Research Company.

Möller, D. W. and P. A. Mueller. 1991. Origin and age of the Mediterranean Messinian evaporites: Implications from Sr isotopes. *Earth and Planetary Science Letters* 107:1–12.

Moore, G. T., D. N. Hayashida, and C. A. Ross. 1993. Late Early Silurian (Wenlockian) general circulation model-generated upwelling, graptolitic black shales, and organic-rich source rocks—An accident of plate tectonics? *Geology* 21:17–20.

Moran, P. A. P. 1950. Notes on continuous stochastic phenomena. *Biometrika* 37: 17–23.

——. 1958. Random processes in genetics. *Cambridge Philosophical Society Proceedings* 54:60–71.

Morris, P. J. 1995a. Coordinated stasis and ecological locking. *Palaios* 10:101–102.

——. 1996. Testing patterns and causes of faunal stability in the fossil record, with an example from the Pliocene Lusso Beds of Zaire. *Palaeogeography, Palaeoclimatology, Palaeoecology* 127:313–338.

Morris, P. J., L. C. Ivany, K. M. Schopf, and C. E. Brett. 1995. The challenge of paleoecological stasis: Reassessing sources of evolutionary stability. *Proceedings of the National Academy of Sciences, U.S.A.* 92:11269–11273.

Morris, S. C. 1995b. Ecology in deep time. *Trends in Ecology and Evolution* 10: 290–294.

Morse, D. R., N. E. Stork, and J. H. Lawton. 1988. Species number, species abundance and body length relationships of arboreal beetles in Bornean lowland rain forest trees. *Ecological Entomology* 13:25–37.

Müller, G. 1990. Developmental mechanisms at the origin of morphological novelty: A side-effect hypothesis. In M. Nitecki, ed., *Evolutionary Innovations,* pp. 99–130. Chicago: University of Chicago Press.

Naeem, S., L. J. Thompson, S. P. Lawler, J. H. Lawton, and R. M. Woodfin. 1994. Declining biodiversity can alter the performance of ecosystems. *Nature* 368: 734–737.

Naeser, C. W., G. A. Izett, and J. D. Obradovich. 1980. Fission-track and K-Ar ages of natural glasses. *United States Geological Survey Bulletin* 1489:1–31.

Nee, S., T. G. Barraclough, and P. H. Harvey. 1996. Temporal changes in biodiversity. In K. J. Gaston, ed., *Biodiversity,* pp. 230–252. London: Blackwell.

Nee, S., P. H. Harvey, and R. M. May. 1991. Lifting the veil on abundance patterns. *Proceedings of the Royal Society of London* B243:161–163.

Nee, S., E. C. Holmes, R. M. May, and P. H. Harvey. 1994a. Extinction rates can be estimated from molecular phylogenies. *Proceedings of the Royal Society of London* 344:77–82.

———. 1995. Estimating extinction from molecular phylogenies. In J. H. Lawton and R. M. May, eds., *Extinction Rates,* pp. 164–182. Oxford: Oxford University Press.

Nee, S., R. M. May, and P. H. Harvey. 1994b. The reconstructed evolutionary process. *Philosophical Transactions of the Royal Society of London* B344:305–311.

Nee, S., A. O. Mooers, and P. H. Harvey. 1992. Tempo and mode of evolution revealed from molecular phylogenies. *Proceedings of the National Academy of Science* 89:8322–8326.

Nei, M. 1987. *Molecular Evolutionary Genetics.* New York: Columbia University Press.

Nei, M. and W.-H. Li. 1979. Mathematical models for studying genetic variation in terms of restriction endonucleases. *Proceedings of the National Academy of Science U.S.A.* 76:5269–5273.

Neigel, J. E. and J. C. Avise. 1983. Clonal diversity and population structure in a reef-building coral, *Acropora cervicornis:* Self-recognition analysis and demographic interpretation. *Evolution* 37:437–453.

Nelson, G. 1989. Cladistics and evolutionary models. *Cladistics* 5:275–289.

Nelson, G. and N. Platnick. 1981. *Systematics and Biogeography: Cladistics and Vicariance.* New York: Columbia University Press.

Newell, N. D. 1949. Phyletic size increase, an important trend illustrated by fossil invertebrates. *Evolution* 3:103–124.

———. 1959. The nature of the fossil record. *Proceedings of the American Philosophical Society* 103:264–285.

———. 1967. Revolutions in the history of life. *Geological Society of America Special Paper No. 89*, pp. 63–91.

Newman, M. E. J. and B. W. Roberts. 1995. Mass extinction: Evolution and the effects of external influences on unfit species. *Proceedings of the Royal Society of London* B260:31–37.

Nichols, J. D. and K. H. Pollock. 1983. Estimating taxonomic diversity, extinction rates, and speciation rates from fossil data using capture-recapture models. *Paleobiology* 9:150–163.

Nicol, D. 1962. The biotic development of some Niagaran reefs—An example of an ecological succession or sere. *Journal of Paleontology* 36:172–176.

Nowak, R. M. 1991. *Walker's Mammals of the World.* Baltimore: Johns Hopkins Press.

Odum, E. P. 1969. The strategy of ecosystem development. *Science* 164:262–270.

Oliver, W. A., Jr. 1990. Extinctions and migrations of Devonian rugose corals in the Eastern Americas Realm. *Lethaia* 23:167–178.

Oliver, W. A., Jr. and A. E. H. Pedder. 1994. Crises in the Devonian history of the rugose corals. *Paleobiology* 20:178–190.

Olson, E. C. 1952. The evolution of a Permian vertebrate chronofauna. *Evolution* 6:181–196.

———. 1983. Coevolution or coadaptation? Permo-Carboniferous vertebrate chronofauna. In M. Nitecki, ed., *Coevolution,* pp. 301–338. Chicago: University of Chicago Press.

O'Neill, R. V., D. L. DeAngelis, J. B. Waide, and T. F. H. Allen. 1986. A Hierarchical Concept of Ecosystems. *Monographs in Population Biology 23.* Princeton: Princeton University Press.

O'Neill, R. V., A. R. Johnson, and A. W. King. 1989. A hierarchical framework for the analysis of scale. *Landscape Ecology* 3:193–205.

Opdam, P. 1990. Metapopulation theory and habitat fragmentation: A review of holarctic breeding bird studies. *Landscape Ecology* 5:93–106.

Orth, C. J., J. S. Gilmore, L. R. Quintana, and P. M. Sheehan. 1986. Terminal Ordovician extinction: Geochemical analysis of the Ordovician/Silurian boundary, Anticosti Island, Quebec. *Geology* 14:433–436.

Otte, D. and J. A. Endler, eds. 1989. *Speciation and Its Consequences.* Sunderland, Mass.: Sinauer.

Overpeck, J. T., P. J. Bartlein, and T. Webb III. 1991. Potential magnitude of future vegetation change in Eastern North America: Comparisons with the past. *Science* 254:692–695.

Paine, R. T. 1966. Food web complexity and species diversity. *American Naturalist* 100:65–75.

——. 1992. Food-web analysis through field measurement of per capita interaction strength. *Nature* 355:73–75.

Palmer, A. R. 1984. The biomere problem: Evolution of an idea. *Journal of Paleontology* 58:599–611.

——. 1986. Inferring relative levels of genetic variability in fossils: The link between heterozygosity and fluctuating asymmetry. *Paleobiology* 12:1–5.

Palmer, A. R. and C. Strobeck. 1986. Fluctuating asymmetry: Measurement, analysis, patterns. *Annual Review of Ecology and Systematics* 17:391–421.

Palmer, M. W. and P. S. White. 1994. Scale dependence and the species-area relationship. *American Naturalist* 144:717–740.

Palumbi, S. R. 1994. Genetic divergence, reproductive isolation, and marine speciation. *Annual Review of Ecology and Systematics* 25:547–572.

Pandolfi, J. M. 1996. Limited membership in Pleistocene reef coral assemblages from the Huon Peninsula, Papua New Guinea: Constancy during global change. *Paleobiology* 22:152–176.

Parrish, J. T. 1987. Palaeo-upwelling and the distribution of organic-rich rocks. In J. Brooks and A. J. Fleet, eds., *Marine Petroleum Source Rocks,* pp. 199–205. London: Geological Society of London.

——. 1993. Climate of the supercontinent Pangea. *Journal of Geology* 101:215–233.

Parsons, P. A. 1993. Stress, extinctions and evolutionary change: From living organisms to fossils. *Biological Reviews* 68:313–333.

Paterson, H. E. H. 1985. The recognition concept of species. In E. S. Vrba, ed., *Species and Speciation,* pp. 21–30. Transvaal Museum Monograph, No. 4.

Patterson, B. D. 1990. On the temporal development of nested subset patterns of species composition. *Oikos* 59:330–342.

Patterson, B. D. and W. Atmar. 1986. Nested subsets and the structure of insular mammalian faunas and archipelagos. *Biological Journal of the Linnean Society* 28:65–82.

Patterson, B. D. and J. H. Brown. 1991. Regionally nested patterns of species composition in granivorous rodent assemblages. *Journal of Biogeography* 18:395–402.

Patterson, C. and A. B. Smith. 1987. Is the periodicity of extinctions a taxonomic artefact? *Nature* 330:248–251.

Patterson, R. T. and A. D. Fowler. 1996. Evidence of self organization in planktic foraminiferal evolution: Implications for interconnectedness of paleoecosystems. *Geology* 24:215–218.

Patzkowsky, M. E. 1994. Coordinated stasis in Middle and Upper Ordovician brachiopod biofacies in Eastern North America. *Abstracts with Programs, GSA Annual Meeting, Seattle,* p. A519.

——. 1995. A hierarchical branching model of evolutionary radiations. *Paleobiology* 21:440–460.

Paul, C. R. C. and S. F. Mitchell. 1994. Is famine a common factor in marine mass extinctions? *Geology* 22:679–682.

Paulay, G. 1990. Effects of late Cenozoic sea-level fluctuations on the bivalve faunas of tropical oceanic islands. *Paleobiology* 16:415–434.

Pearson, P. N. 1992. Survivorship analysis of fossil taxa when real-time extinction rates vary: The Paleogene planktonic foraminifera. *Paleobiology* 18:115–131.

———. 1995. Investigating age-dependency of species extinction rates using dynamic survivorship analysis. *Historical Biology* 10:119–136.

Pease, C. M. 1992. On the declining extinction and origination rates of fossil taxa. *Paleobiology* 18:89–92.

Peitgen, H.-O. and D. Saupe. 1988. *The Science of Fractal Images.* New York: Springer-Verlag.

Penny, D., M. D. Hendy, and M. A. Steele. 1992. Progress with methods for constructing evolutionary trees. *Trends in Ecology and Evolution* 7:73–79.

Perry, D. A. 1995. Self-organizing systems across scales. *Trends in Ecology and Evolution* 10:241–244.

Peters, R. H. 1983. *The Ecological Implications of Body Size.* Cambridge, England: Cambridge.

Peterson, C. H. 1983. The pervasive biological explanation. (Book review.) *Paleobiology* 9:429–436.

Peterson, K. J. and C. R. Marshall. 1995. The Cambrian explosion II: Phylogeny, problematica, and the nature and timing of the metazoan radiation. *Geological Society of America Abstracts with Program* 27:A-269.

Petraitis, P. S., R. E. Latham, and R. A. Niesenbaum. 1989. The maintenance of species diversity by disturbance. *Quarterly Review of Biology* 64:393–418.

Pickett, S. T. A. 1980. Non-equilibrium coexistence of plants. *Bulletin of the Torrey Botanical Club* 107:238–248.

———. 1989. Space-for time substitution as an alternative to long-term studies. In G. E. Likens, ed., *Long-Term Studies in Ecology,* pp. 110–135. New York: Springer-Verlag.

Pickett, S. T. A. and J. N. Thompson. 1978. Patch dynamics and the design of nature reserves. *Biological Conservation* 13:27–37.

Pickett, S. T. A. and P. White, eds. 1985. *The Ecology of Natural Disturbance and Patch Dynamics.* Orlando: Academic Press.

Pimm, S. L. 1984a. The complexity and stability of ecosystems. *Nature* 307: 321–326.

———. 1984b. Food chains and return times. In D. R. Strong, D. Simberloff, L. G. Abele, and A. B. Thistle, eds., *Ecological Communities: Conceptual Issues and the Evidence,* pp. 397–412. Princeton: Princeton University Press.

———. 1991. *The Balance of Nature?* Chicago: University of Chicago Press.

———. 1994. *The Balance of Nature? Ecological Issues in the Conservation of Species and Communities.* Chicago: University of Chicago Press.

Pimm, S. L., J. M. Diamond, T. M. Reed, G. J. Russell, and J. M. Verner. 1993. Times to extinction for small populations of large birds. *Proceedings of the National Academy of Sciences of the U.S.A.* 90:10871–10875.

Pimm, S. L., J. L. Gittleman, G. J. Russell, and T. M. Brooks. 1995. The future of biodiversity. *Science* 269:347–350.

Pimm, S. L., H. Lee Jones, and J. M. Diamond. 1988. On the risk of extinction. *American Naturalist* 132:757–785.

Pitrat, C. W. 1970. Phytoplankton and the late Paleozoic wave of extinction. *Palaeogeography, Palaeoclimatology, Palaeoecology* 8:49–55.

Platt, W. J. and I. M. Weis. 1977. Resource partitioning and competition within a guild of fugitive prairie plants. *American Naturalist* 111:479–513.

Plotnick, R. E. and R. H. Gardner. 1993. Lattices and landscapes. *Lectures on Mathematics in the Life Sciences* 23:129–157.

Plotnick, R. E. and M. L. McKinney. 1993. Ecosystem organization and extinction dynamics. *Palaios* 8:202–212.

Plotnick, R. E., M. L. McKinney, and R. Gardner. 1994. Multiscale ecosystem stasis and multiscale disturbances: The thousand natural shocks that flesh is heir to. *Abstracts with Programs, GSA Annual Meeting, Seattle*, 26(7):A520

Plotnick, R. E. and K. L. Prestegaard. 1995. Fractal and multifractal models and methods in stratigraphy. In C. Barton and P. R. Lapointe, eds., *Fractals in Petroleum Geology and Earth Processes*, pp. 73–96. Mt. Kisco, N. Y.: Plenum Press.

Popp, B. N., T. F. Anderson, and P. A. Sandberg. 1986. Brachiopods as indicators of original isotopic compositions in some Paleozoic limestones. *Geological Society of America Bulletin* 97:1262–1269.

Porter, J. W. and O. W. Meier. 1992. Quantification of loss and change in Floridian reef coral populations. *American Zoologist* 32:625–640.

Potts, D.C. 1984. Generation times and the Quaternary evolution of reef-building corals. *Paleobiology* 10:48–58.

Preston, F. W. 1948. The commonness, and rarity, of species. *Ecology* 29:254–283.

———. 1960. Time and space and the variation of species. *Ecology* 41:785–790.

———. 1962a. The canonical distribution of commonness and rarity: Part I. *Ecology* 43:185–215.

———. 1962b. The canonical distribution of commonness and rarity: Part II. *Ecology* 43:410–432.

Price, P. W. 1984. Alternative paradigms in community ecology. In P. W. Price, C. N. Slobodchikoff, and W. S. Gaud, eds., *A New Ecology: Novel Approaches to Interactive Systems*, pp. 353–383. New York: Wiley.

Prigogine, I. 1980. *From Being to Becoming: Time and Complexity in the Physical Sciences*. New York: W. H. Freeman.

Probert, P. K. 1984. Disturbance, sediment stability, and trophic structure of soft-bottom communities. *Journal of Marine Research* 42:893–921.

Prothero, D. R. 1985. North American mammalian diversity and Eocene-Oligocene extinctions. *Paleobiology* 11:389–405.

———. 1994. *The Eocene-Oligocene Transition*. New York: Columbia University Press.

Prothero, D. R. and T. H. Heaton. 1996. Faunal stability during the Early Oligocene climatic crash. *Palaeogeography, Palaeoclimatology, Palaeoecology* 127:257–284.

Provine, W. B. 1986. *Sewall Wright and Evolutionary Biology.* Chicago: University of Chicago Press.

Pulliam, H. R. 1988. Sources, sinks and population regulation. *American Naturalist* 132:652–661.

Pulliam, H. R. and B. J. Danielson. 1991. Sources, sinks, and habitat selection: A landscape perspective on population dynamics. *American Naturalist* 137: S50-S66.

Purvis, A., J. L. Gittleman, and H.-K. Luh. 1994. Truth or consequences: Effects of phylogenetic accuracy on two comparative methods. *Journal of Theoretical Biology* 167:293–300.

Raff, R. A. 1996. *The Shape of Life. Genes, Development, and the Evolution of Animal Form.* Chicago: University of Chicago Press.

Rahbek, C. 1997. The relationship among area, elevation, and regional species richness in Neotropical birds. *American Naturalist* 149:875–902.

Rahel, F. J. 1990. The hierarchical nature of community persistence: A problem of scale. *American Naturalist* 136:328–344.

Raina, S. N. and Y. Ogihara. 1994. Chloroplast DNA diversity in Vicia faba and its close wild relatives: Implications for reassessment. *Theoretical Applied Genetics* 88:261–266.

Rambaut, A., P. H. Harvey, and S. Nee. 1998. In press. End-Epi: An application for reconstructing population dynamic histories from phylogenies. *Computers in Applied Bioscience.*

Rand, D. A. and H. B. Wilson. 1993. Evolutionary catastrophes, punctuated equilibria and gradualism in ecosystem evolution. *Proceedings of the Royal Society of London* B253:137–141.

Raup, D. M. 1966. Geometric analysis of shell coiling: General problems. *Journal of Paleontology* 40:1178–1190.

——. 1967. Geometric analysis of shell coiling: Coiling in ammonoids. *Journal of Paleontology* 41:43–65.

——. 1968. Theoretical morphology of echinoid growth. *Journal of Paleontology* 42:50–63.

——. 1972. Taxonomic diversity during the Phanerozoic. *Science* 177:1065–1071.

——. 1975a. Taxonomic diversity estimation using rarefaction. *Paleobiology* 1: 333–342.

——. 1975b. Taxonomic survivorship curves and Van Valen's law. *Paleobiology* 1: 82–96.

——. 1976a. Species diversity in the Phanerozoic: A tabulation. *Paleobiology* 2: 279–288.

——. 1976b. Species diversity in the Phanerozoic: An interpretation. *Paleobiology* 2:289–297.

——. 1977a. Stochastic models in evolutionary paleontology. In A. Hallam, ed., *Patterns of Evolution,* pp. 59–78. Amsterdam: Elsevier.

———. 1977b. Species diversity in the Phanerozoic. Systematists follow the fossils. *Paleobiology* 3:328–329.

———. 1978. Cohort analysis of generic survivorship. *Paleobiology* 4:1–15.

———. 1979. Biases in the fossil record of species and genera. *Bulletin of the Carnegie Museum of Natural History* 13:85–91.

———. 1981. Physical disturbance in the life of plants. In M. H. Nitecki, ed., *Biotic Crises in Ecological and Evolutionary Time*, pp. 39–52. New York: Academic Press.

———. 1983. On the early origins of major biologic groups. *Paleobiology* 9:107–115.

———. 1985. Mathematical models of cladogenesis. *Paleobiology* 11:42–52.

———. 1986. Biological extinction in Earth history. *Science* 231:1528–1533.

———. 1991a. A kill curve for Phanerozoic marine species. *Paleobiology* 17:37–48.

———. 1991b. *Extinction: Bad Genes or Bad Luck?* New York: Norton.

———. 1991c. The future of analytical paleobiology. In N. L Gilinsky and P. W. Signor, eds., *Analytical Paleobiology; Paleontological Society Short Courses in Paleontology* 4, pp. 207–216.

———. 1992. Large-body impact and extinction in the Phanerozoic. *Paleobiology* 18:80–88.

———. 1994. The role of extinction in evolution. *Proceedings of the National Academy of Science U.S.A.* 91:6758–6763.

Raup, D. M. and G. E. Boyajian. 1988. Patterns of generic extinction in the fossil record. *Paleobiology* 14:109–125.

Raup, D. M. and S. J. Gould. 1974. Stochastic simulation and evolution of morphology—towards a nomothetic paleontology. *Systematic Zoology* 23:305–322.

Raup, D. M., S. J. Gould, T. J. M. Schopf, and D. S. Simberloff. 1973. Stochastic models of phylogeny and the evolution of diversity. *Journal of Geology* 81:525–542.

Raup, D. M. and D. Jablonski. 1993. Geography of end-Cretaceous marine bivalve extinctions. *Science* 260:971–973.

Raup, D. M. and L. G. Marshall. 1980. Variation between groups in evolutionary rates: A statistical test of significance. *Paleobiology* 6:9–23.

Raup, D. M. and J. J. Sepkoski, Jr. 1982. Mass extinctions in the marine fossil record. *Science* 215:1501–1503.

———. 1984. Periodicity of extinctions in the geologic past. *Proceedings of the National Academy of Sciences* 81:801–805.

———. 1986. Periodic extinction of families and genera. *Science* 231:833–836.

Raymo, M. E. 1991. Geochemical evidence supporting T. C. Chamberlin's theory of glaciation. *Geology* 19:344–347.

Raymond, A. and C. Metz. 1995. Laurussian land-plant diversity during the Silurian and Devonian: Mass extinction, sampling bias, or both? *Paleobiology* 21:74–91.

Rea, D. K., J. C. Zachos, R. M. Owen, and P. D. Gingerich. 1990. Global change at the Paleocene-Eocene boundary: Climatic and evolutionary consequences of tectonic events. *Palaeogeography, Palaeoclimatology, Palaeoecology* 79: 117–128.

Reed, T. M. 1980. Turnover frequency in island birds. *Journal of Biogeography* 7: 329–335.

Reice, S. R. 1994. Nonequilibrium determinants of biological community structure. *American Scientist* 82:424–435.

Renard, M. 1986. Pelagic carbonate chemostratigraphy (Sr, Mg,^{18}O,^{13}C). *Marine Micropaleontology* 10:117–164.

Renne, P. R., Z. Zichao, M. A. Richards, M. T. Black, and A. R. Basu. 1995. Synchrony and causal relations between Permian-Triassic boundary crises and Siberian flood volcanism. *Science* 269:1413–1416.

Rhoads, D.C. and D. K. Young. 1970. The influence of deposit-feeding organisms on bottom-sediment stability and community trophic structure. *Journal of Marine Research* 28:150–178.

Rhodes, M. C. and C. W. Thayer. 1991. Mass extinctions: Ecological selectivity and primary production. *Geology* 19:877–800.

Rhodes, M. C. and R. J. Thompson. 1993. Comparative physiology of suspension-feeding in living brachiopods and bivalves: Evolutionary implications. *Paleobiology* 19:322–334.

Richmond, R. H. 1989. Competency and dispersal potential of planula larvae of a spawning versus a brooding coral. *Proceedings 6th International Coral Reef Symposium* 2:827–831.

Richmond, R. H. and C. L. Hunter. 1990. Reproduction and recruitment of corals: Comparisons among the Caribbean, the tropical Pacific, and the Red Sea. *Marine Ecology Progress Series* 60:185–203.

Richter, F. M., D. B. Rowley, and D. J. DePaolo. 1992. Sr isotope evolution of seawater: The role of tectonics. *Earth and Planetary Science Letters* 109:11–23.

Ricklefs, R. E. 1987. Community diversity: Relative roles of local and regional processes. *Science* 235:167–171.

———. 1991. *Ecology.* New York: W. H. Freeman.

———. 1995. The distribution of biodiversity. In V. H. Heywood, ed., *Global Biodiversity Assessment*, pp. 139–173. Cambridge: Cambridge University Press.

Ricklefs, R. E. and S. S. Renner. 1994. Species richness within families of flowering plants. *Evolution* 48:1619–1636.

Ricklefs, R. E. and D. Schluter. 1993a. Species diversity: Regional and historical influences. In R. E. Ricklefs and D. Schluter, eds., *Species Diversity in Ecological Communities. Historical and Geographical Perspectives*, pp. 350–364. Chicago: University of Chicago Press.

Ricklefs, R. E. and D. Schluter, eds. 1993b. *Species Diversity in Ecological Communities.* Chicago: University of Chicago Press.

Ridley, M. 1993. Analysis of the Burgess Shale. *Paleobiology* 19:519–521.

Riedl, R. 1978. *Order in Living Organisms*. New York: Wiley.

Rieppel, O. 1986. Species are individuals: A review and critique of the argument. *Evolutionary Biology* 20:283–317.

———. 1991. Things, taxa and relationships. *Cladistics* 7:93–100.

———. 1992. Homology and logical fallacy. *Journal of Evolutionary Biology* 5: 701–715.

Robbins, C. S., B. Bruun, and H. S. Zim. 1983. *Birds of North America: A Guide to Field Identification*. New York: Golden Press.

Rodgers, J. 1971. The Taconic Orogeny. *Geological Society of America Bulletin* 82: 1141–1178.

Rogers, C. S. 1985. Degradation of Caribbean and western Atlantic coral reefs and decline of associated fisheries. *Proceedings of the 5th International Coral Reef Congress, Tahiti* 6:491–496.

———. 1990. Responses of coral reefs and reef organisms to sedimentation. *Marine Ecology Progress Series* 62:185–202.

Rohlf, F. J. 1985. *NTSYS. Numerical Taxonomy System of Multivariate Statistical Programs*. Stony Brook, New York: State University.

Rohlf, F. J., W. S. Chang, R. R. Sokal, and J. Kim. 1990. Accuracy of estimated phylogenies: Effects of tree topology and evolutionary model. *Evolution* 44: 1671–1684.

Rohlf, F. J. and M. C. Wooten. 1988. Evaluation of the restricted maximum-likelihood method for estimating phylogenetic trees using simulated allele-frequency data. *Evolution* 42:581–595.

Rosen, B. R. 1984. Reef coral biogeography and climate through the late Cenozoic: Just islands in the sun or a critical pattern of islands? In P. J. Brenchley, ed., *Fossils and Climate*, pp. 201–262. New York: John Wiley.

Rosenzweig, M. L. 1975. On continental steady states of species diversity. In M. L. Cody and J. M. Diamond, eds., *Ecology and Evolution of Communities*, pp. 121–140. Cambridge: Harvard University Press, Belknap Press.

———. 1992. Species diversity gradients: We know more and less than we thought. *Journal of Mammalogy* 73:715–730.

———. 1995. *Species Diversity in Space and Time*. Cambridge: Cambridge University Press.

———. 1997. Tempo and mode of speciation. *Science* 277:1622–1623.

Rosenzweig, M. L. and Z. Abramsky. 1993. How are diversity and productivity related? In R. Ricklefs and D. Schluter, eds., *Species Diversity in Ecological Communities: Historical and Geographical Perspectives*, pp. 52–65. Chicago: University of Chicago Press.

Rosenzweig, M. L. and C. W. Clark. 1994. Island extinction rates from regular censuses. *Conservation Biology* 8:491–494.

Rosenzweig, M. L. and J. L. Duek. 1979. Species diversity and turnover in an Ordovician marine invertebrate assemblage. In G. P. Patil and M. L. Rosenzweig, eds., *Contemporary Quantitative Ecology and Related Ecometrics*, pp. 109–119. Fairland, MD: International Co-operative.

Rosenzweig, M. L. and R. D. McCord. 1991. Incumbent replacement: Evidence for long-term evolutionary replacement. *Paleobiology* 17:202–213.

Rosenzweig, M. L. and J. A. Taylor. 1980. Speciation and diversity in Ordovician invertebrates: Filling niches quickly and carefully. *Oikos* 35:236–243.

Roth, P. H. 1987. Mesozoic calcareous nanofossil evolution: Relation to paleoceanographic events. *Paleoceanography* 2:601–611.

Roth, V. L. 1984. On homology. *Biological Journal of the Linnaean Society* 22:13–29.

——. 1994. Within and between organisms: Replicators, lineages, and homologues. In B. K. Hall, ed., *Homology: The Hierarchical Basis of Comparative Biology*, pp. 301–337. San Diego: Academic Press.

Roughgarden, J. 1989. The structure and assembly of communities. In J. Roughgarden, R. M. May, and S. A. Levin, eds., *Perspectives in Ecological Theory*, pp. 203–226. Princeton: Princeton University Press.

Rousseau, D. D. 1992. Is causal ecological biogeography a progressive research program? *Quaternary Science Reviews* 11:593–601.

Roy, K. 1996. The roles of mass extinction and biotic interaction in large-scale replacements: A reexamination using the fossil record of stroboidean gastropods. *Paleobiology* 22:436–452.

Roy, K., D. Jablonski, and J. W. Valentine. 1995. Thermally anomalous assemblages revisited: Patterns in the extraprovincial latitudinal range shifts of Pleistocene marine mollusks. *Geology* 23:1071–1074.

Roy, K., J. W. Valentine, D. Jablonski, and S. M. Kidwell. 1996. Scales of climatic variability and time averaging in Pleistocene biotas: Implications for ecology and evolution. *TREE* 11(11):458–463.

Roy, K. and P. J. Wagner. 1995. Communities in the fossil record: Coordination or coincidence? *American Paleontologist* 3(1):3–4.

Rummel, J. D. and J. Roughgarden. 1985. A theory of faunal buildup for competition communities. *Evolution* 39:1009–1033.

Runkle, J. R. 1985. Disturbance regimes in temperate forests. In S. T. A. Pickett and P. White, eds., *The Ecology of Natural Disturbance and Patch Dynamics*, pp. 17–34. Orlando: Academic Press.

Ruse, M. 1986. *Taking Darwin Seriously*. Oxford: Blackwell.

Rusler, R. D. 1987. Frequency distribution of mammalian body size analyzed by continent. Unpublished M. S. thesis. Tucson: University of Arizona.

Russell, G. J., J. M. Diamond, S. L. Pimm, and T. M. Reed. 1995. Centuries of turnover: Community dynamics at three timescales. *Journal of Animal Ecology* 64:628–641.

Russell, M. P. and D. R. Lindberg. 1988. Real and random patterns associated with molluscan spatial and temporal distributions. *Paleobiology* 14:322–330.

Ruttenberg, K. C. and R. A. Berner. 1993. Authigenic apatite formation and burial in sediments from non-upwelling, continental margin environments. *Geochimica et Cosmochimica Acta* 57:991–1007.

Rylaarsdam, K. W. 1983. Life histories and abundance patterns of colonial corals on Jamaican reefs. *Marine Ecology Progress Series* 13:249–260.

Sale, P. F. 1977. Maintenance of high diversity in coral reef fish communities. *American Naturalist* 111:337–359.

Salthe, S. N. 1985. *Evolving Hierarchical Systems: Their Structure and Representation.* New York: Columbia University Press.

——. 1993. *Development and Evolution.* Cambridge: MIT Press.

Sammarco, P. W. 1980. *Diadema* and its relationship to coral spat mortality: Grazing, competition, and biological disturbance. *Journal of Experimental Marine Biology and Ecology* 45:245–272.

——. 1982. Echinoid grazing as a structuring force in coral communities: Whole reef manipulations. *Journal of Experimental Marine Biology and Ecology* 61:31–55.

——. 1985. The Great Barrier Reef vs. the Caribbean: Comparisons of grazers, coral recruitment patterns, and reef recovery. *Proceedings 5th International Coral Reef Symposium* 4:391–397.

——. 1987. A comparison of some ecological processes on coral reefs of the Caribbean and the Great Barrier Reef. In C. Birkeland, ed., *Comparison Between Atlantic and Pacific Tropical Marine Coastal Ecosystems: Community Structure, Ecological Processes, and Productivity,* pp. 127–166. Paris: UNESCO.

Sanders, H. L. 1968. Marine benthic diversity: A comparative study. *American Naturalist* 102:243–282.

Sanderson, M. J. and M. J. Donoghue. 1994. Shifts in diversification rate with the origin of angiosperms. *Science* 264:1590–1593.

——. 1996. Reconstructing shifts in diversification rates on phylogenetic trees. *Trends in Ecology and Evolution* 11:15–20.

Savage, D. E. and D. E. Russell. 1983. *Mammalian Paleofaunas of the World.* Reading, Mass.: Addison-Wesley.

Savazzi, E. 1995. Theoretical shell morphology as a tool in constructional morphology. *Neues Jahrbuch für Geologie und Paläontologie Abhandlungen* 195:229–240.

Scheltema, R. S. 1971. Larval dispersal as a means of genetic exchange among geographically separated populations of shallow-water benthic marine gastropods. *Biological Bulletin* 140:284–322.

Schidlowski, M. 1991. Quantitative evolution of global biomass through time: Biological and geochemical constraints. In S. H. Schneider and P. J. Boston, eds., *Scientists on Gaia,* pp. 211–222. Cambridge: MIT Press.

Schindel, D. E. 1982. Resolution analysis: A new approach to gaps in the fossil record. *Paleobiology* 8:340–353.

Schmidt-Nielson, K. 1984. *Scaling: Why is Animal Size So Important?* New York: Cambridge.

Schoener, T. W. 1983. Rate of species turnover decreases from lower to higher organisms: A review of the data. *Oikos* 41:372–377.

——. 1986. Overview: Kinds of ecological communities—ecology becomes pluralistic. In J. Diamond and T. J. Case, eds., *Community Ecology,* pp. 467–479. New York: Harper and Row.

——. 1991. Extinction and the nature of the metapopulation: A case system. *Acta Oecologia* 12:53–75.

Schoener, T. W. and D. A. Spiller. 1987. High population persistence in a system with high turnover. *Nature* 330:474–77.

Schopf, J. W. 1994. Disparate rates, differing fates: Tempo and mode of evolution changed from the Precambrian to the Phanerozoic. *Proceedings of the National Academy of Sciences of the U.S.A.* 91:6735–6742.

Schopf, K. M. 1996. Coordinated stasis: Biofacies revisited and the conceptual modeling of whole-fauna dynamics. *Palaeogeography, Palaeoclimatology, Palaeoecology* 127:157–176.

Schopf, K. M. and L. C. Ivany. 1997. Comment on "Long term faunal stasis without evolutionary coordination: Jurassic benthic marine paleocommunities, Western Interior, United States." *Geology* 25(5):473.

Schopf, T. J. M. 1972. Varieties of paleobiologic experience. In T. J. M. Schopf, ed., *Models in Paleobiology,* pp. 8–25. San Francisco: Freeman, Cooper.

——. 1978. Fossilization potential of an intertidal fauna: Friday Harbor, Washington. *Paleobiology* 4:261–270.

——. 1979. The role of biogeographic provinces in regulating marine faunal diversity through geologic time. In J. Gray and A. J. Boucot, eds., *Historical Biogeography, Plate Tectonics, and the Changing Environment,* pp. 449–457. Corvallis: Oregon State University Press.

Schroeder, M. 1991. *Fractals, Chaos, Power Laws.* New York: Freeman.

Scoffin, T. P. 1992. Taphonomy of coral reefs: A review. *Coral Reefs* 11:57–77.

Scotese, C. R. and W. S. McKerrow. 1990. Revised world maps and introduction. In W. S. McKerrow and C. R. Scotese, eds., *Palaeozoic Palaeogeography and Biogeography, Geological Society of London Memoir* 12:1–21.

Scott, P. J. B., M. J. Risk, and J. D. Carriquiry. 1988. El Niño, bioerosion and the survival of East Pacific Reefs. *Proceedings 6th International Coral Reef Symposium* 2:517–520.

Seagle, S. W. and H. H. Shugart. 1985. Faunal richness and turnover on dynamic landscapes: A simulation study. *Journal of Biogeography* 12:499–508.

Seilacher, A. 1991. Self-organizing mechanisms in morphogenesis and evolution. In N. Schmidt-Kittler and K. Vogel, eds., *Constructional Morphology and Evolution,* pp. 251–271. Berlin, Heidelberg: Springer- Verlag.

——. 1994. Candle wax shells, morphodynamics, and the Cambrian explosion. *Acta Paleontologica Polonica* 38:273–280.

Sepkoski, J. J., Jr. 1976. Species diversity in the Phanerozoic: species-area effects. *Paleobiology* 2:298–303.

——. 1978. A kinetic model of Phanerozoic taxonomic diversity: I. Analysis of marine orders. *Paleobiology* 4:223–251.

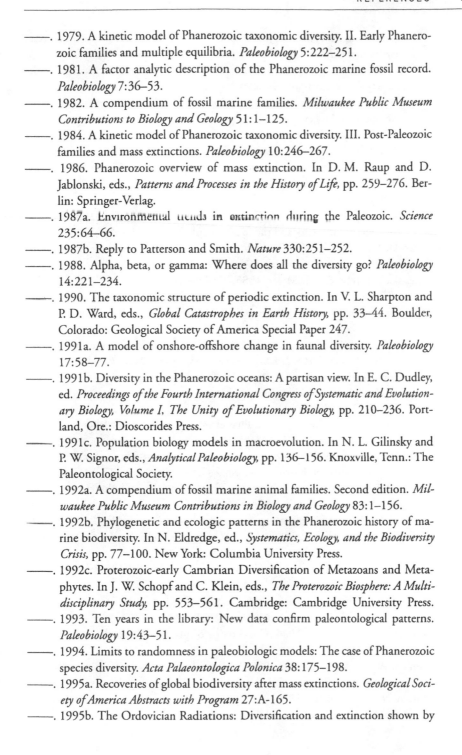

———. 1979. A kinetic model of Phanerozoic taxonomic diversity. II. Early Phanerozoic families and multiple equilibria. *Paleobiology* 5:222–251.

———. 1981. A factor analytic description of the Phanerozoic marine fossil record. *Paleobiology* 7:36–53.

———. 1982. A compendium of fossil marine families. *Milwaukee Public Museum Contributions to Biology and Geology* 51:1–125.

———. 1984. A kinetic model of Phanerozoic taxonomic diversity. III. Post-Paleozoic families and mass extinctions. *Paleobiology* 10:246–267.

———. 1986. Phanerozoic overview of mass extinction. In D. M. Raup and D. Jablonski, eds., *Patterns and Processes in the History of Life,* pp. 259–276. Berlin: Springer-Verlag.

———. 1987a. Environmental trends in extinction during the Paleozoic. *Science* 235:64–66.

———. 1987b. Reply to Patterson and Smith. *Nature* 330:251–252.

———. 1988. Alpha, beta, or gamma: Where does all the diversity go? *Paleobiology* 14:221–234.

———. 1990. The taxonomic structure of periodic extinction. In V. L. Sharpton and P. D. Ward, eds., *Global Catastrophes in Earth History,* pp. 33–44. Boulder, Colorado: Geological Society of America Special Paper 247.

———. 1991a. A model of onshore-offshore change in faunal diversity. *Paleobiology* 17:58–77.

———. 1991b. Diversity in the Phanerozoic oceans: A partisan view. In E. C. Dudley, ed. *Proceedings of the Fourth International Congress of Systematic and Evolutionary Biology, Volume I, The Unity of Evolutionary Biology,* pp. 210–236. Portland, Ore.: Dioscorides Press.

———. 1991c. Population biology models in macroevolution. In N. L. Gilinsky and P. W. Signor, eds., *Analytical Paleobiology,* pp. 136–156. Knoxville, Tenn.: The Paleontological Society.

———. 1992a. A compendium of fossil marine animal families. Second edition. *Milwaukee Public Museum Contributions in Biology and Geology* 83:1–156.

———. 1992b. Phylogenetic and ecologic patterns in the Phanerozoic history of marine biodiversity. In N. Eldredge, ed., *Systematics, Ecology, and the Biodiversity Crisis,* pp. 77–100. New York: Columbia University Press.

———. 1992c. Proterozoic-early Cambrian Diversification of Metazoans and Metaphytes. In J. W. Schopf and C. Klein, eds., *The Proterozoic Biosphere: A Multidisciplinary Study,* pp. 553–561. Cambridge: Cambridge University Press.

———. 1993. Ten years in the library: New data confirm paleontological patterns. *Paleobiology* 19:43–51.

———. 1994. Limits to randomness in paleobiologic models: The case of Phanerozoic species diversity. *Acta Palaeontologica Polonica* 38:175–198.

———. 1995a. Recoveries of global biodiversity after mass extinctions. *Geological Society of America Abstracts with Program* 27:A-165.

———. 1995b. The Ordovician Radiations: Diversification and extinction shown by global genus-level data. In J. D. Cooper, M. L. Droser, and S. C. Finney, eds., *Ordovician Odyssey: Short Papers for the Seventh International Symposium on the Ordovician System,* pp. 393–396. Fullerton, California: Pacific Section of the Society for Sedimentary Geology.

———. 1996. Competition in macroevolution: The double wedge revisited. In D. Jablonski, D. H. Erwin, and J. Lipps, eds., *Evolutionary Paleobiology,* pp. 211–255. Chicago: University of Chicago Press.

Sepkoski, J. J., R. K. Bambach, and M. L. Droser. 1991. Secular changes in Phanerozoic event bedding and the biological overprint. In G. Einsele, W. Ricken, and A. Seilacher, eds., *Cycles and Events in Stratigraphy,* pp. 298–312. Berlin: Springer-Verlag.

Sepkoski, J. J., Jr., R. K. Bambach, D. M. Raup, and J. W. Valentine. 1981. Phanerozoic marine diversity and the fossil record. *Nature* 293:435–437.

Sepkoski, J. J., Jr. and D.C. Kendrick. 1993. Numerical experiments with model monophyletic and paraphyletic taxa. *Paleobiology* 19:168–184.

Sepkoski, J. J. Jr. and A. I. Miller. 1985. Evolutionary faunas and the distribution of Paleozoic benthic communities in space and time. In J. W. Valentine, ed., *Phanerozoic Diversity Patterns,* pp. 153–190. Princeton: Princeton University Press.

Shaffer, M. 1987. Minimum viable populations: Coping with uncertainty. In M. E. Soulé, ed., *Viable Populations for Conservation,* pp. 69–86. Cambridge: Cambridge University Press.

Sheehan, P. M. 1975. Brachiopod synecology in a time of crisis. *Paleobiology* 1:205–212.

———. 1977. Species diversity in the Phanerozoic. A reflection of labor by systematists? *Paleobiology* 3:325–328.

———. 1982. Brachiopod macroevolution at the Ordovician-Silurian boundary. *Proceedings 3rd North American Paleontological Convention* 2:477–481.

———. 1985. Reefs are not so different—They follow the evolutionary pattern of level-bottom communities. *Geology* 13:46–49.

———. 1991. Patterns of synecology during the Phanerozoic. In E. C. Dudley, ed., *The Unity of Evolutionary Biology, Vol. 1,* pp. 103–118. Portland, Ore.: Dioscorides Press.

———. 1996. A new look at ecologic evolutionary units (EEUs). *Palaeogeography, Palaeoclimatology, Palaeoecology* 127:21–32.

Sheehan, P. M. and T. A. Hansen. 1986. Detritus feeding as a buffer to extinction at the end of the Cretaceous. *Geology* 14:868–870.

Shmida, A. and S. Ellner. 1984. Coexistence of plant species with similar niches. *Vegetatio* 58:29–55.

Shmida, A. and M. V. Wilson. 1985. Biological determinants of species diversity. *Journal of Biogeography* 12:1–20.

Shubin, N. and P. Alberch. 1986. A morphogenetic approach to the origin and basic organization of the tetrapod limb. *Evolutionary Biology* 20:319–387.

Shuto, T. 1974. Larval ecology of prosobranch gastropods and its bearing on biogeography and paleontology. *Lethaia* 7:239–256.

Sibley, C. G. and J. E. Ahlquist. 1990. *Phylogeny and Classification of the Birds.* New Haven: Yale University Press.

Sibley, C. G. and B. L. Monroe, Jr. 1990. *Distribution and Taxonomy of Birds of the World.* New Haven: Yale University Press.

———. 1993. *A Supplement to Distribution and Taxonomy of Birds of the World.* New Haven: Yale University Press.

Signor, P. W. 1990. The geologic history of diversity. *Annual Review of Ecology and Systematics* 21:509–539.

———. 1992. Taxonomic diversity and faunal turnover in the Early Cambrian: Did the most severe mass extinction of the Phanerozoic occur in the Botomian Stage? In S. Lidgard and P. R. Crane, eds., *Fifth North American Paleontological Convention, Abstracts and Program,* p. 272. Paleontological Society Special Publication No. 6.

———. 1994. Biodiversity in geological time. *American Zoologist* 34:23–32.

Signor, P. W. and C. E. Brett. 1984. The mid-Paleozoic precursor to the Mesozoic marine revolution. *Paleobiology* 10:229–245.

Signor, P. W. and J. H. Lipps. 1982. Sampling bias, gradual extinction patterns and catastrophes in the fossil record. In L. T. Silver and P. H. Schultz, eds., *Geological Implications of Impacts of Large Asteroids and Comets on the Earth,* pp. 291–296. Geological Society of America Special Paper No. 190.

Signor, P. W. and G. J. Vermeij. 1994. The plankton and the benthos: Origins and early history of an evolving relationship. *Paleobiology* 20:297–319.

Signor, P. W., III. 1978. Species richness in the Phanerozoic: An investigation of sampling effects. *Paleobiology* 4:394–406.

———. 1985. Real and apparent trends in species richness through time. In J. W. Valentine, ed., *Phanerozoic Diversity Patterns,* pp. 129–150. Princeton: Princeton University Press.

Simberloff, D. 1980. A succession of paradigms in ecology: Essentialism to materialism and probabilism. *Synthese* 43:3–39.

Simberloff, D. S. 1969. Experimental zoogeography of islands. A model of insular colonization. *Ecology* 50:296–314.

———. 1972. Properties of the rarefaction diversity measurement. *American Naturalist* 106:414–418.

———. 1974. Equilibrium theory of island biogeography and ecology. *Annual Review of Ecology and Systematics* 5:161–182.

———. 1981. Community effects on introduced species. In M. H. Nitecki, ed., *Biotic Crises in Ecological and Evolutionary Time,* pp. 53–81. New York: Academic Press.

Simberloff, D. S. and E. O. Wilson. 1969. Experimental zoogeography of islands: The colonization of empty islands. *Ecology* 50:278–296.

Simpson, G. G. 1944. *Tempo and Mode in Evolution.* New York: Columbia University Press.

——. 1953. *The Major Features of Evolution.* New York: Columbia University Press.

Sjogren, P. 1994. Distribution and extinction patterns within a northern metapopulation case of the pool frog, *Rana lessonae. Ecology* 75:1357–1367.

Skellam, J. G. 1951. Random dispersal in theoretical populations. *Biometrika* 38: 196–218.

Skelton, P. W. 1993. Adaptive radiation: Definition and diagnostic tests. In D. R. Lees and D. Edwards, eds., *Evolutionary Patterns and Processes*, pp. 45–58. London: Academic Press.

Slatkin, M. 1981. A diffusion model of species selection. *Paleobiology* 7: 421–425.

——. 1985. Gene flow in natural populations. *Annual Review of Ecology and Systematics* 16:393–430.

Slowinski, J. B. 1990. Probabilities of n-trees under two models: A demonstration that asymmetrical interior nodes are not improbable. *Systematic Zoology* 39: 89–94.

Slowinski, J. B. and C. Guyer. 1989. Testing the stochasticity of patterns of organismal diversity: An improved null model. *American Naturalist* 134:907–921.

Smit, J. 1982. Extinction and evolution of planktonic foraminifera after a major impact at the Cretaceous/Tertiary boundary. In L. T. Silver and P. H. Schultz, eds., *Geological Implications of Impacts of Large Asteroids and Comets on the Earth*, pp. 329–352. Geological Society of America Special Publication No. 190.

Smith, A. B. 1984. *Echinoid Paleobiology.* London: Allen & Unwin.

——. 1994. *Systematics and the Fossil Record.* Oxford: Blackwell.

Smith, A. B., D. T. J. Littlewood, and G. A. Wray. 1995. Comparing patterns of evolution: Larval and adult life history stages and ribosomal RNA of post-Paleozoic echinoids. *Philosophical Transactions of the Royal Society of London* B349:11–18.

Smith, C. R. 1994a. Tempo and mode in deep-sea benthic ecology: Punctuated equilibrium revisited. *Palaios* 9:3–13.

Smith, L. 1994b. Fluctuating asymmetry and developmental stability in Cambrian and Ordovician trilobites. *Geological Society of America Abstracts with Program* 26:A-373.

Sneath, P. H. A. and R. R. Sokal. 1973. *Numerical Taxonomy.* San Francisco: Freeman.

Sneppen, K., P. Bak, H. Flyvbjerg, and M. H. Jensen. 1995. Evolution as a self-organized critical phenomenon. *Proceedings of the National Academy of Sciences of the U.S.A.* 92:5209–5213.

Sokal, R. R. and F. J. Rohlf. 1995. *Biometry.* Third edition. New York: W. H. Freeman.

Solé, R. V. and S. C. Manrubia. 1995. Are rainforests self-organized in a critical state? *Journal of Theoretical Biology* 173:31–40.

Solé, R. V., S. C. Manrubia, M. Benton, and P. Bak. 1997. Self-similarity of extinction statistics in the fossil record. *Nature* 388:764–766.

Soulé, D. F. and G. S. Kleppel. 1988. *Marine Organisms As Indicators*. New York: Springer-Verlag.

Soulé, M. E. 1980. Threshold for survival: Maintaining fitness and evolutionary potential. In M. E. Soulé and B. A. Wilcox, eds., *Conservation Biology: An Evolutionary-Ecological Perspective*, pp. 151–169. Sunderland, Mass.: Sinauer.

Sousa, W. P. 1985. Disturbance and patch dynamics on rocky intertidal shores. In S. T. A. Pickett and P. White, eds., *The Ecology of Natural Disturbance and Patch Dynamics*, pp. 101–124. Orlando: Academic Press.

Southward, A. J. 1980. The Western English Channel—an inconstant ecosystem? *Nature (London)* 285:361–366.

Southward, A. J., G. T. Boalch, and L. Maddock. 1988. Fluctuations in the herring and pilchard fisheries of Devon and Cornwall linked to change in climate since the 16th century. *Journal of the Marine Biological Association of the United Kingdom* 68:423–445.

Springer, D. A. and A. I. Miller. 1990. Levels of spatial variability: The "community" problem. In W. Miller III, ed., *Paleocommunity Temporal Dynamics: The Long-Term Development of Multispecies Assemblages*, pp. 13–30. Paleontological Society Special Publication No. 5. Knoxville: University of Tennessee Press.

Staff, G. M., R. J. Stanton, Jr., E. N. Powell, and H. Cummins. 1986. Time-averaging, taphonomy and their impacts on paleocommunity reconstruction: Death assemblages in Texas bays. *Geological Society of America Bulletin* 97:428–443.

Stanley, S. M. 1973a. An ecological theory for the sudden origin of multicellular life in the Late Precambrian. *Proceedings of the National Academy of Sciences of the U.S.A.* 70:1486–1489.

——. 1973b. An explanation for Cope's rule. *Evolution* 27:1–26.

——. 1975. A theory of evolution above the species level. *Proceedings of the National Academy of Sciences U.S.A.* 72:646–650.

——. 1979. *Macroevolution: Pattern and Process*. San Francisco: W. H. Freeman.

——. 1985. Rates of evolution. *Paleobiology* 11:13–26.

——. 1986a. Anatomy of a regional mass extinction: Plio-Pleistocene decimation of the western Atlantic bivalve fauna. *Palaios* 1:17–36.

——. 1986b. Population size, extinction, and speciation: The fission effect in Neogene Bivalvia. *Paleobiology* 12:89–110.

——. 1987. *Extinction*. New York: Scientific American.

——. 1988. Paleozoic mass extinctions: Shared patterns suggest global cooling as a common cause. *American Journal of Science* 288:334–352.

——. 1990a. Adaptive radiation and macroevolution. In P. D. Taylor and G. P. Larwood, eds., *Major Evolutionary Radiations*, pp. 1–16. Oxford: Clarendon Press.

——. 1990b. Delayed recovery and the spacing of major extinctions. *Paleobiology* 16:401–414.

——. 1990c. The general correlation between rate of speciation and rate of extinction: Fortuitous causal linkages. In R. M. Ross and W. D. Allmon, eds., *Causes of Evolution. A Paleontological Perspective*, pp. 103–127. Chicago: University of Chicago Press.

Stanley, S. M., P. W. Signor III, S. Lidgard, and A. F. Karr. 1981. Natural clades differ from "random" clades: Simulations and analyses. *Paleobiology* 7: 115–127.

Stanley, S. M., K. L. Wetmore, and J. P. Kennett. 1988. Macroevolutionary differences between the two major clades of Neogene planktonic formaminifera. *Paleobiology* 14:235–249.

Stanley, S. M. and X. Yang. 1994. A double mass extinction at the end of the Paleozoic Era. *Science* 266:1340–1344.

Stearns, S. C. 1992. *The Evolution of Life Histories*. Oxford: Oxford University Press.

Stebbins, G. L. 1969. *The Basis of Progressive Evolution*. Chapel Hill: University of North Carolina Press.

Stehli, F. G. and S. D. Webb, eds. 1985. *The Great American Biotic Interchange*. Mt. Kisco, N. Y.: Plenum Press.

Stemann, T. A. and K. G. Johnson. 1995. Ecologic stability and spatial continuity in a Holocene reef, Lago Enriquillo, Dominican Republic (abstract). *Geological Society of America Annual Meeting, Abstracts with Programs* 27:A166.

Steneck, R. S. 1988. Herbivory on coral reefs: A synthesis. *Proceedings of the 6th International Coral Reef Symposium, Townsville* 1:37–49.

Stenseth, N. C. 1984. The tropics: Cradle or museum? *Oikos* 43:417–420.

Stenseth, N. C. and J. Maynard Smith. 1984. Coevolution in ecosystems: Red Queen evolution or stasis? *Evolution* 38:870–880.

Storer, J. E. 1989. Rodent faunal provinces, Paleocene-Miocene of North America. *Natural History Museum of Los Angeles County Science Series* 33:17–29.

Stucky, R. K. 1984. Revision of the Wind River faunas, Early Eocene of central Wyoming. Part 5. Geology and biostratigraphy of the upper part of the Wind River Formation. *Annals of Carnegie Museum* 53:231–294.

——. 1990. Evolution of land mammal diversity in North America during the Cenozoic. In H. H. Genoways, ed., *Current Mammalogy, Vol. 2*, pp. 375–432. Mt. Kisco, N. Y.: Plenum Press.

Suchanek, T. H. 1981. The role of disturbance in the evolution of life history strategies in the intertidal mussels *Mytilus edulis* and *Mytilus californianus. Oecologia* 50:143–152.

Sytsma, K. J. and L. D. Gottlieb. 1986. Chloroplast DNA evolution and the phylogenetic relationships in *Clarkia* sect. *peripetasma. Evolution* 40:1248–1262.

Szmant, A. M. 1986. Reproductive ecology of Caribbean reef corals. *Coral Reefs* 5: 43–53.

Tang, C. M. and D. J. Bottjer. 1996. Long-term faunal stasis without evolutionary coordination: Jurassic benthic marine paleocommunities, Western Interior, United States. *Geology* 24(9):815–818.

Tang, C. M. and D. J. Bottjer. 1997. Reply to "Long term faunal stasis without evolutionary coordination: Jurassic benthic marine paleocommunities, Western Interior, United States." *Geology* 25(5):474–475.

Taper, M. L. and T. J. Case. 1992. Models of character displacement and the theoretical robustness of taxon cycles. *Evolution* 46:317–333.

Tappan, H. 1968. Primary production, isotopes, extinctions, and the atmosphere. *Palaeogeography, Palaeoclimatology, Palaeoecology* 4:187–210.

——. 1970. Reply to Phytoplankton abundance and late Paleozoic extinctions. *Palaeogeography, Palaeoclimatology, Palaeoecology* 8:56–66.

——. 1971. Microplankton, ecological succession, and evolution. *Proceedings of the North American Paleontological Convention, Part H*, pp. 1058–1103.

——. 1980. *Paleobiology of Plant Protists.* San Francisco: W. H. Freeman.

——. 1982. Extinction or survival: Selectivity and causes of Phanerozoic crises. In L. T. Silver and P. H. Schultz, eds., *Geological Implications of Impacts of Large Asteroids and Comets on the Earth*, pp. 265–276. Boulder: Geological Society of America.

——. 1986. Phytoplankton: Below the salt at the global table. *Journal of Paleontology* 60:545–554.

Tappan, H. and A. R. Loeblich. 1971. Geobiologic implications of fossil phytoplankton evolution and time-space distribution. In R. Kosanke, R. and A. T. Cross, eds., *Symposium on Palynology of the Late Cretaceous and Early Tertiary*, pp. 247–340. Boulder: Geological Society of America.

——. 1973. Evolution of the oceanic plankton. *Earth Science Reviews* 9:207–240.

——. 1988. Foraminiferal evolution, diversification, and extinction. *Journal of Paleontology* 62:695–714.

Tardy, Y., R. N'Kounkou, and J.-L. Probst. 1989. The global water cycle and continental erosion during Phanerozoic time (570 my). *American Journal of Science* 289:455–483.

Tauber, C. A. and M. J. Tauber. 1989. Sympatric speciation in insects: Perception and perspective. In D. Otte and J. A. Endler, eds., *Speciation and Its Consequences*, pp. 307–344. Sunderland, Mass.: Sinauer.

Tavare, S. 1984. Line-of-Descent and genealogical processes, and their applications in population genetics models. *Theoretical Population Biology* 26:119–164.

Taylor, J. D. 1978. Faunal response to the instability of reef habitats: Pleistocene molluscan assemblages of Aldabra Atoll. *Palaeontology* 21:1–30.

Taylor, P. D. 1988. Major radiation of cheilostome bryozoans: Triggered by the evolution of a new larval type? *Historical Biology* 1:45–64.

Teeter, J. W. 1973. Geographic distribution and dispersal of some recent shallow water marine Ostracoda. *Ohio Journal of Science* 73:46–54.

Templeton, A. R. 1989. The meaning of species and speciation: A genetic perspective. In D. Otte and J. A. Endler, eds., *Speciation and Its Consequences,* pp. 3–27. Sunderland, Mass.: Sinauer.

Terborgh, J. and J. Faaborg. 1973. Turnover and ecological release in the avifauna of Mona Island, Puerto Rico. *Auk* 90:759–779.

Thayer, C. W. 1979. Biological bulldozers and the evolution of marine benthic communities. *Science* 203:458–461.

——. 1983. Sediment-mediated biological disturbance and the evolution of marine benthos. In M. J. S. Tevesz and P. L. McCall, eds., *Biotic Interactions in Recent and Fossil Benthic Communities,* pp. 480–635. Mt. Kisco, N. Y.: Plenum Press.

——. 1992. Escalating energy budgets and oligotrophic refugia: Winners and dropouts in the Red Queen's race. *Fifth North American Paleontological Convention, Abstracts and Program.* p. 290. Paleontological Society Special Publication No. 6.

Thomas, C. D. 1994. Extinction, colonization and metapopulations: Environmental tracking by rare species. *Conservation Biology* 8:373–378.

Thomas, C. D., J. A. Thomas, and M. S. Warren. 1992. Distributions of occupied and vacant butterfly habitats in fragmented landscapes. *Oecologia* 92: 563–567.

Thomas, R. D. K. and W.-E. Reif. 1993. The skeleton space: A finite set of organic designs. *Evolution* 47:341–360.

Thomas, W. A., G. Goldstein, and W. H. Wilcox. 1973. *Biological Indicators of Environmental Quality.* Ann Arbor: Ann Arbor Science Publishers, Inc.

Thompson, J. N. 1994. *The Coevolutionary Process.* Chicago: University of Chicago Press.

Tillman, D. and J. A. Downing. 1994. Biodiversity and stability in grasslands. *Nature* 367:363–365.

Tilman, D. 1982. *Resource Competition and Community Structure.* Princeton: Princeton University Press.

Tipper, J. C. 1979. Rarefaction and rarefiction—the use and abuse of a method in paleoecology. *Paleobiology* 5:423–434.

Toft, C. A. and M. Mangel. 1991. Discussion: From individuals to ecosystems; the papers of Skellam, Lindeman and Hutchinson. *Bulletin of Mathematical Biology* 53:121–134.

Tomascik, T. 1991. Settlement patterns of Caribbean scleractinina corals on artificial substrata along a eutrophication gradient, Barbados, West Indies. *Marine Ecology Progress Series* 77:261–269.

Tomascik, T. and F. Sander. 1985. Effects of eutrophication on reef-building corals I. Growth rate of the reef-building coral *Montastrea annularis. Marine Biology* 87:143–155.

———. 1987a. Effects of eutrophication on reef-building corals II. Structure of scleractinian coral communities on fringing reefs, Barbados, West Indies. *Marine Biology* 94:53–75.

———. 1987b. Effects of eutrophication on reef-building corals III. Reproduction of the reef-building coral *Porites porites*. *Marine Biology* 94:77–94.

Tonn, W. M. and J. J. Magnuson. 1982. Patterns in the species composition and richness of fish assemblages in northern Wisconsin lakes. *Ecology* 63: 1149–116.

Tracy, C. R. and T. L. George. 1992. On the determinants of extinction. *American Naturalist* 139:102–122.

Turcotte, D. L. 1994. Fractal aspects of geomorphic and stratigraphic processes. *GSA Today* 4:201–213.

Ulanowicz, R. E. 1980. An hypothesis on the development of natural communities. *Journal of Theoretical Biology* 85:223–245.

Underwood, A. J., E. J. Denly, and M. J. Moran. 1983. Experimental analyses of the structure and dynamics of mid-shore rocky intertidal communities in New South Wales. *Oceologia* 56:202–219.

Upton, G. J. G. and G. Fingleton. 1985. *Spatial Data Analysis by Example*. Chichester: Wiley.

Valentine, J. W. 1969. Patterns of taxonomic and ecological structure of the shelf benthos during Phanerozoic time. *Palaeontology* 12:684–709.

———. 1970. How many marine invertebrate fossil species? *Journal of Paleontology* 44:410–415.

———. 1973a. *Evolutionary Paleoecology of the Marine Biosphere*. Englewood Cliffs, N.J.: Prentice-Hall.

———. 1973b. Phanerozoic taxonomic diversity: A test of alternate models. *Science* 180:1078–1079.

———. 1985. Diversity as data. In J. W. Valentine, ed., *Phanerozoic Diversity Patterns: Profiles in Macroevolution*, pp. 3–8. Princeton: Princeton University Press.

———. 1990a. The fossil record: A sampler of life's diversity. *Philosophical Transactions of the Royal Society of London* B330:261–268.

———. 1990b. The macroevolution of clade shape. In R. M. Ross and W. D. Allmon, eds., *Causes of Evolution: A Paleontological Perspective*, pp. 128–150. Chicago: University of Chicago Press.

———. 1995. Why no new phyla after the Cambrian? Genome and ecospace hypotheses revisited. *Palaios* 10:190–194.

Valentine, J. W., S. M. Awramik, P. W. Signor, and P. M. Sadler. 1991a. The biological explosion at the Precambrian-Cambrian boundary. *Evolutionary Biology* 25:279–356.

Valentine, J. W. and C. A. Campbell. 1975. Genetic regulation and the fossil record. *American Scientist* 63:673–680.

Valentine, J. W. and D. H. Erwin. 1987. Interpreting great developmental experiments: The fossil record. In R. A. Raff and E. C. Raff, eds., *Development as an Evolutionary Process,* pp.71–107. New York: Alan Liss.

Valentine, J. W., D. H. Erwin, and D. Jablonski. 1996. Developmental evolution of metazoan body plans: The fossil evidence. *Developmental Biology* 173: 373–381.

Valentine, J. W., T. C. Foin, and D. Peart. 1978. A provincial model of Phanerozoic marine diversity. *Paleobiology* 4:55–66.

Valentine, J. W. and D. Jablonski. 1983. Speciation in the shallow sea: Patterns and biogeographic controls. In R. W. Sims, J. H. Price, and P. E. S. Whalley, eds., *Evolution, Time and Space: The Emergence of the Biosphere,* pp. 201–226. Systematics Association Special Volume No. 23. New York: Academic Press.

——. 1986. Mass extinctions: Sensitivity of marine larval types. *Proceedings National Academy of Sciences U.S.A.* 83:6912–6914.

——. 1993. Fossil communities: Compositional variation at many time scales. In R. E. Ricklefs and D. Schluter, eds., *Species Diversity in Ecological Communities: Historical and Geographical Perspectives,* pp. 341–349. Chicago: University of Chicago Press.

Valentine, J. W. and B. Mallory. 1965. Recurrent groups of bonded species in mixed death assemblages. *Journal of Geology* 73:683–701.

Valentine, J. W. and C. L. May. 1996. Hierarchies in biology and paleontology. *Paleobiology* 22:23–33.

Valentine, J. W. and E. M. Moores. 1972. Plate-tectonic regulation of faunal diversity and sea level: A model. *Nature* 228:657–659.

Valentine, J. W., B. H. Tiffney, and J. J. Sepkoski, Jr. 1991b. Evolutionary dynamics of plants and animals. *Palaios* 5:81–88.

Van Valen, L. 1976. Energy and evolution. *Evolutionary Theory* 1:179–229.

——. 1980. One man's view of evolution. *BioScience* 30:620.

——. 1984. A resetting of Phanerozoic community evolution. *Nature* 307:50–52.

——. 1985a. A theory of origination and extinction. *Evolutionary Theory* 7: 133–142.

——. 1985b. How constant is extinction? *Evolutionary Theory* 7:93–106.

——. 1985c. Why and how do mammals evolve unusually rapidly? *Evolutionary Theory* 7:127–132.

——. 1994. Concepts and the nature of selection by extinction: Is generalization possible? In W. Glen, ed., *The Mass Extinction Debates: How Science Works in a Crisis,* pp. 200–341. Stanford: Stanford University Press.

Van Valen, L. M. 1973. A new evolutionary law. *Evolutionary Theory* 1:1–30.

Van Valen, L. M. and V. C. Maiorana. 1985. Patterns of origination. *Evolutionary Theory* 7:107–125.

Van Valkenburgh, B. and C. M. Janis. 1993. Historical diversity patterns in North American large herbivores and carnivores. In R. E. Ricklefs and D. Schluter, eds., *Species Diversity in Ecological Communities: Historical and Geographical Perspectives,* pp. 330–340. Chicago: University of Chicago Press.

Veech, W. A. 1992. Topological dynamics. *McGraw-Hill Encyclopedia of Science and Technology*, 7th ed. 18:428–431. New York: McGraw-Hill.

Vermeij, G. J. 1977. The Mesozoic marine revolution: Evidence from snails, predators and grazers. *Paleobiology* 3:245–258.

———. 1978. *Biogeography and Adaptation: Patterns of Marine Life*. Cambridge: Harvard University Press.

———. 1987a. *Evolution and Escalation: An Ecological History of Life*. Princeton: Princeton University Press.

———. 1987b. The dispersal barrier in the tropical Pacific: Implications for molluscan speciation and extinction. *Evolution* 41:1046–1058.

———. 1989. Geographical restriction as a guide to the causes of extinction: The case of the cold Northern oceans during the Neogene. *Paleobiology* 15:335–356.

———. 1991. Anatomy of an invasion: The trans-Arctic interchange. *Paleobiology* 17:281–307.

———. 1993. Biogeography of recently extinct marine species: Implications for conservation. *Conservation Biology* 7:391–397.

———. 1994. The evolutionary interaction among species: Selection, escalation, and coevolution. *Annual Review of Ecology and Systematics* 25:219 236.

———. 1995. Economics, volcanoes, and Phanerozoic revolutions. *Paleobiology* 21:125–152.

Vitousek, P. M., C. D'Antonio, L. Loope, and R. Westbrooks. 1996. Biological invasions as global environmental change. *American Scientist* 84:468–478.

Vrba, E. S. 1980. Evolution, species and fossils: How does life evolve? *South African Journal of Science* 76:61–84.

———. 1983. Macroevolutionary trends: New perspectives on the roles of adaptation and incidental effect. *Science* 221:387–389.

———. 1985. Environment and evolution: Alternative causes of the temporal distribution of evolutionary events. *South African Journal of Science* 81:229–236.

———. 1987. Ecology in relation to speciation rates: Some case histories of Miocene-Recent mammal clades. *Evolutionary Ecology* 1:283–300.

———. 1989. Levels of selection and sorting with special reference to the species level. *Oxford Surveys in Evolutionary Biology* 6:111–168.

———. 1993. Turnover-pulses, the Red Queen, and related topics. *American Journal of Science* 293-A:418–452.

Vrba, E. S. and S. J. Gould. 1986. The hierarchical expansion of sorting and selection: Sorting and selection cannot be equated. *Paleobiology* 12:217–228.

Wade, M. J. and D. E. McCauley. 1988. Extinction and colonization: Their effects on the genetic differentiation of local populations. *Evolution* 42:995–1005.

Wagner, G. P. 1988. The significance of developmental constraints for phenotypic evolution by natural selection. In G. de Jong, ed., *Population Genetics and Evolution*, pp. 222–229. Berlin, Heidelberg: Springer-Verlag.

———. 1989. The biological homology concept. *Annual Review of Ecology and Systematics* 20:51–69.

———. 1994. Homology and the mechanisms of development. In B. K. Hall, ed., *Homology: The Hierarchical Basis of Comparative Biology,* pp. 273–299. San Diego: Academic Press.

Wagner, P. J. 1995a. Diversity patterns among early gastropods: Contrasting taxonomic and phylogenetic descriptions. *Paleobiology* 21:410–439.

———. 1995b. Testing evolutionary constraint hypotheses with early Paleozoic gastropods. *Paleobiology* 21:248–272.

Wake, D. B. 1991. Homoplasy: The result of natural selection, or evidence of design limitations? *American Naturalist* 138:543–567.

Wake, D. B. and A. Larson. 1987. Multidimensional analysis of an evolving lineage. *Science* 238:42–48.

Waldrop, M. M. 1990. Spontaneous order, evolution, and life. *Science* 247:1543–1545.

Walker, T. D. 1985. Diversification functions and the rate of taxonomic evolution. In J. W. Valentine, ed., *Phanerozoic Diversity Patterns: Profiles in Macroevolution,* pp. 311–334. Princeton: Princeton University Press.

Walker, T. D. and J. W. Valentine. 1984. Equilibrium models of evolutionary species diversity and the number of empty niches. *American Naturalist* 124:887–899.

Wallace, C. C. and B. L. Willis. 1994. Systematics of the coral genus *Acropora:* Implications of new biological findings for species concepts. *Annual Review of Ecology and Systematics* 25:237–262.

Wallace, R. S. and R. K. Jansen. 1990. Systematic implications of chloroplast DNA variation in the genus *Microseris* (Asteraceae: Lactuceae). *Systematic Botany* 15:606–616.

Wang, K., H. H. J. Geldsetzer, and H. R. Krouse. 1994. Permian-Triassic extinction: Organic $\delta^{13}C$ evidence from British Columbia, Canada. *Geology* 22:580–584.

Ward, L. W., R. H. Bailey, and J. G. Carter. 1991. Pliocene and early Pleistocene stratigraphy, depositional history, and molluscan paleobiogeography of the Coastal Plain. In J. W. Horton, Jr. and V. A. Zullo, eds., *The Geology of the Carolinas,* pp. 274–289. Knoxville: University of Tennessee Press.

Ward, L. W. and N. L. Gilinsky. 1993. Biostratigraphic analysis of the Chowan River Formation (upper Pliocene) and adjoining units, the Moore House Member of the Yorktown Formation (upper Pliocene) and the James City Formation (lower Pleistocene). *Virginia Museum of Natural History Memoir 3, part A.* pp. 1–33.

Ward, P. 1980. Comparative shell shape distribution in Jurassic-Cretaceous ammonites and Jurassic-Tertiary nautilids. *Paleobiology* 6:32–43.

Ward, R. D. 1990. Biochemical genetic variation in the genus *Littorina* (Prosobranchia: Mollusca). *Hydrobiologia* 193:53–69.

Watkins, R. and A. J. Boucot. 1978. Temporal pattern of species diversity among some Silurodevonian brachiopods. In M. K. Hecht, W. C. Steere, and B. Wallace, eds., *Evolutionary Biology,* 11 pp. 636–647. Mt. Kisco, N. Y.: Plenum Press.

Watson, H. C. 1835. *Remarks on the Geographical Distribution of British Plants.* London.

Watt, A. S. 1947. Pattern and process in the plant community. *Journal of Ecology* 35:1–22.

Wayne, R. K., R. E. Benveniste, D. N. Janczewski, and S. J. O'Brien. 1989. Molecular and biochemical evolution of the Carnivora. In J. L. Gittleman, ed., *Carnivore Behavior, Ecology, and Evolution*, pp. 465–494. Ithaca: Cornell University Press.

Webb, S. D. 1969. Extinction-origination equilibria in Late Cenozoic land mammals of North America. *Evolution* 23:688–702.

——. 1983. The rise and fall of the Late Miocene ungulate fauna in North America. In M. H. Nitecki, ed., *Coevolution*, pp. 267–306. Chicago: University of Chicago Press.

——. 1984. Ten million years of mammal extinctions in North America. In P. S. Martin and R. G. Klein, eds., *Quaternary Extinctions: A Prehistoric Revolution.* Tucson, Ariz.: University of Arizona Press.

——. 1989. The fourth dimension in North American terrestrial communities. In D. Morris, Z. Abramsky, B. Fox, and M. R. Willig, eds., *Patterns in the Structure of Mammalian Communities*, pp.181–203. Lubbock: Texas Tech University Press.

Webb, S. D. and N. D. Opdyke. 1995. Global climatic influence on Cenozoic land mammal faunas. In Board on Earth Sciences and Resources, National Research Council. *Effects of Past Global Change on Life*, pp. 184–208. Washington, D. C.: National Academy Press.

Webb, T., III. 1987. The appearance and disappearance of major vegetational assemblages: Long-term vegetational dynamics in eastern North America. *Vegetatio* 69:177–187.

West, D.C., H. H. Shugart, and D. B. Botkin, eds. 1981. *Forest Succession: Concepts and Application.* Berlin: Springer-Verlag, Berlin.

Westoby, M. 1993. Biodiversity in Australia compared with other continents. In R. E. Ricklefs and D. Schluter, eds., *Species Diversity in Ecological Communities*, pp. 170–177. Chicago: University of Chicago Press.

Westphall, M. J. 1986. *Anatomy and History of a Ringed-Reef Complex, Belize, Central America.* M. S. thesis. Coral Gables: University of Miami.

Westrop, S. R. 1994. Biofacies stability and replacement during an evolutionary radiation: Sunwaptan (Upper Cambrian) trilobite faunas of North America. *Abstracts with Programs, GSA Annual Meeting, Seattle*, p. A454.

——. 1996. Temporal persistence and stability of Cambrian biofacies: Sunwaptan (Upper Cambrian) trilobite faunas of North America. *Palaeogeography, Palaeoclimatology, Palaeoecology* 127:33–46.

Wethey, D. S. 1985. Catastrophe, extinction, and species diversity: A rocky intertidal example. *Ecology* 66:445–456.

White, M. J. D. 1978. *Modes of Speciation.* San Francisco: Freeman.

White, P. S. 1979. Pattern, process, and natural disturbance in vegetation. *Botanical Review* 45:229–299.

White, P. S. and S. T. A. Pickett. 1985. Natural disturbance and patch dynamics: An introduction. In S. T. A. Pickett and P. S. White, eds., *The Ecology of Natural Disturbance and Patch Dynamics,* pp. 3–16. Orlando: Academic Press.

Whitlock, M. C. 1992. Nonequilibrium population structure in forked fungus beetles: Extinction, colonization, and genetic variation among populations. *American Naturalist* 139:952–70.

Whittaker, R. H. 1960. Vegetation of the Siskiyou Mountains, Oregon and California. *Ecological Monographs* 30:279–338.

——. 1970. *Communities and Ecosystems.* London: Macmillan.

——. 1975. *Communities and Ecosystems,* 2nd Edition. New York: Macmillan.

——. 1977. Evolution of species diversity in land communities. *Evolutionary Biology* 10:1–67.

Wicken, J. 1988. Thermodynamics, evolution, and emergence: Ingredients for a new synthesis. In B. H. Weber, D. J. Depew, and J. D. Smith, eds., *Entropy, Information, and Evolution: New Perspectives on Physical and Biological Evolution,* pp. 139–169. Cambridge: MIT Press.

Wiens, J. A. 1989. Spatial scaling in ecology. *Functional Ecology* 3:385–397.

Wiens, J. A., J. F. Addicott, T. J. Case, and J. Diamond. 1986. Overview: The importance of spatial and temporal scale in ecological investigations. In J. Diamond and T. J. Case, eds., *Community Ecology,* pp. 145–153. Cambridge: Harper & Row.

Wignall, P. B. and A. Hallam. 1992. Anoxia as a cause of the Permian/Triassic extinction: Facies evidence from northern Italy and the western United States. *Palaeogeography, Palaeoclimatology, Palaeoecology* 93:21–46.

Wilcox, B. A. 1978. Supersaturated island faunas: A species-age relationship for lizards on post-pleistocene land-bridge islands. *Science* 199:996–998.

Wilde, P. and W. B. N. Berry. 1984. Destabilization of the oceanic density structure and its significance to marine "extinction" events. *Palaeogeography, Palaeoclimatology, Palaeoecology* 48:143–162.

Wilkinson, C. R. 1987. Sponge biomass as an indication of reef productivity in two oceans. In C. Birkeland, ed., *Comparison Between Atlantic and Pacific Tropical Marine Coastal Ecosystems: Community Structure, Ecological Processes, and Productivity,* pp. 99–103. Paris: UNESCO.

——. 1993. Coral reefs of the world are facing widespread devastation: Can we prevent this through sustainable management practices? *Proceedings of the 7th International Coral Reef Symposium, Guam* 1:11–21.

Wilkinson, C. R. and A. C. Cheshire. 1988. Cross-shelf variations in coral reef structure and function—Influences of land and ocean. *Proceedings 6th International Coral Reef Symposium* 1:227–233.

Williams, A. and J. M. Hurst. 1977. Brachiopod evolution. In A. Hallam, ed., *Pat-*

terns of Evolution, as Illustrated by the Fossil Record, pp. 79–122. Amsterdam: Elsevier Scientific Publishing Company.

Williams, C. B. 1943. Area and the number of species. *Nature* 152:264–267.

——. 1964. *Patterns in the Balance of Nature.* London: Academic Press.

Williams, E. E. 1972. The origin of faunas. Evolution of lizard congeners in a complex island fauna: A trial analysis. *Evolutionary Biology* 6:47–90.

Williams, G. C. 1992. *Natural Selection: Domains, Levels, and Challenges.* New York: Oxford University Press.

Williams, M. R. 1995. An extreme-value function model of the species incidence and species-area relations. *Ecology* 76:2607–2616.

Williamson, M. H. 1983. The land-bird community of Skokholm: Ordination and turnover. *Oikos* 41:378–384.

Wills, M. A., D. E. G. Briggs, and R. A. Fortey. 1994. Disparity as an evolutionary index: A comparison of Cambrian and Recent arthropods. *Paleobiology* 20: 93–130.

Wilson, D. E. and D. M. Reeder. 1993. *Mammal Species of the World: A Taxonomic and Geographic Reference.* Washington, D. C.: Smithsonian.

Wilson, E. O. 1969. The species equilibrium. In Diversity and Stability in Ecological Systems. In G. M. Woodwell and H. H. Smith, eds., *Brookhaven Symposium in Biology 22,* pp. 38–47. Springfield: United States Department of Commerce.

——. 1975. *Sociobiology: A New Synthesis.* Cambridge: Harvard University Press.

Wilson, E. O. and D. S. Simberloff. 1969. Experimental zoogeography of islands: Defaunation and monitoring techniques. *Ecology* 50:267–278.

Wimsatt, W. C. 1986. Developmental constraints, generative entrenchment and the innate-acquired distinction. In W. Bechtel, ed., *Integrating Scientific Disciplines,* pp.185–208. Dordrecht: Martinus Nijhoff.

Wimsatt, W. C. and J. C. Schank. 1988. Two constraints on the evolution of complex adaptations and the means for their avoidance. In M. Nitecki, ed., *Evolutionary Progress,* pp. 231–273. Chicago: University of Chicago Press.

Wing, L. 1943. Spread of the starling and English sparrow. *Auk* 60:74–87.

Wing, S. L., J. Alroy, and L. J. Hickey. 1995. Plant and mammal diversity in the Paleocene to early Eocene of the Bighorn Basin. *Palaeogeography, Palaeoclimatology, Palaeoecology* 115:117–155.

Wise, K. P. and T. J. M. Schopf. 1981. Was marine faunal diversity in the Pleistocene affected by changes in sea level? *Paleobiology* 7:394–399.

Wissel, C. and B. Maier. 1992. A stochastic model for the species-area relationship. *Journal of Biogeography* 19:355–361.

Wood, R. 1993. Nutrients, predation, and the history of reef-building. *Palaios* 8: 526–543.

Woodburne, M. O. and C. C. Swisher, III. 1995. Land mammal high-resolution geochronology, intercontinental overland dispersals, sea level, climate, and vicariance. *Geochronology, Time Scales, and Global Stratigraphic Correlation, SEPM Special Publication* 54:335–364.

Woodley, J. D. 1992. The incidence of hurricanes on the north coast of Jamaica since 1870: Are the classic reef descriptions atypical? *Hydrobiologia* 247:133–138.

Woodruff, F. and S. M. Savin. 1991. Mid-Miocene isotope stratigraphy in the deep sea: High-resolution correlations, paleoclimatic cycles, and sediment preservation. *Paleoceanography* 6:755–806.

World Conservation Monitoring Centre. 1992. *Global Biodiversity: Status of the Earth's Living Resources.* London: Chapman & Hall.

Worsley, T. R., R. D. Nance, and J. B. Moody. 1986. Tectonic cycles and the history of the Earth's biogeochemical and paleoceanographic record. *Paleoceanography* 1:233–263.

Wray, G. A. 1992. The evolution of larval morphology during the post-Paleozoic radiation of echinoids. *Paleobiology* 18:258–287.

——. 1995. Punctuated evolution of embryos. *Science* 267:1115–1116.

Wray, G. A. and A. E. Bely. 1994. The evolution of echinoderm development is driven by several distinct factors. *Development 1994 Supplement* 97–106.

Wray, G. A., J. S. Levinton, and L. H. Shapiro. 1996. Molecular evidence for deep Precambrian divergences among metazoan phyla. *Science* 274:568–573.

Wright, D. H., B. D. Patterson, G. M. Mikkelson, A. Cutler, and W. Atmar. 1998. In press. A comparative analysis of nested subset patterns of species composition. *Oecologia.*

Wright, D. H. and J. H. Reeves. 1992. On the meaning and measurement of nestedness of species assemblages. *Oecologia* 92:416–428.

Wright, S. 1931. Evolution in Mendelian populations. *Genetics* 16:97–159.

——. 1940. Breeding structure of populations in relation to speciation. *American Naturalist* 74:232–48.

——. 1978. *Evolution and the Genetics of Populations. IV. Variability Within and Among Populations.* Chicago: University of Chicago Press.

Wright, S. J. 1981. Intra-archipelago vertebrate distributions: The slope of the species-area relation. *American Naturalist* 118:726–748.

Wu, J. and O. L. Loucks. 1995. From balance of nature to hierarchical patch dynamics: A paradigm shift in ecology. *Quarterly Review of Biology* 70:439–466.

Yule, G. U. 1924. A mathematical theory of evolution based on the conclusions of Dr. J. C. Willis, F. R. S. *Philosophical Transactions of the Royal Society of London* B213:21–87.

Zachos, J. C., M. A. Arthur, and W. E. Dean. 1989. Geochemical evidence for suppression of pelagic marine productivity at the Cretaceous/Tertiary boundary. *Nature* 337:61–64.

Zachos, J. C., J. R. Breza, and S. W. Wise. 1992. Early Oligocene ice-sheet expansion on Antarctica: Stable isotope and sedimentological evidence from Kerguelen Plateau, southern Indian Ocean. *Geology* 20:569–573.

Index

Printed in the USA
CPSIA information can be obtained
at www.ICGtesting.com
JSHW021435221024
72172JS00002B/3

9 780231 104159